AF746058

Particle Technology Series

Volume 25

Series editor
José Manuel Valverde Millán, University of Sevilla, Spain

Many materials exist in the form of a disperse system, for example powders, pastes, slurries, emulsions and aerosols, with size ranging from granular all the way down to the nanoscale. The study of such systems necessarily underlies many technologies/products and it can be regarded as a separate subject concerned with the manufacture, characterization and manipulation of such systems. The series does not aspire to define and confine the subject without duplication, but rather to provide a good home for any book which has a contribution to make to the record of both the theory and applications of the subject. We hope that engineers and scientists who concern themselves with disperse systems will use these books and that those who become expert will contribute further to the series.

The Springer Particle Technology Series is a continuation of the Kluwer Particle Technology Series, and the successor to the Chapman & Hall Powder Technology Series.

More information about this series at http://www.springer.com/series/6433

Henk G. Merkus • Gabriel M.H. Meesters
Editors

Production, Handling and Characterization of Particulate Materials

Springer

Editors
Henk G. Merkus, Emeritus Professor
Chemical Engineering Department
Delft University of Technology
Pijnacker, The Netherlands

Gabriel M.H. Meesters
DSM Food Specialties & Delft University of Technology
Delft, The Netherlands

ISSN 1567-827X
Particle Technology Series
ISBN 978-3-319-20948-7 ISBN 978-3-319-20949-4 (eBook)
DOI 10.1007/978-3-319-20949-4

Library of Congress Control Number: 2015953436

Springer Cham Heidelberg New York Dordrecht London
© Springer International Publishing Switzerland 2016
This work is subject to copyright. All rights are reserved by the Publisher, whether the whole or part of the material is concerned, specifically the rights of translation, reprinting, reuse of illustrations, recitation, broadcasting, reproduction on microfilms or in any other physical way, and transmission or information storage and retrieval, electronic adaptation, computer software, or by similar or dissimilar methodology now known or hereafter developed.
The use of general descriptive names, registered names, trademarks, service marks, etc. in this publication does not imply, even in the absence of a specific statement, that such names are exempt from the relevant protective laws and regulations and therefore free for general use.
The publisher, the authors and the editors are safe to assume that the advice and information in this book are believed to be true and accurate at the date of publication. Neither the publisher nor the authors or the editors give a warranty, express or implied, with respect to the material contained herein or for any errors or omissions that may have been made.

Printed on acid-free paper

Springer International Publishing AG Switzerland is part of Springer Science+Business Media (www.springer.com)

Preface

Particulate products make up around 80 % of all chemical products, from all industry sectors. The solid particles may be amorphous or have different crystalline forms, all of which have different properties. They originate from mining or other natural occurrences or are manufactured in crystallization, precipitation, agglomeration, or atomization processes. Prior to application they may undergo filtration, flotation, or classification and/or be subjected to compaction, extrusion, or comminution in order to bring them in the desired particle size range. Most often, they undergo transport and storage in some kind of equipment. "Particles" may not only be solid but also consist of entities like small droplets (by emulsification or atomization) or gas bubbles in another phase. The physical and mechanical behavior of single particles generally are reasonably well understood.

Base materials and products not only occur in the form of dry powders but also as dispersions, such as suspensions, emulsions, air bubbles in liquids, foams, or aerosols. In addition to the properties of the particles themselves, the properties of these products strongly depend upon the type of system, particulate concentration, and type and viscosity of the continuous phase, as well as upon particle size distribution and particle shape. Base materials and products have a complex nature as a result of the many possible mutual interactions between particles and of particles with chemical components in the system. Although much progress in understanding has been made over the past decades, the physical and mechanical properties of these complex systems are still much less understood than those of gases and liquids. Therefore, adequate product characterization is a challenging task. Sometimes, the products are characterized by techniques that have some more or less direct relationship with the property of interest. Often, however, various parameters derived from measured particle size distributions are linked to the properties of interest through empirical correlations. Selecting the optimum parameter(s) and measurement method(s) for characterization is one of the challenges. Finding the best correlation with behavior is the other challenge.

Although several books and many articles have been written that are specialized to one of the above unit operations or to a specific material or product, hardly any comprehensive book describes the ins and outs of all unit operations in comparison. That is the aim of this book. Specialized authors have written the various chapters. We are very grateful to all of them for their contribution and hope that our book will pay a useful contribution to further optimization of the process equipment applied for particulate materials and products.

Pijnacker, The Netherlands Henk G. Merkus
Delft, The Netherlands Gabriel M.H. Meesters

Contents

Author Information

Dr. Sajid Alavi is a professor at the Department of Grain Science and Industry at the Kansas State University in Manhattan, Kansas, USA. He has close to 20 years of experience with extrusion and other processing technologies. He routinely designs technology and R&D solutions for numerous food, feed, and pet food processors and is involved in processing and food aid-related projects in the USA, Africa, India, and other countries around the world. He has been invited to speak at numerous international forums and institutions in the USA, Italy, South Africa, Brazil, India, Mozambique, and China. He has provided training and networking opportunities to 900 industry leaders from 30 countries spanning all six continents through his internationally reputed short course "Extrusion Processing: Technology and Commercialization" and similar offerings and workshops in other countries such as India, Brazil, and Mozambique.

Dr. R.P. Kingsly Ambrose is an assistant professor (Designer Particulate Products) at the Department of Agricultural and Biological Engineering of Purdue University, West Lafayette, IN, USA. He obtained his B.S. in agricultural engineering and M.S. in agricultural processing from Tamil Nadu Agricultural University, Coimbatore, India, and his Ph.D. in agricultural and biological engineering from Purdue University. Dr. Ambrose's research interests are in the application of particle technology concepts to grain processing. Research areas include milling technologies, bulk material handling, characterization of powders for their particle and flow characteristics, grain dust explosion mitigation, and modeling particulate systems. Dr. Ambrose has authored/coauthored 39 peer-reviewed journal manuscripts, 2 book chapters, 68 conference presentations, and 16 invited talks.

Dr. Grant Ballantyne (Senior Research Fellow, the University of Queensland, Julius Kruttschnitt Mineral Research Centre (JKMRC)) is an early career researcher at the Julius Kruttschnitt Mineral Research Centre (JKMRC). Grant's three research themes are energy reduction, integration of comminution and flotation, and site optimization. His work is focused on the reduction of energy consumed through hard-rock mining, specifically rock-breakage processes. The measurement of energy use and efficiency is intrinsically involved in this objective. Grant is passionate about equipping the next generation of metallurgists through undergraduate lecturing, postgraduate supervision, and professional development courses. Grant has been involved in successful site-based industry research over the past 10 years covering four continents and eight commodities. Grant's research has gained industry recognition, especially through an international non-for-profit group called the Coalition for Eco-Efficient Comminution (CEEC).

Dr. Claire C. Berton-Carabin is assistant professor at Wageningen University in the Netherlands (Department Agrotechnology and Food Sciences, Lab. Food Process Eng.). Her research is focused on the connection between the (micro)structure of food emulsions and their reactivity and functionality. She received her Ph.D. at the University of Nantes and INRA (France) in 2011. Her Ph.D. project, supervised by Prof. Genot, was on the effect of the structure of the oil-water interface on lipid oxidation in emulsions and led to eight peer-reviewed publications, including a recently published review article ("Lipid Oxidation in Oil-in-Water Emulsions: Involvement of the Interfacial Layer," Compr. Rev. Food Sci. Food Safety, 2014). She then conducted a postdoctoral project at the Pennsylvania State University (Food Science Department), in the group of Prof. Coupland, on the use of electron spin resonance to investigate the location and reactivity of small molecules in emulsion-based systems. Then, after a few months at the Danone-Nutricia Research Centre (Utrecht, the Netherlands), she got appointed at Wageningen University. She now teaches several M.Sc. courses and supervises three Ph.D. students and several M.Sc. students. These projects are all related to emulsion science, from characterizing and modeling the formation and structure of the oil-water interface to studying the behavior of complex emulsion-based encapsulates in storage or digestive conditions.

Em. Prof. Norman Chigier was awarded M.A. (1960), Ph.D. (1961), and Sc.D. (1977) degrees from the University of Cambridge. He has held teaching appointments at Sheffield University and Technion-Israel Institute of Technology. He was appointed William J. Brown Professor of Mechanical Engineering at Carnegie Mellon University, Pittsburgh, Pennsylvania, in 1982. Dr. Chigier was a recipient of the ASME's Lewis F. Moody Award (1965), the Institute of Fuel's Lubbock-Sambrook Award (1968 and 1975), and the Tanasawa Award (1988). Prof. Chigier's first book, *Combustion Aerodynamics* (1972), was coauthored with Prof. John Beer of Massachusetts Institute of Technology and was subsequently translated into Japanese (1976). His textbook, *Energy, Combustion and Environment*, was published by McGraw-Hill (1981). He is the author or coauthor of over 300 papers. In 1974, he founded the international review journal, *Progress in Energy and Combustion Science*. In 1992, Professor Chigier was named a fellow of the ASME. He is editor of the *Combustion: An International Series*, published by Hemisphere Publ. Corp. He was one of the founding members of the International

Institute of Liquid Atomization and Spray Systems (ILASS) and founding editor of the archival research journal, *Atomization and Sprays*. In 2006 he received the Arthur Lefebvre Award for distinguished contributions in the field of atomization and sprays.

Dr. Heather Emady is an assistant professor of chemical engineering in the School for Engineering of Matter, Transport, and Energy at Arizona State University. Dr. Emady's Ph.D. at Purdue University was on granule formation mechanisms for single drop granulation. Following her Ph.D., she joined Procter & Gamble as a postdoctoral researcher, working on particle sedimentation in microstructured fluids. Dr. Emady then joined Rutgers University as a postdoctoral researcher for the Catalyst Manufacturing Science and Engineering Consortium, as well as for the Engineering Research Center for Structured Organic Particulate Systems, before joining ASU in January 2015. Her research interests include single drop granule formation, discrete element method modeling, and particulate process and product design in general.

Dr. Tevfik Gemci was awarded B.Sc. (1983) and M.Sc. (1985) degrees in mechanical engineering from Istanbul Technical University. He was a recipient of a research fellowship from the German Academic Exchange Council (DAAD – Deutscher Akademischer Austauschdienst) from 1987 to 1989. He received his Ph. D. (1993) degree in mechanical and environmental engineering from the Technical University of Kaiserslautern in Germany. He was appointed as assistant professor of environmental engineering at Sakarya University in Turkey from 1996 to 1998. In 1999 he joined Carnegie Mellon University, Pittsburgh, Pennsylvania, as a research professor and worked with Dr. Norman Chigier in the Spray Systems Technology Center till 2004. He continued his academic research as research professor in the Mechanical Engineering Department at the University of Nevada Las Vegas from 2004 to 2007. He is the author or coauthor of over 40 papers. From 2008 to 2015 he has worked as an expert consultant in the field of thermal fluids science and spray and aerosol dynamics on a wide variety of industrial projects of mechanical, environmental, pharmaceutical, biomedical, and biotechnological nature. Since 2015 he started to work as a senior quality assurance and validation engineer at B. Braun Medical Inc.

Prof. Karen Hapgood spent 5 years in the US pharmaceutical industry before joining Monash University in 2006, where she cofounded the Monash Advanced Particle Engineering Laboratory. Prof. Hapgood is most well known for her work on nucleation during wet granulation, and her broader research interests include drug delivery via powders, granules, and tablets. She received the AAPS New Investigator in Pharmaceutics and Pharmaceutical Technology Award (2006), is a fellow of IChemE and Engineers Australia, and is an executive editor of *Advanced Powder Technology*.

Dr. Marko Hilden (Senior Research Fellow, the University of Queensland, Julius Kruttschnitt Mineral Research Centre (JKMRC)). Marko's research interests cover a range of topics in mineral processing and comminution including Semi-Autogenous Grinding and High-Pressure Grinding Roll modeling and optimization, mathematical modeling of mineral liberation, conceptual circuit design and simulation, and comminution energy efficiency. In 2007 he was awarded a Ph.D. studying industrial screen separation at the JKMRC. His earlier experience includes a process-improvement engineering role at iron ore operations in Western Australia's Pilbara region and as a coal technology researcher based in Melbourne.

Prof. George E. Klinzing holds degrees in chemical engineering from the University of Pittsburgh and Carnegie Mellon University and a Ph.D. degree from the later. He recently was given an honorary doctor of engineering degree from the University of Newcastle in Australia. He joined the University of Pittsburgh as a professor on its USAID Ecuador project at Universidad Central in Quito. After 3 years he returned to the main campus of the University of Pittsburgh and began his teaching and research career that has span 50 years with summer employment in the energy industries. He helped to produce 54 M.S. theses and 25 Ph.D. dissertations in the areas of mass transfer and particle technology. His research spans a wide spectrum of topics in particle technology mostly in the experimental field addressing challenging problems from pressure behaviors to electrostatic influences. He has written two books on pneumatic conveying. In 1987 he joined the administrative team of the Swanson School of Engineering as the associate dean for research followed by a 17-year stint as the vice provost for research for the University. He helped to establish university-wide interdisciplinary centers: the Center for Energy, the Petersen Institute for NanoScience and Engineering, the Center for National Preparedness, and the Center for Simulation and Modeling. He began several multidisciplinary initiatives on such topics as suffering, geriatrics and ambulatory and cognitive capacity, wisdom and aging, cybersecurity and immunology, business of humanity, nanoparticle safety, and synthetic biology. Prof. Klinzing is a fellow of the American Institute of Chemical Engineering and the American Association for the Advancement of Science. He holds the Life Time Achievement Award from the Particle Technology Forum of the A.I.Ch.E. as well as the Gary Leach Award and the McAfee Award. He serves on the editorial board of *Particulate Science and Technology* and *Powder and Bulk Engineering*. At the Swanson School of Engineering he holds a Whitford Professorship. He has produced over 250 research publications and holds 3 US patents and 6 US copyrights.

Dr. Ing. Ger J.M. Koper is associate professor at the Department of Chemical Engineering of the Delft University of Technology. He became electronics engineer in 1975 and worked as such in the Department of Histochemistry and Cytochemistry of the Leiden University from 1976 to 1985. He then moved to theoretical physics and received his Ph.D. degree from Leiden University in 1990 on *aging in spin glasses*. From 1990 until 2000 he was senior lecturer at the Physical and

Macromolecular Chemistry group. He coauthored more than 150 publications in refereed journals.

His main teaching emphasis lies on applied and advanced thermodynamics, specializing on nonequilibrium phenomena, as well as *interfacial engineering*, focusing on colloidal stability and particle synthesis. He regularly (co-)organizes national and international conferences as well as training schools for graduate students, postdocs, and professionals.

His research field is colloid and interface science with emphasis on aggregation and adsorption phenomena. His experimental expertise includes optical techniques such as transient electro-optical birefringence (Kerr effect), reflectometry, microscopy, and static and dynamic light scattering as well as dielectric spectroscopy and electro-kinetics. His current research topics are in the field of materials research for polyelectrolyte membrane fuel cells, in particular nanostructure formation, and applied thermodynamics.

His topics of particular interest include investigation of aggregation phenomena of colloidal particles and binding of protons and metal ions by polyelectrolytes.

He served in many committees and boards and currently is both member of the management team and working group leader of the COST Action CM1101 *Colloidal Aspects of Nanoscience for Innovative Processes and Materials* and newsletter editor for the *International Association of Colloid and Interface Scientists*.

Dr. Roman Latsuzbaia currently holds a position as scientist at Netherlands Organization for Applied Scientific Research, TNO, in Zeist, the Netherlands. In 2010 he received his M.Sc. degree in applied chemical engineering from the University of Manchester with a thesis on the synthesis of catalyst nanoparticles for direct methanol fuel cells. In 2011 he moved to Delft University of Technology, group of Advanced Soft Matter, and received his Ph.D. under the supervision of Dr. Ger J.M. Koper on a topic of fuel cell catalyst production and regeneration. His main areas of expertise are in fuel cells, electrochemistry, colloid and interface science, and production and characterization of nanomaterials. His current focus is in the field of electrochemical synthesis of specialty chemicals.

Dr. Lian X. Liu (Senior Research Fellow, the University of Queensland, Julius Kruttschnitt Mineral Research Centre (JKMRC)) obtained her bachelor and master's degree in mineral processing from Northeastern University, China, which is the premier institution in mining and minerals. Her master's thesis was focused on comminution processes, specifically on grinding kinetics and modeling. She did her Ph.D. at the University of Queensland in the area of population balance modeling in granulation including particle breakage processes. Dr. Liu has developed a strong research profile and expertise since her Ph.D. in granulation, breakage, and compaction of particles, through many applied research projects funded by the industry, ARC and CSIRO. She is now working at JKMRC in the areas of ore characterization and comminution process modeling.

Prof. Don McGlinchey is the head of the Centre for Industrial Bulk Solids Handling and leader of the Design, Process and Manufacturing Research Group at Glasgow Caledonian University. He has undertaken consultancy projects for both multinational companies and small to medium enterprises (SMEs) and delivered short courses in Europe and the USA. He is a chartered physicist, with a B.Sc. in physics and a Ph.D. in on the subject of the effect of vibration on particulate materials. Don is the editor of two books and has authored over 50 research articles. He is a participant in the European Federation of Chemical Engineers Working Party on the Mechanics of Particulate Solids (EFCE – WPMPS) and a member of the Institution of Mechanical Engineers (IMechE) Bulk Materials Handling Committee. His current research interests include multiphase flow instrumentation and the application of computational fluid dynamics (CFD) techniques to gas-solids conveying.

Dr. Gabriel M.H. Meesters has a B.Sc. and M.Sc. in chemical engineering with a major in bioprocess technology from the Delft University of Technology. He has a Ph.D. in particle technology also from the Delft University of Technology. He worked at biotechnology companies like Gist-Brocades in the Netherlands, as well as for Genencor International and currently at DSM in research and development in the Netherlands. In all these functions he was working on formulation and product development. Since 1996 he holds a part-time position at the Delft University of Technology, as assistant professor at the faculty of Applied Sciences, first in the Particle Technology group, later the Nanostructured Materials Group, and currently in the Product and Process Engineering group. He supervised over 15 Ph.D. students and more than 50 M.Sc. students. He published around 60 refereed papers, holds around 15 patents and patent applications, and is coauthor and coeditor of the book *Particulate Products: Tailoring Properties for Optimal Performance* (2014; Springer). He (co-)organized several international conferences in the field of particle technology and was president of the World Congress on Particle Technology in 2010.

Em. Prof. Dr. Henk G. Merkus graduated in physical organic chemistry at the University of Amsterdam. He worked several years at the Royal Dutch Shell Laboratories in Amsterdam on research in the field of detergents and industrial chemicals, followed by development work on thermal wax cracking for production of C_2–C_{14} olefins and on acid-catalyzed synthesis of carboxylic acids from C_3–C_6 olefins. Then, he made the change to analytical chemistry, involving both measurements and method development with a large variety of techniques and methods, first at Shell's process development department in Amsterdam and later in the chemical engineering department of Delft University of Technology. Gradually, the analytical horizon widened: first surface area and porosity measurements were added to chemical analysis, later followed by particle size analysis. He is author of the book *Particle Size Measurements: Fundamentals, Practice, Quality* (2009 Springer) and many journal articles, as well as coauthor and coeditor of the book

Particulate Products: Tailoring Properties for Optimal Performance (2014 Springer). Moreover, he participates in standardization activities regarding particle size measurement, both in the Dutch NEN and the International Organization for Standardization (ISO) (TC24).

Dr. Sam Palaniandy (Senior Research Fellow, the University of Queensland, Julius Kruttschnitt Mineral Research Centre (JKMRC)). His main research interest is in stirred milling and fine grinding technology. He has experience in conducting performance evaluation of stirred mills and has conducted stirred mill circuit surveys in Australia, South America, Africa, and Europe. His current activities include site surveys, development of ore characterization method for fine particles, small-scale test for stirred mills, and process modeling.

Prof. Dr. Karin Schroën obtained her Ph.D. degree in food process engineering from Wageningen University (still agricultural at the time) in 1995, working on an emulsion membrane bioreactor within the group of Klaas van ‘t Riet. Since then she has worked as a postdoc at University College London (UK) focusing on downstream processing of two-phase bioconversions and in the Biotechnology group of Wageningen University, studying fundamentals and process design for enzyme-catalyzed antibiotic synthesis. She became an assistant professor within the laboratory of food process engineering at Wageningen University in 2001, an associate professor in 2010, and was appointed full professor in September 2012. Her fields of expertise comprise emulsification, microtechnology, encapsulation, membrane separation, modeling, surface modification, and food process engineering in general. She teaches courses at various levels, has more than 120 scientific publications to her name, and holds seven patents.

Prof. Dr. Dietmar Schulze is professor of mechanical process engineering at the Ostfalia University of Applied Sciences Braunschweig/Wolfenbüttel since March 1996. He is an expert in the areas of powder characterization and handling (e.g., silo design). Dietmar Schulze has published more than 100 papers on powder technology including a recent book named *Pulver und Schüttgüter* and its English translation *Powders and Bulk Solids*. He received the Technology Transfer Award from the local Chamber of Industry and Commerce in 1994 and the Arnold Eucken Award from VDI in 1995.

Dietmar Schulze studied mechanical engineering with focus on mechanical process engineering at the Technical University of Braunschweig and graduated in 1985. From 1985 to 1991 he worked as scientific assistant of Prof. Jörg Schwedes at the Institute of Mechanical Process Engineering in Braunschweig. He finalized his thesis on silo stresses and discharge in 1991. After this, he and Prof. Jörg Schwedes founded the consultancy “Schwedes + Schulze Schüttguttechnik” focusing on powder characterization and silo design. Since 1993 Dr. Schulze has been developing and manufacturing ring shear testers for the measurement of flow properties of powders.

Dr. Fengnian Shi (Principal Research Fellow, the University of Queensland, Julius Kruttschnitt Mineral Research Centre (JKMRC)), a Chinese by birth, joined the JKMRC in 1988 and has been associated with the JKMRC for the past 27 years. He was awarded a Ph.D. degree at the University of Queensland in 1995. His research covers wide areas in comminution for the mineral and coal industries, including breakage characterization for ore, coal, and coke, mathematic modeling of impact crusher, SAG mill, ball mill, HPGR and vertical spindle mills, circuit simulation, and plant optimization. He jointly holds two international patents on the rotary breakage tester (JKRBT) and the breakage characterization method, part of which is presented in Section 6.4. He is a team leader of electrical comminution, developing the technology of ore pre-weakening, coarse particle liberation, and ore pre-concentration using high-voltage pulses. He has published more than 100 papers, the majority being in the peer-reviewed journals and refereed conference proceedings.

Dr. Rachel Smith is a lecturer in the Department of Chemical and Biological Engineering at the University of Sheffield, UK. Dr Smith holds a B.Eng. in chemical engineering from the University of Queensland and a Ph.D. in chemical engineering from the University of Queensland, during which she studied wet granulation and granule breakage. From 2008 to 2012 Dr Smith conducted post-doctoral research into wet granule nucleation and particle coating and joined the University of Sheffield in 2012. Dr Smith's research interests cover a range of industrial particulate processes including the study of granulation, particle and powder coating, and powder flow and fluidization, using a combination of experimental and computational simulation tools. Dr Smith was awarded a Royal Society Industry Fellowship in 2015.

Dr. Steve Tarleton is a senior lecturer in the Department of Chemical Engineering at Loughborough University who has more than 30 years' experience of working in filtration and separation. Details of his publications and achievements are available at http://www-staff.lboro.ac.uk/~cgest.

Dr. Pieter Vonk is working as senior scientist in crystallization and solid processing in the Advanced Chemical Engineering Solution Department of DSM in the Netherlands. He has a 14-year experience in supporting DSM in the design, troubleshooting, and optimization of crystallization, solid processing, and solid handling processes. Products supported range from pharmaceutical products (commercial production and pharmaceutics for clinical trials), food specialties (vitamins), and industrial chemicals (melamine, ammonium sulfate, succinic acid).

Before joining DSM he had been working as assistant professor in the Industrial Pharmacy at the University of Groningen, where the focus of the research was on solid processing aspects of pharmaceutical production (fluid bed and high shear agglomeration and coating for controlled-release applications). He holds a Ph.D. in Chemical Engineering from the University of Groningen with the subject of multicomponent diffusion in bio-separations (ultrafiltration and chromatography).

Prof. Dr. Richard Wakeman (B.Sc., M.Sc., Ph.D., D.Tech., C.Eng., FR.Eng., F.I. Chem.E.) is a consultant chemical engineer (www.richardwakeman.co.uk) who has worked internationally in Europe, Asia, and the USA since 1971. He is an emeritus professor at Loughborough University (UK), was previously professor of process engineering at Exeter University, and has been a visiting professor at Pardubice University (Czech), a university professor at Chung Yuan University (Taiwan), and the Golden Jubilee Fellow at the University of Mumbai (India). He was awarded an honorary doctorate of technology by Lappeenranta University of Technology (Finland) in 2012.

He obtained a Ph.D. in chemical engineering from the University of Manchester Institute of Science and Technology (UMIST) after working in the chemical industry as a troubleshooting engineer; he then joined Exeter University where he later became professor of process engineering. His interests are in solid/fluid separation science and technology, membrane processes, and particle processing. He has published some 400 papers, patents, and books, and his work has received recognition through several awards, including the Junior Moulton Medal (1978), Moulton Medals (1991 and 1995) and Arnold Greene Medal (2008) of the Institution of Chemical Engineers, the Suttle Award (1971) and the Gold Medal (1993, 2003, and 2005) of the Filtration Society, and the Chemical Weekly Award of the Indian Institute of Chemical Engineers (2005).

He has been the executive editor of *The Transactions of The Institution of Chemical Engineers* and is an editor of *Filtration*. He is a fellow of the Institution of Chemical Engineers, a chartered engineer, and was elected a fellow of the Royal Academy of Engineering in 1996. He was the honorary secretary of the Filtration Society between 2000 and 2013 and is a past chairman.

Dr. Ralf Weinekötter studied chemical engineering in Karlsruhe (Germany) and Nancy (France). His Ph.D. thesis at the *ETH* Zurich (Swiss Institute of Technology) covered the "continuous mixing of fine particles." Part of the research was executed at the Particle Technology group of DUPONT in Delaware (USA). During his time as senior lecturer at the Institute of Process Engineering *ETH* Zurich), he published with H.R. Gericke the book *Mischen von Feststoffen* (Springer Verlag 1995). This book was followed by *Mixing of Solids,* Kluwer Academic Publishers (2000). In 1994 Weinekötter joined the R&D group of Gericke AG, Switzerland. He became responsible for the development of powder mixers. Over the years he has held various positions within this international company which specialized on feeding, mixing, and conveying installations for powders. Today he is managing director of Gericke AG. He contributed to the latest edition of Perry's *Chemical Engineers' Handbook* with the chapter on solids mixing.

Dr. Mohsen Yahyaei (Research Fellow, the University of Queensland, Julius Kruttschnitt Mineral Research Centre (JKMRC)) obtained his Ph.D. in mineral processing from Shahid Bahonar University of Kerman in 2010. He has more than

13 years' industrial and academic experience through conducting successful industry-based research work. Planning and implementing multidisciplinary projects in comminution, flotation, and thickening for a number of industrial projects provided him a strong desire to conduct applied research underpinned by sound fundamental understanding. Since 2011, when he joined JKMRC, his research is focused mainly on two areas, namely, surface breakage of rocks and liner wear modeling. Although Mohsen is an early career researcher at the University of Queensland, he has a strong track record of publications by publishing more than 10 journal papers and presenting over 35 papers in high-profile international conferences and winning many awards for his research and papers.

Chapter 1
Introduction

Henk G. Merkus

Abstract A great variety of industrial products is based on particulate materials. Behavior and performance of such products depend upon particle size and shape in the base materials, in addition to chemical and structural characteristics. An overview of the major production and handling processes of particulate materials as well as the basic characteristics for their design, monitoring and control is presented in this book. It is meant to facilitate easy comparison and good choices of production and handling processes and of equipment. Topics for particulate production comprise size enlargement – crystallization, precipitation, granulation and extrusion – as well as size reduction, viz. comminution, atomization and emulsification. In the chapters on handling, mixing and segregation, filtration and sedimentation, flotation, classification, storage and transport are discussed. The quality of batches of base materials and products is assessed through measurements in samples. Adequate care of the sampling process to obtain representative samples is required, especially in cases where the batches show segregation, i.e. when the composition is different at different locations in the batch. Then, deviations in non-representative samples may be substantial. The quality of products is often described in terms of PSD parameters in addition to descriptors that are more closely related to behavior. Unfortunately, the choice of PSD parameters is often made without good reasoning, causing the chosen parameters to be sub-optimal. Therefore, suggestions are given for optimum choices. The behavior of bulk powders and concentrated dispersions depends in addition to particle characteristics upon the degree of particle packing and the inter-particle forces, and for dispersions also upon the particulate concentration, the surface type and the zeta-potential. Optimum rheological behavior of such concentrated systems, powders as well as liquid dispersions, is of prime importance, during processing as well as for product quality. Some examples are presented for the parameters that have an empirical relationship to practical situations. The choice of PSD parameters that are typically used for control of processes is easier than that for product quality, since usually only adequate repeatability and instrument robustness are important for good

H.G. Merkus (✉)
Em. Prof., Chemical Engineering Department, Delft University of Technology,
Pijnacker, The Netherlands
e-mail: henkmerkus@hetnet.nl

© Springer International Publishing Switzerland 2016

H.G. Merkus, G.M.H. Meesters (eds.), *Production, Handling and Characterization of Particulate Materials*, Particle Technology Series 25,
DOI 10.1007/978-3-319-20949-4_1

Table 1.1 Examples of particulate materials

Dry powders	Sugar, flour, starch, sand, cement, coal, pigments, polymer beads, peas, diamonds, toner powder, etc.
Liquid mixtures	Emulsions (L/L) + suspensions (S/L): milk, butter, margarine, creams, bacteria, blood, paint, etc.
Solid mixtures (S/S)	Ores, sediments, pharmaceuticals, etc.
Aerosols (L/G + S/G)	Fog, mist, sprays, etc., which may consist of dust, salt, sand, coal, ore, inhalers, etc.
Gas bubbles in medium	Whipped cream (G/L), insulating foam (G/S), etc.

control. This chapter gives an introduction to the general background and the challenges of selecting adequate processes and to the relevant particulate parameters.

1.1 Objective of This Book

Particulate products make up around 80 % of all chemical products. They have in common that they contain one or more particulate base materials, which may come from a wide variety of sources. Sometimes the source is natural, sometimes the materials are man-made through a chemical reaction (see Table 1.1).

Often the particles are solid, sometimes tiny droplets or air bubbles. Always, some kind of production process is involved to give the materials the desired size range and/or shape of the particles, composition and performance quality of the ultimate product.

In addition to the production processes, mechanical processes take care of mixing (to counteract segregation and/or reach a homogeneous composition), transport and storage. Optimum particle size range and shape are essential for best product quality, since these properties relate to product behavior and performance [13].

Many textbooks and articles have been written on the techniques applied for the production and handling of particulate materials. They provide a wealth of detailed information but are nearly always restricted to a single technique, base material or field of application. The objective of this book is to give an overview of all techniques with sufficient background to provide insight and understanding. Thus, it is meant to facilitate comparison of all techniques and equipment applied for improved implementation in design and control of particulate processes.

1.2 Production and Handling Processes

A wide variety of particulate materials and products exists, as illustrated in Table 1.1. Some originate from natural sources, others are man-made through some chemical reaction. Most often some mechanical processing is involved.

Given the vast differences between these materials, it is clear that they require different production techniques. The main production processes can be identified as [16, 17, 19]:

- **Size enlargement**. Particles can be formed from solutions through crystallization and precipitation. Most often, purification in addition to particle formation is the objective.

  Crystallization is executed in both batch and continuous processes. Its main goal is production of crystalline particles of high purity, usually from solutions in which other components are present as well. In batch processes crystals are formed by cooling, by solvent evaporation of saturated solutions, through addition of anti-solvents or by changing the pH; often, seeding with small particles is applied in order to start the crystallization. Continuous processes typically proceed by solvent evaporation. Here the 'seeds' are often the result of attrition of the formed larger particles. The particle size of the product typically approaches the millimeter range (see further Chap. 2).

  Precipitation results from the reaction between two (soluble) ionic compounds, the product of which is insoluble in the medium, usually water. The resulting particle size of the precipitate is very small (nanometer range), since its solubility product is very abruptly overstepped by the addition of the one dissolved compound to the other. Precipitation processes usually also involve agglomeration or aggregation of the primary particles, through which particle size is significantly enlarged (see further Chap. 3). Typically, precipitation occurs in batch processes.

  Polymerization of aqueous monomer emulsions may be considered as a variety of precipitation, since the monomers are very slightly soluble whereas the solubility of oligomers and polymers is zero. The polymerization typically starts in the continuous aqueous phase. At the point that the formed oligomers become insoluble, they form emulsion droplets (primary particles as a dispersed phase).

  During these precipitation and polymerization processes spontaneous agglomeration, aggregation or coalescence of primary particles often occurs already at early stages of particle formation and growth. This is especially true if the primary particles have little or no electric charge or are insufficiently stabilized by surfactants.

  Solid particles in the micrometer range can be enlarged to millimeter sizes by agglomeration/aggregation, granulation and extrusion. These processes are very helpful to reach the desired properties – e.g. flowability and particle shape – for the end product.

Agglomeration is the formation of clusters having point contacts between the primary particles. It often occurs not only in dry powders when the particles are smaller than about 10 μm or are humid (by formation of liquid bridges), but also in liquids when the particles have little or no surface charge. Here, it is induced by Brownian motion and stirring. The bonding strength can be substantially increased by the formation of solid bridges coming from dissolved material.

Aggregation is the formation of clusters in which the particles have a stronger bonding than in agglomerates due to the presence of side contacts. Aggregation typically takes place in liquid dispersions. Here too, solid bridges can be formed coming from dissolved material.

Note that agglomeration and aggregation may be counterproductive when they occur unintentionally as they can give rise to undefined lumps.

Granulation and tabletting is applied to improve the flow characteristics of products as well as to homogenize mixtures of ingredients. Granulation is applied for a wide variety of products. It is typically executed in equipment where the ingredient particles are continuously mixed and sprayed with binder liquid, the combination of which leads to clustering and compaction of the particles. The compacts may be used as such, but they can act also as precursors for a tabletting stage, in which they are further compressed for better strength and shape (see further Chap. 4).

Extrusion was first introduced in the processing of plastic granules. Soon after, its application expanded to metals and rubber. Gradually, it has also become very popular for food applications in e.g. cereals, pastas and snacks, since subsequent steps in the preparation process can be executed in the same extruder. In this technique use is made of the plastic deformability of products. Sufficient plasticity is usually reached at somewhat elevated temperatures, for some products in combination with the available water content in the feed (see further Chap. 5).

- **Size reduction**. Often, the dry solid particles produced are too large for their envisaged application in a product. One example is when they result from mining of ores. Another reason may be that the drying process involved in particle production has caused the presence of large lumps that consist of agglomerates and/or aggregates. Then, particle size must be reduced by *comminution (crushing or milling)*, which involves dry or wet processes of breakage, attrition and de-agglomeration. Comminution operations usually consume a lot of energy. Besides differences in equipment, this causes that the total comminution operation often is split into several stages for better economy (see further Chaps. 6 and 11).

  Crushing involves usually dry breakage of large pieces from up till about 1 m into cm-sized particles. Crushers of various types are typically applied for (fairly) hard materials, cutters for tough ones.

  Milling is executed in both dry and wet state to result in millimeter-sized or smaller particles. Dry milling usually is performed in roller mills, wet milling in ball mills or bead mills. The advantage of wet processes is that the resulting small (micrometer-sized) particles can be stabilized to prevent re-agglomeration and aggregation.

Dispersion of dry powders in a liquid is also often called milling. It is applied to break up agglomerates and when stable particulate suspensions are required. It involves application of energy and dispersants/surfactants to reach sufficient de-agglomeration and suspension stability.

Atomization, spraying and nebulization are processes of producing droplets from a liquid in an air stream. If solid particles are aimed at, then the liquid is a melt or a solution, and the droplet formation is followed by cooling or drying. The resulting size of droplets and particles may vary from millimeters to smaller than micrometers (see further Chap. 7).

- **Emulsification**. Here, the particles are in liquid form. They are dispersed in another, immiscible liquid, typically through application of high energy/shear in various types of equipment and stabilized by means of surfactants/emulsifiers. This is a special way of size reduction for liquids. Particle size is usually in the nano- to micrometer range. Typical applications of emulsions are in food, pharmaceuticals and lubricants. Adequate stabilization is required to guarantee a long shelf life (see further Chap. 8).

In addition to production processes, some kind of mechanical processing is often involved, in which particle size is not intentionally altered:

- **Mixing**. Mixing is an important process step for particulate materials. Its goal is to reach a homogeneous product. For single particulate materials, the reason for mixing is the existence of (undesired) differences in particle size distribution at different positions within a product batch. This may be caused by process fluctuations, but more often it is mainly the result of segregation (see below). Always, the degree of segregation shall be checked in batches of particulate materials.

Segregation is the opposite process from mixing. It usually occurs in dry, free-flowing materials in relation to differences in particle size and/or density. It is induced by vibration of the materials during transport or by aeration upon storage in silos. The detrimental effect of such heterogeneity on product quality can be severe, since product properties relate to particle size and its distribution. Note that some degree of segregation especially in free-flowing materials has to be accepted in practice since ideal mixtures do not exist in practice.

Mixing of a segregated material batch is required to reach a sufficiently homogeneous size distribution within the batch. Moreover, it is essential for particulate products when they are composed of different ingredients. For such products, mixing is necessary to reach a homogeneous product composition, in addition to particle size distribution. Especially for products, which are composed of small quantities of a given component and a majority of other particulate material(s), such as active material and excipients for pharmaceutical pills, adequate mixing before further processing is essential. Decreasing the particle size to below about 20 μm or making the powder humid can be used to counteract segregation during mixing. Both measures increase the particle-particle interactions and make powders cohesive, thus resulting in lower or no sensitivity to segregation. Note that cohesive powders may greatly challenge steady transport and adequate silo charge and discharge. Thus, these powders require adequate choices in the design and operating conditions of the equipment (see further Chap. 9).

- **Filtration/sedimentation**. Filtration and sedimentation are typically applied to separate solid particles from a liquid or air, in which they are produced or dispersed. Sometimes, the goal is collection of the particles, sometimes it is obtaining a clear liquid or clean air. This may be the case for protection of the environment or for obtaining a clean room. Typically, the efficiency of collection is expressed in a mass percentage of collected particles in comparison to originally present particles. The quality of clearing of a liquid is usually expressed in a degree of remaining turbidity or percentage transmission, for clean gas or clear liquids in the mass or number of particles smaller than a given size per unit volume, e.g. *PM10*.

  Filtration uses the porosity of mechanical filters or particulate filter beds for particle collection or removal from suspensions. Typically, removal of particles from suspensions concerns particle sizes larger than about 1 μm; in air, it is usually below 1 μm. Mechanical filters in the form of sieves can only remove particles if they have smaller aperture sizes than the size of the particles. Therefore, industrial sieves are limited to collection and removal of larger particle sizes. An advantage of sieves is that the particles can be easily separated from the sieves. Filter cloth and particulate filter beds remove particles based on the existing void sizes and the possibilities for adsorption. Because of the presence of very small voids and the adsorption possibilities, also submicrometre particles can be removed. Well known examples of application of filter beds are beer filtration and application of sand beds for filtration of drinking water. Filter cloth is e.g. applied for obtaining clean air. Recovery of the particles from the cloth or the beds is usually incomplete (see further Chap. 10).

  Sedimentation is the result of positive density differences between particles and suspension liquid, which cause settling of particles when they are larger than about 1 μm (depending on density). Sedimentation of smaller particles or when the density difference is small can be enhanced in centrifuges. The opposite case – particle density smaller than that of the liquid – is named *creaming*. A well known example is the creaming of milk. Sedimentation can be applied for clearing of liquids as well as for particle collection. The latter typically requires some kind of filtration following sedimentation, and drying of the resulting cake. Various types of sedimentation are also applied in classification (see below and also Chap. 10).

- **Flotation** is another way to separate solid particles from a liquid. It is usually applied in cases where specific minerals are to be separated, or Brownian motion prevents particles from adequate sedimentation, or the density difference between particles and liquid is small. In this process, various types of chemicals (pH adjusters, surfactants and ionic substances) are applied to make the surface

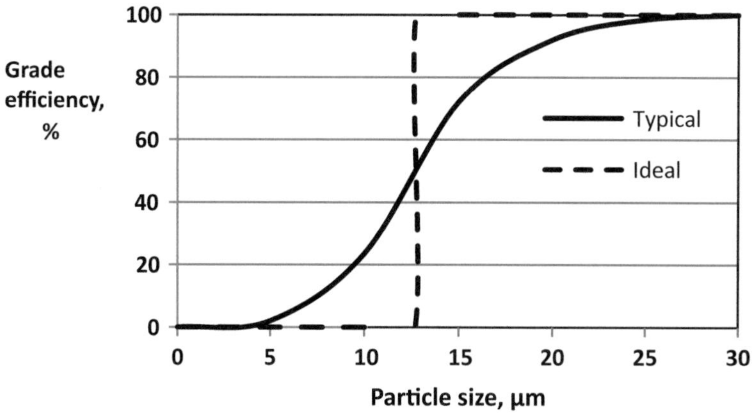

Fig. 1.1 Grade efficiency curve for a gas cyclone, showing typical and ideal separation (Adapted from [16])

of target (mineral) particles hydrophobic and to generate air bubbles to which the hydrophobic particles are to be attached. This air causes the particles to float as a froth at the liquid surface, which facilitates their removal from the liquid. This process is e.g. widely applied in the production of valuable minerals from ores and in waste water treatment (see further Chap. 11).

- **Classification**. Classification of produced particulate materials is applied if their size range does not match the desired size range, by removing fines or coarse particles, by separating particles having a different density, or to remove particles completely from a gas or liquid stream. Typically, (hydro)cyclones, sieves or gravitational or centrifugal sedimentation units are applied, in particulate mixtures with air or liquid. For dry mixtures, also magnetic and electrostatic separators are used. (Hydro)cyclones typically operate at particle sizes in the range of about 1–500 µm (depending on particle and fluid density). The grade efficiency for separating particles having a size distribution into fractions of fine and coarse product is often used to sketch the separation efficiency against particle size. As an example, a typical grade efficiency or Tromp curve for a gas cyclone is presented in Fig. 1.1 (see further Chaps. 10 and 12).
- **Storage and powder discharge**. Intermediate storage in bins, silos or heaps is required when production and usage are not in a direct sequence but at different times or locations. Silos have the advantage that they provide for a closed environment so that effects by humidity or oxygen can be avoided. Note that bulk storage always leads to some kind of consolidation of the lower parts of the stored material, which increases its bulk density and may decrease its flowability. Note also that segregation often occurs during the total of feeding, storage and discharge. Moreover, poor silo and hopper design typically leads to problems upon discharging the particulates. Then, this discharge from the silo may proceed only for a small part while the product near the silo walls remains stagnant. In order to avoid such problems, flow properties and bulk density of the

product to be stored should be determined and hoppers and silos, including inlet, outlet and additional equipment, be designed in relation to potential stresses and segregation. This is essential for optimum operation (see further Chap. 13).

- **Transport/conveying**. Often, production, storage and further usage do not occur at the same location but at some distance. Then, transportation of the material is necessary. Different types of equipment are applied, which all have their economical and technical pros and cons. Pneumatic and hydraulic conveying in pipelines is often favored for short and medium long distances (up till a few kilometers). If the transport distance is short, belts are used also. For long distances, product batches are transported in bags, trucks, wagons and ships. Note that segregation and dusting often occur during transport and, thus, should be taken care of. Note further that generation of electrostatics in powder flows may be strong, which can cause dust explosions resulting in dramatic casualties and losses of property. Prevention of dust explosions makes adequate care for good grounding and nonconductive piping, etc. in pneumatic conveying essential (see further Chap. 14).

Note: One of the founding fathers of particle technology, Professor Hans Rumpf of the University of Karlsruhe, named the above unit operations mechanical process technology (mechanische Verfahrenstechnik), because they deal with the transformation of material systems by predominantly mechanical operations [18]. Still, they are the subject of R&D and education in the corresponding departments of many German technical universities. In many other countries, at least some of the unit operations are studied and taught in other departments, such as chemical, civil and equipment engineering. Major companies typically form project groups of experts from different departments for the task to make optimum choices for all processing equipment and the appropriate operating conditions for a given material or product.

Typically, the processing of particulate materials is considered as very complex in view of the many parameters that can play a role, such as different phases (solid, liquid and gas), particle size distributions, different particle shapes, often high concentrations causing particle-particle interactions, etc. It is impressive to see that, despite this complexity, significant progress has been made in the development of modeling and simulation packages that approach reality, although they are still mainly based on empirical equations (see Chaps. 2, 3, 6, 12, and 14).

1.3 Characterization of Particulate Bulk Products and Processes

As shown in Table 1.1, particulate materials in a wide variety form the basic materials for many industrial products, in almost all industrial areas. Typically, each product requires a specific particle size range, particle shape as well as bulk properties for its base material(s) in addition to specific other components in the end products. The reason is that the performance quality of particulate materials and products depends upon particle size, size distribution, shape and bulk properties, since these relate to the way that the particles interact with each other and the surrounding medium. Quality aspects of bulk products involve for example [13, 19]:

- rheological properties of powders (important for steady flow and dosage)
- fluidization behavior (important if a fluid bed is applied in its application)
- dusting behavior (related to health hazards and explosion risks)
- particle packing density (related to e.g. bulk density, concrete strength, resistance of filter beds to fluid flow and rheological behavior)
- rheological behavior of dispersions, especially at high particulate concentrations (important during processing as well as with respect to e.g. taste or paint application)
- optical properties (related to color, gloss, transparency and hiding power of paint and sunscreen)
- surface properties (chemical and physical) and zeta-potential
- dissolution rate
- sensorial characteristics (e.g. taste of food, sweets and beverages).

Characterization of industrial materials is required in order to test whether they comply to their required quality and specifications, or for monitoring and control of the process, or with respect to environmental requirements in view of health and explosion behavior. This characterization usually involves both the particles and the bulk product. Measurement of particle size distribution (PSD), shape and/or porosity of the particles is often the first approach as it is fairly easy. Characterization of the bulk properties is usually second; it is more complex and product specific, but at least as important. It regards bulk density and rheological properties, which are essential in relation to handling and processing (see further Chaps. 5, 13, and 15).

Typically, test samples are used to characterize the quality of material batches. Always, such samples shall be representative for the product batch and contain a sufficient number of particles over the full size distribution. Most often such representativeness can only be obtained, due to segregation and/or process fluctuations, by collecting samples at different times during particulate flow or from different locations in a bulk, mixing them to reach a composite sample and, if necessary, dividing this composite sample to reach a smaller test sample. Note that sometimes very small test samples are used for analysis. In such small samples, the number of particles present may severely limit the quality of the results. ***Especially highly segregated materials require adequate quality of the sampling procedure, which always should be checked***. For compliance testing, the materials should preferably be analyzed after production as well as before application, in view of potential segregation during transport and storage (see Chaps. 9 and 15).

The quality criteria for the total analytical method, including sampling, sample preparation and measurement, are dictated by the performance requirements of the ultimate product. In general, the analysis aims at characteristics of the particles (size distribution, particle shape and porosity), at bulk solids properties as bulk density, flowability and wall friction, and for dispersions at particulate concentration and rheological behavior.

Measurement of the PSD is often the first approach for product quality control in view of the ease of execution and its relationship with both bulk properties and product quality. Its measurement method requires, besides obtaining a representative test sample, also adequate dispersion, concentration range, size range, analysis time, precision (both repeatability and reproducibility), accuracy, resolution and sensitivity. Note that inadequate sampling and dispersion as well as incompetent analysts, who deviate from given instructions or overlook unexpected behavior during analysis, may strongly deteriorate the quality of the analysis results [12]. Moreover, the optimum control of many processes requires very fast availability of measurement results, for which PSD parameters are often adequate. To reduce human involvement, various kinds of commercial samplers and on- and in-line PSD measurement instruments have been developed for application in the control of these industrial processes. On the other hand, analysis of the bulk properties is essential when particle-particle interactions play an important role in the product performance.

The properties of single particles strongly depend upon particle composition, type and size. For near-spheres, the relationships are generally well understood. For example, color and transparency strongly depend, in addition to chemical nature, upon both crystal type and particle size, especially when the size is in the same range as the wavelength of the light. Good examples are pigments in paints and sunscreens [13]. For non-spherical particles, however, the understanding is hampered by the influence of particle shape.

The behavior of bulk particulate materials and products may be very complex since the inter-particle distance is often small and mutual interactions of the particles and with the surrounding medium play a role. Mechanical interactions between particles dominate in dry powders. If the particles are regular and have sizes larger than about 20–50 μm in a (medium-) narrow size distribution, the powders show a behavior that depends mostly upon bulk density and size distribution (degree of particle packing and strength of attractive forces). They are usually free flowing, since the inter-particle attractive forces are small in comparison to particle mass. The consequences are that they can move easily but also segregate easily.

An example of the influence of particle size is sketched in Fig. 1.2 for the fluidization behavior of powders [6, 13].

If the particles in a powder show a broad size distribution, which is the case for most industrial products, then the bulk density strongly depends upon the history of aeration, consolidation and vibration. Aeration during storage often leads to relatively small bulk densities especially when the particles take their first position while leaving a lot of empty space. Tapping, vibration and consolidation generally result in significantly greater bulk density than in the loose state after pouring. This is because the smaller particles may fill the voids in between the larger ones, but only after breaking the attractive inter-particle forces by the induced external forces. Higher bulk density also means that bed porosity is significantly smaller and mechanical interactions larger. Near maximum packing the mobility of the particles is strongly reduced due to mutual hindrance.

Note that particles having a substantial static charge at their surface behave in the same way as with a high degree of aeration since repulsive forces between the

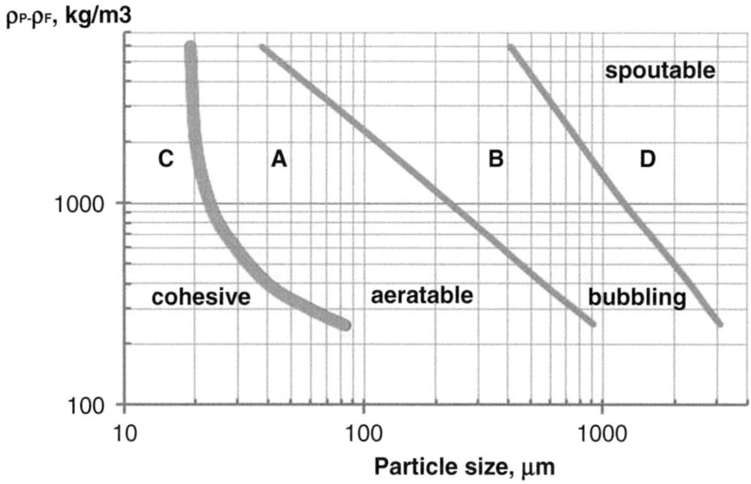

Fig. 1.2 Geldart fluidization diagram [13] (Adapted from [6]); (Copyright Springer; reproduced with permission)

particles dominate in the behavior. An example is polymer powder, which can build up charge by friction. This charge causes that particles repel each other and cannot reach close proximity.

Dry powders, where the particles have sizes smaller than about 20 μm and/or have shapes like plates or fibers, behave quite differently from powders containing larger, regular particles. They are cohesive already in fairly loose packing, since the inter-particle forces are large in comparison to particle mass and, thus, dominate in the behavior. This makes such powders very resistive to flow, especially after some kind of consolidation (as e.g. occurs during storage), when inter-particle distances are decreased. Presence of small amounts of water in a powder enhances this phenomenon of cohesivity.

Several performance quality aspects of products, for example cohesivity, flowability, rheological behavior, shear strength and taste, can be measured more or less directly through standardized measurement methods (see further Chaps. 5, 13, and 15).

The rheological behavior of bulk powders is usually characterized through shear tests and measurements of bulk density (free and tapped), angle of repose and wall friction. Taylor et al. [22] have concluded through principal component analysis that a combined flowability index based on measurements of critical orifice, compressibility and angle of repose provides a better product characterization than the individual tests. The different facets of the flow properties of powders also can be illustrated in so-called spider diagrams, based, for example, on measured characteristics of bulk density, wall friction, shear strength, Hausner ratio and some equipment characteristics such as hopper wall angle for mass flow and critical outlet size. Such diagrams present a nice overview of powder flow behavior and equipment characteristics through ranging the various parameters in terms of easy, modest and poor flow. Thus, they also highlight specific aspects that deserve more attention. This is advantageous for application in hopper design (see further Chap. 15).

Liquid dispersions – emulsions and suspensions – exist having different particulate concentrations. In addition to the mechanical interactions between particles, electrical interactions also play a role. Thus, particulate concentration and particle characteristics are important for e.g. rheological behavior, which is a property of major importance in liquid systems. At low concentration and fairly regular particles, the behavior is Newtonian, similar to the corresponding continuous phase. It means that the required force to maintain flow is proportional to the shear rate, with the viscosity as the proportionality constant. In other words, the viscosity is a measure of the resistance of a fluid to flow. Presence of particles influences viscosity in direct relation to their concentration. Such dilute systems, however, form a minority of the industrial products. Most suspensions and emulsions have a medium to high particulate concentration. Then, the inter-particle distance is decreased and the attractive or repulsive forces (dependent upon the presence of particle charge, usually expressed in terms of zeta-potential) between the particles increased. These forces also strongly depend upon the type of continuous phase (ionic or not) and the presence of dispersant molecules (ionic or steric stabilizers). Above particulate concentrations of about 5 % v/v, the viscosity increases strongly, in a non-linear way with particulate concentration. Above concentrations of about 30 % v/v, the rheological behavior changes from Newtonian to non-Newtonian, where the viscosity depends upon shear rate and sometimes also upon exposure time to shear. Moreover, also a certain minimum shear stress – the yield stress – may be required to start movement, i.e. to break up structures that are formed by the particles. Non-Newtonian behavior is named pseudoplastic, shear-thinning, dilatant/shear-thickening, Bingham plastic, thixotropic or rheoplastic, depending upon its nature. Changes of rheological behavior are most extreme when the particulate concentration comes close to that of maximum packing of the particles. Then, products may show viscoelastic behavior having both viscous and elastic properties [13]. Rheological behavior is usually of prime importance in relation to processing and product quality and stability.

Typically, the rheological behavior is determined in steady-state procedures, which may include some pretreatment (pre-shearing) during standardized time periods to break-up particulate structures. Viscoelasticity is usually determined in dynamic measurement techniques that result in a complex elastic modulus, from which an elastic and a viscous component can be derived [13].

Note that the standardized measurements of the performance quality are executed at different conditions from real life. Moreover, they are usually costly as they require experience and knowledge and/or are time-consuming. Thus, the results should always be handled with due care for application in practical situations.

In addition to the characterization of these bulk properties, regular assessment of the quality of a particulate material is often executed through measurement of particle size distributions and/or particle shape characteristics, typically by off-line analysis. Parameters of PSD and shape are relevant because they relate to aspects of the performance quality of a material. Mostly, empirical correlations are used for these relationships, or the results are simply compared with prior data. Measurement is easy and many different techniques and methods are available, the quality of which can be guaranteed by frequent qualification of both instruments and methods. Moreover, off-line analysis allows optimization of the particulate concentration for the measurement technique applied. Many different PSD and shape parameters can be produced through using different instruments. Since different PSD and shape parameters relate to different product quality aspects, usually several (about 3–10) parameters are derived from the measurements for assessment and verification of product quality. Note that specification of a large number of parameters in product specifications may be superfluous as well as costly.

The objective of monitoring and control of particulate processes is different from that of quality assessment of products. It is to maintain a constant level for product quality within stated limits at minimum costs, maximum throughput, minimum off-spec material, minimum waste and minimum energy consumption. In addition to product analyses at the end of the process, two approaches may be applied for this purpose; often, the two are applied in conjunction. The first uses several input parameters, often in some kind of a standard recipe of e.g. feed rates, residence time and energy consumption; this approach is based on experience. It is often used for simple batch processes. In complex processes like granulation, different input variables often have similar or opposite effects on the same output parameters. Then, there is a demand for integrated process control, meaning that several input variables are to be combined for the process control (see Chap. 4). Such combinations are frequently used in chemical engineering. Good examples are the characteristic dimensionless numbers named after Bodenstein, Prandtl, Reynolds, Weber, etc.

The second control approach is based on measurement – during processing – of (output) parameters that relate to product quality; for particulate materials, often one or two selected PSD parameters are chosen. This is sufficient as long as the same base material is applied in the equipment at the same conditions and, thus, the same type of process behavior can be expected. For batch processes and continuous processes which show slow fluctuations, conventional off-line product analysis can be applied for control and monitoring, which is usually less costly in terms of investment but requires capable manpower. It means that samples are taken at various time intervals and measured in a laboratory at some distance from the production site. The total required time may be substantial (several hours). This dead time can be shortened by installing the measurement instrument near the production line but following the same sequence of sampling, dispersion and analysis (typically manually by operators or automated: at-line analysis). For control of continuous processes having fast fluctuations, however, still smaller dead times are required, leading to more or less continuous measurements in often automated and more expensive on-line or in-line measurement equipment. In general, the main requirement for the total dead time – the period of time in between sampling up till availability of analytical results – is that the frequency of measurement results meets the process dynamics, meaning that the PSD parameters required for control are obtained fast in comparison to process fluctuations and potential disturbances. Note that some processes are slow, i.e. having a characteristic time in the order of hours, but that others have very small characteristic times, in the order of seconds to minutes. Further requirements for analytical instrumentation applied for monitoring and control are: adequate particulate concentration range, particle size range, repeatability, stability and robustness [12]. The latter factors are important since the industrial environment typically is much less clean and controlled than the analytical laboratory, in addition to the fact that there is much less opportunity for instrument calibration and qualification.

Note that inadequate process control may lead to poor product quality or large recycle streams and, thus, to poor economics.

1.4 Particulate Characteristics in Relation to Production and Handling Processes

Characterization of particulate products typically involves PSD and porosity measurement as well as bulk density and rheological behavior of the bulk product. For the selective production of minerals from ores, good knowledge of the surface properties is essential. Product quality requirements are usually laid down in product specifications. Particle size may be in the nanometer, micrometer, millimeter or even centimeter range. Typically, product properties relate to particle size and size distribution. Some particulate products have a narrow size distribution. Crystal sugar is a good example of a narrow size distribution near 1 mm that allows easy flow, in contrast to icing sugar containing particles of about 1 μm. Most industrial materials, however, have a broad size distribution. If all particles contribute to the behavior and performance of a product, then some kind of mean size, based on number, area or volume (see later) is to be taken for the relationship with properties. In some products, however, this is less meaningful. This is the case when small amounts of large or small particles, if present in the bulk, deteriorate performance. For example, few large particles in polishing powder cause scratches in the substrate, and in chocolate a gritty taste. On the other hand, small amounts of relatively small particles, which correspond to large particle numbers, may hamper filtration through blockage of the filter, or may cause dusting which may result in health problems for humans or dust explosions.

Fineness of particles and width of particle size distributions (PSDs) are defined differently in different fields of application. In my opinion, it seems best to standardize to:

Fineness (D_{90})		PSD width (D_{90}/D_{10} ratio)	
Nanoparticles	<0.1 μm	Monosized	<1.02 (ideally = 1.00)
Ultrafine	0.1–1 μm	Ultra narrow	1.02–1.05
Fine	1–10 μm	Narrow	1.05–1.5
Medium	10–1000 μm	Medium	1.5–4
Coarse	1–10 mm	Broad	4–10
Very coarse	>10 mm	Very broad	>10

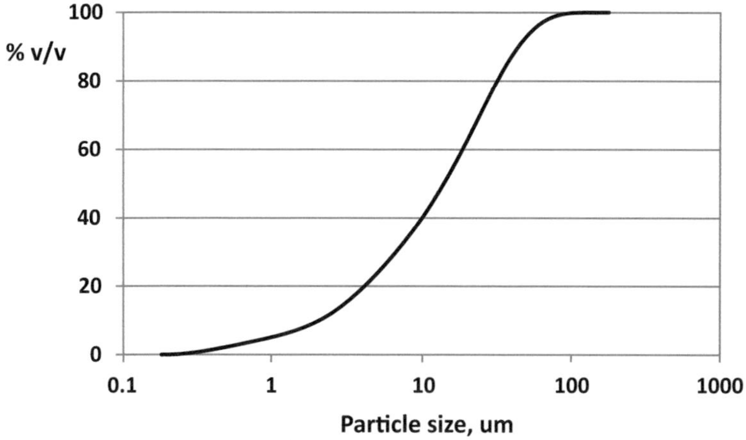

Fig. 1.3 Cumulative size distribution of Portland cement (laser diffraction results)

As an example, the size distribution of Portland cement is presented in Fig. 1.3. This cement clearly shows a very broad, monomodal PSD: $D_{10;3} = 2$ μm, $D_{50;3} = 14$ μm and $D_{90;3} = 45$ μm. Thus, 90 % by volume of this distribution is larger than 2 μm and 10 % v/v larger than 45 μm. Furthermore, the PSD shows some tailing towards the small size end.

Particle shape of most industrial materials shows almost equal length, breadth and thickness, that is have an aspect ratio close to 1. These particles are called isometric or regular. Some particles are even spherical, like droplets in air or in emulsions. But there are exceptions with extreme shapes, like fibers, needles or platelet particles; for example, many pharmaceuticals crystallize in needles. Then, the effect of particle shape may dominate that of size, for example by causing specific optical effects, by decreasing flowability or by providing for decreased (needles) or enhanced (fibers) product strength. Another peculiarity is when the particles contain pores, which decrease apparent particle density and, if accessible, largely increase the specific surface area. This is important for e.g. catalysis.

The size of a sphere is easily identified in terms of its diameter. For non-spherical particles, the size is typically given in terms of the diameter of an equivalent sphere, which depends upon measurement principle and instrument. Table 1.2 gives for some examples of equivalent diameters. Only for particles and other objects having a simple and more or less identical geometry, length, breadth and/or thickness are used for representation.

For measurement of particle size distributions and shape, many instruments, based on different principles, are available [12]. If particle shape deviates largely from being spherical, the use of equivalent sphere diameters may cause significant differences between PSD measurement results coming from techniques based on different principles. Moreover, the basis of PSD results may differ: sometimes the

Table 1.2 Examples of equivalent diameters ('particle size')

Aerodynamic particle size	(Stokes') diameter of a sphere with density 1000 kg/m^3 having the same aerodynamic property as the particle
Equivalent light scattering diameter	Diameter of a sphere (or set of spheres) having the same light scattering properties
Equivalent projected area diameter	Diameter of a circle having the same area as the particle's projection
Equivalent sieve diameter	Equivalent volume diameter of particles that just pass through the apertures of a sieving medium (near-mesh size)
Equivalent volume diameter	Diameter of a sphere having the same volume as the particle
Hydraulic particle size	Stokes' diameter of a (sediment) particle, while assuming a density of 2650 kg/m^3
Stokes' diameter	Diameter of a sphere that has the same density and settling rate as the particle under conditions of Stokes' law (viscous flow conditions)

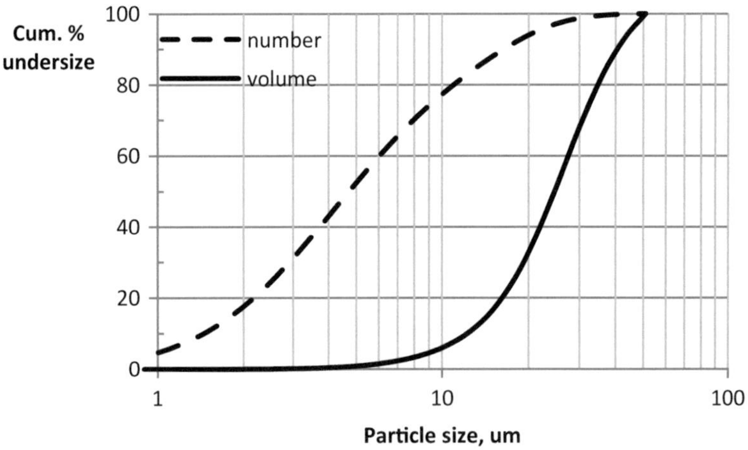

Fig. 1.4 Coulter Counter PSD analysis of chocolate particles; number data taken from [14] and recalculated to volume data [13] (Copyright Springer; reproduced with permission)

particles are measured one by one and the PSD is based on numbers (e.g. in case of microscopy), sometimes the particles are separated into size fractions and the mass of these fractions measured (e.g. sieving) and sometimes the collective behavior is interpreted through some model based on volume (e.g. laser diffraction) or fluctuating scattered light intensity (e.g. dynamic light scattering). These differences also cause large differences between PSD curves for broad distributions since the number, etc. per unit mass largely depends upon particle size.

As an example, the vast differences between number- and volume-based PSD results of chocolate particles are shown in Fig. 1.4. In this product, more than 50 %

n/n is involved in the 1 % v/v of smallest particles and only about 0.2 % n/n represents the upper 10 % v/v. Further differences between measured results may be caused when the measurement technique involves some preference for orientation of particles having large aspect ratios, or when instruments using the same principle differ in hardware or software.

At this moment, typical shape characterization concentrates on only one or two of the main shape features. Generally, quantitative surface area, porosity and pore size distribution features are used for application. The other shape characteristics are mostly described in a qualitative or semi-quantitative way, e.g. (short) fibers or sharp, angular particles.

Particle shape characteristics involve [10, 12]:

- macro-shape features, related to the 3-dimensional form of the particles (length, breadth and thickness), expressed e.g. as aspect ratio
- meso-shape features, related to the general aspects of roundness and angularity
- micro-shape features, related to rugosity and smoothness as well as porosity, pore size distribution and other structural heterogeneities.

The quality of a material in relation to specifications for particle size and shape can be expressed as follows:

- PSD type, viz. monomodal, bimodal, ..., multimodal
- particle size range, often expressed in D_{10} and D_{90} (with an extra subscript related to the PSD basis, being number, area, volume, etc.)
- a weighted mean size
- the maximum allowable amount oversize (above a stated size), or undersize (below a stated size), or the minimum and/or maximum amount in one or more stated size classes
- the parameters of a modeled distribution, such as the normal, log-normal, Rosin-Rammler or Gaudin-Schumann distribution
- surface area per unit volume or mass
- porosity and pore size distribution
- specified particle shape and shape homogeneity, e.g. spheres, cubes, rounded, angular, fibers, sometimes with a parameter, e.g. aspect ratio, elongation, flakiness, roundness.

The most appropriate identifiers and their measurement method(s) should preferably be based on literature, standards or own investigations. Often, additional properties are laid down in specifications, for example bulk density, color, degree of contamination, flowability and dissolution time.

Note that the frequently used median size, D_{50}, is a statistical parameter of a PSD without any direct relation to product quality. For example, it does not take the shape and width of the PSD into account.

It is very unfortunate that PSD parameters are often chosen – without good reasoning – for relations with product performance and for representation in characteristic, dimensionless numbers, such as the Reynolds or the Weber number. Note that the type of the PSD parameter is hidden in these numbers, if it is not mentioned explicitly. The above mentioned median size is an example of such misuse. Moreover, far from always mentioning is made of the type of equivalent sphere measured, which may greatly depend upon measurement principle and instrument if the particles have an irregular shape. In relation to this dependency, it is important to choose an optimum measurement technique and instrument for determining the optimum PSD characteristic of a given product. The link with optimum product performance is strong and, thus, precision, resolution, sensitivity and stability of the measurement instrument should agree with product quality requirements.

This problem is much less severe in the case of process control, since this is executed in a single instrument at the production location while process feed is fairly constant and process conditions are kept within narrow margins. At those conditions, almost any PSD parameter will show more or less similar responses to the small process variations with respect to PSD shifts. Here, also the median size can be used successfully. Note that symmetrical widening of the PSD cannot be seen in the mean and median size.

Statistical moments can be used for defining mean sizes, standard deviations and shape characteristics of particle size distributions. In general, the l-th moment of a random variable x around the origin is the expected value of x^l:

$$\mathrm{E}\left(x^l\right) = \Sigma\left(x_i^l \cdot \mathrm{f}(x)\right) \tag{1.1}$$

Here, f(x) represents the normalized probability of occurrence of x_i as a weighting and l the power of x involved; the subscript i refers to the respective size classes. These moments are denoted with a prime, M'.

The zero-th moment, denoted as M'_0, represents the total probability of occurrence of x. By convention, the occurrence is normalized: $\Sigma(\mathrm{f}(x)) = 1$.

If the distribution does not come from the full population but from a sample, containing N elements, of that population, then the normalization, for example of the first moment, denoted as M'_1, is done by dividing the moment by N.

$$M'_1 = \mathrm{E}(x)/N = \Sigma\,(x_i \cdot \mathrm{f}(x))/\Sigma(n_i) \tag{1.2}$$

Thus, E(x) equals the arithmetic mean value of x (= $<x>$).

Central moments relate to the distribution of x^l around the mean value. They give an indication of the shape of the distribution around the mean value and are denoted without a prime. For example, the second *central* moment is the variance of the distribution (standard deviation squared), indicating its width:

$$M_2 = \mathrm{E}(x - <x>)^2 = \Sigma(x_i - <x>)^2 \cdot \mathrm{f}(x) \tag{1.3}$$

Higher central moments relate to the shape of the distribution: the third moment to skewness (or asymmetry), the fourth one to kurtosis (or flatness/peakedness).

The relevance of the moments in case of particle size distributions can be easily identified:

$M'_0 = \Sigma(n_i)$ equals the total number of particles in the sample $(= N = \Sigma(n_i))$
$M'_1 = \Sigma(n_i \cdot D_i)$ equals the total length of particles in the sample
$M'_2 = \Sigma(n_i \cdot D_i^{\,2})$ relates to the total area of particles in the sample $(= \Sigma n_i \cdot \pi \cdot D_i^{\,2})$
$M'_3 = \Sigma(n_i \cdot D_i^{\,3})$ relates to the total volume of particles in the sample $(= \Sigma^1/_6 \cdot n_i \cdot \pi \cdot D_i^{\,3})$.

In the *Moment-Ratio notation*, mean particle sizes are derived from the ratio of two moments [1, 2, 3, 9]:

$$< D_{p,q} >= (M'_p/M'_q)^{1/(p-q)} = \{\Sigma(n_i \cdot D_i^p)/\Sigma(n_i \cdot D_i^q)\}^{1/(p-q)} \tag{1.4}$$

where:

D = particle size (i.e. a named equivalent size)
D_i = mean particle size in size class i
n_i = number of particles in size class i
p, q = real numbers ($p \neq q$); they relate to the type of quantity, e.g.: *0* to number, *1* to length, *2* to area and *3* to volume.

The moment in the numerator expresses the product of the particle quantity of interest and the weighting quantity; the moment in the denominator gives the weighting.

Some examples of mean particle sizes are:

$< D_{1,0} >= M'_1/M'_0 = \Sigma(n_i \cdot D_i)/\Sigma(n_i)$
arithmetic mean size; in other words, it represents the size of the particles which constitute the bulk of the sample by number

$< D_{3,0} >= (M'_3/M'_0)^{1/3} = (\Sigma(n_i \cdot D_i^{\,3})/\Sigma(n_i))^{1/3}$
mean volume size; in other words, it represents the size that corresponds to the mean volume or size cubed (D^3) of a size distribution

$< D_{3,2} >= M'_3/M'_2 = \Sigma(n_i \cdot D_i^{\,3})/\Sigma(n_i \cdot D_i^{\,2})$
surface-weighted mean size or Sauter mean diameter (also named SMD) [20]; in other words, the size that has the same volume/surface ratio as the entire sample; thus, it is inversely related to the volume specific surface area

$< D_{4,3} >= M'_4/M'_3 = \Sigma(n_i \cdot D_i^4)/\Sigma(n_i \cdot D_i^{\,3})$
volume-weighted mean size or De Brouckere mean size; it represents the size of the (larger) particles which determine the bulk of the sample volume

$< D_{6,5} > = M'_6/M'_5 = \Sigma(n_i \cdot D_i^{\,6})/\Sigma(n_i \cdot D_i^{\,5})$
scattered light intensity-weighted, harmonic mean or Z-average diameter; it comes from the cumulants PSD evaluation in dynamic light scattering (DLS) and, since it is intensity-weighted, relates to the stronger light-scattering particles.

Differences between the various mean values, surface area and volume characteristics are illustrated for mixtures of 10 and 100 nm non-porous spheres in Table 1.3.

The table clearly shows the differences between the various mean sizes. Also, it illustrates that a small number of relatively large particles represents a significant contribution to the total volume. On the other hand, large numbers of small particles give only a small contribution to the volume but a significant contribution to the total surface area.

The *Moment notation* to derive mean particle sizes, which is by origin German, is used as an alternative for the Moment-Ratio notation in various countries and fields of particulates application. Typically, it uses x as the symbol for particle size, with k and r as subscripts. The mean size in this Moment notation is defined as:

$$< x_{k,r} >= \left(M'_{k,r}\right)^{1/k} \tag{1.5}$$

Here, k and r have the same function as p and q in the Moment-Ratio notation. It can be shown that:

$$(x_{k,r} >)^k = M'_{k,r} = (M'_{k+r,0})/(M'_{r,0}) \;=\; (< D_{k+r,r} >)^k \tag{1.6}$$

and, thus:

$$< x_{k,r} >=< D_{p,q} > , \quad \text{where} \quad p = k + r \quad \text{and} \quad q = r \tag{1.7}$$

Hence, e.g. following mean sizes are equivalent:

$$< D_{1,0} >=< x_{1,0} >= M'_{1,0} \tag{1.8}$$

$$< D_{3,0} >=< \; x_{3,0} >= \left(M'_{3,0}\right)^{1/3} \tag{1.9}$$

$$< D_{3,2} >=< \; x_{1,2} >= M'_{1,2} \;=\; M'_{3,0}/M'_{2,0} \tag{1.10}$$

$$< D_{4,3} >=< x_{1,3} >= M'_{1,3} = M'_{4,0}/M'_{3,0} \tag{1.11}$$

Some examples of application of these mean values are:

$<D_{1,0}>$ the arithmetic mean size is advised when the number-based size distribution of particles is important, as in health effects or contamination of surfaces.

$<D_{3,2}>$ the Sauter mean diameter is advised in calculations where the active surface area is important in a relationship, as in fuel combustion [21], droplet evaporation rate [2], atomizers [23], explosion behavior [5, 7, 11] and heat and mass transfer [4, 15]. Spherical droplets in sprays dominate in these relationships of particle size with behavior, although the Sauter mean size is also frequently used for materials containing non-spherical particles.

Table 1.3 Comparison of characteristic values in mixtures of 10 and 100 nm non-porous spheres (density 1000 kg/m^3)

	Number	Number								
Size	10 nm	100 nm	$\langle D_{1,0} \rangle$ nm	$\langle D_{3,0} \rangle$ nm	$\langle D_{3,2} \rangle$ nm	$\langle D_{4,3} \rangle$ nm	$\langle D_{6,5} \rangle$ nm	Surf. area m^2	Volume m^3	Spec. surf. area m^2/gram
	1	0						3.14E-16	5.24E-25	600
	0	1						3.14E-14	5.24E-22	60
	1	1	55.0	79.4	99.1	99.9	100.0	3.17E-14	5.24E-22	61
	10	1	18.2	45.1	91.8	99.1	100.0	3.46E-14	5.29E-22	65
	100	1	10.9	22.2	55.0	91.8	99.9	6.28E-14	5.76E-22	109
	1000	1	10.1	12.6	18.2	55.0	99.1	3.46E-13	1.05E-21	330
	1000	2	10.2	14.4	25.0	70.0	99.6	3.77E-13	1.57E-21	240

<$D_{3,0}$> the mean volume size allows estimation of the number of particles in a sample from the sample mass (if particle density is known), which may be important in relation to e.g. dose, health effects or contamination of surfaces.

<$D_{4,3}$> the volume-weighted or De Brouckere mean size is advised for situations where the larger particles are important, as in combustion equilibrium [15].

Examples of other PSD parameters used are:

$D_{90;3}$ *or amount of material larger than a stated size* to indicate few large particles in a sample, e.g. for gritty taste of chocolate or glossy paint quality (the advantage of using the volume-based parameter is that it is much more sensitive to the presence of few large particles than the corresponding number-based parameter).

$D_{10;3}$ *or amount of material smaller than a stated size* to indicate the presence of small particles in a sample, e.g. for slimy taste of chocolate, filtration behavior or catalyst abrasion. Note, however, that a number-based parameter is more sensitive to the presence of small amounts of small particles (see example presented in Fig. 1.3). Still, the volume-based parameter is often chosen in view of easier measurement.

PM_{10} mass concentration (μg/m^3) of particles in air having an aerodynamic size equal to or smaller than 10 μm, related to health effects of fine dust (similarly: $PM_{2.5}$).

Note 1: The number of particles is important in health effects through inhalation if each particle may cause an effect, e.g. asbestos fibers or insoluble nanoparticles. The mass concentration is important for the total dose in health effects.

Note 2: The number of relatively large particles in a stated sample of e.g. paint is also measured without determination of the full PSD by means of the Hegman gauge, or by a special method using the electrical sensing zone technique [12].

Note: The application of these parameters primarily relates to spherical particles, e.g. droplets, in which case the particle size distributions and their characteristic parameters can be easily converted to each other. Such conversion is also often applied when the particles have similar length, breadth and thickness; sometimes, a shape factor has to be added in the conversion. However, when particles differ largely in length, breadth and/or thickness such as fibers or flakes, shape effects may dominate in the application as well as may cause significant differences between size distributions coming from measurement techniques based on different principles. Then, conversion of PSD's and characteristic parameters is neither straight forward nor relevant for the application.

In the past, PSD data were often compressed by fitting them to a 2-parameter model distribution, e.g. to evaluate the progress of production processes (e.g. grinding and milling) or the quality of narrowly sized materials (e.g. reference materials) [12]. These models have in common that they contain one parameter expressing some characteristic size value (e.g. mean or maximum size), whereas the other parameter relates to the width of the distribution (e.g. (geometrical) standard deviation). With the use of special graph paper, these functions can be linearized, which allows easy access to the parameters. A drawback is that their description usually does not hold for the full PSD. Since modern computers allow easy handling of large amounts of data as provided by modern PSD measurement instruments, this approach of model distributions has now decreased to only few cases.

Different PSD parameters may be used for the determination of a single aspect of product quality, since they will reflect in some way to PSD changes. However, some parameters are more sensitive to changes in product quality than others. The same holds for measurement techniques. Moreover, different aspects of product quality are usually best represented by different parameters. Thus, good consideration should be given to the choice of both technique and parameters in order to cover all product quality aspects in an optimum manner.

For emulsions and suspensions, good knowledge of the surface properties and ways to adapt them is essential. One reason is for establishing adequate stability of dispersions through application of surfactants and/or ionic substances (see Chap. 8). Another reason is for the selective and economic extraction of valuable minerals from ores by flotation (see Chap. 11).

The application of Quality Function Deployment/House of Quality (QFD) offers good possibilities for optimizing the design not only of products but also of processes, as explained in our earlier book [13]. In this QFD approach, the characteristics of potential production and handling processes are linked in an explicit and transparent manner to the required performance parameters of products. To this end, following steps are to be made by a team of engineers, in good contact with the company's management:

I. Prepare a clear, concise **mission statement** on the desired material and its expected production process.
II. Generate **ideas** for the quantity of the envisaged material to be produced and handled.
III. Collect **basic data** for the desired/expected quality of the material, viz. composition (both chemical and physical, i.e. dry particulate, emulsion, suspension), purity, particle size and shape, flowability, hygroscopicity, rheological properties and sensitivity to particle attrition, breakage, agglomeration and aggregation.
IV. **Define** specifications for the material.
V. Generate **ideas** for the manufacturing and handling processes of the material as well as for equipment.
VI. Collect **basic data** for the processes and equipment that may be applied.
VII. **Relate** the potential of the processes to the desired and expected material properties, where possible, in a quantitative manner.
VIII. **Select** the most promising processing possibilities in relation to type and properties of the material, process complexity, safety and costs.

IX. **Evaluate** the most promising ideas, **test** potential equipment, **set up** possible full process schemes in the form of flow sheets and estimate the total process costs.
X. **Decide** upon the best process scheme, equipment, sampling points and process control.

A nice example of considerations for setting up a process design is given in Chap. 2 for crystallization processes.

For process control, repeatability, stability and robustness are the main quality requirements for the PSD parameter applied and, thus, also for the measurement instrument, especially in the sometimes harsh factory conditions. The same PSD parameters as for product quality can be used, but also statistical parameters such as the median size may be applied successfully.

Many processes, however, are not controlled by PSD parameters but by other means, like energy consumption of stirrers or processing time. Here, experience plays an important role, but the experience of one operator may lead to other settings and results than that of another. One of the reasons is that some processes are very complex (e.g. granulation, Chap. 4) and in fact require multiple in- or on-line measurements, which are strongly hampered in an environment of particles where dust and liquid cause severe fouling of many sensors.

1.5 Set Up of the Book

The objective of this book is to facilitate comparison of all processes and specific equipment applied for particulate materials through providing an overview of all major techniques with sufficient background and insight for good understanding. This should allow improved and systematic implementation in set up and control of the unit operations in particulate processes. To this end, all processes for production and handling of particulate materials are described in this book by experts in the respective fields.

Following this introduction, the chapters deal with **size enlargement techniques** viz.

- *Industrial Aspects of Crystallization* in Chap. 2
- *Wet Colloid Synthesis: Precipitation and Dispersion* in Chap. 3
- *Granulation and Tabletting* in Chap. 4
- *Particulate Flow and Agglomeration in Food Extrusion* in Chap. 5.

Next, **size reduction techniques** are described:

- *Comminution* in Chap. 6
- *Atomization, Spraying and Nebulization* in Chap. 7
- *Emulsification, Established and Future Technologies* in Chap. 8.

Then, chapters follow related to **handling of particulate materials:**

- *Mixing of Solid Materials* in Chap. 9
- *Particle Separations by Filtration and Sedimentation* in Chap. 10
- *Flotation* in Chap. 11
- *Classification* in Chap. 12
- *Storage and Discharge of Bulk Solids* in Chap. 13
- *Solids Transport and Handling* in Chap. 14.

The final Chap. 15 deals with *Sampling and Characterization of Bulk Particulate Materials and Products.*

1.6 Definitions, Abbreviations and Symbols

Aerodynamic size	diameter of a sphere with density 1000 kg/m^3 having the same aerodynamic property as the particle of interestdiameter of a sphere with density 1000 kg/m^3 having the same aerodynamic property as the particle of interest
Agglomerate	assemblage of primary particles with intermediate attractive forces (sometimes named aggregate)
Aggregate	assemblage of primary particles with strong attractive forces (sometimes including or named agglomerate)
Aspect ratio	ratio of maximum to minimum Feret diameter of a particle (often taken as ratio of length to breadth; the inverse value is often used to limit the value between 0 and 1 [10])
Attrition	reduction in size of particles through loss of relatively small fragments by collision with other particles or surfaces
Breakage	reduction in size of particles by collision with other particles or surfaces, leading to about equal fragments
Cohesivity	stickiness of a powder, caused by strong attractive forces between individual particles
Dead time	time evolved between sampling and availability of required measurement results, including sampling, sample transport, dispersion, measurement, data interpretation and communication of data
Equivalent diameter	diameter of an equivalent sphere
Equivalent sphere	sphere that has the same property as the observed particle in relation to a given measurement principle
Floc	assemblage of loosely bound primary particles in a liquid
Flocculation	formation of flocs in suspensions through decreasing the surface charge (zeta-potential) of the particles to smaller than $\vert 30 \vert$ mV (absolute value)
Grade efficiency	mass ratio of coarse solids separated as product to that present in the feed, in the same size class

Grade efficiency curve	grade efficiency against particle size of particle separation equipment
Hausner ratio	ratio of tapped density of a powder to its freely settled density
Laser diffraction	technique for estimating particle size distributions from the measured scattering pattern by an ensemble of dispersed particles using laser light
Lump	generic name for agglomerated or aggregated particles
Mean size	mean size of a size distribution, weighted according to number, area, volume or light intensity
Particle	discrete piece of material (gas bubble, droplet or solid)
Particle size	diameter of a defined equivalent sphere (see e.g. Table 1.2)
Tromp curve	grade efficiency curve

Characteristic, dimensionless numbers

Bodenstein number	Bo	see Péclet number
Capillary number	*Ca*	ratio of viscous forces to interfacial tension: $Ca = \eta_L \cdot \nu_L/\gamma$
Deborah number	*De*	ratio of relaxation time of a visco-elastic material (characteristic time for material to adjust to applied stresses) to the characteristic time scale of observing the response of the material; it characterizes the fluidity of a material generated by its non-constant stress history: $De = \tau_c/\tau_p$
Ohnesorge number	*Oh*	ratio of viscous forces to the square root of the product of inertia and surface tension; indicator of jet or sheet stability: $Oh = \mu_L/(\rho_L \cdot \sigma \cdot d_0)^{0.5} = We^{0.5}/Re$
Péclet number	*Pe*	ratio of advective to diffusive transport rate; relevant for studies of transport phenomena in a continuum; for mass transfer in liquid: $Pe_L = Re_L \cdot Sc$
Prandtl number	*Pr*	ratio of momentum diffusivity (kinematic viscosity) to thermal diffusivity: $Pr = \nu/\alpha = C_p \cdot \eta_L/k$
Reynolds number	*Re*	ratio of inertial forces to viscous forces; its value relates to the degree of turbulence in a fluid; low values – laminar flow, high values – turbulent flow; for a particle: $Re_p = \rho_L \cdot D_p \cdot u_p/\eta_L$
Schmidt number	*Sc*	ratio of momentum diffusivity (viscosity) to mass diffusivity; relevant in fluid flows in which there are simultaneous momentum and mass diffusion: $Sc = \eta_L/\rho_L \cdot Đ$

Weber number	We	ratio of inertial effects to interfacial tension: $We = \rho_L \cdot v^2 \cdot D_{drop}/\gamma$
Weissenberg number	Wi	ratio of relaxation time of a visco-elastic fluid to a specific process time; it indicates the degree of anisotropy or orientation generated by simple stress (e.g. deformation): $Wi = \tau_c \cdot (d\gamma/dt)$

G	gas
G/L	dispersion of gas bubbles (gaseous 'particles') in a liquid
G/S	dispersion of gas bubbles in a solid
L	liquid
L/L	emulsion; dispersion of liquid droplets in an immiscible liquid
PM	particulate matter
PSD	particle size distribution
QFD	Quality Function Deployment (House of Quality)
S	solid
S/L	suspension; dispersion of solid particles in a liquid
Bo	Bodenstein number (see Péclet number, *Pe*)
Ca	capillary number $Ca = \eta_L \cdot v_L/\gamma$
C_p	specific heat
D_{drop}	droplet size
D_p	particle size
$D_{10;0}$	10th percentile of a cumulative, undersize, number-based PSD (first subscript gives percentile value; second subscript the distribution basis: *0* number, *1* length, *2* area, *3* volume) [8]
$D_{50;3}$	median size of a volume-based PSD
$<D_{1,0}>$	arithmetic mean size (based on number)
$<D_{3,0}>$	mean volume size
$<D_{3,2}>$	surface area-weighted mean size; Sauter mean diameter
$<D_{4,3}>$	volume-weighted mean size
$<D_{6,5}>$	Z-average; intensity-weighted mean size
De	Deborah number $De = \tau_c/\tau_p$
Đ	diffusion coefficient
k	thermal conductivity
Oh	Ohnesorge number $Oh = \mu_L/(\rho_L \cdot \sigma \cdot d_0)^{0.5} = We^{0.5}/Re$
% n/n	percentage by number
% v/v	percentage by volume
Pe	Péclet number $Pe_L = Re_L \cdot Sc$
$PM_{2.5}$	mass concentration (μg/m^3) of particles in air having an aerodynamic size equal to or smaller than 2.5 μm
PM_{10}	mass concentration (μg/m^3) of particles in air having an aerodynamic size equal to or smaller than 10 μm
Pr	Prandtl number $Pr = \nu/\alpha = C_p \cdot \eta_L/k$
Re	Reynolds number $Re_p = \rho_L \cdot D_p \cdot u_p/\eta_L$

Sc	Schmidt number $Sc = \eta_L/\rho_L \cdot Đ$
u_p	linear velocity particle
v_L	characteristic velocity of liquid
We	Weber number $We = \rho_L \cdot {v_L}^2 \cdot D_{drop}/\gamma$
α	thermal diffusion $= k/\rho_L \cdot C_p$
γ	interface tension
$d\gamma/dt$	shear rate
η_L	dynamic viscosity liquid (μ also used as symbol)
ν	kinematic viscosity $= \eta_L/\rho_L$
ρ_L	liquid density
τ_c	stress relaxation time
τ_p	characteristic time of observation

Acknowledgement The author gratefully acknowledges the contribution of Dr. Maarten Alderliesten to the text on the Moment-Ratio notation for the derivation of mean particle sizes.

References

1. Alderliesten, M.: Mean particle diameters. Part IV: Empirical selection of the proper type of mean particle diameter describing a product or material property. Part. Part. Syst. Charact. **21**, 179–196 (2004)
2. Alderliesten, M.: Mean particle diameters. Part V: Theoretical derivation of the proper type of mean particle diameter describing a product or process property. Part. Part Syst. Charact. **22**, 233–245 (2005)
3. Alderliesten, M.: Mean particle diameters. From statistical definition to physical understanding. J. Biopharm. Stat. **15**, 295–325 (2005)
4. Estes, K.A., Mudawar, I.: Correlation of Sauter mean diameter and critical heat flux for spray cooling of small surfaces. Int. J. Heat Mass Transf. **38**, 2985–2996 (1995)
5. Föster, H.: Properties of flammable mists; Chap. 5 in [7] (2004)
6. Geldart, D. (ed.): Gas Fluidization and Technology. Wiley, New York (1986)
7. Hattwig, M., Steen, H.: Handbook of Explosion Prevention and Protection. Wiley-VCH, Weinheim (2008)
8. ISO 9276-1: Representation of results of particle size analysis – graphical representation. International Organization for Standardization (1998/2004)
9. ISO 9276-2: Representation of results af particle size analysis – calculation of average particle sizes/diameters and moments from particle size distributions. International Organization for Standardization (2014)
10. ISO 9276-6: Representation of results of particle size analysis – particle shape and morphology. International Organization for Standardization (2008)
11. Lemkowitz, S.M., Pasman, H.J.: Chap. 4, Assessment and control of fire and explosion hazards and risks of particulates; in [13] (2014)
12. Merkus, H.G.: Particle Size Measurements – Fundamentals, Practice, Quality. Particle Technology Series, vol. 17. Springer, New York (2009)
13. Merkus, H.G., Meesters, G.M.H. (eds.): Particulate Products – Tailoring Properties for Optimal Performance. Particle Technology Series, vol. 19. Springer, New York (2014)

14. Mohos, F.A.: Confectionery and Chocolate Engineering: Principles and Applications. Wiley-Blackwell, Ames (2010)
15. Mugele, R.A., Evans, H.D.: Droplet size distribution in sprays. Ind. Eng. Chem. **43**, 1317–1324 (1951)
16. Rhodes, M.: Introduction to Particle Technology. Wiley, Chichester (1998)
17. Richardson, J.F., Harker, J.H., Backhurst, J.R.: Coulson and Richardson's Chemical Engineering. Particle Technology and Separation Processes, vol. 2, 5th edn. Butterworth & Heinemann, Oxford (2002)
18. Rumpf, H.: Mechanische Verfahrenstechnik. Carl Hanser Verlag, Munich (1975)
19. Rumpf, H.: Particle Technology. Particle Technology Series, vol. 1. Chapman & Hall (English translation of [18] by Dr. F.A. Bull) (1990)
20. Sauter, J.: Die Grössenbestimmung der im Gemischnebel von Verbrennungskraft-maschinen vorhandenen Brennstoffteilchen. Forschungsarbeiten auf dem Gebiete des Ingenieurswesens, Heft. 279. VDI-Verlag, Berlin (1926)
21. Sauter, J.: Untersuchung der von Spritzvergasern gelieferten Zerstäubung. Forschungsarbeiten auf dem Gebiete des Ingenieurswesens, Heft. 312. VDI-Verlag, Berlin (1928)
22. Taylor, M.K., Ginsburg, J., Hickley, A.J., Gheyas, F.: Composite method to quantify powder flow as a screening method in early tablet or capsule formulation development. AAPS PharmSciTechn. **1**(3), E18 (2000), doi:art 18
23. Williams, A.: Combustion of Liquid Fuel Sprays. Butterworths & Co. Ltd., London (1990)

Chapter 2
Industrial Aspects of Crystallization

Pieter Vonk

Abstract Industrial crystallization is the production of crystalline particulate material from a liquid solution. Because of the production of crystalline material the technique has a high purification potential. Proper control of the crystallization circumstances gives the opportunity to tune the particle size and to a limited extent the shape. These characteristics lead to the fact that crystallization is one of the most applied purification and formulation techniques in industry.

Crystallization is applied to all kinds of industries like industrial chemicals (sodium chloride, terephthalic acid, ammonium sulphate, melamine, adipic acid), food chemicals (dextrose, lactose, benzoic acid, vitamins), pharmaceutical (antibiotics, Paracetamol, Atorvastatin, Estradiol). Production takes place on small scales (e.g. 100 g in early stages of clinical trials of pharmaceutics) as well as very large scales (e.g. alumina, 10,000 m^3 crystallizers).

Development of crystallization processes requires the assessment of the feasibility of formation of crystals (including polymorphs), determination of the thermodynamics and crystallization kinetics, all in view of the required product specifications. The thermodynamics of the solid and the origin of the product usually determine the nature of the crystallization process. The scale of the production determines the complexity of the process and equipment. Small scale, high value products (like pharmaceutics) are usually produced in multi-product environments with relative straight forward equipment (e.g. stirred vessels), but may require a complex operation schedule (multiple solvent switches, cooling/heating cycles,. . .). Large scale commodities usually are produced in dedicated complex equipment (multi-functionalities in equipment like mixing, evaporation, heat-integration) enabling the efficient production of the crystals.

Process modeling can be used to accelerate process development (small and large scale production) and to optimize processes. The key factor in the control of crystallization processes is control of the super-saturation. This is done by controlling the liquid composition, mixing, cooling and evaporation in the process. Super-saturation and solubility are important factors for determining the particle size of

P. Vonk (✉)
Crystallization and Solid Processing, DSM Advanced Chemical Engineering Solution, Geleen, The Netherlands
e-mail: Pieter.vonk@dsm.com

© Springer International Publishing Switzerland 2016

H.G. Merkus, G.M.H. Meesters (eds.), *Production, Handling and Characterization of Particulate Materials*, Particle Technology Series 25,
DOI 10.1007/978-3-319-20949-4_2

the product. High-super-saturation processes and low solubility usually lead to small particles (<100 μm). High solubility and low-super-saturation processes have the ability to form large crystals (>100 μm). The shape of crystals is generally an intrinsic property of the solid material and can be influenced to a limited extent. Changing needle shape crystals into cubic crystals is generally not possible, but reducing the aspect ratio (length/width) might be possible.

2.1 Introduction

The target of industrial crystallization is to obtain crystalline particles of high purity having a specified, usually fairly narrow size distribution from a solution. Crystallization is achieved by cooling, evaporation of the solvent, addition of salt, acid/base or anti-solvent. Cooling crystallization is applied to a large extent because this type of crystallization operation gives the best control of the process. Typical examples are melamine and dextrose. Evaporation is done when the product has a high solubility. Typical examples are ascorbic acid, sodium chloride and ammonium sulphate. This way of operation requires more energy compared to cooling crystallization, but application of multi-effect evaporation and vapor recompression of low quality steam from a process reduces the energy usage. Anti-solvent crystallization and pH-shift are often done to prepare a powder with small particles (i.e. fast dissolution).

The particle size and shape are important in view of the intended application. Powders with particles below 20 μm (fines) need to be handled with care, because they dust easily (safety issues) and tend to have bad flow behavior. On the other hand fines dissolve quite quickly if the wetting is good. Needles break easily and flow badly, especially when they are thin and are very elongated.

Eighty to ninety percent of the products produced in the chemical, fine-chemical and pharmaceutical industry are solid under ambient conditions. In order to make handling of the material easy, powder/granulates are often requested as final product formulation. Therefore crystallization is in many cases a part of the production process, because of the formulation aspects of the technique and the purification power. Economic aspects of crystallization process depend on the kind of product produced. High value pharmaceuticals require robustness in production of the registered material without a lot of attention to the operational costs. Commodities require the balancing of producing the right material and low investment and operational costs.

2.1.1 Organization of the Chapter

The chapter is organized as follows.

- First a general guideline is given on how to develop a crystallization process. It gives the main points of attention, but the reader should be aware of the fact that every crystallization is different.
- The theoretical background of crystallization and process modeling of crystallization describes the thermodynamics of a crystallizing system that are important for the development of the process. The thermodynamics determine to a large extent the kind of crystallization process chosen. Process modeling of crystallization processes can be applied to understand the crystallization process using limited amount of information and accelerate the development using experimental design. Process optimization by modeling is possible once the process is running.
- The industrial aspects section describes the types of equipment used in industry and gives a number of product examples. The examples range from pharmaceuticals to industrial chemicals. Also the production of nano-crystals by microfluidic devices in mentioned, although the usage in industry is limited.

The content of this chapter is by no means sufficient to act as a complete overview of crystallization. Further information can be found in the advised literature [2, 13, 21, 24, 28].

2.2 Development of Crystallization Processes

The development of crystallization processes is not a straight-forward process. Depending on the application of the product to be crystallized and the characteristics of the product and process itself, different steps have to be taken. The first thing to realize is that each crystalline product and each crystallization process is different. There are general rules, e.g. Ostwald rule of stages, but exceptions on these rules can easily be found. In general the following steps have to be taken:

1. Check the feasibility to form crystals of the desired product. For existing products this is quite evident, but for new chemical components (e.g. pharmaceuticals) this is not the case. An important aspect is the possibility to form different polymorphs [22]. Note that the metastable form is often formed first during crystallization. Especially for pharmaceuticals extensive screening is required for application, registration and IP purposes, but for other products limited screenings are still needed in order to identify bottlenecks in the process development.
2. Once crystals have been prepared, a first idea is obtained of the possible crystal shape of the final product. This giving indications what kind of challenges arise during the process development and production. Cubical crystals like sodium chloride and ammonium sulphate are easy to process (filtration and drying). Many pharmaceuticals, however, crystallize as thin needles that easily lead to inclusion of liquid during filtration, extended drying times, insufficient purity

and easy breakage. Changing the shape of a crystal is possible to a limited extent (Lovette et al. [19]).

3. Define the begin and end point of the process. Usually the source of the product is known as is the intended application of the product. This may set limitations to the process under development. Crystallization of a commodity type product from an aqueous solution gives less freedom to change the solvent, whereas during the development of a new pharmaceutical compound the reaction scheme is still under development associated with some freedom in the solvent system.
4. Define the product specification in terms of the chemical and physical quality of the product. Chemical quality is mostly related to the level and type of impurities present in the product. Monomers for polymers (e.g. caprolactam) often require a very high purity (>99.9 %) whereas the purity demands for pharmaceutical (e.g. paracetamol) are slightly more relaxed (>99 %). Specific demands with respect to color or toxicity may set boundaries to the level of specific impurities. The physical quality of the product is defined by the required particle size and shape of the product. The size and shape of a product is usually controlled to a large extent by the product itself and can be changed to a limited amount. Therefore, post-processing of the product (milling or agglomeration) may be required after the crystallization process.
5. Screening of the solubility and crystallization behavior in different relevant solvents. The solubility behavior is important information for making decisions on the steps to be taken during the crystallization process. When the solubility is high and not very dependent on the temperature, cooling crystallization is usually not an option. In this case evaporation of the solvent or addition of a different solvent to lower the solubility (anti-solvent) might be the way forward. When the component contains an acid or base group, pH-changes or formation of a salt by the addition of sodium hydroxide, hydrogen chloride, oxalic acid,... can be done.
6. Detailed investigation of the intended crystallization process. Here the complete thermodynamics and crystallization kinetics have to be investigated. This enables the fine-tuning of the process with respect to temperature, addition profiles, equipment volumes, stirring rates, and other process parameters that need to be determined. Also stress tests with respect to extreme situations (temperatures, impurity levels,...) need to be performed to determine their influence on the process and design mitigating actions.
7. Re-evaluate the design process between each development step, because the development requires a number of iteration loops. Factors that influence the final process are:

    - Market developments may set new product specifications.
    - Polymorphs or process steps are protected by competitors in new patents.
    - Impurities are difficult to remove requiring a new design or multiple process steps.

As mentioned before, these steps are general guidelines and adaptation of one of the steps is required on a regular basis.

2.3 Theoretical Background and Process Modeling of Crystallization

2.3.1 Crystalline Solids and Polymorphism

The target in industrial crystallization is to obtain crystalline solid particles from dissolved material in a fluid. In most cases this is a liquid, although gas of super-critical phases are also used. These particles have a variety of shapes and sizes. The sizes are in the range of a few nanometers (zeolites) to several mm (ammonium sulphate), the shapes are predominantly acicular, cubic, needles, plates and rounded. The morphology is determined by the crystal structure and crystallization conditions (driving force, impurities and fracture mechanics). Several crystal structures are presented in Fig. 2.1.

Figure 2.2 gives a microscope picture of two industrial products from a crystallizer. The material on the left is succinic acid, which is a typical bulk product that is produced in large equipment and where process economics are of utmost importance. The crystals are large (about 200–500 μm) and thereby sensitive to attrition. This results in the edges of the crystals being rounded and in the production of fines. Several particles have a somewhat irregular shape because they are made from two or more particles, which is an indication of agglomeration.

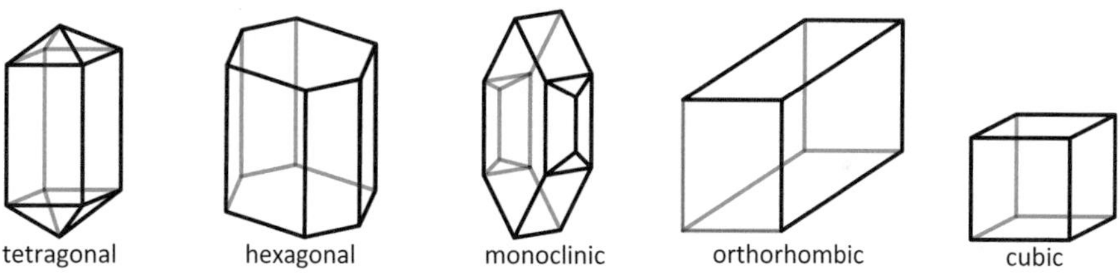

Fig. 2.1 Examples of crystalline structures (crystal lattice)

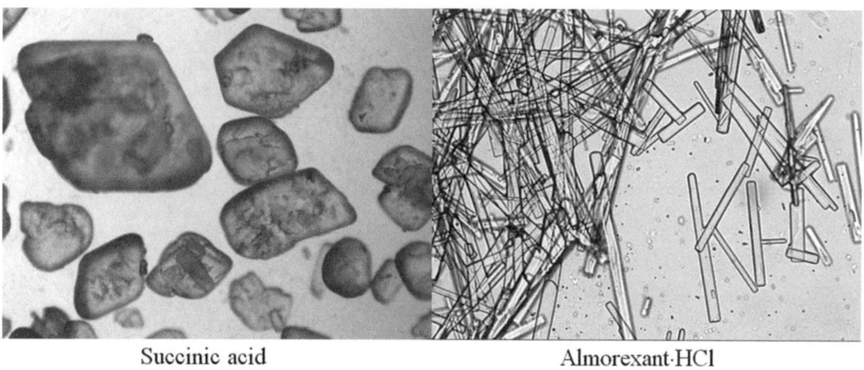

Fig. 2.2 Industrial crystalline products

The material on the right is Almorexant·HCl, a typical pharmaceutical product. This product often crystallizes as needle-shaped particles. Processing and application of needle-shaped particles poses several challenges. Long, thin needles tend to capture large amounts of solvent or form gels during processing. This gives filtration and drying problems. Also, thin needles are very vulnerable to breakage. Finally, flow properties of needles are not good, which gives problems during the formulation (e.g. dosing and tableting).

Changing the shape of crystals is possible to a limited extent. The solvent from which the crystallization is done is the first factor that influences the shape. The growth rate of different crystal surfaces depends on the surface energy of the crystal surface [19]. The surface energy is influenced by the solvent and thereby it influences the shape. Additives (or certain impurities) in the solvent may (partially) block the growth rate of specific crystal surfaces. This will influence the shape of the crystal as well. However, additives are often not allowed and in general they slow down the process. Super-saturation control can also be used to influence the shape of a crystal. This is caused by the fact that many crystal surfaces have a different dependence of the growth rate on the super-saturation. For example, this can be used to increase the thickness of a needle-shape crystal in order to improve the flow behavior and reduce the attrition sensitivity. Transformation of a needle-shape crystal towards a bulky spherical crystal is in general not possible.

In crystalline solids the molecules or ions are organized in a structured way (crystal lattice). The organized structure leads to strong interactions between the molecules. This strong interaction makes it difficult for other kind of molecules to enter the lattice, unless they have great resemblance in molecular structure. This makes crystallization a powerful purification technology.

Many molecules and ionic species have multiple crystal lattices that can be formed, i.e. polymorphs. Figure 2.3 gives two structures of Paracetamol. Form I (monoclinic structure) is the stable form under ambient conditions and has a

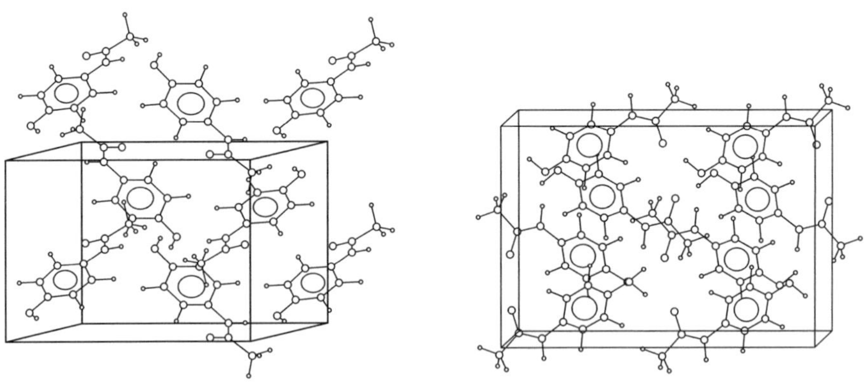

Form I Form II

Fig. 2.3 Polymorphic forms of paracetamol

melting temperature of 170 °C [11]. Form II (orthorhombic structure) is the metastable form, which has a melting temperature of 157 °C. The difference in melting temperature shows the difference in physical properties between polymorphs, which in view of the intended application can be important (e.g. difference in solubility for pharmaceuticals).

Another aspect of polymorphism is the fact that during crystallization, the metastable form is often formed first (Ostwald rule of stages). The general assumption is that metastable phases have faster growth kinetics and win out over the stable phases in the short term. This is, for example, shown by Davey et al. [7]. Whether or not the metastable form transforms into the stable form depends on factors like time, structure difference, temperature and solvent.

2.3.2 *Solubility of a Solid*

The solubility of a solid depends on the properties of the solid (melting point, heat of fusion and heat capacity difference between the liquid and solid) and the properties of the liquid (activity coefficients). The general solubility of a solid is given by the extended Van ‘t Hoff equation:

$$ln(\gamma x_{sat}) = \frac{\Delta H_f}{R}\left(\frac{1}{T_f} - \frac{1}{T}\right) - \frac{\Delta C_p}{R}\left(1 - \frac{T_f}{T} + ln\left(\frac{T_f}{T}\right)\right) \tag{2.1}$$

where:

γ = activity coefficient of the solute in solution
x_{sat} = mole fraction of the solute at saturation
ΔH_f = heat of fusion
R = gas constant
T_f = temperature of fusion
T = temperature
ΔC_p = difference in heat capacity between the liquid and solid solute.

In general, the required properties (γ, ΔH_f, ΔC_p) to describe the solubility cannot be predicted a-priori with sufficient accuracy. Therefore experimental data is required to get a good description. The activity coefficients can be omitted in the case of a single solvent, by using the simple Van ‘t Hoff equation:

$$ln(x_{sat}) = \frac{\Delta H_{sol}}{R}\left(\frac{1}{T_{sol}} - \frac{1}{T}\right) \tag{2.2}$$

where:

ΔH_{sol} = apparent heat of solution
T_{sol} = apparent temperature of solution.

The heat and temperature of fusion have been replaced by the apparent heat and temperature of solution. Other descriptions for solubility are possible (e.g. polynomial functions containing temperature and concentration), but they are less predictive when other solutes are present.

In the case a mixture of solvents is used, activity coefficient models like NRTL or UniQuac [29] can be applied to describe this effect. More predictive models like the NRTL-SAC [5] can also be used, but experience shows that solubility prediction without prior experimental knowledge should be done with great caution.

A more recent development in the field of the description of solubility is the application of the SAFT equation of state. Here two different approaches are under development:

- Perturbed Chain approach: Here, the molecules are described as a chain of hard spheres where the characteristics are obtained from pure component properties like vapor pressure and density (Prudic et al. [27]).
- Group-contribution approach: Here, the molecules are described using the different chemical groups in the molecule [8].

These two methods have the ability to predict the solubility in a broad range of solvents using a limited set of experimental data.

2.3.2.1 Solubility Examples in Relation to Crystallization Processes

Figure 2.4 shows the solubility of three common chemicals in water. It indicates the clear distinctions that exist between the solubility behavior of these compounds. 6-ACA (ε-amino-caproic acid, hydrolysis product of caprolactam) has a high

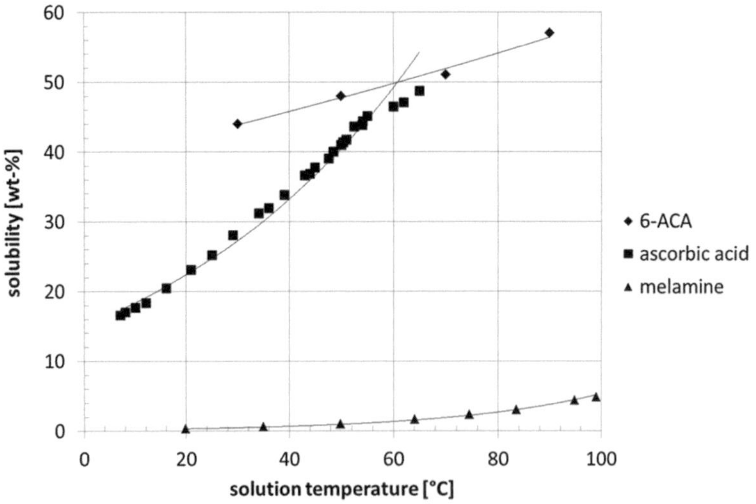

Fig. 2.4 Solubility of some common chemicals in water

solubility in water throughout the whole temperature range. This product is best crystallized with evaporative crystallization (removal of solvent). Ascorbic acid (vitamin C, [32]) has a high solubility with a reasonable dependence on temperature. Here a crystallization process that combines cooling and evaporation would be feasible. Melamine [4] has a low solubility at ambient circumstances that increases with rising temperature, enabling a cooling crystallization process that starts at sufficiently high temperature.

Figure 2.5 shows the solubility of ascorbic acid in different solvents (vitamin C, [32]). For this component, the solubility can be significantly lowered by changing the solvent. Cooling crystallization becomes easier if the preceding process step (e.g. reaction to ascorbic acid) is performed in ethanol or IPA. Table 2.1 gives the van 't Hoff parameters for the solubility of ascorbic acid.

The apparent heat of solution of IPA is significantly higher than the apparent heat of solution of the other alcohols. This indicates that IPA is a good solvent for the crystallization. If the process would start at 80 °C and the final temperature would be 20 °C, a yield of 93 % could be achieved. The yield in ethyl acetate is even higher (95 %), but the low starting solubility would mean that large quantities of solvent need to be handled. This makes this solvent economically less attractive.

Figure 2.6 gives the solubility of two components in water/ethanol mixtures at 25 °C. It is a clear demonstration that solvent composition is an important

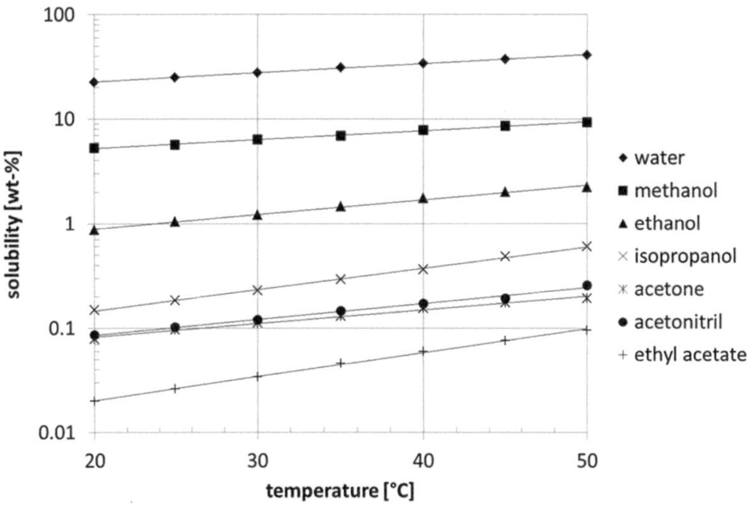

Fig. 2.5 Solubility of ascorbic acid in different solvents

Table 2.1 Van 't Hoff parameters of the solubility of ascorbic acid in different solvents

Solvent	Water	MeOH	EtOH	IPA	Acetone	Acrylonitril	EtAc	Unit
ΔH_{sol}	21.9	16.2	25.7	36.8	23.7	27.6	41.7	kJ/mol
T_{sol}	485	957	692	591	1,889	1,193	635	K

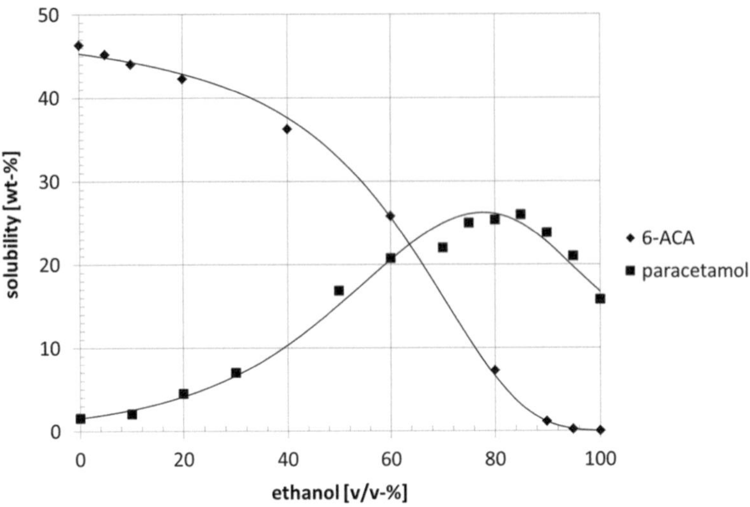

Fig. 2.6 Effect of solvent composition of solubility of 6-ACA and paracetamol at 25 °C

parameter to consider during the development of a crystallization process. In the case of 6-ACA [20] the crystallization from an aqueous solution is possible by adding pure ethanol to the solution (anti-solvent crystallization). Mass balance investigations show that the volume fraction of ethanol needed to crystallize 6-ACA in sufficient amount should be 80 % v/v or more. Because 6-ACA is a commodity chemical, this process option is not likely to be economically feasible.

Paracetamol [31] shows another interesting behavior, namely a maximum in solubility. This means that a process can be optimized by applying the right solvent mixture to increase the solubility. The water/ethanol solvent system is probably not the right choice, because the solubility at ambient conditions is quite high, but other solvent mixtures can be used that have lower solubility's at ambient conditions.

Shift of the pH is a fourth method to induce crystallization (or purification if dissolution is promoted). Figure 2.7 gives the solubility of DL-methionine as a function of the pH [10].

In the case of amphoteric materials, like amino acids, solubility increases both at low and high pH. The dependence of solubility is given by:

$$x_{sat}(T, pH) = x_{sat}(T)\left[1 + 10^{pK_{a,1} - pH} + 10^{pH - pK_{a,2}}\right] \tag{2.3}$$

where:

pH = acidity of the solution
pK_a = dissociation coefficient of the acid/base groups.

The crystallization can be performed starting at a low and high pH. Which option is best needs to be experimentally determined and depends on the kind of crystals and purity obtained.

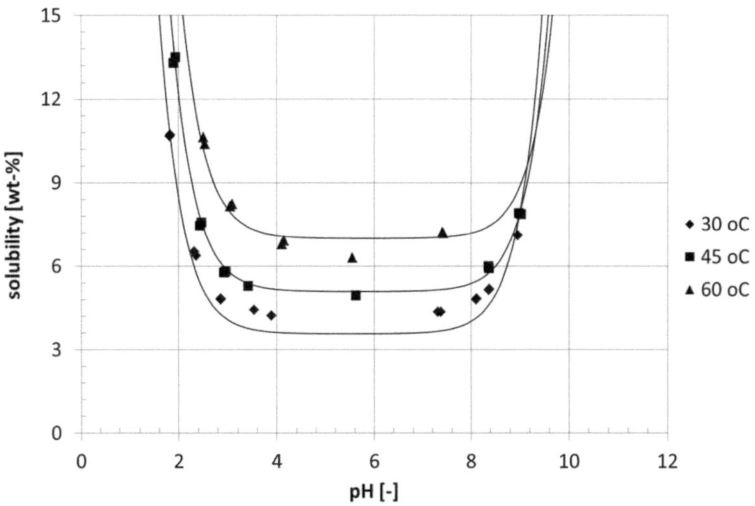

Fig. 2.7 Solubility of DL-methionine as a function of the pH

2.3.3 Nucleation, Crystal Growth, Agglomeration and Attrition

The production of crystals during a crystallization process takes place by a series of related events:

- First of all, super-saturation needs to be generated for crystals to be formed or to grow. This can be done by evaporation, cooling, anti-solvent or acid/base addition.
- Formation of new crystals. This can either be done by spontaneous formation of crystals from the super-saturated liquid (primary nucleation) or by formation of new crystals from existing crystals (secondary nucleation). It should be mentioned that at present a large deal of academic work is ongoing in the field of nucleation. One of the reasons for this attention is the fact that in many cases where primary nucleation plays a role, also secondary nucleation acts at the same time. This makes it difficult to separate the two phenomena.
- Addition of crystals (seeding). Primary nucleation is a statistical process. This may give unwanted variations in the product quality. Seeding a crystallizer with crystals can reduce this variability if the seeds are pre-conditioned (sieving to remove fines tail and dispersion in the solvent before adding them to the crystallizer). When seeds are prepared by milling, mechanical stresses in the crystals causes uncontrolled (partial) dissolution of the seeds. How much crystals are dissolved depends on the actual saturation of the solution and the amount of mechanical stress stored in the crystals and is difficult to predict. Pre-conditioning (e.g. slurrying in a saturation solution) removes the mechanical

stress and enables reproducible seeding which leads to production of crystals of the same size each time the process is run.

- Growth of crystals. In general two mechanisms that operate at the same time are important in industrial processes of crystal growth. Mass transfer from the bulk of the solution to the crystal surface is a first step, which may be limited when crystals become sufficiently large. Integration of molecules/ions in the crystal lattice at the surface is an equally important second step. Two surface integration mechanisms are well identified in literature: spiral or Burton-Cabrera-Frank (BCF) growth and the Birth and Spread mechanism.
- Agglomeration of crystals. Brownian motion of small crystals (<~1 μm) and liquid velocity variations (usually caused by turbulence) may cause crystals to collide with each other. The collisions can lead to agglomeration of crystals if sufficient binding forces are present to withstand the disruptive forces in the system. Binding between particles can be achieved by interaction forces like van der Waals force or an electric double-layer or when the contact time between particles is sufficiently large to grow a solid bridge between the particles. Disruption of the agglomerates is possible by the same forces that brings crystals together or by the action of an impeller/pump.
- Mechanical interactions are almost always present in crystallization and auxiliary equipment. Crystals may collide with walls, stirrers or in pumps leading to attrition of the crystals. Depending on the saturation and the mechanical stress stored in the fragments, these attrition fragments may dissolve or only partially dissolve and act as new nuclei for crystal growth.

A large variety of models can be found in literature for each of the four mechanisms described (nucleation, crystal growth, agglomeration and attrition). Some of the descriptions are purely empirical, some of them are based on first principles. Here, a limited overview is given of some of the descriptive models for the mechanisms with the focus on first principle models with a limited number of freely adjustable parameters. In this way the number of parameters that need to be estimated from experimental data is limited and good predictive models can be obtained.

2.3.3.1 Saturation and Super-Saturation

Saturation level (S) and super-saturation (σ) gives the description of the state of a solute compared to the saturated solution. Many different definitions exist. Very common is the definition based on the concentration:

$$\sigma = S - 1 = \frac{C_{act} - C_{sat}}{C_{sat}} \tag{2.4}$$

where:

σ = super-saturation
S = saturation level

c_{act} = actual solute concentration
c_{sat} = solute concentration at saturation.

This definition is appropriate when dealing with a single solid in a single liquid. However, when large quantities of byproducts are present or the solvent composition changes during the process, the definition on the basis of the mole fraction x is better:

$$\sigma = S - 1 = \frac{x - x_{sat}}{x_{sat}} \tag{2.5}$$

Here x is the actual solute mole fraction.

2.3.3.2 Primary Nucleation

Formation of clusters of molecules/ions in the liquid is in many cases the first step in the spontaneous formation of new crystals/nuclei. The process of formation and disintegration of molecular clusters is a continuous process that is statistical in nature and is controlled by the balance between the generation of a new phase (solid, lower in Gibbs free energy) and generation of new surface (higher in Gibbs free energy). As soon as the gain in Gibbs free energy (high saturation) is higher than the loss in Gibbs free energy, clusters can act as nuclei for crystal to grow. This process has been described in the past using traditional thermodynamics leading to the following nucleation relation:

$$B_{p,0} = ASexp\left[-\frac{16\pi\gamma_s^3 V_m}{3k_b^3 T^3 ln^2 S}\right] \tag{2.6}$$

where:

$B_{p,0}$ = primary nucleation rate
A = pre-exponential coefficient
S = saturation level
γ_s = surface energy of the nuclei
V_m = molar volume of the solute
k_b = Boltzmann constant
T = temperature.

The primary nucleation description contains two important parameters: pre-exponential factor A and the surface energy of the nuclei γ_s. Both parameters have to be estimated from experiments because ab-initio estimations of the pre-exponential factor are not yet developed enough and impurities/heterogeneous particle often play a decisive role in the surface energy of nucleating crystals.

2.3.3.3 Secondary Nucleation

Secondary nucleation is the generation of new nuclei, from existing crystals. Here two mechanisms play a role:

- Activated surface nucleation; small asperities that growth on top of the surface of the crystals are sheared of from the surface due to turbulent action.
- Attrition of crystals; collision of crystals with walls and stirrers cause chipping off of fragments from the crystals. This will be discussed later.

Activated surface nucleation depends on the turbulence in the liquid and the amount of nuclei growing on the surface. The amount of nuclei is proportional to:

$$N_E^{''} = \frac{Đ}{d_m^2} exp\left[-\frac{\pi \gamma_{CL}^2 d_m^4}{(k_b T)^2 ln(S)} \right] \tag{2.7}$$

where:

$N_E^{''}$ = number concentration of available nuclei on the surface of the crystal
$Đ$ = diffusion coefficient of the solute
d_m = molecular diameter
γ_{CL} = surface energy of the crystals.

The rate at which the nuclei are released from the surface depends on the turbulent movement in the solution:

$$B_{s,0} = k_s \frac{\gamma}{6} m_3 \sum \eta_i d_i^2 N_i \tag{2.8}$$

where:

$B_{s,0}$ = secondary nucleation rate
k_s = rate coefficient
γ = turbulent deformation rate of the liquid
m_3 = volume fraction of crystals in the crystallizer
η_I = shearing efficiency of the removal of nuclei from the surface
d_i = particle size
N_i = number concentration of the crystals.

The shearing efficiency of nuclei removal from the surface depends on the maximum viscous force in the liquid that is exerted on the crystal:

$$\sigma_i = \frac{\sigma_{i,0}}{\Psi_i exp[1 - \Psi_i]} \tag{2.9}$$

where:

σ_I = maximum shear force exerted on the surface of the crystal

$\sigma_{i,0}$ = minimum shear force
Ψ_i = particle Stokes number.

2.3.3.4 Crystal Growth

Two different growth mechanisms are generally described in literature for growth of crystals:

- Transport of molecules/ions from the bulk of the solution to the surface of the crystal by diffusion.
- Surface integration of the molecules/ions in the crystal lattice.

Diffusion limitation is often described using Fick's law for diffusion and boundary layer theory. The mass transfer coefficient would typically be something like:

$$\frac{k_d d_p}{Đ} = 2 + 0.8 Re_p^{0.6} Sc^{0.33} \tag{2.10}$$

where:

k_d = mass transfer coefficient
d_p = particle size
$Đ$ = diffusion coefficient
Re_p = particle Reynolds number
Sc = Schmidt number.

In the case of ionic solutions and melts, multi-component diffusion theory (i.e. Maxwell-Stefan diffusion, [36]) might be needed to deal with the presence of other species. A typical example would be the presence of metal ions like potassium and magnesium during the crystallization of sodium chloride.

Surface integration can be described by either spiral growth/BCF theory or the Birth-and-Spread model. Spiral growth theory has the advantage that it can be approximated by a quadratic and a linear relation (depending on the saturation) that can be easily combined with the diffusion description for crystal growth. The crystal growth rate at low super-saturations is given by:

$$\frac{G}{2k_d} = \frac{\Delta c}{c_s} + \frac{k_d}{2k_g}\left(\frac{c_{sat}}{c_s}\right)^2 - \sqrt{\left(\frac{k_d}{2k_g}\right)^2\left(\frac{c_{sat}}{c_s}\right)^4 + \frac{k_d}{k_g}\left(\frac{c_{sat}}{c_s}\right)^2\frac{\Delta c}{c_s}} \tag{2.11}$$

where:

G = growth rate of the crystals
k_d = mass transfer coefficient
Δc = difference between the actual solute concentration and saturation concentration
c_s = solute concentration in the solid phase

k_g = surface integration coefficient
c_{sat} = saturation concentration.

The BCF/spiral surface integration coefficient is given by:

$$k_g = 2.25 \cdot 10^{-3} f_g \frac{Đ}{d_m} \frac{(c_{sat}/c_s)^{4/3}}{ln(c_s/c_{sat})} \tag{2.12}$$

where:

f_g = growth rate coefficient
$Đ$ = diffusion coefficient
d_m = molecular diameter
c_{sat} = saturation concentration
c_s = solute concentration in the solid phase.

2.3.3.5 Agglomeration

Agglomeration of crystals is usually important when high super-saturations are reached during the process. This is typically the case during anti-solvent and pH-shift processes. The rate of agglomeration of two particles is given by:

$$R_{i,j} = \beta_{i,j} \psi_{i,j} N_i N_j \tag{2.13}$$

where:

$R_{i,j}$ = rate of agglomeration of particles of size i and j
$\beta_{i,j}$ = intrinsic collision kernel
$\psi_{i,j}$ = sticking probability
N = number concentration of crystals.

The agglomeration rate contains two parameters:

- Intrinsic collision kernel $\beta_{i,j}$; the rate coefficient at which particles collide.
- Sticking probability $\psi_{i,j}$; the probability that two colliding particles stay together.

For both parameters a broad range of relations can be found in literature, depending on the situation dealt with. The intrinsic collision kernel has two main situations. The most important one is the flow induced agglomeration (ortho-kinetic agglomeration). The kernel in this case is given by:

$$\beta_{i,j} = \frac{\gamma}{6}(d_i + d_j)^3 \tag{2.14}$$

Here γ is the turbulent deformation rate and d the particle size of the colliding particles. The second important kernel is diffusion induced agglomeration (peri-kinetic agglomeration). The kernel in this case is given by:

$$\beta_{i,j} = \frac{2\,k_b T}{3\,\eta_{liq}}\left(d_i + d_j\right)\left(\frac{1}{d_i} + \frac{1}{d_j}\right) \tag{2.15}$$

where:

k_b = Boltzmann constant
T = temperature
η_{liq} = dynamic viscosity of the liquid
d = diameter of the colliding particles.

The description of the sticking probability depends on the binding mechanism involved in the agglomeration. In crystallization, the mechanism given by Hounslow et al. [12] is often interesting, because it relates the sticking probability to the mixing intensity and the crystal growth rate. The controlling parameter is the strength number which is the ratio between the strength of the solid bridge formed between two particles and the disruptive turbulent forces.

$$M = \frac{\sigma G^2}{\eta_{liq} d_{i,j}^2 \dot{\gamma}^3} \tag{2.16}$$

where:

M = strength number
σ = yield strength of the material
G = growth rate of the crystals
η_{liq} = dynamic viscosity
d = size of the agglomerate
γ = turbulent deformation rate of the liquid.

2.3.3.6 Attrition of Crystals

Attrition of crystals is related to the mechanical impact of stirrers/pumps on crystals. The volume of fragments that are released from the crystals depends on the severity of the impact, the size of the particle (mass equivalent diameter) and the material properties of the crystals:

$$\frac{V_{att}}{V_p} = \frac{20}{3}\left(\frac{\pi}{6}\right)^{1/3} \frac{\sigma_{col}^{4/3} d_p}{W_c^{1/3} H} \tag{2.17}$$

where:

V_{att} = volume of fragments that is removed from the particle by mechanical impact
V_p = volume of the particle
σ_{col} = stress of the mechanical impact
d_p = particle size

W_c = critical energy of deformation
H = hardness of the material.

The particle size distribution of fragments is often given by a power law:

$$\frac{dn}{dl} = \frac{c}{l^m} \tag{2.18}$$

where:

n = population density of the fragments
l = size of the fragments
c = coefficient that depends on the amount of material released from the particle as fragments
m = coefficient that depends on the material.

Coefficient c depends on the total volume of the attrition fragments, and the minimum and maximum size of the fragments. A complicating aspect of attrition is the fact that fragments often have a tendency to (partially) dissolve. This leads to growth rate dispersion, i.e. the crystal growth rate of crystals of a certain size has a distribution. This phenomena was nicely shown by Virone et al. [35] and is being investigated quantitatively by Ochsenbein et al. [25].

2.3.4 *Modeling of Crystallization Processes*

The modeling of crystallization processes requires the description of a number of process related phenomena:

- The solubility of the requested solute, impurities and the influence of non-solvent components.
- The dynamic description of the crystal size distribution of the crystalline product.
- Distribution of solid particles in process equipment.

2.3.4.1 Dynamic Description of the Crystal Size Distribution

The dynamic description of the crystal size distribution is done with the so-called population balance. When spacial distribution of the crystals is omitted (compartment description), the population balance becomes:

$$\frac{\partial n}{\partial t} + \sum\nolimits_i \frac{\partial Gn}{\partial l_i} = B - D \pm \sum \Phi_{in/out} \tag{2.19}$$

where:

n = population density of the crystals
t = time
G = growth rate
l = size of the crystals in the population balance
B = birth rate of the crystals
D = death/disappearance rate of the crystals
Φ = in/out flow rate of the crystals from/to the compartment.

A broad range of methods can be found in literature to solve this equation [28]. Many of the methods reduce the multi-dimensionality of the balance equation by using a single dimension for characterization of the particle size (usually mass/volume equivalent diameter). Such methods have been implemented in commercial software packages like Aspen plus (steady state situations) and gProms/gCrystal (dynamic situations, [14]). A simple straight forward implementation that can be easily implemented in general programming platforms is the use of finite differences on a logarithmic size grid. The partial differential equation is transformed into a set of ordinary differential equations using the number of particles in a size class as main variable. In the case of crystallization (positive growth rate) the population balance becomes:

$$\frac{dN_i}{dt} = \frac{aG_iN_i + bG_{i-1}N_{i-1}}{l_i} + B_i - D_i \tag{2.20}$$

where:

$$\begin{bmatrix} a \\ b \end{bmatrix} = \frac{1}{r^2 - 1}\begin{bmatrix} -2 \\ 2r \end{bmatrix} \tag{2.21}$$

and:

$$r = \frac{l_{i+1}}{l_i} = \frac{l_i}{l_{i-1}} = 2^{\frac{1}{2q}} \tag{2.22}$$

where:

N = number concentration
a, b = growth coefficients
G = growth rate
l = particle size in population balance
B = birth rate
D = death/disappearance rate
r = size ratio
q = an integer parameter controlling the discretization.

The accuracy of the result depends on the size ratio used. A good description can in generally be obtained when q is set to 3 ($r = 1.1225$). When dissolution occurs, the finite derivative has to be changed in direction (upwind differences). The implementation of the birth and disappearance of crystals from a size class depends on the phenomena taken into account [18]. Primary and secondary nucleation are usually applied to the smallest crystal sizes considered in the calculation. Agglomeration usually influences all size classes because agglomeration causes small crystals to disappear and large crystals to be formed (birth). Attrition acts opposite to agglomeration. Here large crystals disappear due to mechanical action and small crystal (fragments) are formed.

2.3.4.2 Distribution of Solids in Equipment

Distribution of solids in equipment is an important parameter to consider during the design and optimization of a crystallization process. Two aspects are important, viz. sedimentation of particles and removal of a surplus of fines that may act as seeds.

Sedimentation of particles on the bottom or walls of a crystallizer should be prevented in order to prevent fouling. Fouling reduces the productivity of a process due to lowering the heat transfer or leading to blockage. Mitigation of fouling problems is usually done by cleaning/boil-out of the equipment; however, this gives additional cost (e.g. energy) and a lower productivity. Prevention of unwanted sedimentation can be predicted using the so-called off-bottom criteria. A well-known relation for this is the Zwittering-relation:

$$N_{js} = s\frac{\nu^{0.1} d_p^{0.2} (g\Delta\rho/\rho_l)^{0.45} X^{0.13}}{D^{0.85}} \tag{2.23}$$

where:

N_{js} = stirring speed required to just suspend the crystals
s = a stirrer dependent parameter
ν = kinematic viscosity
d_p = particle size
g = gravitational acceleration
$\Delta\rho$ = difference between the density of the solid and liquid
ρ_l = liquid density
X = volume fraction of solids
D = diameter of the stirrer.

Another aspect of solids distribution is fines destruction. Fines destruction might be applied when the particle size of a product from a crystallizer is too small (resulting in e.g. bad flow behavior) due to the excessive generation of nuclei. When during the process a part of the small crystals is removed and dissolves, the remainder of the crystals can grow to larger size and thereby increase the particle size of the final product. Special sedimentation zones in a crystallizer enable the partial removal of fines and subsequent heating of the diluted slurry leads to dissolution of the fines and thereby enabling larger product crystals.

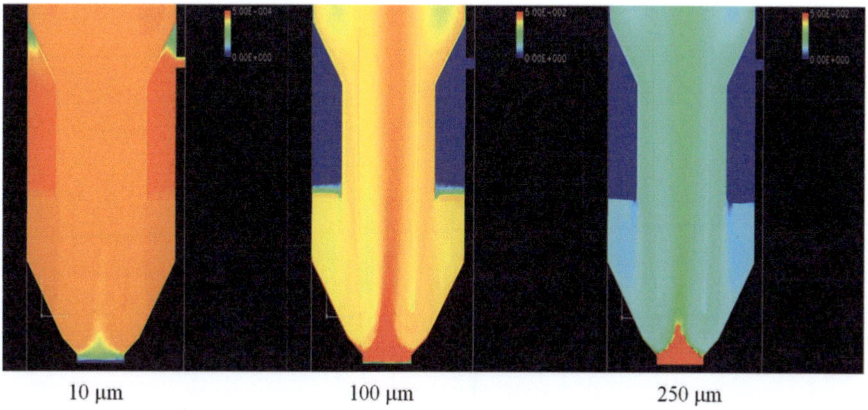

Fig. 2.8 Prediction of solids distribution of 10, 100 and 250 μm crystals in a crystallizer with fines removal. Color indicators range from 0 vol-% (*blue*) to 5 vol-% (*red*) for the two drawings on the right and until 0.05 vol-% (*red*) for the drawing on the left

These sedimentation zones can be designed on the basis of sedimentation theories like Richardson and Zaki [30], Barnea and Mizrahi [1] or Beetstra [3]. The use of Stokes sedimentation is usually not sufficient, because this assumes laminar flow, which is not often the case. The actual distribution of solid in crystallizers and sedimentation zones can be predicted using CFD calculations. Figure 2.8 gives the results of a multi-fluid prediction of a Draft-Tube-Baffled crystallizer with fines removal. In this case three size classes of crystals (10, 100 and 250 μm) and one liquid are considered. It is interesting to see that the fines concentrate partially beneath the skirt (fines removal section), which will influence to efficiency of fines removal.

Another aspect that can be learned from CFD calculations is the fact that the situation becomes quite dynamic. This was shown experimentally by Mitsutania et al. [23] and mathematically by Lettieri et al. [17]. Figure 2.9 gives the results of the separation performance of a classification zone of particles around the cut-off size of the crystallizer settling zone. The separation performance gives the fraction of particles in the outflow of the classification zone compared with the amount present in the crystallizer. The graph shows that the outflow of the classification zone shows short-term dynamics that may influence the overall performance of the system.

2.4 Industrial Aspects of Crystallization

2.4.1 Crystallization Equipment

The kind of equipment used in crystallization strongly depends on the characteristics of the process:

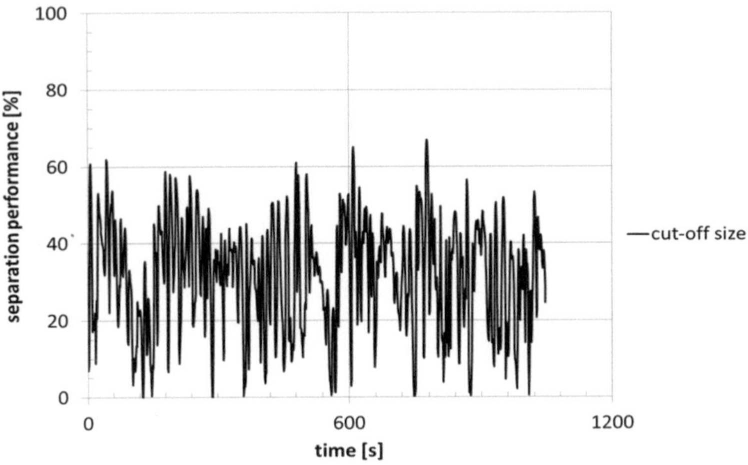

Fig. 2.9 Dynamic separation performance of a classification zone

- Mode of operation: batch or continuous
- Generation of super-saturation: cooling, evaporative, anti-solvent of pH-shift

Many pharmaceutical and fine-chemical products are produced in multi-product plants using batch operations. Often stirred vessels equipped with a Retreat Curve Impeller (RCI) are used. The walls of these vessels are often coated with glass in order to deal with aggressive media used during organic synthesis. These vessel can be fitted with a distillation column to accommodate solvent switches and evaporative processes.

When the number of products to be processed becomes less, more dedicated equipment can be used both in batch as well as continuous mode. Historically the following types of equipment are used:

- Forced circulation crystallizers
- Draft tube crystallizers with fines destruction
- Fluid bed crystallizers

New developments within the field of crystallization are:

- Oscillating baffled crystallizers
- Air lift crystallizers

Figure 2.10 gives the schematic drawings of the three main crystallizer types used in dedicated production facilities. A forced circulation (FC) crystallizer uses an external pump for the circulation of the crystallizer contents. The circulation rate needs to be relatively high in order to secure sufficient mixing in the system. The high circulation leads to high secondary nucleation and particle attrition, leading to smaller crystals in comparison with the other two configurations. Larger crystals can generally be achieved with draft-tube-baffle (DTB or Swenson) crystallizers. In this case mixing in the crystallizer is done using a stirrer located in a draft-tube and

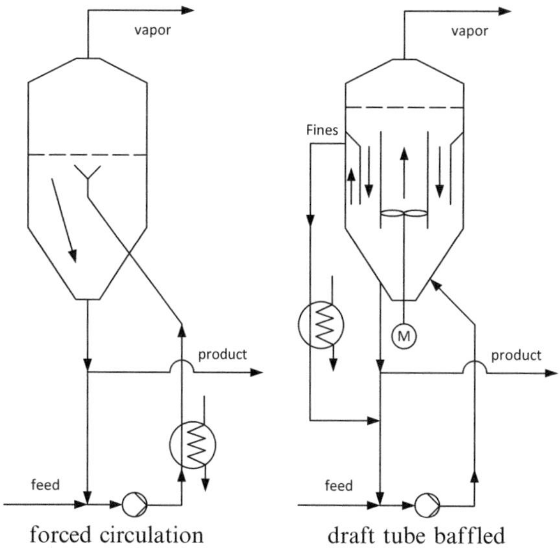

Fig. 2.10 Main crystallizer types for dedicated production facilities

fines can be removed by means of the baffle/skirt. The skirt creates a settling zone that preferentially leaves large crystals in the crystallizer. In case even bigger crystals are required, crystallization can be done in so-called Oslo or fluid bed crystallizers. In this case a large circulation stream is used to generate a fluid bed in the crystallizer. The fluid bed acts as a large settling zone, resulting in an almost crystal-free circulation stream. This reduces the nucleation of new crystals and leads to large crystals if the residence time in the fluid bed is sufficient.

Next to the crystallizers additional equipment is needed to produce the final product. Two important pieces of equipment are solid-liquid separators (filters or centrifuges) and dryers (flash dryers, fluid bed dryers, contact dryers,. . .). The amount of liquid handled by a solid-liquid separation step can be reduced by using hydro-cyclones or thickeners. Storage vessels for a feed/recycle stream and the final slurry can be included. Storage of the final slurry before the solid-liquid separation should be done with caution. The operation of filters and centrifuges often requires the continuous circulation of the slurry. This may lead to additional attrition of the product and excessive production of fines. Classification of the product in different product grades or for the removal of fines can be done using sieves after the drying step. The fines can be dissolved and returned to the process to increase the yield.

Figure 2.11 shows a schematic drawing of the oscillating flow baffled (OFB) crystallizer.

Traditional equipment like the FC and the DTB crystallizer often have inhomogeneous mixing intensity. The mixing in the pump is usually high, leading to high secondary nucleation rates, while the mixing in the body is low leading to lower growth due to diffusion limitation. The OFB crystallizer generates a homogeneous mixing field using a pulsating flow through a baffled pipe. The OFB crystallizer also

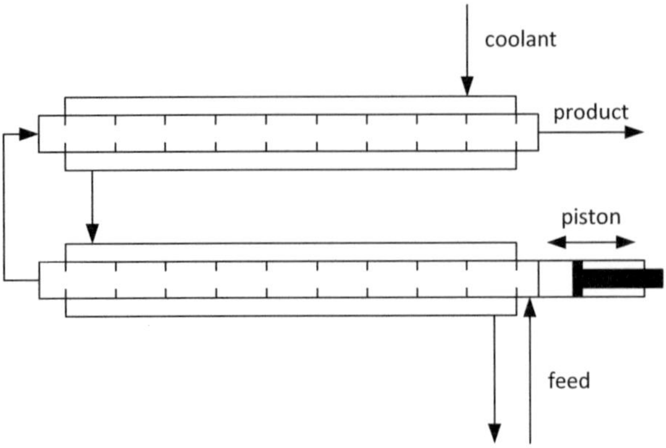

Fig. 2.11 Oscillating flow baffled crystallizer

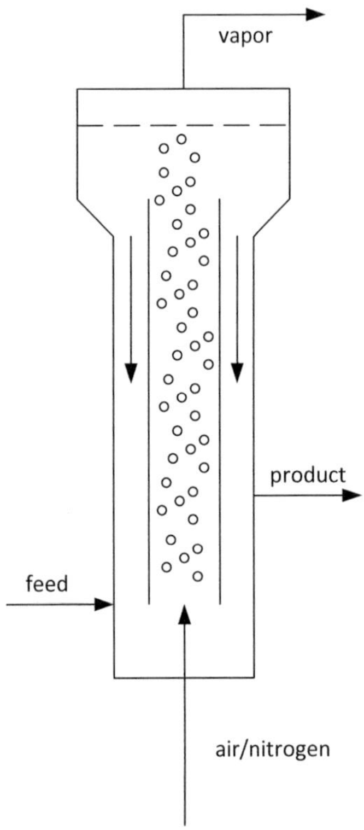

Fig. 2.12 Air lift crystallizer

has plug-flow characteristics, which can be advantageous if crystals have to be prepared with a narrow particle size distribution [16]. The disadvantage of this system is the fact that the temperature profile of the wall needs to be properly designed and controlled, because the OFB crystallizer has a relatively large wall surface area, making it sensitive to fouling of the wall.

A second way to generate a homogeneous mixing field is the use of air/nitrogen to induce mixing. The air lift reactor (see Fig. 2.12) has been developed for

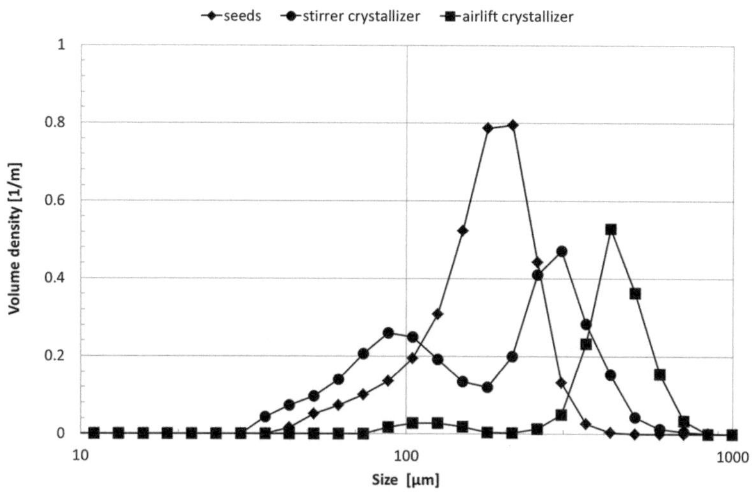

Fig. 2.13 Comparison between airlift and stirrer tank crystallizer using vitamin C (Modified from Lakerveld et al. [15])

fermenters, where the use of air reduces the power input by the stirrer, thus causing less energy usage and lower shear on the microorganisms. This concept has also been investigated by Lakerveld et al. [15] for crystallization. The concept of the airlift crystallizer is further developed within the EU-OPTICO project as part of the Task-based design of crystallization systems. The big idea behind this concept is the de-coupling of the main crystallization tasks (generation of super-saturation, nucleation and growth) in order to improve the control of the system and generate higher quality products. The application of gas also gives the opportunity of temperature control without a heat transfer surface that may get fouled.

Figure 2.13 gives a nice illustration of the separation of tasks using an airlift crystallizer, compared to a conventional stirrer tank crystallizer. Figure 2.13 gives the particle size distribution of vitamin C crystals prepared in two comparable seeded batch cooling crystallization experiments. The seeds in the stirred tank crystallizer show a limited outgrowth and a large peak of fines. This indicates that secondary nucleation has occurred. The seeds in the airlift crystallizer have grown larger, without the formation of many fines. This shows that the secondary nucleation is strongly suppressed in this type of crystallizer, focusing it on crystal growth. However, the drawback of this equipment is that additional measures are needed (continuous seeding or recycle of fines) to enable a stable continuous operation.

Which equipment is chosen, depends on a number of factors like process economics and required particle size. Forced Circulation crystallizers usually apply a high energy input and short residence time. This results in a high productivity operation and small particles. Draft Tube Baffles crystallizers use lower energy input and longer residence time. Therefore the particle size increases in comparison with a Force Circulation crystallizer, but the productivity is lower.

Table 2.2 Characteristic of traditional crystallization equipment

Crystallizer	Energy input	Residence time	Particle size
Forced circulation	10 W/kg	1–2 h	0.2–0.8 mm
Draft tube baffles	0.1–1 W/kg	2–6 h	0.8–2.5 mm
Fluid bed crystallizers	<0.1 W/kg	>3 h	1.5–2.5 mm

Data from Beckmann [2]

Further reduction of energy input and increase of residence time and particle size is achieved in the Fluid Bed crystallizer. Table 2.2 give some typical number for the energy input, residence time and particle size (data taken from Beckmann [2]). The high productivity of the Forced Circulation crystallizer usually means that the investment costs are low compared to the other options, but the operational costs are usually higher.

Oscillating Baffled Crystallizers require a more complex operation. Therefore this operation is chosen when improved control of the particle size is required for pharmaceutical purposes. Traditionally these products are produced in batch crystallizers and continuous operation can lead to cost reduction. Air lift crystallizers aren't used yet in industry, but offer the opportunity to reduce operational costs (both for mixing as well for drying) combined with good product control.

2.4.2 *Design and Control of Crystallizers*

The design and control of crystallizers are determined by the required product quality. Product quality is quantified in different ways. Purity of the product (chemical quality) is in almost all the cases one of the key parameters in the design of the process. Particle size distribution and particle shape are two parameters that are important in view of the ease of processing and in view of the intended application. Influencing the particle size is done by adjusting the saturation at which the process is run and the mixing/stirring intensity used.

Control of saturation depends on the crystallization process used. In batch cooling crystallization the saturation is influenced by the rate at which the crystallizer is cooled. In continuous cooling crystallization saturation is influenced by the difference in temperature between the feed and the crystallizer. For evaporative processes the rate of evaporation is the key parameter. For pH-shift and anti-solvent the addition rate of the precipitation agent is important.

The mixing/stirring intensity has to be controlled in such a way that differences in saturation level are removed efficiently, but attrition of the crystals is avoided (in most cases). Sometimes attrition is wanted in order to get smaller crystals.

The key parameters that control the product quality are the productivity of the process (kg/m^{3}h), the amount of solids present and the saturation level at which the process is run. The saturation level of the process (especially at the mixing point of the feed) controls the growth and nucleation of the crystals. The amount of solids controls the nucleation of the crystals. These two parameters together give the productivity of the process.

Small crystals are generally difficult to filter (solid-liquid separation) and dry. Needle shape crystals have the tendency to capture large amounts of mother lye during filtration. This increases the drying time and may give purity problems. Good handling characteristics (flow behavior and bulk density) are promoted by large, compact particles. Good dissolution characteristics are promoted by small particles. Product purity and particle size are to a large extent controlled by the starting conditions, degree of super-saturation and the mixing of the process. Therefore these parameters should be investigated during the design of the process.

2.4.2.1 Batch Crystallization

Batch crystallization is often applied when dealing with small production amounts that are typically produced in multi-product plants. This makes the production flexible and control of product quality is more easy by taking samples and performing off-line analysis. Another reason for using batch crystallization is the stability of the product. In a continuous plant build-up of by-products may prevent the formation of high-purity product and reduce the economics of the process.

The solute concentration at the start of a batch cooling crystallization is one of the parameters that needs to be investigated experimentally. Figure 2.14 gives the effect of the starting solute concentration of a fine chemical on the purity and the yield after crystallization.

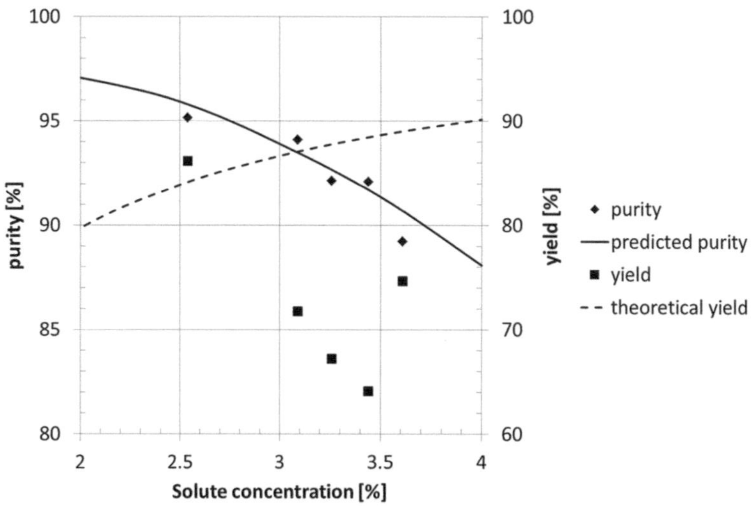

Fig. 2.14 Experimental purity and yield of the crystallization of a fine chemical

The purity shows a decreasing line with solute concentration. This is expected, because the level of impurities also rises with solute concentration. The purification of a crystallization step can be described using the purification factor, P:

$$P = \frac{c_{\text{product, final}}\, c_{\text{impurity, start}}}{c_{\text{product, start}}\, c_{\text{impurity, final}}} \tag{2.24}$$

where:

c = concentration
P = purification factor

In the case of the fine chemical shown in Fig. 2.14, a nearly linear relation was found between the initial concentration of the impurities and the purification factor, enabling the prediction of the final purity.

The yield behaves unexpectedly. The dashed line gives the prediction of the yield based on the solubility of the solute. The experimental yield shows a complete opposite behavior. The impurities cause the solubility of the product to increase.

Designing the super-saturation for a batch cooling crystallization is done through the cooling profile applied. Batch cooling crystallization is often done with a natural or linear cooling profile. Natural cooling leads to an exponential temperature profile with a large cooling rate at the beginning of the process and a low cooling rate at the end. This gives a high saturation level and growth rate in the beginning of the process, making the crystals more susceptible for liquid inclusion (decreased purity) and rough crystals (decreased washing behavior/purity and excessive initial nucleation). Alternatively, linear cooling profiles are also used. However this doesn't improve the situation with respect to inclusions and rough crystals, because initially not much crystal surface is present leading to a high growth rate.

Active control of the cooling rate can be obtained by applying in-line sensors like refractive index or FT-IR. The objective of the control is to enable a constant super-saturation during the process. Active growth control of the crystals by means of particle size measurements is investigated by several academic groups and reported in literature. However, this is not yet common practice in industry because in-line particle size sensors are often troubled by calibration, fouling and drift. Also, robust growth control algorithms are still an academic field of investigation.

The optimization of the cooling profile can also be done using the population balance approach for process development purposes.

Figure 2.15 gives the results of such an optimization. The optimization has been performed for a pharmaceutical intermediate with a molar mass of approximately 0.7 kg/mol that is crystallized using a seeded batch cooling crystallization. Screening crystallization experiments showed that the crystallization was slow due to the large molecular weight and complex nature of the molecule. This resulted in a crystallization process that required an average cooling rate of 5 °C/h in order to reduce the inclusion of mother lye. Optimization of the cooling profile was performed by estimating the solubility and crystallization parameters on the basis of 11 solubility measurements on a 1 mL scale and 1 crystallization experiment on 15 mL scale.

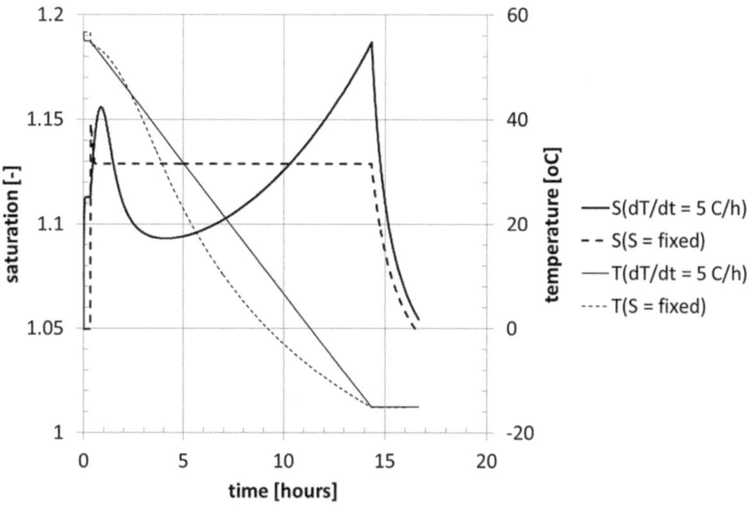

Fig. 2.15 Effect of cooling profile on saturation profile during crystallization of pharmaceutical intermediate

Figure 2.15 shows that optimization of the cooling profile can remove the overshoot of the super-saturation at the beginning and end of the process. Further optimization of the process is possible (e.g. seed amount and seed size) as was shown by Koyama et al. [14]. The optimization of the cooling profile of a crystallization process was already discussed by Nyvilt et al. [24]. Based on a temperature independent growth rate (only saturation level) he concluded that the ideal cooling profile should be a cubic function with a slow cooling in the beginning and a fast cooling at the end. In the above mentioned case this temperature profile is not found. The reason is the fact that temperature effects on the growth rate are taken into account. The low temperatures cause the diffusion of the molecule to decrease, leading to a decreasing cooling profile at the end of the crystallization. An interesting part of this kind of optimization is the fact that it can be applied in an iterative mode for process optimization. This has been shown by Forgione et al. [9].

Control of mixing is important for processes that operate at relative high supersaturations. An example is the crystallization of amphoteric materials. In this crystallization process charged molecules are neutralized by the addition of acid (or base, depending on the initial pH). Figure 2.16 gives the experimental relation between the particle size, impurity level and energy dissipation of an anti-biotic building block. The data show that the particle size and the level of impurities of the crystalline product depends on the energy dissipation of crystallizer, i.e. the mixing behavior of the system.

The mixing of the acid stream is not only influenced by the energy dissipation (i.e. stirring rate), but also by the rate at which the acid is dosed to the system. Here, meso-mixing plays a role [13]:

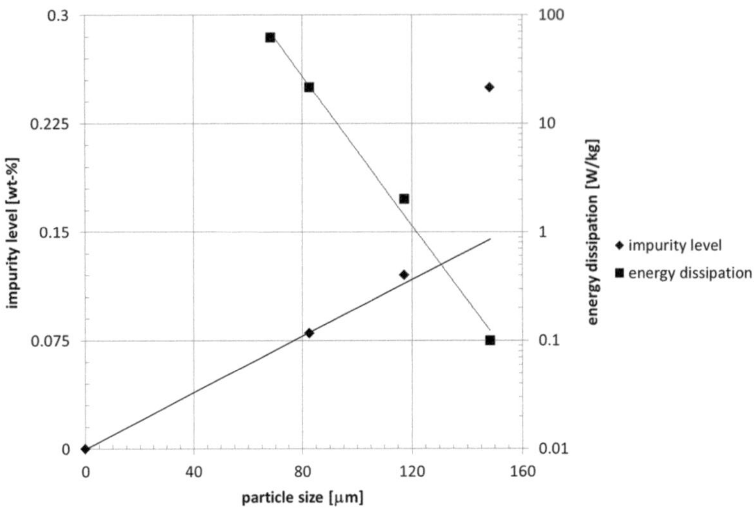

Fig. 2.16 Experimental relation between particle size, impurity level and energy dissipation for the pH-shift crystallization of an anti-biotic building block

$$\tau_{meso} = A \frac{\varepsilon_{avg}}{\varepsilon_{local}} \frac{Q_{feed}^{1/3}}{N^{4/3} D_s} \tag{2.23}$$

where:

τ_{meso} = characteristic time for meso-mixing
A = mixing coefficient
s_{avg} = average energy dissipation
s_{local} = local energy dissipation
Q_{feed} = feed rate
N = stirrer speed
D_s = stirrer diameter

The meso-mixing can be used to develop a multi-compartment population balance model, the segregated feed model (SFM [13]), to describe the formation of the particles. Figure 2.17 gives the configuration of the SFM-model. This model has been used to investigate the effect of the addition rate. The kinetic parameters of the SFM-model were estimated on the basis of standard production circumstances. The objective of the simulations was to find circumstances were smaller crystals were prepared. Initially a small increase of the feed flow was tested (experiment 2), not leading to a decrease of the particle size. Simulations showed that a threshold value exists for the acid feed flow in order to change the particle size. Experiment 3 has been done with 50 % increase of the average acid feed rate, reducing the addition time with 30 %. Here an effect of the feed rate change is seen. The particle size of the simulations and the actual production plant were in agreement with each other (Table 2.3).

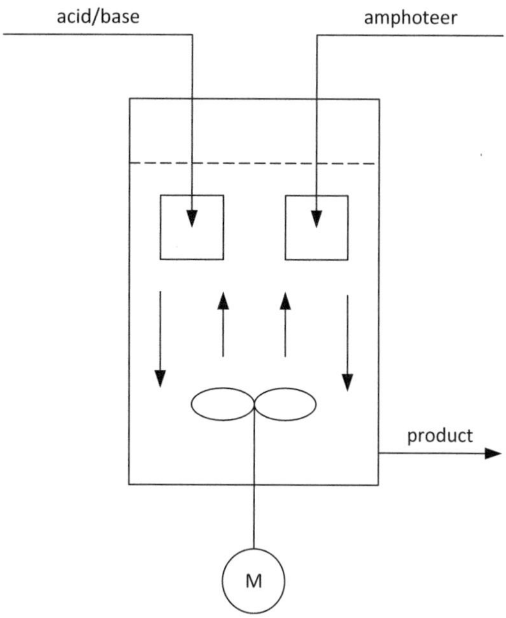

Fig. 2.17 Compartment model for the crystallization of a pH-shift or anti-solvent crystallization

Table 2.3 Effect of acid addition feed rate of particle size of pH-shift crystallization of anti-biotic building block. D50 and D95 are the 50 and 95 % quantiles of the particle size distribution

Parameter	Base case	Exp 2	Exp 3	
Acid feed flow	Standard	+5 %	+50 %	
D50 plant	22.1	23.9	14.1	[μm]
D50 model	22.1	20.5	13.9	[μm]
D95 plant	59.3	71.7	51.1	[μm]
D95 model	55.7	59.6	45.2	[μm]

2.4.2.2 Continuous Crystallization

Continuous crystallization is the preferred process option when economic production of (large quantities of) material is essential. Therefore optimization of operational costs (feed stock, energy usage, reuse of material) is one of the ways to lower the costs.

Control of super-saturation in continuous crystallization is usually the key aspect of the operation. The saturation level is often related to the mixing of the feed of the system with the content of the crystallizer. A known example is the crystallization of melamine.

Figure 2.18 gives the simplified crystallization process of melamine on the basis of US patent 4408046 [34]. The crystallization section is fed with a melamine slurry from a previous process step. Build-up of by-products may occur because mother lye from the product is recycled to an adsorption section (not drawn for simplicity reasons). Therefore, the conditions in the system are such that the by-products are

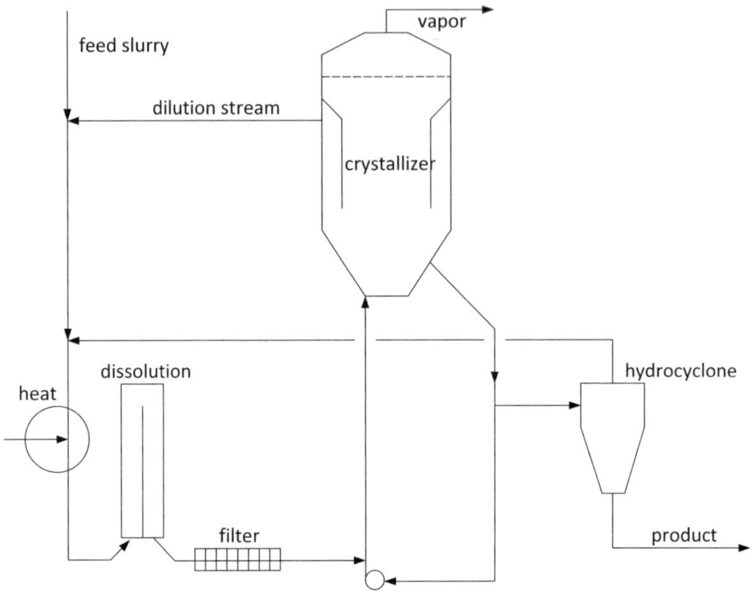

Fig. 2.18 Crystallization of melamine on the basis of USP 4408046

solid in the feed slurry and can be removed from the feed stream by filtration. The feed stream is diluted with an almost clear stream from the fines withdrawal part of the crystallizer and the top stream of the hydrocyclone. This stream is heated to such an extent that the melamine dissolves in the dissolution loop, otherwise the excess of melamine crystals would clog the filter.

However, the amount of solids in the dilution stream from the settling zone of the crystallizer and the hydrocyclone must be limited because this amount of melamine adds to the amount of melamine in the feed stream and increases the super-saturation at the mixing point of the crystallizer. When the super-saturation in this point becomes too high, excessive nucleation takes place, leading to smaller particles and even more material in the dilution loop.

Figure 2.19 gives an example of a three-stage process combining energy integration and consecutive feeding of crystallizers. The multi-stage approach enables the efficient use of energy resources and the production of coarse grain material. This is an approach that is typically applied for inorganic salts like Sodium Chloride, Sodium Sulphate and Ammonium Sulphate. The shown example uses three stages and so-called feed-forward feeding (feed of stage 2 comes from stage 1, etc.). Other configurations are also possible. Darwish et al. [6] investigates a number of different configuration, including the application of vapor recompression to improve the energy content of a vapor stream from a crystallizer. Which configuration is optimal, depends on the situation, i.e. material and costs and availability of energy. It should also be mentioned that removal of solvent by means of eutectic freeze crystallization is also becoming interesting for the crystallization of highly soluble inorganic salts.

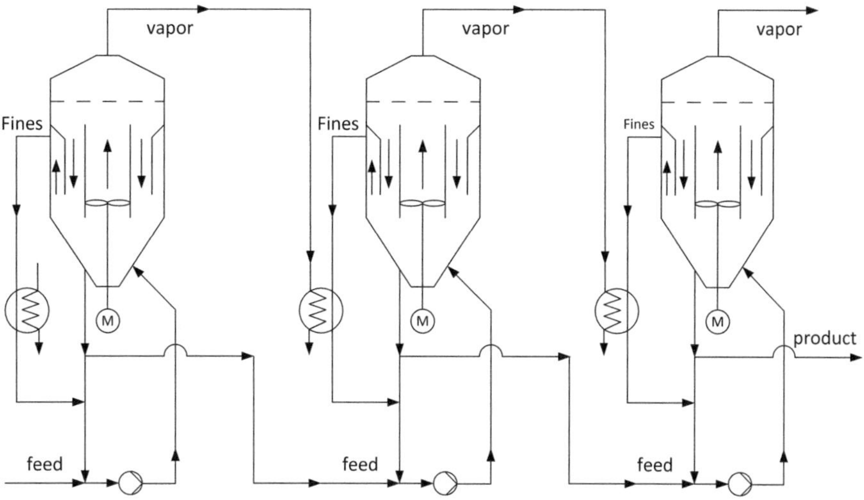

Fig. 2.19 Multi-effect configuration for energy integration during evaporative crystallization

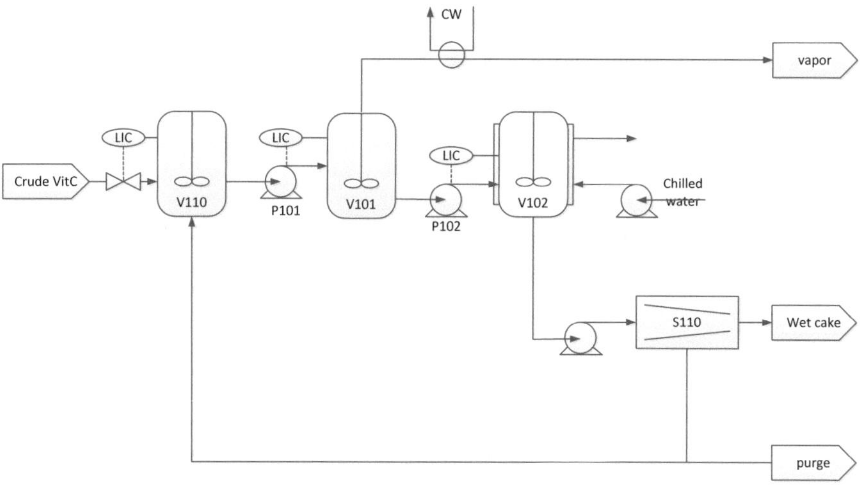

Fig. 2.20 Combined two-step process for crystallization of vitamin C

Multi-stage processes can also be used to increase the yield of a continuous crystallization process. An example is the crystallization of vitamin C. The solubility of vitamin C is such that a combination of cooling and evaporative crystallization can be used to increase the yield of the process.

Figure 2.20 gives the process flow sheet for a two-step crystallization process of vitamin C, with a combination of an evaporative crystallization and cooling crystallization. The first step uses evaporation to concentrate the feed liquid. Crystals are grown at relative high temperature, resulting in high growth rates and high

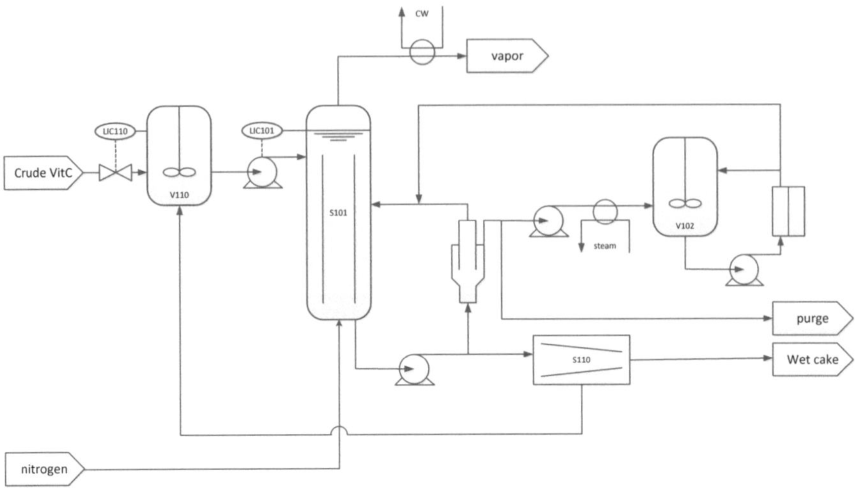

Fig. 2.21 Airlift and membrane assisted crystallization process

productivity. The second stage cools the liquid to sufficient low temperatures to increase the yield. The lower growth rate is counteracted by the high crystal surface area in the crystallizer.

A drawback of evaporative/cooling crystallization process for vitamin C is the fact that cooling to low temperatures is required to get sufficient yield. Application of membrane-assisted super-saturation generation is a method to overcome this drawback.

Figure 2.21 gives a crystallization process containing an airlift crystallizer and membrane assisted super-saturation generation than can be used to crystallize components that require a combination of evaporation and deep cooling to get sufficient yield.

2.4.2.3 Micro-Fluidic Crystallization Processes

Crystallization of nanomaterials can be done in so-called micro-fluidic devices, i.e. micro-reactors. In the case of reactive crystallization, direct mixing of the dissolved reactants can be done in T- or Y-mixers. Panagiotou et al. [26] gives the example of the production of Norfloxacin nanoparticles (antibacterial agent) by means of impinging jets technology. Figure 2.22 gives the principle of the impinging jet technology. Norfloxacin has a high solubility in DMSO, but a low solubility in water. Therefore, Norfloxacin can be crystallized by anti-solvent crystallization. A drawback of anti-solvent crystallization is the dependence on mixing. A way to minimize the effect of mixing is to mix two liquid jets (water and Nosfloxacin/DMSO) at a high velocity in a T-mixer. When the mixing is sufficient, single crystals are formed. Figure 2.23 gives the particle size distribution of the formation

Fig. 2.22 Impinging jet principle for nano crystallization

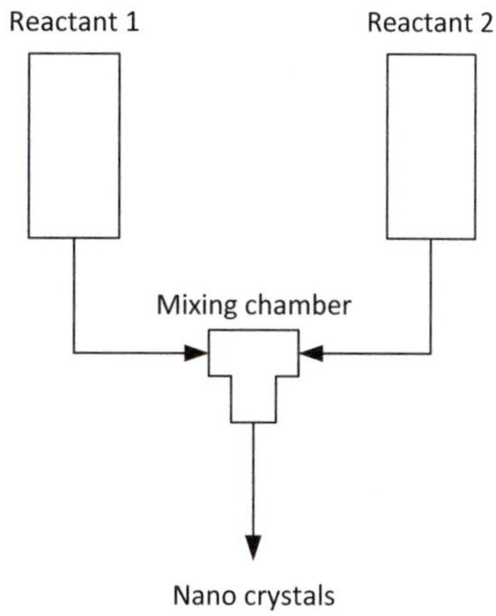

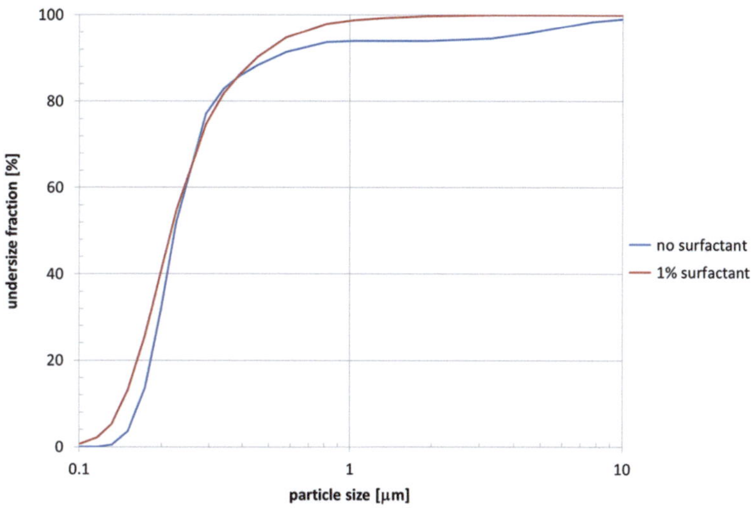

Fig. 2.23 Effect of surfactant on nano crystallization of Norfloxacin (Modified from Panagiotou et al. [26])

of Norfloxacin crystals. In the case of pure solvents 5 % of the material has a particle size between 3 and 10 μm. This indicates that some agglomeration has occurred. When 1 % of surfactant is added, the agglomerated particles disappear.

Preparation of nano-crystals by mixing of fluids that don't de-mix due to liquid-liquid de-mixing is possible for reactive and anti-solvent crystallization. When nano-crystals are to be prepared through cooling crystallization this is not a good option. Normally cooling has to be applied though the wall. This generates additional super-saturation near the wall, leading to fouling of the wall (Fig. 2.24) and blockage of the system.

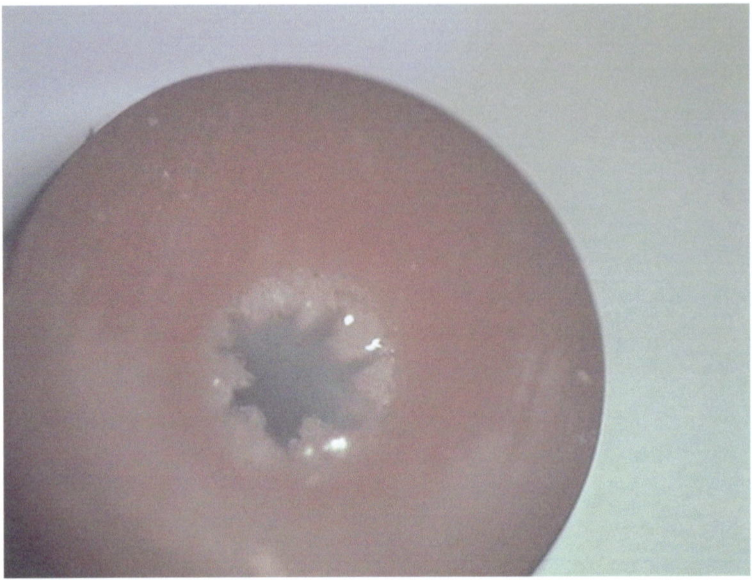

Fig. 2.24 Fouling of micro-fluidic device during cooling crystallization using a homogenous liquid

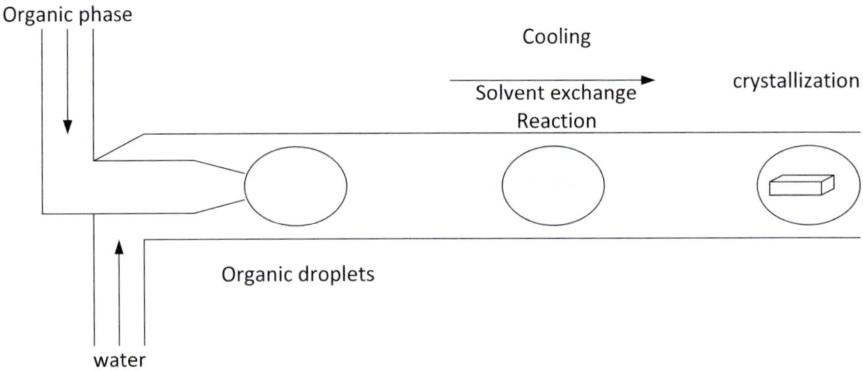

Fig. 2.25 Micro-fluidic principle for nano-crystallization using dispersed droplets

Fouling of micro-fluidic devices during cooling crystallization can be prevented by dispersing the mother lye as droplets in another non-miscible liquid. This can be done by preparing emulsions in stirred vessels or by dispersing the liquid droplets in micro-fluidic devices. Teychené and Biscans [33] have applied this methodology for the crystallization of Eflucimibe. Figure 2.25 gives the general principle of the methodology. The two immiscible liquids are mixed in a small chamber that is formed in such a way that nice droplets are formed. The droplets are transported through a long channel (e.g. etched in a glass plate). The channel gives residence time and the possibility to perform crystallization actions (cooling, solvent exchange or reaction).

One of the critical issues with micro-fluidic devices is the scale-up. Here, two methodologies are used. In the case where fouling is not an issue, transfer to static mixers is a possibility. The throughput of the droplet technology is usually done by parallelization, i.e. applying multiple channels (see for example Panagiotou et al. [26]).

2.5 Definitions, Abbreviations and Symbols

6-ACA	ε-amino-caproic acid
BCF	Burton-Cabrera-Frank
CFD	Computational fluid dynamics
DMSO	Di-methyl sulfoxide
DTB	Draft tube baffle
EtAc	Ethyl acetate
EtOH	Ethanol
FC	Forced circulation
FT-IR	Fourier transform infrared
IPA	Isopropanol
MeOH	Methanol
NRTL	Non-random-two-liquid
NRTL-SAC	Non-random-two-liquid segment activity coefficient
OFB	Oscillating flow baffled
RCI	Retreat curve impeller
SAFT	Statistical associating fluid theory
SFM	segregated feed model
UniQuac	Universal Quasi Chemical

A	#/m^{3}s	Pre-exponential coefficient of primary nucleation
A	m^2	Area
B	#/m^{3}s	Birth rate
$B_{p,0}$	#/m^{3}s	Primary nucleation rate
$B_{s,0}$	#/m^{3}s	Secondary nucleation rate
ΔC_p	J/mol	Heat capacity difference between solid and liquid solute
c_{act}	mol/m^3	Actual concentration
c_s	mol/m^3	Solute concentration in the solid
c_{sat}	mol/m^3	Solute concentration in the liquid at saturation
Δc	mol/m^3	Difference between actual and saturation concentration
$Đ$	m^2/s	Diffusion coefficient

(continued)

D	#/m^{3}s	Death rate
D_s	m	Stirrer diameter
d_m	m	Molecular diameter
d_i	m	Particle size of crystals from size class i
d_p	m	Particle size
f_g	–	Growth rate coefficient
G	m/s	Crystal growth rate
g	m/s^2	Gravitational acceleration
H	Pa	Hardness of a material
ΔH_f	J/mol	Heat of fusion
ΔH_{sol}	J/mol	Apparent heat of solution
k_b	J/K	Boltzmann constant ($1.38{\cdot}10^{-23}$)
k_d	m/s	Mass transfer coefficient
k_g	m/s	Surface integration growth coefficient
k_s	–	Secondary nucleation rate coefficient
l	m	Particle size in population balance
M	–	Strength number
N_E''	#/m^2	Nuclei surface concentration
N_i	#/m^3	Number concentration
N	rpm	Stirring rate
N_{js}	rpm	Just suspension stirring rate
n	1/m^4	Population density
P	–	Purification factor
pH	–	Acidity of solution
pK_a	–	Dissociation coefficient of acid group
Q_{feed}	m^3/s	Volumetric feed rate
R	J/mol K	Gas constant (8.314)
$R_{i,j}$	#/m^{3}s	Collision rate
Re_p	–	particle Reynolds number $= \frac{\rho_l d_p u_p}{\eta_{liq}}$
r	–	Size ratio
S	–	Saturation level
Sc	–	Schmidt number $= \frac{\eta_{liq}}{\rho_{liq} D}$
T	K	Temperature
T_f	K	Temperature of fusion
T_{sol}	K	Temperature of solution
t	s	Time
V_{att}	m^3	Volume of attrition fragments
V_m	mol/m^3	Molar volume
V_p	m^3	Particle volume
W_c	J	Critical deformation energy
X	–	Volume fraction solids
x	–	Mole fraction
x_{sat}	–	Solute mole fraction at saturation
$\beta_{i,j}$	m^3/s	Intrinsic agglomeration rate kernel

(continued)

ε_{avg}	W/kg	Average power dissipation
ε_{local}	W/kg	Local power dissipation
Φ	#/s	In/out flow rate of particles
γ	–	Activity coefficient
γ	1/s	Liquid deformation rate
γ_s	N/m	Surface energy of nuclei
γ_{CL}	N/m	Surface energy of crystals
η_{liq}	Pa·s	Dynamic viscosity
ν	m^2/s	Kinematic viscosity
$\Delta\rho$	kg/m^3	Density difference between solid and liquid
ρ_l	kg/m^3	Density of liquid
σ	–	Super-saturation
σ	Pa	Yield stress
σ_{col}	Pa	Collision stress
σ_i	Pa	Hydrodynamic surface stress
τ_{meso}	s	Meso mixing time
Ψ_i	–	Stokes number $= \frac{8\gamma\Delta\rho d_p^2}{18\eta_{liq}}$
$\psi_{i,j}$	–	Collision sticking efficiency

References

1. Barnea, E., Mizrahi, J.: A generalized approach to the fluid dynamics of particulate systems part 1 general correlation for fluidization and sedimentation in solid multi-particle systems. Chem. Eng. J. **5**, 171–189 (1973)
2. Beckmann, W.: Crystallization, Basic Concepts and Industrial Applications. Wiley-VCH, Weinheim (2013). ISBN 978-3-527-32762-1
3. Beetstra, R., van der Hoef, M.A., Kuipers, J.A.M.: Numerical study of segregation using a new drag force correlation for polydisperse systems derived from lattice-Boltzmann simulations. Chem. Eng. Sci. **62**, 246–255 (2007)
4. Chapman, R.P., Averell, P.R., Harris, R.R.: Solubility of melamine in water. Ind. Eng. Chem. **35**, 137–138 (1943)
5. Chen, C.-C., Crafts, P.A.: Correlation and prediction of drug molecule solubility in mixed solvent systems with the nonrandom two-liquid segment activity coefficient (NRTL-SAC) model. Ind. Eng. Chem. **45**, 4816–4824 (2006)
6. Darwish, M.A., Aly, S.E., Fathalah, K.A.: Comparative study between conventional and vapor compression multi-effect desalination systems. Desalination **49**, 153–167 (1984)
7. Davey, R.J., Blagden, N., Potts, G.D., Docherty, R.: Polymorphism in molecular crystals: stabilization of a metastable form by conformational mimicry. J. Am. Chem. Soc. **119**, 1767–1772 (1997)
8. Dufal, S., Papaioannou, V., Sadeqzadeh, M., Pogiatzis, T., Chremos, A., Adjiman, C.S., Jackson, G., Galindo, A.: Prediction of thermodynamic properties and phase behavior of fluids and mixtures with the SAFT-γ Mie group-contribution equation of state. J. Chem. Eng. Data **59**, 3272–3288 (2014)

9. Forgione, M., Mesbah, A., Bombois, X., Van den Hof, P.M.J.: Iterative learning control of supersaturation in batch cooling crystallization. 2012 American control conference, pp. 6455–6460 June 2012
10. Fuchs, D., Fischer, J., Tumakaka, F., Sadowski, G.: Solubility of amino acids: influence of the pH value and the addition of alcoholic cosolvents on aqueous solubility. Ind. Eng. Chem. Res. **45**, 6578–6584 (2006)
11. Granberg, R.A., Rasmuson, A.C.: Solubility of paracetamol in pure solvents. J. Chem. Eng. Data **44**, 1391–1395 (1999)
12. Hounslow, M.J., Mumtaz, H.S., Collier, A.P., Barrick, J.P., Bramley, A.S.: A micro-mechanical model for the rate of aggregation during precipitation from solution. Chem. Eng. Sci. **56**, 2543–2552 (2001)
13. Jones, A.G.: Crystallization Process Systems. Butterworth-Heinemann, Oxford (2002). ISBN 0-7506-5520-8
14. Koyama, T., Ueda, H., Oguchi, Y., Mumtaz, H., Kashiwaya, S.: Modelling, Validation and Optimisation of a Lab and Bench Scale Batch Crystallization Process, p. 19. ISIC, Toulouse (2014)
15. Lakerveld, R., van Krochten, J.J.H., Kramer, H.J.M.: An air-lift crystallizer can suppress secondary nucleation at a higher supersaturation compared to a stirred crystallizer. Cryst. Growth Des. **14**, 3264–3275 (2014)
16. Lawton, S., Steele, G., Shering, P., Zhao, L., Laird, I., Ni, X.-W.: Continuous crystallization of pharmaceuticals using a continuous oscillatory baffled crystallizer. Org. Proc. Res. Dev. **13**, 1357–1363 (2009)
17. Lettieri, P., Di Felice, R., Pacciani, R., Owoyemi, O.: CFD modelling of liquid fluidized beds in slugging mode. Powder Technol. **167**, 94–103 (2006)
18. Litster, J.D., Smit, D.J., Hounslow, M.J.: Adjustable discretized population balance for growth and aggregation. AIChE J. **41**, 591–603 (1995)
19. Lovette, M.A., Robben Browning, A., Griffin, D.W., Sizemore, J.P., Snyder, R.C., Doherty, M. F.: Crystal shape engineering. Ind. Eng. Chem. Res. **47**, 9812–9833 (2008)
20. McMeekin, T.L., Cohn, E.J., Weare, J.H.: Studies in the physical chemistry of amino acids, peptides and related substances; VII. A comparison of the solubility of amino acids, peptides and their derivatives. J. Am. Chem. Soc. **58**, 2173–2181 (1936)
21. Mersmann, A.: Crystallization Technology Handbook. Marcel Dekker, New York (2001). ISBN 0-8247-0528-9
22. Miller, J.M., Collman, B.M., Greene, L.R., Grant, D.J.W., Blackburn, A.C.: Identifying the stable polymorph early in the drug discovery–development process. Pharm. Dev. Technol. **10**, 291–297 (2005)
23. Mitsutania, K., Grace, J.R., Limb, C.J.: Residence time distribution of particles in a continuous liquid–solid classifier. Chem. Eng. Sci. **20**, 2703–2713 (2005)
24. Mullin, J.W.: Crystallization. Butterworth-Heinemann, Oxford (2001). ISBN 0-7506-4833-3
25. Ochsenbein, D.R., Schorsch, S., Vetter, T., Mazzotti, M., Morari, M.: Growth rate estimation of β L-glutamic acid from online measurements of multidimensional particle size distributions and concentration. Ind. Eng. Chem. Res. **53**, 9136–9148 (2014)
26. Panagiotou, T., Mesite, S.V., Fisher, R.J.: Production of norfloxacin nanosuspensions using microfluidics reaction technology through solvent/antisolvent crystallization. Ind. Eng. Chem. Res. **48**, 1761–1771 (2009)
27. Prudic, A., Ji, Y., Sadowski, G.: Thermodynamic phase behavior of API/polymer solid dispersions. Mol. Pharm. **11**, 2294–2304 (2014)
28. Ramkrishna, D.: Population Balances. Academic, San Diego (2000). ISBN 0-12-576970-9
29. Reid, R.C., Prausnitz, J.M., Poling, B.E.: The Properties of Gases and Liquids. McGraw-Hill, New York (1989). ISBN 0-07-100284-7
30. Richardson, J.F., Zaki, W.N.: Sedimentation & fluidisation: part 1. Trans. Inst. Chem. Eng. **32**, 35–53 (1954)

31. Romero, S., Reillo, A., Escalera, B., Bustamante, P.: The behavior of paracetamol in mixtures of amphiprotic and amphiprotic-aprotic solvents. Relationship of solubility curves to specific and nonspecific interactions. Chem. Pharm. Bull. **44**, 1061–1064 (1996)
32. Shalmashi, A., Eliassi, A.: Solubility of L-(+)-ascorbic acid in water, ethanol, methanol, propan-2-ol, acetone, acetonitrile, ethyl acetate, and tetrahydrofuran from (293 to 323) K. J. Chem. Eng. Data **53**, 1332–1334 (2008)
33. Teychen, S., Biscans, B.: Microfluidic device for the crystallization of organic molecules in organic solvents. Cryst. Growth Des. **11**, 4810–4818 (2011)
34. US patent 4408046, van Hardeveld, R.: Process of Preparing Melamine. Stamicarbon, B.V., Geleen (1982)
35. Virone, C., ter Horst, J.H., Kramer, H.J.M., Jansens, P.J.: Growth rate dispersion of ammonium sulphate attrition fragments. J. Cryst. Growth **275**, e1397–e1401 (2005)
36. Wesselingh, J.A., Vonk, P., Kraaijeveld, G.: Exploring the Maxwell-Stefan description of ion exchange. Chem. Eng. J. Biochem. Eng. J. **57**, 75–89 (1995)

Chapter 3
Wet Colloid Synthesis: Precipitation and Dispersion

Ger J.M. Koper and Roman Latsuzbaia

Abstract Recent developments have significantly altered our understanding of (nano)particle wet synthesis. The most salient feature is that after nucleation and limited growth the "primary" particles aggregate and subsequently coalesce into larger secondary particles of which the size can be limited in the traditional way such as by surface active agents. Hence, the final size and shape of the secondary particles can be controlled at will through the controlled assembly of the primary particles. The aim of this chapter is to use this in the design of wet synthesis methods for both organic and inorganic particles where a particular example for Pt catalysts is discussed in detail. In order to achieve this, colloidal behavior is presented in a colloquial form and interactions, stability, nucleation and growth are reviewed. Subsequently, the use of surfactant as a template is discussed after which the reader is prepared to embark on a description of a synthesis procedure that was recently developed in our group. The importance cannot be stressed: finally one is able to design a synthesis procedure although many challenges remain that are reviewed in the outlook at the end of the chapter.

3.1 Introduction

A wide variety of methods exist to synthesize nanometer or micrometer sized particles of various shapes and symmetry from liquid solutions [19, 33, 42, 56, 70, 79, 81]. However, each of these methods is special with regards to ingredients and processing conditions. This chapter aims at the development of a processing method by which many different particles can be created with one single procedure. The basis of design of this method is the very recent finding that nanoparticle

G.J.M. Koper (✉)
Department of Chemical Engineering, Delft University of Technology, Delft, The Netherlands
e-mail: g.j.m.koper@tudelft.nl

R. Latsuzbaia
Netherlands Organization for Applied Scientific Research, TNO, Zeist, The Netherlands
e-mail: romanlatsuzbaia@gmail.com

© Springer International Publishing Switzerland 2016
H.G. Merkus, G.M.H. Meesters (eds.), *Production, Handling and Characterization of Particulate Materials*, Particle Technology Series 25,
DOI 10.1007/978-3-319-20949-4_3

synthesis essentially involves the coalescence of small nuclei [9, 20, 59–61, 78], typically not larger than a few nanometer only. By controlling the assembly of these nuclei into larger aggregates before coalescence takes place, one may dictate the size and possibly shape of the resulting particles [45].

For a long time, the canonical view was held that the wet synthesis of colloidal particles is controlled by the Kelvin equation, i.e. by a Gibbs energy balance between the positive surface formation contribution and the negative contribution of bulk solid phase formation [56, 71, 79]. Once, when by fluctuation a nucleus larger than the critical radius is formed, its growth is from there on only limited by mass transport. The Gibbs energy barrier is hence the determining factor in the nucleation rate. The modern view is that this nucleation model, with all its present refinements, still holds but only for the very early stages of particle formation. At later stages, in particular when the monomeric building blocks are depleted, growth continues with the aggregation and subsequent coalescence of the nuclei. The final particle size is determined by capping agents or surface active agents that have the effect of strongly reducing the growth rate of the particles beyond a certain size [35, 51].

In a very recent review article, Wang et al. describe the above sketched growth mechanism for inorganic particles [77]. Almost at the same time, our group contributed to this field by demonstrating that also the formation of polymer colloids by the very classical emulsion polymerization technique evolves along the same lines [9]. In short, after initiation monomers polymerize to oligomers. At some point, the oligomers become insoluble and nucleate into small primary particles. The primary particles subsequently aggregate into larger secondary particles where the polymer continues to grow within and on the surface to result in uniformly sized particles, see Fig. 3.1 (top) for a sketch of the process. An interesting aspect of the aggregation process is that the primary particles carry a significant electrical charge which should prevent them from aggregating. The origin of the charge is either the ionic initiator or, in case of a nonionic initiator, the charge that naturally develops at hydrophobic surfaces in the presence of water [8]. It is the still present radical chains at the surface of the particles that are hypothesized to be responsible for the initial linkage between the particles as illustrated in Fig. 3.1 (bottom). The secondary particles are initially very soft with many voids so that while curing they become perfectly spherical and their surface becomes very smooth.

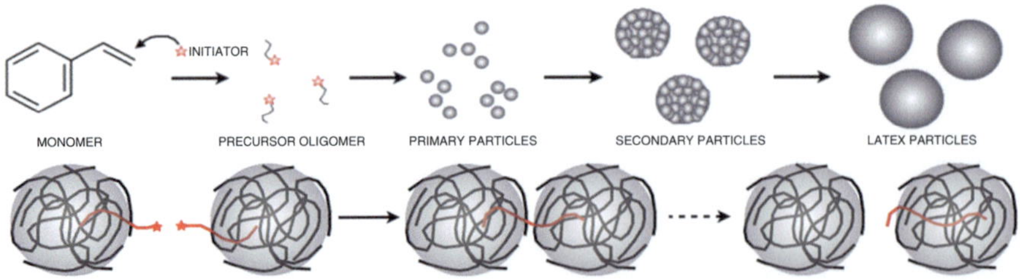

Fig. 3.1 *Top*: Sketch of polymerization process involving nucleation, polymerization, nucleation, coalescence and growth; *Bottom*: visualization of the aggregative growth process by which latex spheres can coalesce reversibly [9] (Copyright American Chemical Society; reprinted with permission)

The rest of the chapter is organized as follows. In Sects. 3.2 and 3.3 the basics of colloidal stability and interactions, respectively, are discussed in order to establish a solid theoretical basis for the further discussion on particle nucleation and growth in Sect. 3.4. In Sect. 3.5 surfactant templating methods are dealt with to be subsequently applied in Sect. 3.6 where the design method is discussed and results are presented for a contemporary example. In Sect. 3.7 we reflect on the present state-of-the-art and discuss some future developments.

3.2 Colloidal Stability[1]

A major concern in wet colloid synthesis and processing is colloidal stability. The term is used whenever a colloidal dispersion loses its homogeneity and clearly shows signs of phase separation such as sediment on the bottom of the container or a layer on top of the container. It is important to realize that almost always the result is the net effect of two intricately interplaying processes: sedimentation or creaming and flocculation. Therefore, before embarking on a discussion of colloidal stability the two separate processes will be discussed.

3.2.1 Sedimentation and Creaming

When a particle is forced to move through a liquid, the velocity v it acquires is proportional to the applied force F. This is essentially Stokes' law and it reads $v = F/f$. The proportionality factor f is called the friction factor which for a sphere[2] is given by $f = 3\pi\eta d$ with d the diameter of the particle and η the viscosity of the liquid which is about 1 mPas for water. The friction that the particle experiences in the liquid arises from the hydrodynamic momentum transfer which in itself is mediated by collisions of liquid molecules with the moving particle surface. There is, however, an additional effect of these collisions. Even at rest, there is a fluctuating momentum transfer from the liquid molecules to the particle and as a consequence the particle exhibits irregular motion, the so called thermal motion or Brownian motion. The time average of the momentum transfer vanishes, but there is a noticeable effect on the position of the particle. In a given time lapse Δt the uncertainty in the center-of-mass position of the particle has increased by an amount $\Delta r = \sqrt{6Đ\Delta t}$. The proportionality constant $Đ$ in the above equation is the diffusion coefficient of the particles, which is related to its friction factor by the Einstein relation $Đ = k_B T/f$, in which $k_B = 1.38 \cdot 10^{-23}$ J/K is the Boltzmann

[1] Sections 3.2 and part of 3.3 are adapted from *An Introduction to Interfacial Engineering* by the author [30].

[2] Even though many (colloidal) objects are not spheres, one still uses expressions for spheres to obtain the correct order of magnitude in the absence of a more accurate value.

Fig. 3.2 Sedimentation in a colloidal dispersion (Copyright VSSD; reprinted with permission)

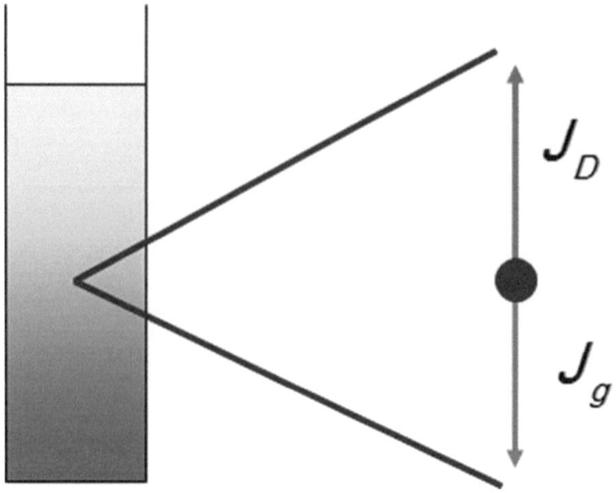

constant or molecular gas constant and T is the absolute temperature of the liquid, typically 300 K. It is important to note that Brownian motion is a chance process and hence there is no way to predict the actual distance covered by a particle. The only statement one may make is that after a time lapse Δt there is a high probability to find the particle within a sphere of radius $\sqrt{6Ð\Delta t}$. Colloquially one says: the particle has "diffused" over a distance $\sqrt{6Ð\Delta t}$.

It is easy to see that the Brownian motion of a tennis ball will not be too impressive, but for a 6 nm sphere in water the diffusion coefficient is about $73 \cdot 10^{-12}\ \mathrm{m^2/s}$ so that after a second it "diffuses" a distance of approximately 20 μm which is about 3500 times its size!

Now, consider a liquid colloidal dispersion in a transparent vessel as in Fig. 3.2 where the dispersed phase has a higher mass density than the liquid. The particles will gain a velocity as discussed above and after some time, a density gradient will set in so that the dispersed phase is denser at the bottom of the vessel than at the top. In the extreme case all the dispersed mass is at the bottom, which is called *sedimentation*. The situation where the dispersed phase has a lower mass density is called *creaming* because the density gradient is opposite. The density gradient will cause the particles to diffuse back to equalize the gradient. The balance of the two opposing fluxes, J_g because of gravity and J_D because of diffusion, is a density profile $c(z)$ that has the same functional form as the barometric height distribution $c(z) = c(z_0)\exp\{-(z - z_0)/L_g\}$ with characteristic length scale L_g. It is evaluated as $L_g = k_B T/(gV_p \Delta\rho)$ in which g is the gravitational constant, $\Delta\rho$ is the mass density difference between dispersed phase and continuous phase, and V_p is particle volume. This length scale is to be compared to the size of the container; we assume a flask of 10 cm length (z – direction) here. For values of the length scale L_g less than a few millimeters, a sediment or cream layer will form. For values larger than tenths of centimeters, the dispersion will appear homogeneous. Values in between will exhibit an appreciable density gradient which for many systems is easily visible whereas for others spectroscopic techniques such as light scattering are required.[3]

3.2.2 Flocculation

In a colloidal system, particles move because of Brownian motion and external forces, such as shear; see Fig. 3.3. Once in a while, two particles collide and then it depends on the interaction between the particles whether they will stick together or will separate again due to for instance shear. Particles may also separate due to Brownian motion albeit that in the latter case this is usually a minor effect. The process that sticks particles together continues and with time larger and larger flocs or aggregates are formed. The larger a floc, the slower it diffuses because of a relatively involved hydrodynamic effect. Simply stated, it is due to rotational

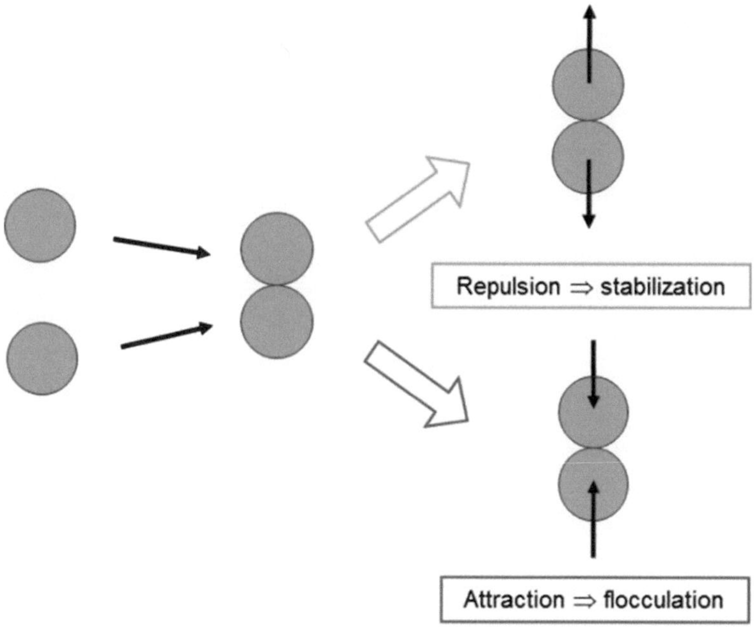

Fig. 3.3 Schematic representation of the flocculation process (Copyright VSSD; reprinted with permission)

[3] It is important to realize that the diffusion rate determines the time scale over which the sedimentation profile establishes itself. The shape of the profile however is independent of the kinetics; as a signature: the profile does not contain the viscosity which is the rate determining liquid property here.

diffusion of the particles during which the liquid enclosed in the solid cavities remains stationary with the particle and hence adds up to its effective volume. In shear flow, particles also rotate which leads in the same way to slower motion. Therefore, the effective mass density of the flocs changes which in the end leads to creaming or sedimentation. Furthermore, especially in the case of liquid-liquid dispersions the sticking of particles can lead to the merging of droplets, a process commonly termed coalescence.

As discussed above, the collisions between particles may either arise from Brownian motion, in that case leading to perikinetic flocculation,[4] or from external forces, leading to orthokinetic flocculation.

The time scale of aggregation is an important product design quantity, for instance in paint the time scale of aggregation should be below that of drying as a film. The fastest aggregation is obtained when there is no interaction between the particles except that they aggregate upon collision. This is not a purely academic exercise as it is often a good model when only van der Waals forces act between the particles: due to the short length scale over which these forces act the particles behave as sticky balls. Initially, aggregation is a second order rate process and hence the time scale of aggregation is inversely proportional to the initial particle concentration, c_0 (in particles per cubic meter). It was Smoluchowski[5] who derived expressions for perikinetic flocculation, $\tau_p = 1/(4\pi Ð d c_0)$, and for orthokinetic flocculation, $\tau_o = 3/\left(2\pi G d^3 c_0\right)$ in which G is the shear rate. During flocculation, the total mass of particles and hence its volume fraction $\phi = \pi d^3 c_0/6$ remains constant so that during orthokinetic flocculation the flocculation time scale is independent of the actual progress of flocculation.

In order to control or prevent flocculation, one needs to tune the interactions between the colloidal particles. In this way, one obtains longer time scales for flocculation and increases the *stability ratio*, i.e. the ratio between the actual flocculation time scale and the fast time scale. In some particular cases, this common route is not available for instance because during processing a dispersion experiences strongly varying pH levels. In those cases, affecting the fast flocculation time scale by increasing the viscosity or introducing a yield stress might be the solution.

---

[4] The word coagulation is also used to denote flocculation or aggregation.

[5] The Smoluchoski family preferred the Polish name as given in the text. German texts, such as his PhD thesis, refer to him as Marian, Ritter von Smolan Smoluchowski; the word *Ritter* stands for Knight. In contemporary texts, confusion dominates and one also finds Maryan von Smoluchowsky.

3.3 Colloidal Interactions

In most text books, the electrical double layer or DLVO-interactions [22] enjoy more attention than the steric interactions whereas in practice their relevance is quite comparable. Therefore, this subsection devotes roughly the same amount of attention to both kinds of interactions.

3.3.1 *Van der Waals Interactions*

The origin of the van der Waals interactions is in the distribution of electrons around the nuclei of the molecules. It is intrinsically quantum mechanical in nature, but a simple picture may be sketched based on the classical Bohr model of atoms and molecules. Consider first a mono-atomic molecule such as Argon. The nucleus represents the center of mass of the positive charges in this molecule. The center of mass of all the negative charges, the electrons, on average coincides with the center of mass of the positive charges. The electrons, however, orbit the nucleus and as a consequence their center of mass fluctuates around the nuclear center of mass. This gives rise to a fluctuating dipole that radiates an electromagnetic field. The electron clouds around molecules in the neighborhood respond to this fluctuating dipole and themselves radiate an electromagnetic field that is partly correlated with the received field. The net average of this fluctuating dipole-fluctuating dipole interaction is attractive. For more involved molecules there are also dipole contributions from vibrational motions inside the nuclei and even rotational motions of the molecules as a whole. A detailed description of van der Waals forces can be found in the book by Israelachvili [28].

To find the van der Waals interaction between surfaces or between spherical particles, summation rules are invoked. For spheres of diameters d_1 and d_2 of a material α in a medium β, the van der Waals interaction can be written as

$$w(h) = \frac{A_{\alpha\beta\alpha}}{3h}\frac{d_1 d_2}{d_1 + d_2} \tag{3.1}$$

with h the distance between the surfaces.

The importance of this expression is that the van der Waals interaction between spheres varies inversely with the separation and is proportional to the "average" size. The strength of the interaction is determined by the Hamaker coefficient $A_{\alpha\beta\alpha}$. Its value can be related to the interfacial tension of the materials. In general one finds, that for particles of the same material α that are dispersed in a liquid β, the Hamaker constant is always positive but that for particles of different materials α and γ it could be negative. An example is poly(methyl methacrylate) or PMMA-particles in water, for which the Hamaker constant is $7.11\,10^{-20}$ J, whereas the Hamaker constant for the interaction between PMMA-particles and air bubbles in water is attractive, it is $-1.23\,10^{-20}$ J.

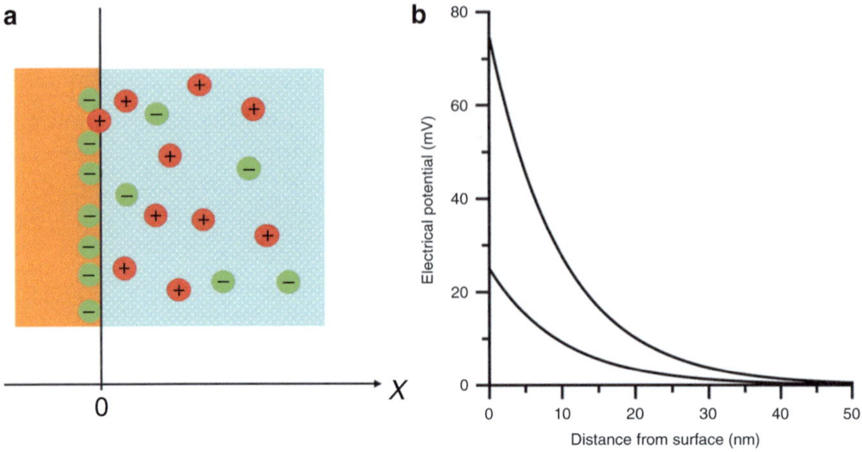

Fig. 3.4 Charged surface in contact with an electrolyte (**a**) and electrical potential in the liquid (**b**) in which both a high and a low surface potential case at the same ionic strength are shown (Copyright VSSD; reprinted with permission)

3.3.2 Electrical Double Layer Interactions

A surface in contact with an electrolyte, see Fig. 3.4a, may acquire a surface charge by several mechanisms. One of the most frequently occurring mechanisms is the dissociation of surface groups and another mechanism is the specific adsorption of some ion type. With all mechanisms, the effect is that there are restricted ions at or within the surface and mobile ions in the electrolyte. The system as a whole is neutral, so that there is a balance between the counter ions and the surface charge. In addition, the electrolyte may contain more ions of different kinds that charge compensate amongst themselves.

The organization of the counter ions will be such that the Gibbs energy is minimized, which implies that (i) the surface charge is compensated on the shortest possible length scale, (ii) the effect of repulsion between the ions is minimal to minimize the enthalpy, and (iii) the ions are as much dispersed as possible to maximize the entropy. The presence of other ions assists in two ways:

(i) By intermingling positive and negative ions the effect of interactions between the ions is balanced and (ii) more ions allow for more distributions which maximize the entropy. The organization of the ions near the surface leads to an electric potential as depicted in Fig. 3.4b. The resulting electrical potential is well approximated by

$$\phi(h) = \phi_{\text{eff}} e^{-\kappa h} \tag{3.2}$$

Table 3.1 Some values of the Debye screening length for a 1–1 salt such as NaCl

[NaCl] M	10^{-7}	10^{-4}	10^{-3}	1
κ^{-1} nm	960	39.4	9.6	0.3

There are two characteristics of this potential that need to be discussed. The first is the exponential decay dictated by the inverse Debye length scale that varies with the square root of the ionic strength[6] of the liquid phase, see Table 3.1. The above equation (3.2) summarizes the experience that the more ions there are in the system, the more effective their screening effect. This screening is essentially the same as in ionic solutions where ions of the one kind organize themselves around ions of the other kind so as to minimize the electric fields that arises because of their presence. The ionic distribution in front of the surface effectively forms a diffuse double layer with a thickness that scales with the Debye screening length.

The second characteristic of the potential is the effective surface potential. In Fig. 3.4 the relation between the effective potential and the actual surface potential as can be determined by for instance potentiometric titration. The picture clearly shows, that for low actual potentials, the effective potential and the surface potential are equal. For high actual potentials, the effective potential levels off to a constant value, $4k_BT/e$ (with e proton charge) in the case of a simple 1–1 electrolyte solution, see Fig. 3.5. The explanation for this behavior lies in the ionic distribution. For low surface potentials the diffuse double layer effectively screens the surface charge, but for high potentials the attraction between the free ions and the surface charge would become too high if not a significant amount of the counter ions reside very close to the surface. This *charge condensation* is a very local effect close to the surface, so that further away from the surface the effective charge that is felt in the double layer and beyond is as if the larger part of the surface charge is not present. Hence the term effective surface potential.

The relation between the surface charge and the effective surface potential or *zeta-potential* is rather involved [26]. For not too large surface potentials it reads

$$\sigma = \varepsilon_0 \varepsilon \kappa \phi_{\text{eff}} \tag{3.3}$$

where κ is the Debye screening length, $\varepsilon_0 = 8.85\ 10^{-12}\ \text{C}^2/\text{Jm}$ the dielectric permittivity of vacuum and ε the relative dielectric permittivity of the solvent (about 80 for water). The surface charge of a particle is a quantity that can be measured by recording its motion under the influence of an electric field. By means of such electrophoretic measurements one hence obtains the zeta potential of the surface. More information on this and similar methods can be found in the books of Hunter, one of the pioneers of the technique [25–27].

---

[6] The ionic strength of a 1–1 electrolyte is equal to its concentration, for more involved electrolytes it can be found in various text books [22].

Fig. 3.5 Effective potential as a function of surface potential (Copyright VSSD; reprinted with permission)

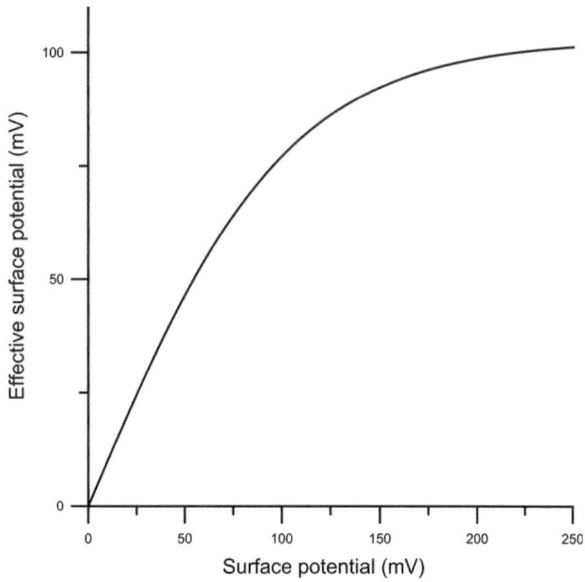

3.3.3 DLVO Potential

When two charged particles approach one another within the range of the electrical double layer, which means that the separation between the surfaces is less than a few times the Debye screening length, repulsive electrical interactions arise with a strength that increases with distance. At closer separation, the attractive van der Waals interactions become noticeable. The resulting interaction potential is depicted in Fig. 3.6. The characteristics of the DLVO-potential are as follows:

- For low ionic strength values, the potential is purely repulsive and exhibits an energy barrier that prevents particles to enter the attractive regime.
- For high ionic strength values, the potential is purely attractive.
- For intermediate ionic strength values the energy barrier is not too large and there may even be a secondary minimum in which particles may temporarily be trapped.

3.3.4 Steric Interactions

All particles and even molecules occupy some space that cannot be occupied or penetrated by other particles or molecules. This is called the excluded volume of a particle or molecule. In gases it is the high temperature value of the second virial coefficient that in essence is the mutually excluded volume of the gas molecules. Completely analogously, the second virial coefficient of the osmotic pressure of a particulate solution is determined by the mutually excluded volume of the particles.

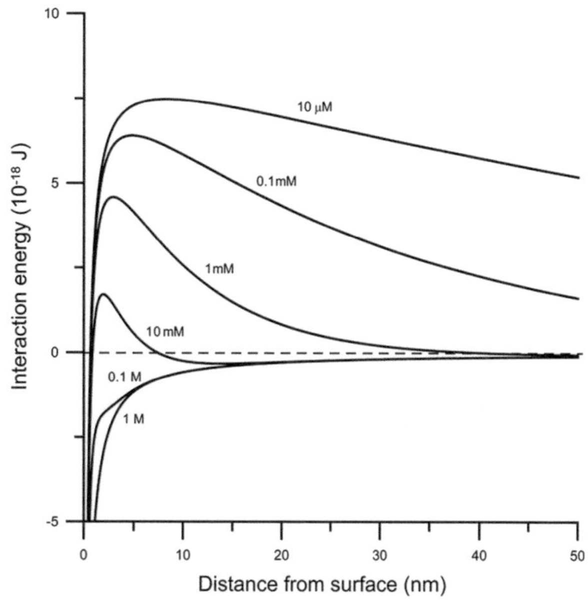

Fig. 3.6 DLVO potential for particles of 0.5 µm with a zeta-potential of 75 mV and a Hamaker coefficient of 1.4 10^{-19} J in a 1–1 electrolyte (Copyright VSSD; reprinted with permission)

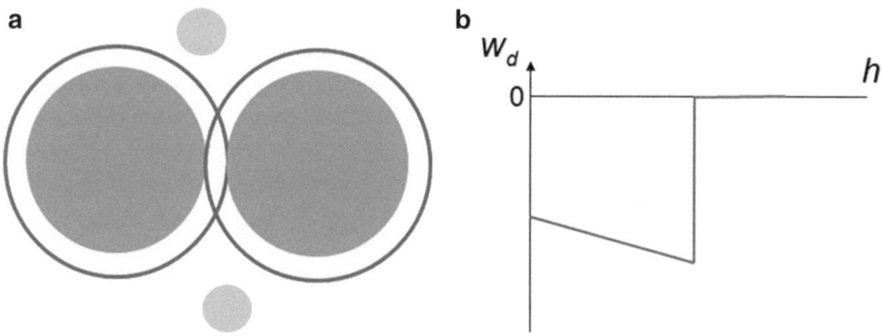

Fig. 3.7 Two colloidal particles in a bath of small particles (**a**) and depletion interaction potential (**b**) (Copyright VSSD; reprinted with permission)

The cause of these so called steric interactions lies in the Born-repulsion of molecules that in themselves originate in the Pauli Exclusion Principle that states that electronic shells cannot be occupied by more than one electron with the same spin. In the following two different manifestations of steric interactions will be discussed, namely depletion interaction and adsorbed layer interaction. Most of the templated particle synthesis routes discussed later in this chapter utilize layer interaction by adsorbed surfactant.

3.3.5 Depletion Interaction

Consider Fig. 3.7a, which depicts two large colloidal particles in a bath of small particles. Around the colloidal particles a shell is drawn within which the centers of

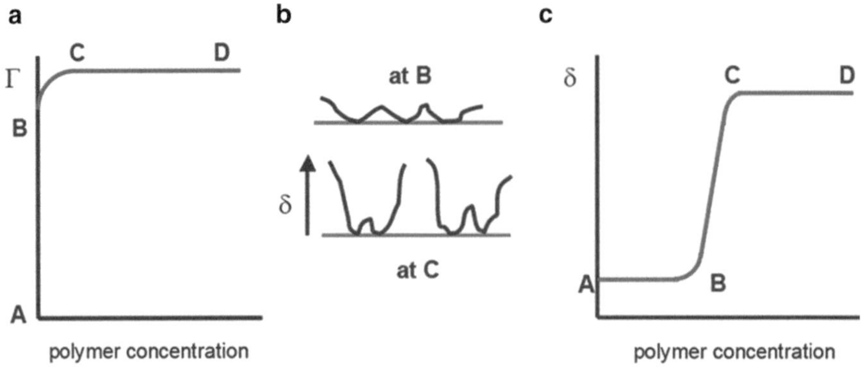

Fig. 3.8 Adsorbed polymer layer properties: (**a**) adsorbed mass, Γ, (**b**) conformation and (**c**) layer thickness, δ (Copyright VSSD; reprinted with permission)

the small colloidal particles cannot penetrate. As long as the separation h between the big particles is larger than the diameter d of the small particles, the small particles may collide from all sides with the big particles. Since any collision on one side of the big particles is equally probable as a collision on the opposite side, the big particles will not experience a net force in one particular direction. However, when the separation $h < d$ the small particles cannot enter the *overlap volume* v_o. This implies that small particle collisions on the opposite side of the big particle are not "compensated" by collisions from the side where now is the overlap volume. The net effect of this is, that the big particles will experience a force that brings them together. In terms of the osmotic pressure Π_s of the dispersion due to the small particles the depletion interaction potential can be written as:

$$w_d = \begin{cases} -\Pi_s v_o & \text{for } h < d \\ 0 & \text{otherwise} \end{cases} \tag{3.4}$$

and has the simple form depicted in Fig. 3.7b.

3.3.6 *Adsorbed Layer Interaction*

Polymers may be made to adsorb on surfaces, being either flat surfaces or particle surfaces. Whether polymers will adsorb or not depends on the polymer-surface interactions and on the solvent. Many times one uses polymers with ionisable groups, so called polyelectrolytes, that are selected such that they preferentially bind to surface groups. In Fig. 3.8 the typical aspects of polymer adsorption are depicted [14]. Figure 3.8a is the graph of surface adsorption versus polymer concentration, Fig. 3.8c depicts the polymer layer thickness as a function of concentration and in Fig. 3.8b are sketches of what the polymer conformations are at varies points.

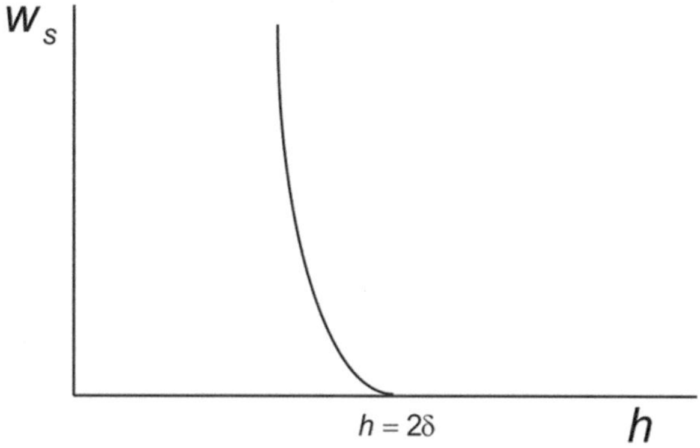

Fig. 3.9 Adsorbed polymer layer interaction (Copyright VSSD; reprinted with permission)

At low concentrations, the regime A–B, there is barely polymer adsorption and those polymers that are adsorbed spread out over the surface at will. For a given polymer, there is a competition between the adsorbed polymer segments and those in solution. In other words, there is a balance between the enthalpy gained with the adsorbed segments and the entropy gained with the segments that are free in solution. At B the adsorption reaches a maximum in the sense that polymers adsorbed in the flat conformation occupy the entire available surface. There can be much more polymers on the surface, but then the conformation needs to change from the flat state to the standing state. This is why the layer thickness increases so drastically going from B to C compared to the increase in adsorption. In the regime C–D the surface gets more and more saturated which is hardly noticeable on the adsorption or on the layer thickness. For all practical purposes the surface can be considered as full.

The quality of the solvent plays an important role in this respect which is expressed in the Flory polymer-solvent interaction parameter χ. For $\chi > 0.5$ the solvent quality is good and the polymer conformations are extended. In the other case, for $\chi < 0.5$ the solvent quality is considered to be bad which leads to compact conformations.

Now consider two particle surfaces that are not fully saturated with adsorbed polymer, see Fig. 3.9. When the separation is larger than twice the polymer layer thickness there is no interaction noticeable. Upon closer separation, the polymer layers are touching. This gives rise to a repulsive interaction when the solvent is also a good solvent for the polymer, because the polymer segments start to interpenetrate which leads to a loss of entropy. When the solvent quality is bad, the interaction is attractive. With polymer adsorption it is well possible to tune the interaction between colloidal particles. There are some pitfalls that one has to beware of:

- If the solvent is too good, adsorbed polymers extend far into the solvent and in fact polymers may bridge between two particles. This phenomenon leads to weak flocculation.
- Free polymer coils in the solution may lead to depletion flocculation because of the above described attractive depletion interactions.

In order to circumvent the problems mentioned above, one often chooses the more expensive block co-polymers with which one may tailor the one block to the adsorption on the surface and the other block to the optimal solvent behavior.

3.4 Nucleation and Growth

The first step in the wet synthesis of (nano)particles is the creation of a supersaturated solution. For instance in the case of fine metal particles such as gold this is classically done by mixing aqueous solutions of a suitable metal precursor such as chloroauric acid ($HAuCl_4$) and a reducing agent such as sodium borohydride ($NaBH_4$) [74]. In such a supersaturated solution atoms experience a strong tendency to aggregate thereby attaining a lower free energy state compared to the non-aggregated state. However, for small nuclei there is a strong surface energy component arising from the fact that the surface atoms partially remain in contact with the liquid phase. The competition between the bulk energy gain and surface energy increase upon aggregation is modelled by the Kelvin equation and the resulting Gibbs energy as a function of nucleus size is given in Fig. 3.10.

The important aspect of this balance is that there is an energy barrier to nucleation: there is a critical radius over which nucleus growth starts. This corroborates with the experimental finding that a certain degree of supersaturation is required in order to obtain a measurable rate of nucleus formation, i.e. to be able to discriminate a density fluctuation from a growing nucleus. Impurities in the solution that prefer to reside in the nuclei rather than in the solution modify the energy balance slightly and may affect that the energy barrier exhibits a weak local minimum by which nuclei can become relatively long-lived and create a meta-stable dispersion [63]. Where for

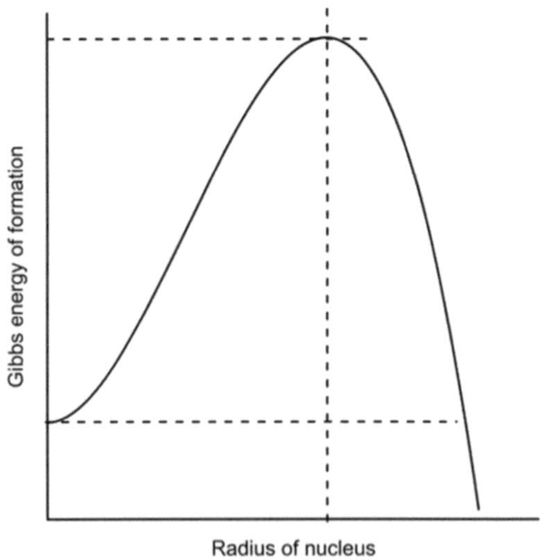

Fig. 3.10 Gibbs energy for a nucleus as a function of size

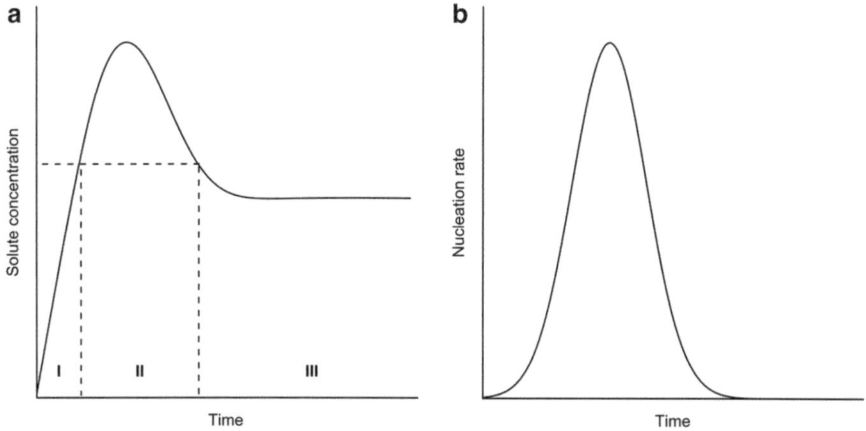

Fig. 3.11 LaMer diagram (**a**) and corresponding nucleation function (**b**) for the formation process of monodisperse particles

most liquids the Gibbs energy barrier is significant this is hardly ever the case for solid materials. Hence, there is hardly a barrier to nucleation for most processes involving solid nucleation. In actual fact, nucleation occurs so fast after creating supersaturation in a solution that the whole process is also termed a *precipitation reaction* albeit that nucleation in itself is (physical) phase separation and only the formation of the solid atoms is a chemical reaction. Nevertheless, the concentration at which nucleation sets in is usually significantly larger than the equilibrium saturation. Similar to what has been discussed in Sect. 3.2 for colloidal aggregation, nucleation kinetics also involves two aspects: the probability to collide and a reduction of the number of collisions that are successful through an energy barrier. A critical concentration needs to be overcome to guarantee sufficient collisions to generate significant amounts of stable growing nuclei.

A first quantitative description of the method to produce fine particles by the precipitation reaction has been provided by LaMer and Dinegar [38]. The main idea is depicted in Fig. 3.11a where the concentration of solute is sketched as a function of time. In the pre-nucleation stage, indicated by a I in the figure, the solute is created by the formation reaction and hence its concentration increases until it reaches a critical concentration. At that concentration the nucleation stage, indicated by a II, sets in and gradually solute is consumed faster than what is being produced. Hence, the concentration drops again also because the already nucleated particles begin to grow. When the concentration drops below the critical concentration, nucleation stops while the nuclei continue to grow in the final stage indicated by III. The concentration subsequently drops further until a stationary value is reached where the growth rate and the consumption balance.

If the concentration follows the LaMer diagram, nuclei are formed only during the nucleation stage, II. A modern development is to actually measure the rate of nucleus formation [78], see Fig. 3.11b for a sketch. There are two aspects to the nucleation function. Its integral, i.e. the area under the curve, gives the total number of nuclei formed; a good estimate is half the product of the peak value and the foot width of the curve. The polydispersity of the formed nuclei is related to the width of the curve because after formation growth immediately sets in so that the first formed nuclei already grow while new ones are formed. Although the curves are typically slightly asymmetric, they can be fitted to a Gaussian for most practical situations.[7]

For the synthesis of core shell particles or for the purpose of obtaining very monodisperse particles one invokes *heteronucleation*. Particles of an affine material then act as nucleation sites for the solute. When the affinity of the solute for the particulate material is high, core shell particles will be formed whereas when the affinity is relatively low buds are created on the surface of the particles that grow until complete coverage is achieved. The nucleation rate is significantly affected by the presence of the particles.

3.4.1 Ostwald Ripening

The last stage, III, in the LaMer diagram is entered when nucleation stops – because the solute concentration has dropped below the critical concentration – and particle growth persists until the solute is virtually depleted. Let us first analyze the situation where growing nuclei do not aggregate. This could be the case when precursor ions adsorb on the particle surface and render the particle charged as is the case in many metal particle synthesis processes [75] or when particles are effectively immobile in a surfactant matrix [55]. Particle growth is usually a solute diffusion controlled process and the driving force is the solute concentration being higher than the solubility. Due to the significant contribution of the surface to the formation energy of small particles, the solute solubility of a suspension of small particles is larger than for large ones, a phenomenon described by the Kelvin equation [22]. The Lifshitz-Slyozov-Wagner (LSW) theory takes this higher solubility as a local effect near the surface of the small particles to predict the average volume – or mass for that matter – of the particles to grow linearly with time [3]. The time scale is set by the diffusive properties of the solute and by surface energetics. A significant drawback of the theory is that the resulting particle size distribution is predicted to be antisymmetric where experimentally it has been found to be symmetric. In addition, the experimental particle size distributions are found to be composition, time and temperature dependent while the LSW distribution is universal and independent of these parameters. There have been many attempts to amend these discrepancies [3] but up till now the more successful explanation appears to be by Senkov who takes into account the effect of local concentration fluctuations

[7] There is no immediate argument as to why the functional form of the nucleation function should be a Gaussian.

[68, 69]. Indeed, concentration fluctuations are expected to be large in practical systems due to the relatively fast reaction times to produce the atoms compared to timescales typically involved in mixing reactant systems.

An important consequence of local solute solubility being dependent on particle size is that in a polydisperse dispersion it favors the dissolution of small particles in favor of the growth of the larger particles, an effect generally known as Ostwald ripening or *coarsening* [22]. The driving force is the Gibbs energy difference between various sized particles [72]. The growth process described here implies that the particle sizes homogenize to a uniform distribution and that the average particle size depends on the amount of solute initially present and the amount of nuclei initially formed [71]. This accentuates the importance of the nucleation function described above of which the integral gives the number of formed nuclei. For a more quantitative description of these processes, the reader is referred to the relevant literature such as the detailed modeling by Van Hyning et al. [75].

3.4.2 Aggregative Growth

Largely due to the high resolution imaging equipment available nowadays, it has been established now that most wet synthesis procedures providing particles larger than some tens of nanometers involve aggregative growth. In Fig. 3.12A–D an example from our own laboratory [9] where intermediate steps of the polystyrene particle synthesis with a non-ionic initiator clearly exhibit a raspberry structure indicative of aggregative growth. In Fig. 3.12E–G an example from Matijevic' laboratory exhibiting the same [40]. The most striking feature of these experimental data is the fact that the primary particles that develop after nucleation from growth and Ostwald ripening remain relatively stationary in size for a time lapse long enough to warrant imaging. Apparently, the primary particles are metastable with a size depending on material properties (Fig. 3.2). For the polystyrene particles this is believed to be due to the natural charge as develops at all hydrophobic surfaces in aqueous media of moderate pH and ionic strength. The aggregation is believed to develop because of combining free radical surface chains on nearby particles [9]. This renders the synthesis procedure very sensitive to ionic strength as indeed observed [8]. Similarly for solid particles where typically the precursor to the particle forming solute is ionic and preferentially adsorbs on the surface of the particles rendering the primary particles charged. However, upon depletion of the precursor, also the surface charge diminishes because of the ongoing reduction reaction [75]. Also here, strong sensitivity to the exact formulation becomes evident: extreme sensitivity to ionic strength and initial reducing agent concentration.

In the past decade many workers, but mainly Privman, have studied the somewhat surprising experimental result that aggregative growth following solute nucleation and subsequent growth eventually leads to rather monodisperse particles in a size range from tens of nanometers up to several tens of micrometers and

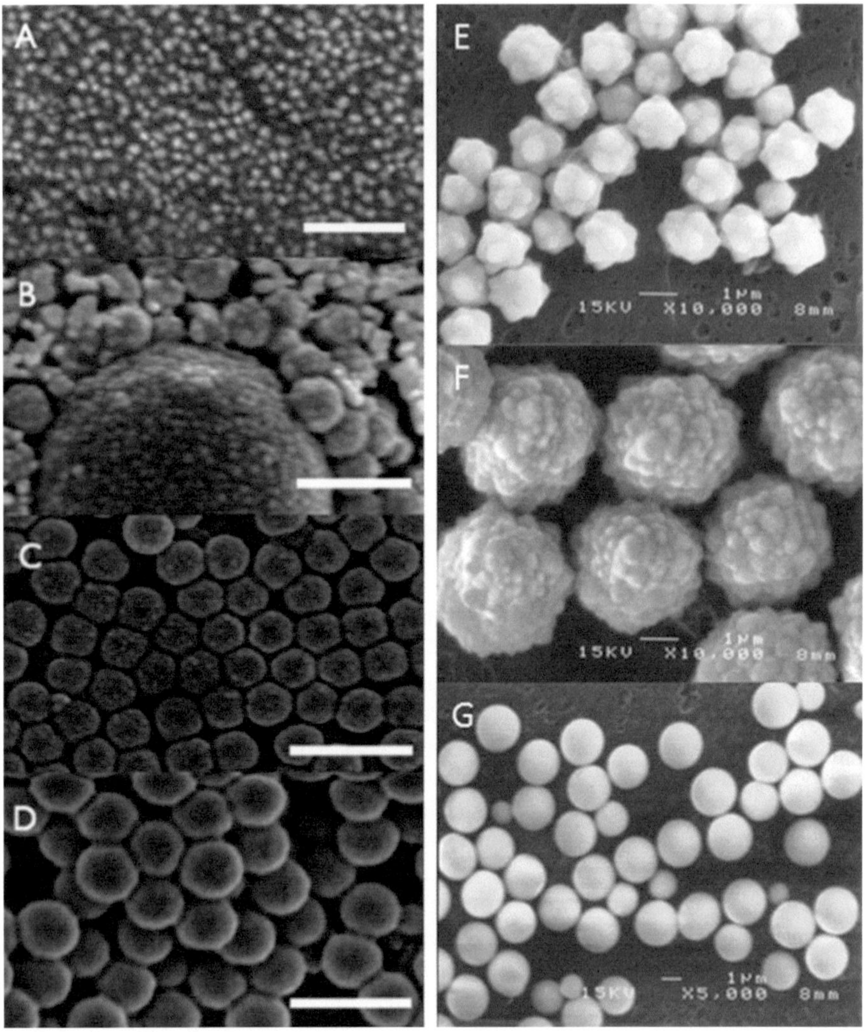

Fig. 3.12 Evidence for aggregative growth (**A–D**) from non-ionically initiated styrene polymerization and (**E–G**) for the formation process of CdS nanoparticles. The micrographs of freeze-dried samples in (**A–D**) were taken during a surfactant-free polymerization process at successive time lapses after initiation. (**A**) 15, (**B**) 45, (**C**) 120, (**D**) 180 min. (**A** and **B** scale bars are 200 nm. C and **D** scale bars are 500 nm). Those of CdS particles were obtained by the controlled double-jet precipitation technique: (**E**) after 8 s; (**F**) after 20 s; and (**G**) of the final product (Copyright ACS; reprinted with permission from **A–D** Dobrowolska et al. [9] and **E–G** Libert et al. [40])

sometimes larger: over three orders of magnitude in size! By careful analytical, experimental and numerical work it has become clear under which conditions fine, monodisperse particles can be created. There are two reasons to give for this: restructuring of the primary particles is feasible as these are typically very small, sometimes less than a nanometer, and additional material deposition is favored in corners: a phenomenon akin to sintering; yet another manifestation captured by the

Kelvin equation. Most of the analyses have focused on a *burst nucleation* process as already envisaged by LaMer et al. [38]. An important criterion, satisfied by this process, is that the production of primary particles should not be continuous but exponentially decaying in time; see the recent review by Gorshkov and Privman [18] and references therein. This stresses the importance of the nucleation function introduced above albeit that experimental techniques for its ready evaluation are still under development.

3.4.3 Restructuring

In actual fact, the wet synthesis of particles as described here is a hierarchical series of agglomeration processes. The first one is where the solute particles assemble into nuclei and subsequently grow by solute particle addition. The resulting small sized primary particles are most often crystalline which is most likely due to the fact that the time required to restructure the particle after solute addition is relatively short compared to the time scale involved in the diffusional nucleation and growth process. In other words, the solute can readily find an optimal position on the surface. The second agglomeration process is the one where the primary particles aggregate into secondary particles. In general, aggregation processes render structures that can be very dendritic if the first attachment is definite. The reason for this is that most of the subsequent particles will hit the tips of the dendrites before reaching the interior. In successful particle synthesis procedures, the aggregation process does allow for restructuring at the surface and a much more compact raspberry structure results. There is ample evidence in the literature of such structures, from our own work as represented in Fig. 3.12 as well as from others. Subsequent primary particle additions often serve to assist in the restructuring process albeit that amorphous walls between otherwise crystalline regions often persist. An important signature of the restructuring process at this stage is the polycrystalline nature of the secondary particles with crystallite domain sizes close to the primary particle size [78]. When most of the primary particles are consumed, a suitable protection against aggregation of the seconday particles is to be supplied. If not, one more aggregation process may set in to form agglomerates of secondary particles. When the secondary particles are not yet fully settled, some restructuring may still take place but in general the resulting structures are flocs of a more fractal nature.

3.4.4 Morphology

The secondary aggregation process in principle allows for the formation of many different particle shapes, often by the choice of suitable agents that assist in the assembly of the primary particles; the master of this is Matijevic and there is a wealth of literature from his group [17, 48–50]. Other methods utilize external fields [45]. Although a clear understanding of the involved processes is lacking, there have been some successful attempts to elucidate the mechanism from the work of Privman and coworkers [18]. A proper understanding is lacking, largely because in actual fact only the end result of the processes can be studied whereas for an understanding of the mechanism also intermediate stages need to be visualized. However, systematic studies are available such as by the group of Pileni [57, 58].

3.5 Surfactant Templates

Surfactants are amphiphilic molecules that consist of a hydrophilic part, usually called head group, and a hydrophobic part, usually an aliphatic chain or tail. These molecules have the tendency to reside at a water/air or water/oil interface and hence they are termed surface active molecules or surfactants for short. A related, but not identical, property of surfactants is their ability to form micelles, associative structures that form at concentrations higher than a critical value called the critical micelle concentration (cmc).

The Gibbs diagram, Fig. 3.13, gives the phase behavior of typical oil-surfactant-water systems. Our interest is in the relatively low surfactant concentration regime, until above the bicontinuous phase. When the surfactant is of relatively low

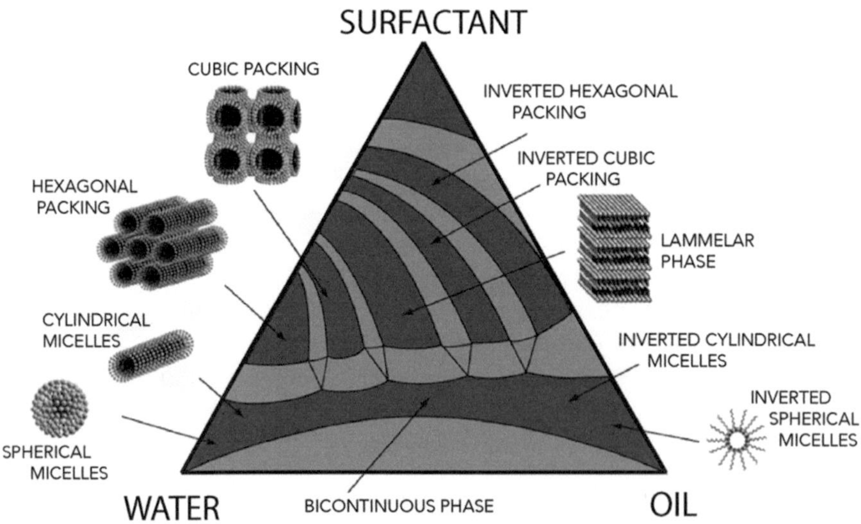

Fig. 3.13 Schematic Gibbs diagram of a typical oil-surfactant-water system. *Dark regions* indicate single phases and *grey regions* are multiple phases. Of particular interest is the single phase region connecting the water and oil corners: near these corners there are spherical micelles while in the middle this is the bicontinuous phase with interconnected water and oil channels separated by a continuous surfactant layer

molecular mass the interactions between the associative structures are mild and the indicated phases over the bicontinuous phase hardly exist. For high molecular weight surfactant the high concentration regimes exhibit the more complex phase behavior which is not of interest to the present discussion. Along the edges starting from the bottom corners of the Gibbs diagram one finds the micellar solutions discussed above: beyond a given cmc value micelles develop both in oil and in water albeit at different concentrations. At low concentrations small globular micelles are found while at larger concentrations wormlike micelles may develop or denser globular micelle systems. At yet higher surfactant concentrations, the associative structures interact significantly and more complex structures such as liquid crystalline phases develop. Such systems are of interest to the synthesis of porous inorganic structures such as SB15 [52]. At low surfactant concentrations one finds a 2-phase region, see Fig. 3.13, where the water-rich phase contains oil swollen micelles and the oil-rich phase contains water-swollen micelles. This is the region where emulsions are formulated and although their equilibrium behavior would be a two-phase system, they are kinetically stabilized to remain for a very long time – in many cases practically forever – in the emulsion state.

The continuous single phase region reaching from the oil corner to the water corner is the one that is of interest to nanoparticle synthesis. The general name for systems in this regime is microemulsions and these distinguish themselves from the above discussed (macro)emulsions in the fact that they form thermodynamically stable systems. In the corners of the Gibbs diagram there are either the oil-continuous regime and the water-continuous regime in which there are globular, surfactant coated water droplets or oil droplets respectively. The size of the pools of dispersed liquid in the globules can be estimated from the formulation, for instance the molar, or molecular, water/surfactant ratio w_0 determines the average diameter of the water pools in an water-in-oil microemulsion according to $d \approx 6w_0 v_w/a_0$ with as molecular parameters v_w the molecular volume of the water and a_0 the molecular area per surfactant molecule; sizes range from 1 nm up till hundreds of nanometers [28]. What is neglected is the usually tiny concentration of free surfactant, typically around the cmc, and the molecular contributions such as the head group hydration of the surfactant. Note that this relation can also be used to formulate (macro)emulsions. The unknown surfactant parameter in this simple rule-of-thumb is the area per surfactant molecule in the interface. For low curvature interfaces, such as a macroscopically flat interface, it can be determined using the Gibbs' adsorption equation. $\Gamma = -(1/(k_B T))d\gamma/d\ln c$, from the dependence of the interfacial tension γ as a function of surfactant concentration [22], the molar surfactant area being equal to the inverse of the adsorption Γ.

Upon increasing the dispersed fluid volume, the globules swell and become worm-like. See Fig. 3.14 for a simulation result. Near the middle of the single phase region, the worms have grown and coalesced to form another continuous phase and the result is a bicontinuous phase with both oil- and water continuity. It is important to realize that both fluid domains, loosely termed oil and water, may actually contain other solutes such as alcohols or salts, that modify their properties.

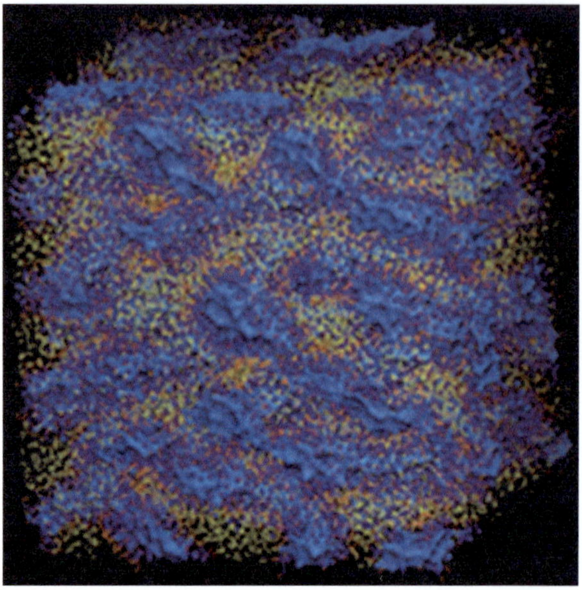

Fig. 3.14 Snapshot at the end of 1 μs simulations for the system water–Na-AOT–n-heptane with 20 % water content using a course grained simulation model. Water: *light blue*, n-heptane: *yellow*, Na^+: *blue*, AOT head group: *purple*, AOT tail: *orange* (Modified from Ref. [54])

3.5.1 Surfactant Parameters

The actual structures that are formed with different concentrations of oil, surfactant and water largely depend on the surfactant. Depending on technological or academic field, different surfactant parametrization schemes have been developed. The most powerful in actually predicting phase behavior has been the concept of curvature free energy as introduced by Helfrich [21, 28]. It is formulated on the basis of an expansion of the local free energy in terms of geometrical curvature: deviations of the actual local curvature from the *spontaneous curvature* are counteracted by an elastic-like rigidity. An important situation arises when the spontaneous curvature vanishes which is usually at a particular temperature for nonionic surfactants and at a particular ionic strength of the aqueous phase for ionic surfactants. This is the situation when bicontinuous microemulsions are formed, where locally both positive and negative curvatures towards the water domains exist such that the overall average curvature vanishes. Although successful in predicting phase behavior, its original formulation lacked a clear link to molecular structure.

In technology, the concept of *hydrophilic-lipophilic balance* (HLB) is still the most frequently used and all surfactant suppliers provide the values for their products – as these are simply related to the mass ratio of hydrophobic part over hydrophilic part – despite the fact that such a quantity must to a certain extent depend on the actual composition of the oil and the water phase [24]. Recently, this and some related concepts were unified into the *hydrophilic-lipophilic deviation* (HLD) by Salager and coworkers [53]. This HLD measures the deviation of the formulation from the optimal formulation which is the same state as described above where the spontaneous curvature vanishes.

In academic circles, the more geometrical concept of packing parameter is more popular. It is defined as the ratio $P = a_s/a_0$ of the minimal surface area a_s that a surfactant can occupy and the actually occupied surface area a_0 defined above. The minimal area is the sterically realizable optimal packing value $a_s = v_s/\ell$, with v_s the surfactant molecular volume and ℓ the realized length of the surfactant; for alkanes it is about 80 % of the extended length as found by molecular simulations. The optimal formulation where bicontinuity can be found is the so called *balanced state* where the packing parameter equals unity. For values around 1/3, spherical oil-in-water droplets are preferred and for a value around 3 it is the water-in-oil droplets. As already alluded to, the three concepts can be unified [37] albeit that more work is needed before the unification leads to technological use. Nevertheless, these concepts help in understanding the phenomenology observed in formulating templating precipitation systems.

3.5.2 Nanoreactors

Around the 1980s it was realized that droplet phase water-in-oil microemulsions could act as a system of *nanoreactors* [5, 13] for precipitation reactions to synthesize metal nanoparticles [66]. One almost always uses water-in-oil systems although in principle oil-in-water systems are possible as well. The size of the water pools, determined as above described by simple geometrical considerations involving properties of the surface active molecules and of the dispersed phase [28], could then be used to predetermine nanoparticle size. There usually appears to be some kind of relation between the size of the water droplets and of the nanoparticles, so that for some particular examples the term microemulsion template is appropriate. Nowadays, it is generally understood that such a relation in general cannot be only understood from the composition of the microemulsion [44]. The complicating factor is that material exchanges between the nanoreactors [15], certainly when their contents vary in composition. The exchange of reactants depends on various parameters that are controlled by composition, in particular droplet concentration and surfactant film flexibility [73]. The inescapable consequence of reactant exchange is that the synthesis of the nanoparticles is not limited to the reactors themselves but in actual fact is a collective phenomenon not unlike the situation in a simple sol albeit at rates that are – at least partly – controlled by the microemulsion composition. Also, nanoparticles are exchanged during or after nucleation and growth and this phenomenon is even enhanced under conditions such as low surfactant film flexibility [44, 66] which leads to nanoparticle sizes ranging far beyond the water pool size or even size distributions becoming bimodal [67], i.e. small micelles without particle and large micelles with particle. Nevertheless, a great advantage of droplet phase microemulsion templating is that in most cases the resulting particles are relatively monodispersed and well protected against aggregation [23, 41, 43, 57, 62, 66]. The use of droplet phase microemulsions has also the advantage, that the morphology of the nanoparticle can be changed by varying the type of surfactant or the microemulsion composition [10, 58]. An important factor from a technological point of view is the yield of these synthesis methods that is relatively low for this method due to the low droplet concentrations used. For dispersed phase fractions beyond 10 %, droplet interactions become important and the water pools aggregate reversibly [31].

3.5.3 Bicontinuous Microemulsion Templating

The microemulsion composition for which optimal conditions can be found for nanoparticle synthesis turns out to be the composition region where bicontinuous microemulsions are preferred, see Fig. 3.13. Due to the water continuity, reactants can diffuse freely within the system so that diffusion is only weakly controlled by microemulsion composition [32, 55]. Although small molecules and ions may move freely within the water continuous structures, larger particles experience strong hindrance due to the structure. This arrests the nanoparticles that are hence no longer capable of aggregating, not even at higher concentrations. Dense microemulsions, particularly the bicontinuous type, have been largely overlooked in the preparation of discrete nanoparticles [46]. This is surprising, because quite some studies have shown them to be superior to droplet type microemulsions by allowing a higher nanoparticle production. Krauel et al. have demonstrated (alkylcyanoacrylate) nanoparticle synthesis using interfacial polymerization in bicontinuous microemulsions. It was claimed that the availability of larger fractions of water for the solubilization of the hydrophilic macromolecules is a great advantage of bicontinuous microemulsions over water-in-oil microemulsions [34]. Esquivel et al. were the first to report an increased yield of nanoparticles produced with bicontinuous microemulsions. These authors were able to reach concentrations of almost 1.2 % of 8 nm nanoparticles per unit mass of microemulsion. This is to be compared to a maximum of 0.4 % for water-in-oil microemulsions [12]. A correlation between the ratio of water to surfactant molecules in the bicontinuous microemulsion to the morphology of the PbS nanoparticles was obtained by Xu et al. and they could selectively choose between nanospheres, nanowires or nanotubes by going from low to high water to surfactant ratios [47, 80]. The claims of the aforementioned authors were reinforced by Reyes et al. who reported that bicontinuous microemulsions are not only necessary for higher yields of nanoparticles but also for higher conversions of the nanoparticle precursors. A nearly 1.7 % of 3 nm silver nanoparticles was achieved and the ratio of nanoparticle precursor to the reducing agent was designated as the critical parameter in controlling the nanoparticle concentration and morphology [64]. Crystallinity of the produced nanoparticles is often not high but no systematic study has been performed so far [29]. A disadvantage of high surfactant loadings that is frequently mentioned is that this significantly increases the cost. However, as surfactant can be fully recovered this is not a real issue.

3.6 Design of a Synthesis Procedure

In this section, a general approach to the synthesis of small and uniform metal nanoparticles in dense microemulsions will be presented. The method has been illustrated using various types of surfactants to control particle size in the nanometer range [39]. The flexibility in the choice of surfactant allows us to employ up to 40 wt.% of water phase for synthesis, which is exceptionally high for microemulsion based synthesis but delivers high yields of ultrafine nanoparticles (up to 3 wt.%).

The requirements for the design of the synthesis procedure are (1) a high yield of mono-dispersed nanoparticles of predefined size, (2) a robust, yet flexible and safe production method with (3) an efficient and safe extraction technique. Such a design can in principle be done in a single series of four steps, graphically depicted in Fig. 3.15, albeit that some optimization cycles may be necessary to attain the desired result. The procedure will be exemplified for platinum nanoparticles as catalyst for the oxygen reduction reaction as takes place in a polyelectrolyte membrane fuel cell [39].

1. *Selection of the surfactant* follows three criteria. The first is that the surfactant is to be in the balanced state, i.e. the packing parameter has to have a value of around 1 or slightly more to guarantee water continuity. This corresponds to HLB-values of about 10, typically between 8 and 12. The composition of the microemulsion does weakly affect this value but in a first design stage this may

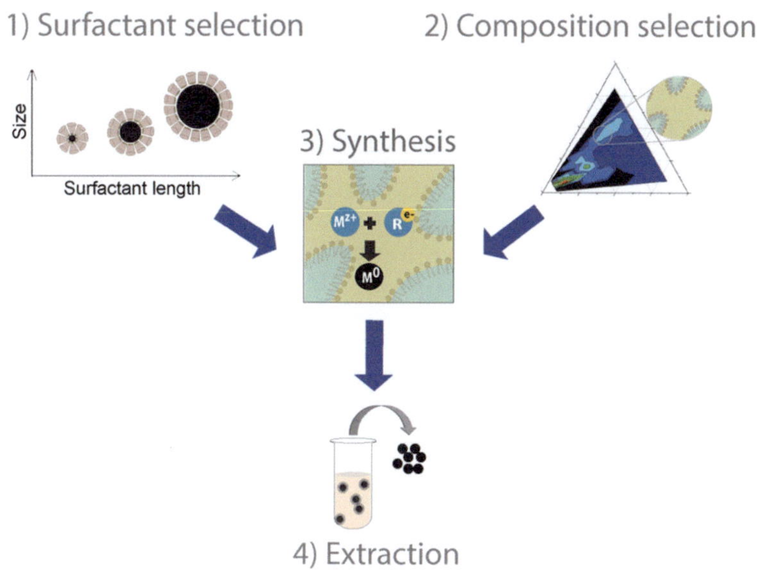

Fig. 3.15 General approach to nanoparticle synthesis in bicontinuous microemulsion systems consisting of four steps: surfactant selection, composition selection, synthesis and extraction

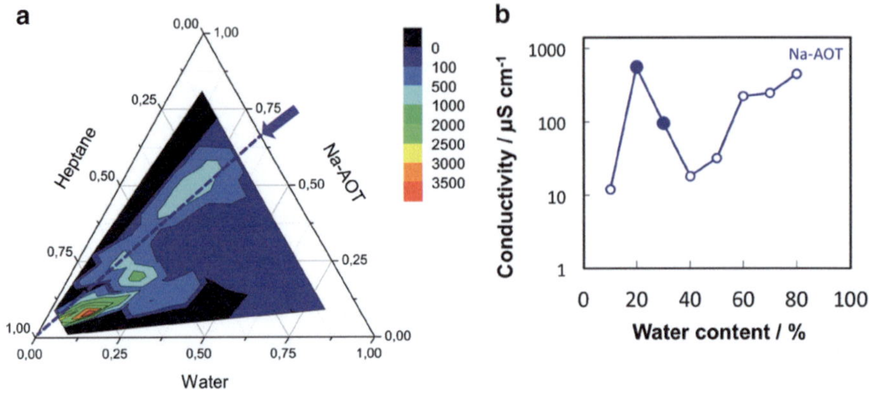

Fig. 3.16 Phase diagram (**a**) based on conductivity measurements, color maps based on relative electrical conductivity; *white regions* were not measured, and *black regions* represent phase-separated samples. (**b**) Conductivity values taken from samples with compositions along the *dashed line*; the bicontinuous phase samples are indicated by *closed circles*, the other circles belonging samples in other phases are *left open*

not be considered. The second criterion is the effective surfactant length. Unfortunately, such values are generally not tabulated albeit that some can be found in the literature. A good first estimate can be obtained using Tanford's regression [28] involving the maximum chain length. A third and fourth criterion, affinity of the head group towards the metal surface and the inertness with respect to the chemicals used, are difficult ones to predict although some information is available in the literature [65]. For the present example of Pt nanoparticle synthesis, we have selected an anionic surfactant head group, Na-AOT, that has a high affinity to Pt, has a tail with a length of 0.9 nm and a packing parameter 1.2.

2. *Formulation of the microemulsion composition* is the next step. The oil type is largely dictated by the choice of the surfactant but some variation is usually possible. For instance, for Na-AOT one can choose alkanes from hexane to hexadecane with an optimum near heptane [55]. This variation allows for some flexibility for instance to fine-tune the particle size. The aqueous phase will be an electrolyte with an ionic composition similar to what is expected to be the liquid state after the precipitation reaction has completed. Various techniques exist to perform the above described composition selection method as signatures of the bicontinuous phase are high conductivity, high viscosity, etc. [6]. In Fig. 3.16 the phase diagram is presented for Na-AOT. The color scheme used is based on electrical conductivity with which regions of high and low conductivity are mapped out. Using a dilution line from the water corner through the approximate location of the bicontinuous phase, one can deduce from the measured conductivity values the optimal composition. These compositions were confirmed by phase calculations based on geometrical considerations [1].

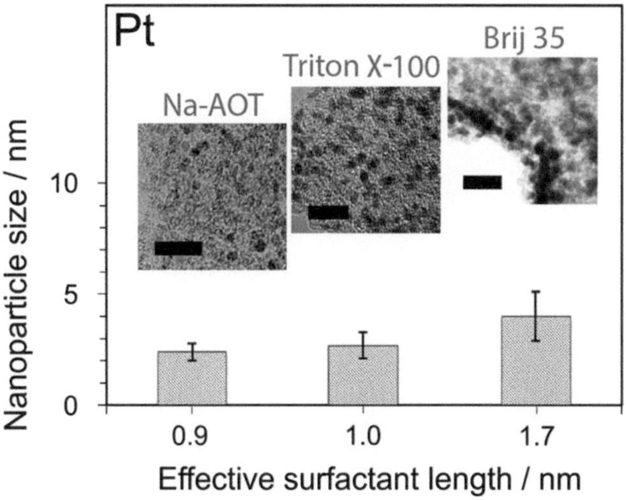

Fig. 3.17 Size dependence of nanoparticles of platinum on the effective length of the surfactant used to form the templating microemulsion. Micrographs are inserted to inspect the particle size and distribution; scale bar 10 nm

3. *Synthesis of the nanoparticles* is achieved by mixing two microemulsions, one with the reducing agent dissolved in the aqueous phase and the other with the dissolved metal precursor. Other schemes exist [11] but will not be discussed here in detail. The reaction concludes typically within a few hours [39] and in principle it can be followed by UV-Vis spectroscopy albeit that very quickly the particle concentration is so high that the absorbance cannot be negotiated by a standard UV-Vis spectrometer [55]. The resulting particles are usually studied by means of Transmission Electron Microscopy (TEM) on small aliquots of the produced liquid strongly diluted to prevent as much as possible the immediate aggregation of the nanoparticles upon the breaking of the bicontinuous phase.

   Figure 3.17 gives results for the systems studied here for the synthesis of platinum, more can be found in the work by Latsuzbaia et al. [39]. The correlation of nanoparticle size with surfactant length is clear; as predicted, it was the main parameter to control the nanoparticle size. Binding to Pt is strong for Na-AOT and Triton X-100 so that small and monodisperse nanoparticles with Relative Standard Deviation (RSD) of 17 and 22 % are obtained. This is not the case for Brij 35 with a large hydrophilic part and short hydrophobic part, which does not allow steric stabilization so that polydisperse nanoparticles are formed (RSD 28 %).

   No other parameters influence the size of nanoparticles significantly in these dense microemulsion systems. Various microemulsion compositions or reagent concentrations were used for the synthesis of nanoparticles in the same system, which did not result in a different nanoparticle size proving a robustness of the method [39]. The method allows synthesis of various metals such as Pd, Fe, Co, Ni and metal alloys such as Pt-Ru-Co also reported by us previously for Na-AOT system [76].

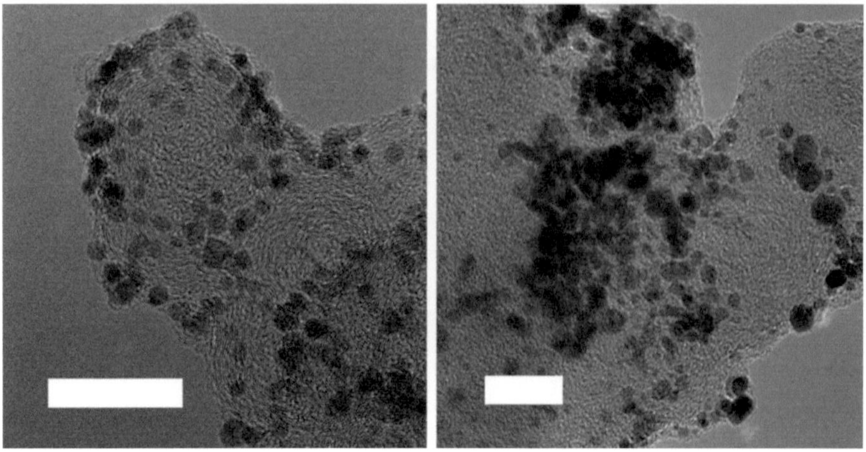

Fig. 3.18 Nanoparticle coverage of carbon support showing before and after heat treatment; scale bar 20 nm. Nanoparticles synthesized in Triton X-100/Toluene/Water

4. *Extraction* methods that can be employed strongly depend on the later application of the nanoparticles. For catalysis, adsorption of the nanoparticles onto a carrier is the preferred method [2]. After binding the nanoparticles to the substrate, these can easily be separated from the production liquid that hence becomes available for recycling: only the remainder of the salt needs to be extracted from the aqueous phase. No heat treatment with associated nanoparticle agglomeration or surfactant decomposition are required, see Fig. 3.18. After adsorption on the carbon support surfactant can be removed by washing and the platinum surface was clean for catalysis as demonstrated elsewhere [39]. It is well known that the alyphatic chains of surfactants exhibit a high affinity towards hydrophobic carbon surfaces, which implies that they promote molecular physisorption rather than solvation of nanoparticles [7]. This phenomenon can be used to advantage when extracting carbon supported metal nanoparticles.

The above demonstrates a robust synthesis route that is available for the wet synthesis of uniformly sized various metal nanoparticles that is scalable yet flexible for further optimization. The dense microemulsion system allows change of reactant concentration and microemulsion composition to increase the yield of the process, while none of these parameter influences the final size of the nanoparticles. The process design and system selection procedure are theoretically underpinned by considering the surfactant parameter, the effective surfactant size, the microemulsion phase diagram and the influence of formulation variables [39]. The size range that can be addressed has not been explored so far, but sizes from 1 to 10 nm are within easy reach.

3.7 Outlook

The important result that is reported on in this chapter is that it has for the first time been possible to design a synthesis route. So far, nanoparticle synthesis has only been able on an empirical level. The range of sizes that can be obtained with the presented method is rather limited and estimated as 1–10 nm, one decade. It is to be expected that this example will soon be followed so that more schemes become available.

To that purpose, quite some issues with other synthesis – and also the present one – need to be dealt with. The largest uncertainty is associated with the affinity of the surfactant for the material to be synthesized. Little has been investigated systematically but some recent developments are expected to quickly yield more insight [4, 16, 36]. Until that time, the wet colloid synthesis involving precipitation and dispersion will involve some degree of empiricism.

3.8 Definitions, Abbreviations and Symbols

3.8.1 Definitions

The Hydrophilic-Lipophilic Balance (HLB) and Hydrophilic-Lipophilic Deviation (HLD) are measures for the relation of the hydrophilic and hydrophobic groups of a surfactant. There is neither a clear definition of either of these, nor is there an unambiguous expression available. A discussion is given by Salager and coworkers [53].

Abbreviations

ACS	American Chemical Society
cmc	Critical micelle concentration
HLB	Hydrophilic-lipophilic balance
HLD	Hydrophilic-lipophilic deviation
RSD	Relative standard deviation
VSSD	Vereniging voor Studie en Studentenbelangen Delft, NL

3.8.2 Symbols

a	Surface (m^2)
A	Hamaker constant (J)
c	Concentration (m^{-3})
d	Diameter (m)
χ	Flory interaction parameter
δ	Layer thickness (m)
ε	Relative dielectric permittivity
ε_0	Vacuum dielectric permittivity, $8.85\ 10^{-12}\ C^2/Jm$

$Đ$ Diffusion coefficient (m^2/s)
e Elementary charge $1.59 \cdot 10^{-19}$ C
f Friction coefficient (Ns/m)
F Force (N)
g Gravitational constant, $9.81 \cdot 10^{-23}\ m/s^2$
G Shear rate (s^{-1})
h Separation (m)
J Flux ($m^{-2}\ s^{-1}$)
k_B Boltzmann constant, $1.38 \cdot 10^{-23}\ J/K$
ℓ Effective surfactant length (m)
L Length (m)
P Surfactant parameter
r Position (m)
t Time (s)
T Temperature (K)
v Velocity (m/s)
Specific volume (m^3)
V Volume (m^3)
w Interaction strength (J)
w_0 Water/surfactant ratio
z Vertical position (m)
γ Surface tension (J/m^2)
ϕ Volume fraction
Electrical potential (V)
κ Inverse screening length (m^{-1})
η Viscosity (Pas)
ρ Mass density (kg/m^3)
σ Surface charge (C/m^2)
τ Time (s)
Δ Small increment
Γ Adsorption (m^{-2})
Π Osmotic pressure (Pa)

Acknowledgement The Chemistry department of Cambridge University is thanked for hosting G.K. in 2014.

References

1. Andre, P., Ninham, B.W., Pileni, M.P.: Mesostructured fluids: a geometrical model predicting experimental data. New J. Chem. **25**(4), 563–571 (2001). doi:10.1039/b009079o
2. Astruc, D., Lu, F., Aranzaes, J.R.: Nanoparticles as recyclable catalysts: the frontier between homogeneous and heterogeneous catalysis. Angew. Chem. Int. Ed. **44**(48), 7852–7872 (2005)
3. Baldan, A.: Review progress in Ostwald ripening theories and their applications to nickel-base superalloys – Part I: Ostwald ripening theories. J. Mater. Sci. **37**(11), 2171–2202 (2002)
4. Ben-Yaakov, D., Andelman, D., Podgornik, R., Harries, D.: Ion-specific hydration effects: extending the Poisson-Boltzmann theory. Curr. Opin. Colloid Interface Sci. **16**(6), 542–550 (2011). doi:10.1016/j.cocis.2011.04.012
5. Boutonnet, M., Kizling, J., Stenius, P.: The preparation of monodisperse colloidal metal particles from micro-emulsions. Colloids Surf. **5**(3), 209–225 (1982). doi:10.1016/0166-6622(82)80079-6
6. Bumajdad, A., Eastoe, J.: Conductivity of water-in-oil microemulsions stabilized by mixed surfactants. J. Colloid Interface Sci. **274**(1), 268–276 (2004)

7. Ciesielski, A., El Garah, M., Haar, S., Kovaricek, P., Lehn, J.M., Samori, P.: Dynamic covalent chemistry of bisimines at the solid/liquid interface monitored by scanning tunnelling microscopy. Nat. Chem. **6**(11), 1017–1023 (2014). doi:10.1038/nchem.2057
8. Dobrowolska, M.E., Koper, G.J.M.: Optimal ionic strength for nonionically initiated polymerization. Soft Matter **10**(8), 1151–1154 (2014)
9. Dobrowolska, M.E., van Esch, J.H., Koper, G.J.M.: Direct visualization of "Coagulative Nucleation" in surfactant-free emulsion polymerization. Langmuir **29**(37), 11724–11729 (2013)
10. Eastoe, J., Hollamby, M.J., Hudson, L.: Recent advances in nanoparticle synthesis with reversed micelles. Adv. Colloid Interface Sci. **128**, 5–15 (2006)
11. Eriksson, S:, Nylen, U., Rojas, S., Boutonnet, M.: Preparation of catalysts from microemulsions and their applications in heterogeneous catalysis. Appl. Catal. A-Gen. **265** (2), 207–219 (2004)
12. Esquivel, J., Facundo, I.A., Trevino, M., Lopez, R.G.: A novel method to prepare magnetic nanoparticles: precipitation in bicontinuous microemulsions. J. Mater. Sci. **42**(21), 9015–9020 (2007)
13. Fendler, J.H.: Atomic and molecular clusters in membrane mimetic chemistry. Chem. Rev. **87** (5), 877–899 (1987)
14. G.J. Fleer, M.A. Cohen Stuart, J.M.H.M. Scheutjens, T. Cosgrove and B. Vincent: Polymers at Interfaces. Chapman and Hall, London (1993)
15. Fletcher, P.D.I., Howe, A.M., Robinson, B.H.: The kinetics of solubilisate exchange between water droplets of a water-in-oil microemulsion. J. Chem. Soc. Faraday Trans. I **83**, 985–1006 (1987)
16. Friedman, R.: Electrolyte solutions and specific ion effects on interfaces. J. Chem. Educ. **90**(8), 1018–1023 (2013). doi:10.1021/ed4000525
17. Goia, D.V., Matijevic, E.: Preparation of monodispersed metal particles. New J. Chem. **22**(11), 1203–1215 (1998). doi:10.1039/a709236i
18. Gorshkov, V., Privman, V.: Models of synthesis of uniform colloids and nanocrystals. Physica E **43**(1), 1–12 (2010)
19. Grzelczak, M., Vermant, J., Furst, E.M., Liz-Marzan, L.M.: Directed self-assembly of nanoparticles. ACS Nano **4**(7), 3591–3605 (2010). doi:10.1021/nn100869j
20. Harada, M., Kamigaito, Y.: Nucleation and aggregative growth process of platinum nanoparticles studied by in situ quick XAFS spectroscopy. Langmuir **28**(5), 2415–2428 (2012). doi:10.1021/la204031j
21. Helfrich, W.: Elastic properties of lipid bilayers – theory and possible experiments. Z. Naturforsch. C **28**(11–1), 693–703 (1973)
22. Hiemenz, P.C., Rajagopalan, R.: Principles of Colloid and Surface Chemistry, 3rd edn, revised and expanded. Undergraduate Chemistry: A Series of Textbooks. Marcel Dekker, New York (1997)
23. Hollamby, M.J., Eastoe, J., Mutch, K.J., Rogers, S., Heenan, R.K.: Fluorinated microemulsions as reaction media for fluorous nanoparticles. Soft Matter **6**(5), 971–976 (2010)
24. Holmberg, K., Jönsson, B., Kronberg, B., Lindman, B.: Surfactants and Polymers in Aqueous Solution. Wiley, New York (2002)
25. Hunter, R.J.: Zeta Potential in Colloid Science: Principles and Applications. Academic, New York (1981)
26. Hunter, R.J.: Introduction to Modern Colloid Science. Oxford Science Publications. Oxford University Press, Oxford (1993)
27. Hunter, R.J.: Foundations of Colloid Science. Foundations of Colloid Science. Oxford University Press, Oxford (2001)
28. Israelachvili, J.N.: Intermolecular and Surface Forces. McGraw-Hill Publishing Co., Japan (2011)
29. Jia, C.J., Schuth, F.: Colloidal metal nanoparticles as a component of designed catalyst. PCCP **13**(7), 2457–2487 (2011). doi:10.1039/c0cp02680h

30. Koper, G.J.M.: An Introduction to Interfacial Engineering. VSSD, Delft (2007)
31. Koper, G.J.M., Sager, W.F.C., Smeets, J., Bedeaux, D.: Aggregation in oil-continuous water sodium bis(2-ethylhexyl)sulfosuccinate oil microemulsions. J. Phys. Chem. **99**(35), 13291–13300 (1995)
32. Kowlgi, K., Lafont, U., Rappolt, M., Koper, G.: Uniform metal nanoparticles produced at high yield in dense microemulsions. J. Colloid Interface Sci. **372**(1), 16–23 (2012)
33. Kraft, D.J., Hilhorst, J., Heinen, M.A.P., Hoogenraad, M.J., Luigjes, B., Kegel, W.K.: Patchy polymer colloids with tunable anisotropy dimensions. J. Phys. Chem. B **115**(22), 7175–7181 (2011). doi:10.1021/jp108760g
34. Krauel, K., Davies, N.M., Hook, S., Rades, T.: Using different structure types of microemulsions for the preparation of poly(alkylcyanoacrylate) nanoparticles by interfacial polymerization. J. Control. Release **106**(1–2), 76–87 (2005)
35. Krommenhoek, P.J., Wang, J., Hentz, N., Johnston-Peck, A.C., Kozek, K.A., Kalyuzhny, G., Tracy, J.B.: Bulky adamantanethiolate and cyclohexanethiolate ligands favor smaller gold nanoparticles with altered discrete sizes. ACS Nano **6**(6), 4903–4911 (2012). doi:10.1021/nn3003778
36. Kunz, W.: Specific ion effects in colloidal and biological systems. Curr. Opin. Colloid Interface Sci. **15**(1–2), 34–39 (2010). doi:10.1016/j.cocis.2009.11.008
37. Kunz, W., Testard, F., Zemb, T.: Correspondence between curvature, packing parameter, and hydrophilic-lipophilic deviation scales around the phase-inversion temperature. Langmuir **25** (1), 112–115 (2009)
38. Lamer, V.K., Dinegar, R.H.: Theory, production and mechanism of formation of monodispersed hydrosols. J. Am. Chem. Soc. **72**(11), 4847–4854 (1950)
39. Latsuzbaia, R.: Increasing the Lifetime of Fuel Cell Catalysts. PhD-thesis, Delft University of Technology (2015)
40. Libert, S., Gorshkov, V., Privman, V., Goia, D., Matijevic, E.: Formation of monodispersed cadmium sulfide particles by aggregation of nanosize precursors. Adv. Colloid Interface Sci. **100**, 169–183 (2003). doi:10.1016/s0001-8686(02)00056-8
41. Liveri, V.T.: Controlled synthesis of nanoparticles in microheterogeneous systems. Springer, New York (2006)
42. Lohse, S.E., Murphy, C.J.: The quest for shape control: a history of gold nanorod synthesis. Chem. Mater. **25**(8), 1250–1261 (2013). doi:10.1021/cm303708p
43. Lopez-Quintela, M.A.: Synthesis of nanomaterials in microemulsions: formation mechanisms and growth control. Curr. Opin. Colloid Interface Sci. **8**(2), 137–144 (2003)
44. Lopez-Quintela, M.A., Tojo, C., Blanco, M.C., Rio, L.G., Leis, J.R.: Microemulsion dynamics and reactions in microemulsions. Curr. Opin. Colloid Interface Sci. **9**(3–4), 264–278 (2004)
45. Ma, F., Wu, D.T., Wu, N.: Formation of colloidal molecules induced by alternating-current electric fields. J. Am. Chem. Soc. **135**(21), 7839–7842 (2013). doi:10.1021/ja403172p
46. Malheiro, A.R., Varanda, L.C., Perez, J., Villullas, H.: The aerosol OT plus n-butanol plus n-heptane plus water system: phase behavior, structure characterization, and application to Pt70Fe30 nanoparticle synthesis. Langmuir **23**(22), 11015–11020 (2007)
47. Martinez-Rodriguez, R.A., Vidal-Iglesias, F.J., Solla-Gullon, J., Cabrera, C.R., Feliu, J.M.: Synthesis of Pt nanoparticles in water-in-oil microemulsion: effect of HCl on their surface structure. J. Am. Chem. Soc. **136**(4), 1280–1283 (2014). doi:10.1021/ja411939d
48. Matijevic, E.: Preparation and properties of uniform size colloids. Chem. Mater. **5**(4), 412–426 (1993). doi:10.1021/cm00028a004
49. Matijevic, E.: Uniform inorganic colloid dispersions – achievements and challenges. Langmuir **10**(1), 8–16 (1994). doi:10.1021/la00013a003
50. Matijevic, E., Goia, D.: Formation mechanisms of uniform colloid particles. Croat. Chem. Acta **80**(3–4), 485–491 (2007)
51. Muhammed, M.A.H., Aldeek, F., Palui, G., Trapiella-Alfonso, L., Mattoussi, H.: Growth of in situ functionalized luminescent silver nanoclusters by direct reduction and size focusing. ACS Nano **6**(10), 8950–8961 (2012). doi:10.1021/nn302954n

52. Naik, B., Ghosh, N.N.: A review on chemical methodologies for preparation of mesoporous silica and alumina based materials. Recent Patents Nanotechnol. **3**(3), 213–224 (2009)
53. Nardello, W., Chailloux, N., Poprawski, J., Salager, J.L., Aubry, J.M.: HLD concept as a tool for the characterization of cosmetic hydrocarbon oils. Polym. Int. **52**(4), 602–609 (2003). doi:10.1002/pi.1012
54. Negro, E., Latsuzbaia, R., de Vries, A.H., Koper, G.J.M.: Experimental and molecular dynamics characterization of dense microemulsion systems: morphology, conductivity and SAXS. Soft Matter **10**(43), 8685–8697 (2014)
55. Negro, E., Latsuzbaia, R., Koper, G.J.M.: Bicontinuous microemulsions for high yield wet synthesis of ultrafine platinum nanoparticles: effect of precursors and kinetics. Langmuir **30**(28), 8300–8307 (2014)
56. Park, J., Joo, J., Kwon, S.G., Jang, Y., Hyeon, T.: Synthesis of monodisperse spherical nanocrystals. Angew. Chem. Int. Ed. **46**(25), 4630–4660 (2007). doi:10.1002/anie.200603148
57. Pileni, M.P.: The role of soft colloidal templates in controlling the size and shape of inorganic nanocrystals. Nat. Mater. **2**(3), 145–150 (2003)
58. Pileni, M.P.: Reverse micelles used as templates: a new understanding in nanocrystal growth. J. Exp. Nanosci. **1**(1), 13–27 (2006)
59. Polte, J., Ahner, T.T., Delissen, F., Sokolov, S., Emmerling, F., Thunemann, A.F., Kraehnert, R.: Mechanism of gold nanoparticle formation in the classical citrate synthesis method derived from coupled in situ XANES and SAXS evaluation. J. Am. Chem. Soc. **132**(4), 1296–1301 (2010)
60. Polte, J., Erler, R., Thunemann, A.F., Sokolov, S., Ahner, T.T., Rademann, K., Emmerling, F., Kraehnert, R.: Nucleation and growth of gold nanoparticles studied via in situ small angle X-ray scattering at millisecond time resolution. ACS Nano **4**(2), 1076–1082 (2010). doi:10.1021/nn901499c
61. Polte, J., Tuaev, X., Wuithschick, M., Fischer, A., Thuenemann, A.F., Rademann, K., Kraehnert, R., Emmerling, F.: Formation mechanism of colloidal silver nanoparticles: analogies and differences to the growth of gold nanoparticles. ACS Nano **6**(7), 5791–5802 (2012). doi:10.1021/nn301724z
62. Rao, C.N.R., Vivekchand, S.R.C., Biswas, K., Govindaraj, A.: Synthesis of inorganic nanomaterials. Dalton Trans. **34**, 3728–3749 (2007)
63. Reiss, H., Koper, G.J.M.: The Kelvin relation – stability, fluctuation, and factors involved in measurement. J. Phys. Chem. **99**(19), 7837–7844 (1995)
64. Reyes, P.Y., Espinoza, J.A., Trevino, M.E., Saade, H., Lopez, R.G.: Synthesis of silver nanoparticles by precipitation in bicontinuous microemulsions. J. Nanomater. Article ID 948941, (2010)
65. Rosen, M.J., Kunjappu, J.T.: Surfactants and Interfacial Phenomena. Wiley, Hoboken (2012)
66. Sager, W.F.: Microemulsion templating. In: Zvelindovsky, A. (ed.) Nanostructured Soft Matter. NanoScience and Technology, pp. 3–44. Springer, Dordrecht (2007)
67. Santra, S., Tapec, R., Theodoropoulou, N., Dobson, J., Hebard, A., Tan, W.H.: Synthesis and characterization of silica-coated iron oxide nanoparticles in microemulsion: the effect of nonionic surfactants. Langmuir **17**(10), 2900–2906 (2001)
68. Senkov, O.N.: Particle size distributions during diffusion controlled growth and coarsening. Scr. Mater. **59**(2), 171–174 (2008)
69. Senkov, O.N.: Particle size distributions during diffusion controlled growth and coarsening (vol 59, pg 171, 2008). Scr. Mater. **62**(2), 122–122 (2010)
70. Singh, V., Khullar, P., Dave, P.N., Kaur, G., Bakshi, M.S.: Ecofriendly route to synthesize nanomaterials for biomedical applications: bioactive polymers on shape-controlled effects of nanomaterials under different reaction conditions. ACS Sustain. Chem. Eng. **1**(11), 1417–1431 (2013). doi:10.1021/sc400159x
71. Sugimoto, T.: Underlying mechanisms in size control of uniform nanoparticles. J. Colloid Interface Sci. **309**(1), 106–118 (2007)

72. Tang, L., Li, X., Cammarata, R.C., Friesen, C., Sieradzki, K.: Electrochemical stability of elemental metal nanoparticles. J. Am. Chem. Soc. **132**(33), 11722–11726 (2010). doi:10.1021/ja104421t
73. Tojo, C., de Dios, M., Barroso, F.: Surfactant effects on microemulsion-based nanoparticle synthesis. Materials **4**(1), 55–72 (2011). doi:10.3390/ma4010055
74. Turkevich, J., Stevenson, P.C., Hillier, J.: A study of the nucleation and growth processes in the synthesis of colloidal gold. Discuss. Faraday Soc. **11**, 55 (1951)
75. Van Hyning, D.L., Klemperer, W.G., Zukoski, C.F.: Silver nanoparticle formation: predictions and verification of the aggregative growth model. Langmuir **17**(11), 3128–3135 (2001)
76. Latsuzbaia, R., Negro, E., Koper, G.: Bicontinuous microemulsions for high yield, wet synthesis of ultrafine nanoparticles: a general approach. Faraday Discuss. **181**, 37–48 (2015). doi:10.1039/c5fd00004a
77. Wang, F., Richards, V.N., Shields, S.P., Buhro, W.E.: Kinetics and mechanisms of aggregative nanocrystal growth. Chem. Mater. **26**(1), 5–21 (2014). doi:10.1021/cm402139r
78. Wang, F.D., Richards, V.N., Shields, S.P., Buhro, W.E.: Kinetics and mechanisms of aggregative nanocrystal growth. Chem. Mater. **26**(1), 5–21 (2014)
79. Xia, Y., Xiong, Y., Lim, B., Skrabalak, S.E.: Shape-controlled synthesis of metal nanocrystals: simple chemistry meets complex physics? Angew. Chem. Int. Ed. **48**(1), 60–103 (2009). doi:10.1002/anie.200802248
80. Xu, W., Shen, Y., Xie, A., Chen, Y., Liu, T., Du, R.: Synthesis and characterization of PbS nanotubes in bicontinuous microemulsion system. Colloid J. **72**(2), 274–278 (2010)
81. Yethiraj, A.: Tunable colloids: control of colloidal phase transitions with tunable interactions. Soft Matter **3**(9), 1099–1115 (2007). doi:10.1039/b704251p

Chapter 4
Granulation and Tabletting

Heather Emady, Karen Hapgood, and Rachel Smith

Abstract This chapter provides an overview of the granulation and tabletting processes, with a focus on wet granulation. The emphasis is on the applications of wet granulation, including practical advice for the design, operation, and control of the equipment, as well as specific industrial applications. Although the mechanisms are fairly well understood, achieving better product control remains a challenge, and therefore is an active area of research.

4.1 Introduction

Granulation and tabletting are two common particle size enlargement processes, used to combine small particles or multiple powder ingredients into a structured particle assembly, to produce pharmaceuticals, catalyst carriers, detergents, and food products, to name a few. Generally, the size of the ingredient particles is rarely large enough or uniform enough to achieve reliable powder flow and avoid segregation of the ingredients. Therefore, granulation and tabletting are used to agglomerate the ingredient particles together into a larger sized granule, in order to:

- Improve the flow properties.
- Adequately mix the ingredients and reduce the risk of subsequent segregation.
- Produce compacts with sufficient strength.

H. Emady (✉)
School for Engineering of Matter, Transport and Energy, Arizona State University, Tempe, AZ, USA
e-mail: Heather.Emady@asu.edu

K. Hapgood
Department of Chemical Engineering, Monash University, Clayton, VIC, Australia
e-mail: Karen.hapgood@monash.edu

R. Smith
Department of Chemical and Biological Engineering, University of Sheffield, Sheffield, UK
e-mail: Rachel.smith@sheffield.ac.uk

© Springer International Publishing Switzerland 2016

H.G. Merkus, G.M.H. Meesters (eds.), *Production, Handling and Characterization of Particulate Materials*, Particle Technology Series 25,
DOI 10.1007/978-3-319-20949-4_4

Product specifications and quality vary widely between industries, but broadly granules with sufficient size, shape, composition, and a suitable internal structure are required to meet the product specifications of flow, strength, disintegration, dissolution or other attributes [66]. Each industry has little flexibility in "active" ingredients, but there is generally a "handbook" for each industry of commonly used formulation components and ingredients, and their regulatory approval details (if applicable) (e.g., [2, 3, 77, 79, 90, 91, 102]).

Examples of more specific product performance goals that can be achieved by granulation and/or compaction are:

- Ensuring all the powdered ingredients are uniformly mixed in view of strict composition regulation (e.g., pharmaceutical tablets and capsules).
- Increase of bulk density so that the product can fit a defined mass into a defined volume (e.g., consumer packaged granules).
- Reduction of dust generation during handling and packaging.
- Production of compacts that have adequate disintegration/dissolution properties (e.g., detergent granules or tablets).
- Production of compacts that have a high specific surface area and an open and well-designed pore structure (e.g., adsorbents).
- Production of pellets that cause little pressure drop in fixed beds (e.g., catalysts).

Granulation is the process of agglomerating the powder particles into a larger aggregate, or granule. There are two key types of granulation processes: *wet granulation*, where a liquid is used to agglomerate the powder, and *dry granulation* or roller compaction, where the powders are compressed together to form granules. Both types of granulation can be operated as batch or continuous processes, depending on the industry and scale of the product produced.

Tabletting is the process of using punches to compress particles contained within a tablet die into a single large agglomerate, usually called a pellet or a tablet. Tabletting can be used to compress powders directly into tablets ("direct compression"), or an intermediate granulation step can be used to prepare the powders prior to tablets.

4.2 Granulation and Compaction Processes

4.2.1 Wet and Dry Granulation

Wet granulation is a particle size enlargement process where small particles are agglomerated using a liquid binder. The liquid is either sprayed, poured, or in some cases heated and melted, into the agitated powder bed. There are numerous different designs of granulation equipment used, which is selected broadly depending on the required manufacturing throughput, whether batch production or continuous production is required, the level of acceptable dust generation and/or product cross contamination, the sophistication of process control, and the capital and operating cost that can be supported by the product margin [62]. These criteria – particularly product margin and throughput – generally restrict each industry application to a few potential equipment choices.

Roller compaction (also known as *dry granulation*) is an alternate method to produce granules, where the powders are fed into a small gap between two rotating rollers, and squeezed to form a thin "ribbon", which is then milled into granules and discharged [54]. Although there are some similarities between roller compaction and extrusion (see Chap. 5), the formulations used are different and the roller-compacted ribbon is only constrained at the upper and lower sides – the edges are free and there is no extrusion die at the exit. It is commonly used to reduce the segregation potential and/or increase the bulk density of a powder prior to tabletting; it is also used where the formulation is moisture-sensitive, as particle agglomeration is performed without liquid addition. Roller compaction is inherently a continuous process, which enables reasonable real-time control of roll speed, feeder speed, roll gap and roll pressure, as well as milling conditions, and is amendable to direct online, real-time monitoring of ribbon density, final granule size and overall throughput [36]. The capital and operating cost of a roller compaction operation is much cheaper than the cost of an equivalent through-put wet granulation process, and is therefore used in detergents [29], foods, and increasingly in pharmaceutical manufacture [108, 109]. Since the process control of roller compaction is fairly well developed [20, 36, 54, 108], it is much easier to scale-up than wet granulation, where process control has historically been fairly rudimentary. This chapter will therefore focus primarily on wet granulation processes.

4.2.2 *Wet Granulation Mechanisms*

There are three main stages in wet granulation [42]:

1. *Nucleation and wetting*: the initial stage of forming the granules, when the liquid being added to the granulator first contacts the agitated powder in the granulator.
2. *Growth and consolidation*: the agitation of the powder results in collisions between the granules, particles and equipment, which results in an increase in granule size and a reduction in the internal porosity of the granules.
3. *Breakage and attrition*: the powder agitation forces exceed the granule strength and the granule fractures into several pieces or the outer layer of particles is abraded from the granule.

4.2.2.1 Nucleation

Several nucleation mechanisms of how granules form from the spray droplets have been identified, based on the relative sizes of the spray drops to the powder particles.

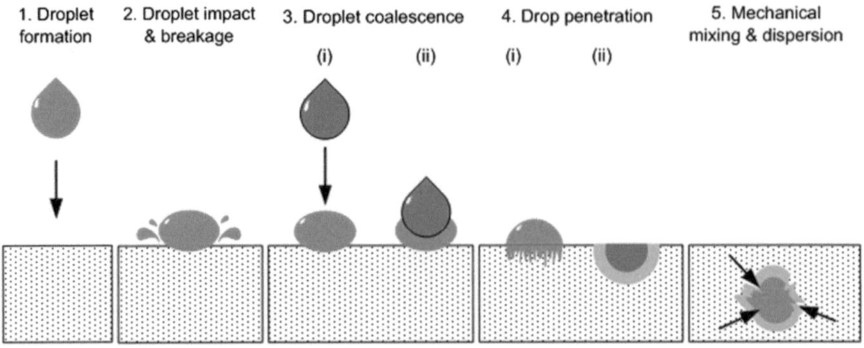

Fig. 4.1 Five possible steps in immersion nucleation: (*1*) Droplet formation; (*2*) Droplet impact and possible breakage on the powder bed; (*3*) Droplet coalescence upon contact with other droplets at high spray flux; (*4*) Drop penetration into the powder bed (*i*) to form a nucleus granule (*ii*); (5) Mechanical mixing and dispersion of the liquid and powder [27, 75] (Reproduced with permission of Elsevier)

If the drop size is larger than the powder size, then nucleation can occur via *immersion nucleation* [1], as shown in Fig. 4.1. The droplet lands on the powder surface and may begin to wet and flow into the powder pores, forming a highly saturated nucleus granule. However, the powder is usually being agitated and ideally the fluid needs to penetrate quickly into the powder bed. This process is described by the drop penetration time, t_p [27], as the equation to predict t_p is shown in Table 4.1. When immersion nucleation occurs via drop penetration into the powder bed, the size of the nucleus formed is directly proportional to the drop size [27, 75, 93, 95, 112]. This is known as the *drop controlled regime* [27]. Within this regime, three different formation mechanisms have been identified, *Tunneling*, *Spreading*, and *Crater Formation*, which can be linked to different nuclei granule shapes as a function of the powder bed porosity and the modified granular Bond number [16–18]. However, if the drop penetration time is slow, or if the liquid is poorly distributed, the fluid will be dispersed and distributed via the mechanical agitation in the granulator. This is known as the *mechanical dispersion* nucleation regime [27].

Distribution nucleation occurs when fine drops are used to coat and nucleate larger powder particles, as shown in Fig. 4.2 [1, 7, 51]. The fraction of the particle surface covered with fine drops can be estimated via the dimensionless particle coating number, Φ_p [51], which is provided in Table 4.1.

The spray zone conditions are also crucial for controlled nucleation. The fluid is atomized and sprayed onto the powder bed. Ideally, the drops would land on the powder surface without significant overlap, and the newly wetted powder areas would be quickly removed and replaced by dry powder, ready for the next stream of landing drops in the spray zone. This is described by the dimensionless spray flux Ψ_a [60], which is a dynamic measure of the density of drops falling on the powder surface compared to the rate of fresh powder surface renewal. At low spray flux (Ψ_a $<<1$), the fresh surface will ensure drop footprints do not overlap and each drop

Table 4.1 Key dimensionless groups and parameters relevant to wet granulation mechanisms

Dimensionless group	Parameters	Description
Drop penetration time t_p $t_p = 1.35 \frac{V_d^{2/3}}{\varepsilon_{eff}{}^2 R_{eff}} \frac{\mu}{\gamma_{lv} \cos\theta}$	V_d = drop volume (m^3) ε_{eff} = effective bed porosity (−) R_{eff} = effective pore radius (m) μ = viscosity of liquid (mPa*s) γ_{lv} = liquid surface tension (N/m) θ = dynamic contact angle (−)	Describes the time taken for a drop of binder fluid to completely sink into the powder bed. Indicates nucleation kinetics. Fast times are desirable for drop controlled nucleation [27]
Spray flux Ψ_a $\Psi_a = \frac{3Q}{2vWd_d}$	Q = liquid flowrate (m^3/s) v = powder velocity (m/s) W = spray width (m) d_d = drop diameter (m)	Describes liquid distribution and spray density in the spray zone [60] for immersion nucleation
Particle coverage Φ_p $\Phi_p = \frac{3w}{2d_d \rho_s \rho_l A_{SA}}$	w = liquid mass per mass dry powder (kg/kg) d_d = drop diameter (m) ρ_l = liquid density (kg/m^3) ρ_s = solid density (kg/m^3) A_{SA} = total surface area of powder (m^2)	Describes wetness of particle surface as a function of particle size/area for distribution nucleation [51]
Viscous Stokes' number St_v $St_v = \frac{2mu_0}{3\pi\mu a^2}$	m = granule mass (kg) u_0 = initial relative granule collision velocity (m/s) μ = viscosity of liquid (mPa*s) a = granule radius (m)	Ratio of initial kinetic energy to viscous dissipation from the collision of two granules [19]
Critical viscous Stokes' number St_v^* $St_v^* = \left(1 + \frac{1}{e}\right) \ln\left(\frac{h}{h_a}\right)$	e = coefficient of restitution (−) h = liquid layer thickness (m) h_a = height of surface asperities (m)	Describes the threshold at which the particle velocity is zero at the edge of the liquid layer [19]
Deformation number St_{def} $St_{def} = \frac{\rho_g U_c^2}{2Y_g}$	ρ_g = granule density (kg/m^3) U_c = granule collision velocity (m/s) Y_g = granule yield strength (kPa)	Ratio of kinetic collision energy to granule strength [42]. Related to granule growth, consolidation and breakage

(continued)

Table 4.1 (continued)

Dimensionless group	Parameters	Description
Saturation level s $s = \frac{w\rho_s(1-\varepsilon)}{\rho_l \varepsilon}$	w = liquid mass per mass dry powder (kg/kg)	Proportion of granule pore space occupied by liquid [42]
	ρ_s = solid density (kg/m^3)	
	ρ_l = liquid density (kg/m^3)	
	ε = granule porosity (−)	

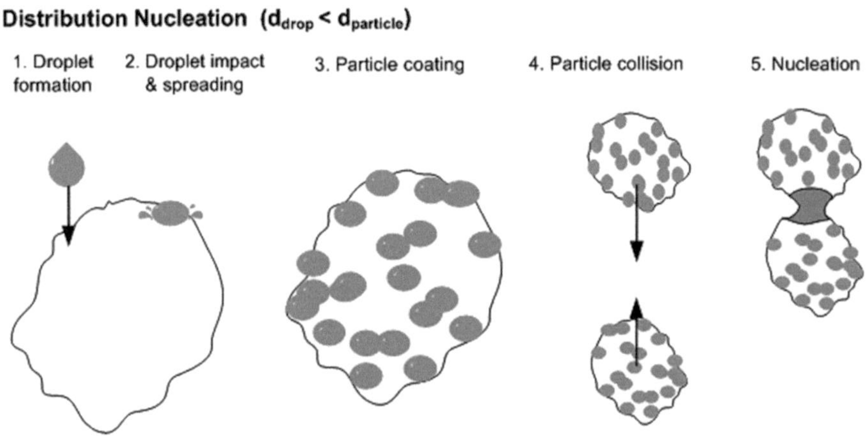

Fig. 4.2 The five steps of distribution nucleation, where the drops are smaller than the particles [51] (Reproduced with permission of Elsevier)

will form a separate nucleus. However, at high spray flux (Ψ_a ~1), there will be significant overlap of drops hitting the powder bed. Nuclei granules formed will be much larger and their size will no longer be a simple function of the original drop size [60, 61].

4.2.2.2 Growth and Consolidation

As nuclei and granules are agitated in a granulator, they experience forces which lead to consolidation, or densification. The rate at which consolidation occurs is dependent on powder and binder properties [44, 46]. It has been observed that as consolidation continues and granule porosity reduces, interstitial liquid can be pushed to the surface of the granules, and the granules become surface wet [120].

Granule growth occurs through two main mechanisms: layered growth, which is driven by the accumulation of dry powder on the outside of surface wet or adhesive granules, and coalescence, which occurs when two or more granules join together to form a larger granule. Several models have been proposed to describe the coalescence process. Ennis et al. [19] proposed a theoretical model for the coalescence of hard spheres. This model assumes the granules act as rigid spheres, and have a thin surface liquid layer of thickness h, and surface asperities of height h_a. The authors proposed a criterion for coalescence based on the viscous Stokes' number, St_v (see Table 4.1). Coalescence will be successful for values of St_v less than a critical viscous Stokes' number, St_v^* (see Table 4.1).

This model assumes that there is little or no deformation of the granules themselves. A further, more complex, model was proposed by Liu et al. [63], to describe the coalescence of deformable granules. Two scenarios were considered for impact and coalescence: surface dry deformable granules, and surface wet deformable granules (see Fig. 4.3). In the case of surface wet granules, two types of coalescence were identified: Type I coalescence, where the inter-granule velocity reduces to zero before granule surfaces touch in the initial approach stage, and type II coalescence, where the inter-granule velocity becomes zero after the surfaces of the granules have come into contact. Granules which do not achieve a zero velocity will rebound fully and not coalesce.

A granule growth regime map [41, 45] has been developed which describes granule growth behavior as a function of the Stokes' deformation number, St_{def}, and the maximum pore saturation level, s (see Table 4.1). This regime map can be found in Fig. 4.4. In this regime map, several qualitative descriptions for the granule growth behavior have been defined. Of particular interest are the steady growth and induction growth zones. In the steady growth zone, granule size increases steadily, and granule size is therefore relatively easy to control. In the induction growth zone, there is a period of little granule growth, during which time the granules are consolidating. This is typically followed by a rapid increase in granule size.

4.2.2.3 Breakage

The mechanism of wet granule breakage is most prevalent in granulators where the forces are large, in particular high shear mixer granulators. The rate at which granule breakage occurs is difficult to predict, due in part to the complex nature of the forces experienced by the granules.

A wide range of experimental studies has been conducted to study the strength and deformation of single granules [43, 44, 46–48, 104, 105]. Compression of granular pellets has been used to identify two main modes of granule deformation [47, 104]: pseudo-plastic failure and semi-brittle fracture. Pseudo-plastic deformation is characterized by a paste-like flow of the granule, with cracks appearing only at high strain. This mode of failure occurs in formulations which demonstrate a peak stress at strains of greater than 5 %, or which do not demonstrate a peak stress [104]. Semi-brittle pellets develop a central vertical crack, and show a peak stress at strains less than 5 %. Examples of pseudo-plastic and semi-brittle failure are shown in Fig. 4.5.

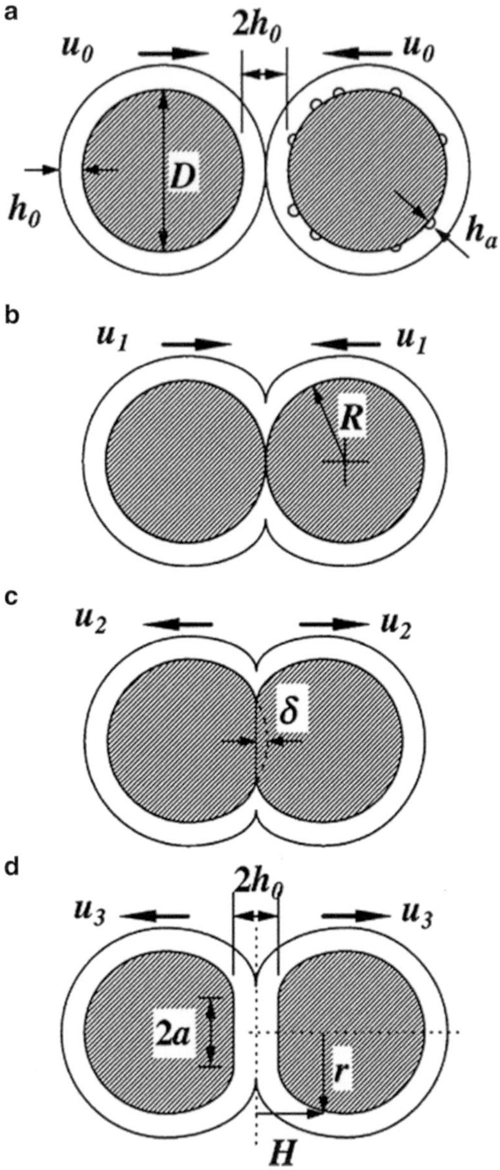

Fig. 4.3 A schematic of the coalescence of surface wet granules. (**a**) Approach stage. (**b**) Deformation stage. (**c**) Initial separation stage. (**d**) Final separation stage [63] (Reproduced with permission from Wiley)

In addition to the study of individual granules and pellets, a variety of experimental studies has been performed to study the breakage of tracer granules in high shear mixer granulators [14, 64, 83, 106, 111]. Experimental results in both tracer and full granulation experiments have shown that granule breakage increases with both increasing impeller speed and granule size [49, 55, 64, 83, 88, 105, 111]. Impeller geometry has also been shown to have a significant effect on granule breakage [106].

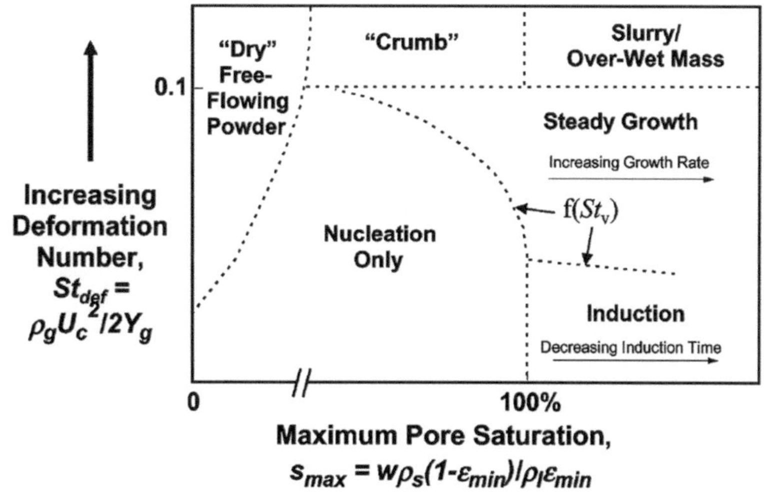

Fig. 4.4 Granule growth regime map [41] (Reproduced with permission of Elsevier)

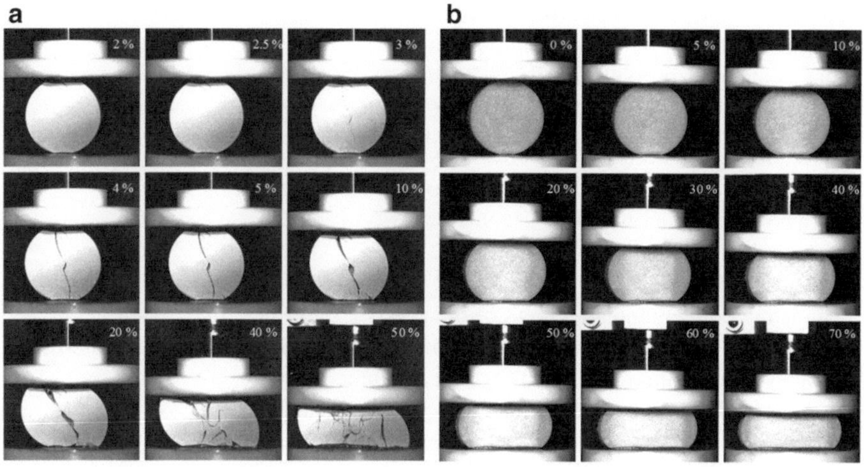

Fig. 4.5 Wet granule deformation. (**a**) Semi-brittle failure. (**b**) Pseudo-plastic failure [104] (Reproduced with permission of Elsevier)

4.2.3 Tabletting and Compression

Granules are often used as a precursor feed to a tabletting machine, which compresses the granules into a solid, shaped tablet (also known as a pellet or compact), which has higher density, well-defined shape, embossing/debossing, and in more sophisticated products may have two or more layers. The tablets are then ready for packaging or direct use in a variety of consumer applications. Although tablets can be formed directly from un-granulated powder, it is common to pre-granulate to avoid segregation of the formulation ingredients and to ensure reliable flow and controlled bulk density when the powder enters the tablet die.

Tabletting is a high-speed compaction operation, where the powder formulation is filled into a small receptacle, called a “die”, and is then compressed as two “punches” – one from above and usually also one from below – enter the die and compress the powder into a small compact. The details of the punch motion during compression (e.g., displacement vs. time) determine the force actually applied to the tablet during compression, although the upper force is usually “set” (or, more accurately, “limited”) by the available hydraulic load applied to the press punches.

During the initial stages of compression, the force applied to the powder causes the particles to rearrange slightly and increase their bulk density. As the punches descend further into the tablet die, the particles begin to deform or fracture, and begin to form interparticle bonds. Further compression increases the strength until the upper punch reverses out of the die, and the lower punch moves up to eject the tablet. Tabletting is a high-speed operation, with multiple punches and dies on a high-speed rotating frame. The typical timescale for a pharmaceutical tablet to be produced is a fraction of a second, and the typical production rate is around 10,000 tablets per minute [80].

The two key tablet properties of interest in most applications are tablet tensile strength and tablet disintegration/dissolution behavior. The tensile strength of the tablet is affected by both the formulation and the mechanics of the compression process. Important formulation properties include:

- granule size distribution and specific surface area,
- the brittle or plastic mechanical properties of each ingredient,
- granule structure (including granule porosity),
- uniformity of flow into the die,
- bulk density in the die,
- moisture content and stickiness of the powder,
- deaeration characteristics (as a lot of air may need to leave the tablet in order to achieve the required reduction in tablet volume), and
- whether the formulation has a tendency to segregate.

Commercial tabletting presses include their own control system, which is able to maintain and control the key parameters of tablet weight and hardness, provided that the flow and bulk density of the incoming powder or granules is consistent. Tablet weight is maintained by adjusting the depth of the lower punch, which adjusts the die volume available to be filled [10]. Tablet hardness is controlled via the punch motion during compression (i.e., the displacement versus time curve traced by the upper punch as it descends into the die). This determines the overall stress-strain history of the compact. Provided that the incoming powder or granule flow is relatively consistent, and that the powder does not segregate, the tablet press can be run for long periods with little intervention besides periodic checks of weight and hardness. Modern pharmaceutical presses can sample and test these parameters automatically. Hence, problems in tabletting are often caused by problems with the formulation, and changes in granulation conditions (e.g., impeller speed) are often reflected in the tablet properties [76].

Generally, tablets produced on different model tabletting machines from an identical powder formulation will have slightly different tablet properties, as the details of the force-displacement curves, punch velocities, tolerances, control systems and ejection systems will all vary slightly, and affect the stress-strain history experienced by the tablet. A recent review of tabletting by Tho et al. [109] includes an extensive list of potential factors affecting tabletting.

4.3 Granulation Equipment Design, Operation and Control

All forms of wet granulation equipment have two common aspects: a method of mixing and turning over the powder, and a method of introducing liquid binder to the powder mass. The designs of granulators vary widely [79], and granulator design and operation has a strong effect on the final granule properties, including size distribution, porosity, and compressibility in tabletting processes.

4.3.1 Drum Granulators

Drum granulation is a well established unit operation [50, 74], commonly used in the manufacture of fertiliser granules [12, 113], specialty products [72], as well as in minerals processing (e.g., calcining kilns) [39]. Drum granulation generally produces large granules, in the 1–10 mm size range, and the granules are spherical in shape due to the rolling action of the drum.

Granules are produced by spraying the binder fluid onto the incoming dry powder at the entry of the drum. As the drum rotates, the particles grow and consolidate as they move slowly towards the drum exit. Granules are sieved into three size classes and split into separate streams. The in-spec granules (e.g., granules between 2 and 5 mm) continue on to the drying step, whilst the under-sized granules (e.g., <2 mm) are mixed with the incoming feed powder at the entry of the drum. The over-sized granules (e.g., larger than 5 mm) are usually crushed before also being returned to the feed stream. An example of a drum granulator is provided in Fig. 4.6.

The first ever granulation study was performed on a drum granulator [74], but despite concerted industrial and academic research effort over 50 years, drum granulators are still problematic with high energy requirements, poorly controlled process conditions, and often produce sub-optimal product [8, 12, 25, 31, 44–46, 81, 84, 113, 120]. High recycle ratios [114] and unstable or surging recycle flows [12] often mean that drum granulators operate at well below the theoretical design capacity [114]. New control strategies for continuous drum granulation circuits are still being developed [25, 84].

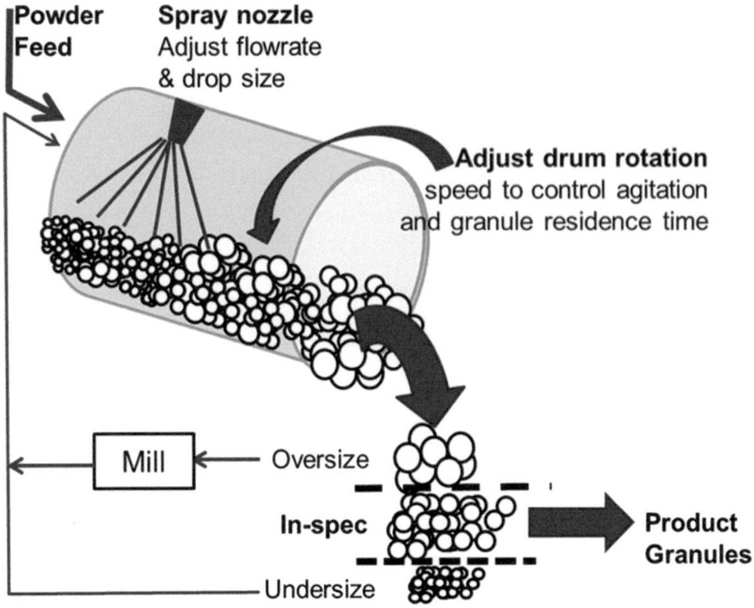

Fig. 4.6 Schematic of a drum granulator, showing size classification and recycle circuit

4.3.2 Mixer Granulators

Mixer granulators are a category of granulators which agitate and mix the powder using rotating impellers or paddles. These granulators are commonly used in the pharmaceutical, detergent and consumer goods, and food industries. Mixer granulators can be classified as either high shear (the most common form) or low shear, depending on the designed impeller rotation speed. The impellers are generally either horizontal axis (i.e., the impeller rotates and sweeps across the floor of the granulator) or a vertical axis (i.e., the impellers lift and drop material vertically in a drum shaped granulator). There is a great deal of variety in the shape of the granulator vessel and impeller, and also in the impeller shaft design, with both fixed shaft and planetary mixers available. An example of a vertical-axis mixer granulator typically used in pharmaceutical granulation is shown in Fig. 4.7.

In addition to the main impeller, mixer granulators often have an additional small high-speed rotating impeller, commonly called a "chopper." The aim of the chopper is nominally to break up over-large granules and aid in densifying granules; however, the actual effect of the chopper on granule properties is unclear [33, 34, 68, 96]. Typical operation of mixer granulators involves a liquid addition stage, where liquid binder is added to the granulator, followed by a mixing stage, known as "wet massing."

Liquid binders used in high shear granulators range from low viscosity liquids to high viscosity liquids, slurries, pastes and melts. They can be aqueous or solvent based (e.g., ethanol), and guides to liquids and binders are found in the various

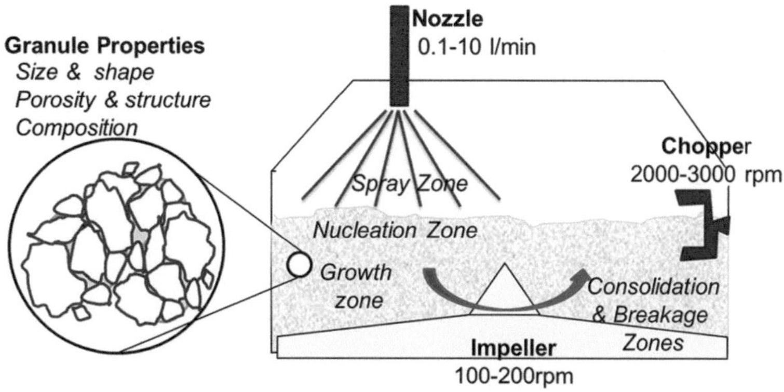

Fig. 4.7 Schematic of a typical vertical-axis high shear mixer, used for pharmaceutical granulation

reference guides for each industry (e.g., [2, 3, 77, 79, 90, 91, 102]). The liquid binder is added to the granulator through the lid via either atomizing nozzles, or an injection or pouring port. In cases where the liquid is not sprayed, the mechanical forces in the granulator (i.e., the impeller, powder flow, and chopper) are used to distribute the liquid binder. The method of liquid addition has a significant effect on granule nucleation, and it has been shown that spraying the liquid during the liquid addition stage can result in smaller granules, with a narrower size distribution [4, 95].

Generally, high-shear mixer granulators produce denser [21, 58], and relatively small granules between 200 and 1000 μm [68], although larger lumps and balls can also form [28, 68]. The final properties of the granules tend to be very dependent on the design and operation of the equipment, and the difficulty of matching the dynamics of the many simultaneous processes occurring in the granulator at any given time. This causes difficulty in producing consistent granules across different equipment and scale without several trial runs [22, 30, 61, 68, 71, 87, 88].

4.3.3 Fluidised Bed Granulators

In contrast to high shear granulators, fluidised bed granulators provide powder motion and mixing by fluidising the powders with a gas. Most fluidised bed granulators use air as the fluidising medium, although in the case of flammable materials such as sugars or cellulose, nitrogen is often used to reduce the risk of explosion. The ability to achieve effective and uniform fluidisation of the powder particles is a pre-requisite for the use of this equipment, and this is a function of particle size and density [57, 101]. Unlike in mixer granulators, liquid binders are exclusively added to fluidised bed granulators as sprays. The droplet size distribution tends to be equal to or smaller than the fluidised particles [51, 95], and ranges around 200–500 μm [1, 4, 32, 51, 95, 100].

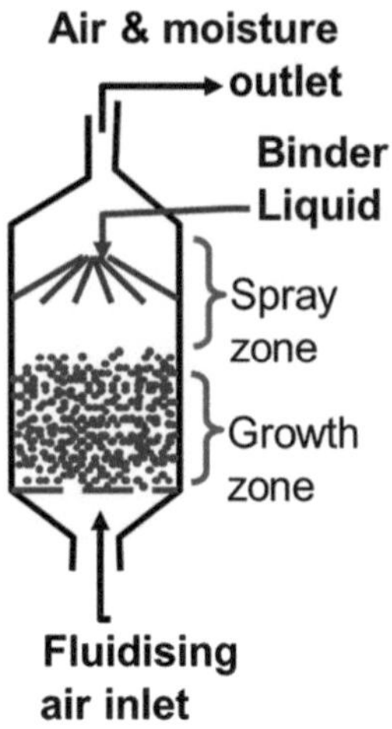

Fig. 4.8 Schematic diagram of a typical top-spray fluidised bed

Operation of fluidised bed granulators starts with fluidisation of powders, followed by the initiation of the liquid binder spray. Due to the drying action of the fluidising air, granule growth is limited to the spray period [53], while the granules are still wet. This is generally followed by a period of no-spray fluidisation, during which the granules are dried. One of the obvious advantages of fluidised bed granulation is the built-in drying of the product.

The vast majority of fluidised bed granulator designs consist of a vertical vessel, roughly generally cylindrical or square in shape; however, more complicated vessel geometries are in use, particularly for continuous fluidised bed granulation processes. The primary points of difference amongst fluidised bed granulators are the vessel shape, and the orientation of the spray. A typical fluidised bed is shown in Fig. 4.8, which shows the fluidisation air entry, the locations of the spray zone, granule growth (& drying) zone, and the freeboard where fine airborne particles may be carried out of the bed and captured by downstream air filters. The majority of fluidised bed granulators are operated with a top-down spray orientation; however, bottom-up and side-spray systems are also used [24, 79].

Fluidised bed granulators tend to produce very porous granules, and are often used to produce granules with good dissolution properties. The design of a typical fluidised bed granulation process is similar to that of a fluidised bed dryer, with the additional complexity of a liquid spray. Granule growth competes with the drying action of the air.

One of the main challenges of fluidised bed granulator design and operation is the fluidisation behavior of the powder and granules. As the granules grow, the minimum fluidisation velocity increases. This can serve as a self-classifying action, as the large granules move out of the spray zone and cease to grow. However, if liquid addition occurs too quickly, the bed can defluidise and turn into a large wet mass, in a process known as “quenching.”

4.3.4 Control of Wet Granulation Processes

The main parameter of interest to be controlled during granulation is the particle size distribution of the granules. There are many potential options for measuring particle size online [67], including:

- Size classification via continuous sieving or screening
- Image analysis [115, 117, 118]
- Laser diffraction, although it is more commonly used off-line
- Laser analysis methods, including beam reflectance [37, 38, 68, 103, 110, 121]
- Other methods such as acoustic methods [110]

In any granulation process, regardless of the equipment, there are several input parameters that need to be controlled. These include:

- the feed powder flowrate (for a continuous process)
- the liquid delivery flowrate [98]
- the total amount of liquid delivered (for a batch system)
- the spray pressure (if using a two fluid nozzle) to control the drop size [94, 97]
- the binder fluid concentration (if an ingredient is being delivered via the liquid phase)

Additionally, it is feasible to measure the moisture content of a granulator powder mass in real time using NIR sensors, and these have been widely applied in pharmaceutical fluid bed granulation, where they are also used to determine the drying endpoint [115–117, 119]. At-line loss on drying is also commonly used.

The level of agitation of the granulating powder mass also requires control. This is usually achieved indirectly by controlling and/or manipulating the agitator speed, the fluidisation velocity or the drum speed in mixers, fluidised beds and drum granulators, respectively. With the exception of fluidisation air, most granulation processes run at pre-set agitation conditions. Only fluidisation is frequently adjusted, because the consequences of defluidisation would most likely be problematic. Direct measurement of the powder agitation and motion would be desirable, but is often difficult. For example, image analysis camera systems to measure powder velocity often encounter issues with window fouling, or dust visibility problems. Powder agitation measurements have been performed at lab scale [61, 89] and in large scale production [28, 85].

Typical parameters controlled for continuous drum granulation, fluidised bed granulation and batch mixer granulation are shown in Tables 4.2, 4.3, and 4.4. The tables also indicate the main variable which is manipulated to achieve control.

Some granulation processes are also prone to building up heat due to exothermic reactions, or simply due to friction and work over long process times. Temperature control can be exerted via cooling jackets on mixer granulators, or using fan air in fluidised beds or drums. Control of the environmental temperature and humidity is generally desirable in all granulation processes, but can usually only realistically be

Table 4.2 Typical control parameters for continuous drum granulators

Controlled parameter	Sensor	Manipulated variables
Total throughput Q_{solids}	Weight feeders	Total throughput Q_{solids}
Recycle flowrate $Q_{recycle}$	Weight feeders	Drum speed s
Exiting granule size distribution D_{exit} (rare)	Image analysis (*advanced control examples only*)	Liquid flowrate Q
Online bulk density (rare)	Mass of dry granules in a cup of known volume (P&G only)	Recycle flowrate R (perhaps with some buffering to reduce surges)
Drum rotation speed	Motor speed	Residence time and holdup

Table 4.3 Typical control parameters for fluidised bed granulators

Controlled parameter	Sensor	Manipulated variables
Fluidising air temperature	Inlet air temperature	Heater power
Fluidisation intensity	Visual observation	Fluidising air velocity
Drying rate	Exit air temperature/relative humidity	Inlet air flowrate and/or temperature
Granule size distribution	Particle size (image analysis, or FBRM)	Batch time (i.e., stop when granules are large enough)
Granule moisture content	NIR moisture probe, or loss on drying of small samples	Total drying time, plus inlet air temperature/flowrate
Pressure drop	Pressure gauges across filter bags	Manual filter bag shake

Table 4.4 Typical control parameters for batch mixer granulators

Controlled parameter	Sensor	Manipulated variables
Batch mass M_{solids}	Weight feeders	N/A
Liquid mass added V_{liquid}	Scales	N/A
Liquid flowrate Q	Flowmeter	N/A
Granule size distribution D_{exit} (rare)	Particle size (FBRM or image analysis) (*advanced control only*)	Granulation batch time, or impeller power consumption
Impeller rotation speed	Motor speed	Impeller speed

achieved in pharmaceutical manufacturing suites where the ambient temperature and relative humidity can be controlled within tight tolerances.

A realistic control scheme for a commercial scale drum granulator was published by P&G [73], which involved measuring the exiting granule size by image analysis and measuring the bulk density of the granules (as a proxy for granule density) at the conclusion of the drying stage. This scheme, shown in Fig. 4.9, is able to

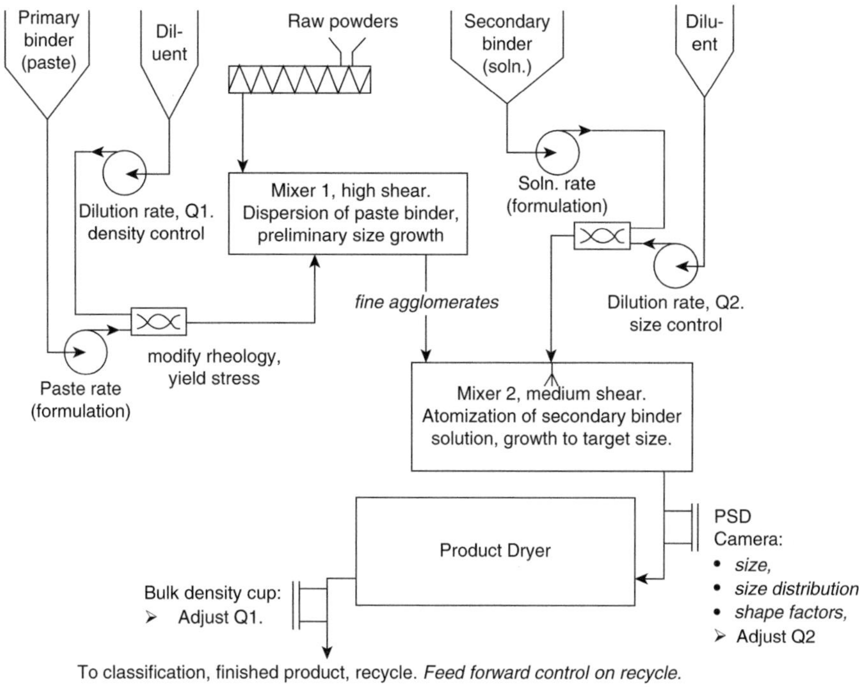

Fig. 4.9 Control scheme for a continuous detergent agglomeration process [73] (Reproduced with permission from Elsevier)

successfully control the steady-state operation, but is unlikely to be able to handle short term transients due to the long time lag between measuring the granule properties at the end of granulation and feeding this back to the start of the process.

4.4 Industrial Applications

4.4.1 Pharmaceuticals

4.4.1.1 Overview of the Current State in the Industry

The pharmaceutical industry is currently undergoing a transition from recipe-driven, batch processing with fixed conditions, towards continuous processing with real time monitoring. This shift is largely due to the industry's Quality by Design (QbD) initiative, introduced in 2006 to improve the efficiency of pharmaceutical manufacturing based on science and engineering [52]. In alignment with QbD, regime maps employing dimensionless groups are gaining traction in the experimental design stage [52] as well as in troubleshooting [28] of pharmaceutical wet granulation processes, which is an improvement over a purely statistical experimental design and trial and error. However, the end goal of fully predictive models for wet granulation remains an active area of research.

Granulation is used in pharmaceutical manufacturing to create uniform dosage forms, to densify, to improve flowability and appearance, to lessen dust, and to aid in dispensing of doses [80]. Common granulators in this industry include high shear mixers and fluidised beds. Twin screw granulation is a current extensive area of research, seen as a way to continuously produce granular products which were traditionally formed via batch high shear mixers [56].

4.4.1.2 Product Specifications and Quality

Important pharmaceutical granule characteristics include content uniformity, dissolution properties, bioavailability, chemical stability, and flowability [80]. Content uniformity is of utmost importance in this industry, where it is imperative that the solid dosage form contain the prescribed amount of active pharmaceutical ingredient (API). Flowability of the granules is also essential, since granulation is usually a precursor to tabletting.

4.4.1.3 Raw Material Selection

Pharmaceutical solid dosage forms are generally comprised of an API and excipients. Excipients can include inert fillers, binders, disintegrants, lubricants, colorants, and stabilizing agents [80]. Examples of different excipients are provided in Table 4.5 [80].

The API and excipients are usually dry mixed prior to addition of the granulating liquid, which can be water, ethanol, or a binder solution. The particulate binder is often dissolved in water or ethanol to be used as the granulating liquid.

Table 4.5 Summary of typical pharmaceutical excipients

Excipient type	Common examples
Filler	Lactose, dicalcium phosphate, starch, microcrystalline cellulose
Binder	Sugars such as glucose, sucrose, and sorbitol; natural polymers such as gelatin, starch, acacia, sodium alginate, and alginic acid; synthetic polymers such as methylcellulose, hydroxypropylcellulose, hydroxypropylmethylcellulose, ethylcellulose, sodium carboxymethylcellulose, polyethylene glycol, and polyvinylpyrrolidone
Disintegrant	Starch, microcrystalline cellulose, croscarmellose, crospovidone, sodium starch glycolate, hydroxypropylcellulose
Lubricant	Magnesium stearate, stearic acid, calcium stearate, talc, starch, wax

4.4.2 Catalysts

4.4.2.1 Overview of the Current State in the Industry

Catalysts and absorbents are used in many industries, including chemicals, polymers, refining, environmental, and fine chemicals [35]. Despite their prevalence, there is a surprising lack of information regarding the processes involved in the formation of catalysts, adsorbents, and supports. The industry seems to prefer to keep most manufacturing knowledge in-house, and academics tend to focus more on the catalysis chemistry rather than processing [35, 69, 70]. In terms of manufacturing, there is some broad knowledge available about different types of formation processes, desired granule attributes, equipment, and raw materials used.

In the manufacture of catalysts and adsorbents, granulation is typically used as a forming process to produce granules of a specific shape, size, pore structure, and strength. Mixers, fluid beds, and rotating granulators (e.g., pans) are all employed [35]. The industry tends to prefer pan granulators, such as the Eirich model, and often refer to the granulation process as "snowballing" [9, 13, 65, 69]. Other forming processes include extrusion, compaction, tabletting, spray drying, and sol-gel [13, 35]. Extrusion is the cheapest and most widely employed method of manufacturing catalysts and supports [9]. After the forming stage, granules are typically calcined [9, 13, 40, 69].

4.4.2.2 Product Specifications and Quality

Catalysts are produced to increase the rate, efficiency and selectivity of desired chemical reactions. They typically consist of an active material on a porous support. These supports are usually synthesized as micron sized particles. However, in order to improve handling and processing, these support or catalyst powders are often agglomerated and calcined into various millimetric shapes, depending on the application. It is expected that the original properties be maintained through the forming stage. A major challenge towards this end is in maintaining the pore structure [35].

Granule attributes that are important in this field include pore volume, pore size and size distribution, strength, specific surface area, bulk density, attrition resistance, particle size and size distribution, and shape [13, 35, 65, 99, 107]. Large specific surface area and pore volume, as well as a narrow pore size distribution, are desirable for optimum catalytic behavior [13]. High porosity and strength are essential, and it is difficult to achieve both of these opposing requirements. Unlike detergents and pharmaceuticals, catalysts and absorbents are not designed to disintegrate. Rather, they need to withstand in the process for many years at high pressures and temperatures, quick changes in process conditions, and should have high attrition resistance and long shelf lives [35].

Catalysts are used in fixed beds, fluid beds, and slurry reactors [9, 59, 65]. The desired granule size depends on the reaction rate and the fluid used [35]. For fixed beds, granules are usually 3–6 mm, and a narrow particle size distribution is desirable to reduce the pressure drop [35, 107]. In slurry reactors, granules are typically 50–1500 μm [35]. Fluidised beds require relatively large particle sizes, since small particles are cohesive and therefore difficult to fluidise [9, 59]; typical sizes are about 20–100 μm [9, 66].

Some common catalyst shapes include pellets, extrudates, spheres, beads, granules, and microspheres [9, 13]. The desired catalyst size and shape depends on the type of process and reactor, as well as the hydrodynamics and mass and heat transfer limitations within the reactor, and the resistance to attrition [9, 70]. Spherical shape results in favorable transport properties, both internally for the reactions, and externally for the bulk flow of material [40]. Spherical granules can be formed via spray drying, sol-gel, and granulation [13]. With granulation, spherical granules can be formed via layering in fluidised beds, such as by starting with inert core seeds and layering the active shell on top [99].

A promising way to make smooth, strong, uniformly sized, spherical granules is via oil-drop or sol-gel granulation, though this route is fairly expensive [13, 40]. - Sol-gel formed granules are proportional in size to the size of the droplet needle gauge, with a uniform size distribution, similar to granules formed in the drop controlled regime in wet granulation [13, 17, 26]. Granules less than 3 mm in size can be viably produced via this method, but it is difficult to maintain the spherical shape of granules larger than this, which tend to deform [13].

4.4.2.3 Raw Material Selection

Some popular catalyst and catalyst support materials include alumina, diatomite, and zeolites. Gamma alumina is a widely used support material due to its thermal stability and high surface area [13, 40]. Diatomite, which is primarily composed of silica, is a natural porous substance often employed as a catalyst support [65]. Zeolites are also very common catalyst materials due to the possibility to design their structured porosity [69].

Both inorganic and organic binders can be employed for granulation, depending on the processing. Examples of inorganics include alumina, cement, clay, silica, and silicates; organics include acrylic, cellulose, gum, polyethylene glycol, starch, sugar, and vinyl [35, 69]. Other additives employed in addition to binders include fillers, lubricants, modifiers, peptisers, plasticisers, and porogens [70]. Any added organics are usually removed via calcination, creating an internal pore structure [70].

4.4.3 Detergents

4.4.3.1 Overview of the Current State in the Industry

Granulation is a key step in the manufacturing of many detergent powders [5, 6]. The inclusion of the granulation process to the detergent industry is a relatively recent innovation, driven in part due to the demand for compact detergent powders and tablets. Until the late 1980s, the manufacture of detergent powders relied primarily on spray drying, producing powders with low bulk density. The introduction of granulation allowed the industry to produce compact formulations with high bulk densities.

The agglomeration process in detergent manufacture varies across detergent powder types, and across companies. Often, a combination of mixer granulators are used [79]. The initial granulation step occurs in a high shear mixer, where the highly viscous binder (typically linear alkybenzene sulphonic acid, LAS) is distributed throughout the powder mass. This is followed by a lower shear mixer granulator, where the binder is further distributed, to create a more homogenous granular bed. In addition to mixer granulators, fluidised bed granulators are also commonly used.

In recent years, there has been a rise in the use of single dosage detergents, which are convenient for the consumer, and also reduce overdosing. For powders, these take the form of compacted tablets, which can often have multiple layers and complex structures [79]. These tablets tend to be loosely compacted when compared with pharmaceutical tablets, as quick dissolution of the tablet is key to the final product performance.

4.4.3.2 Product Specifications and Quality

There are many characteristics of detergent powders which are controlled to ensure a consistent product. The complex nature of the constituent materials can make controlling the detergent manufacturing process a challenge.

A key product specification is granule size. Detergent granules typically average around 500–710 μm [6]. Additionally, the bulk density of detergent granules are monitored carefully, as detergents are usually dosed using a volume measure; by controlling bulk density, the mass of detergent in each dose is consistent. Bulk density can vary between products from 300 to 900 kg/m^3.

Product homogeneity is carefully controlled, as it is important to ensure that each dose delivers the desired active ingredients, and performs within narrow limits. Disintegration and dissolution are also very important in detergent manufacture, as they can have a significant effect on the performance of the detergent. Granule porosity has a significant effect on disintegration and dissolution, and should be controlled.

Powder flow issues and caking are common problems in the detergent industry, and can have an effect on handling and packaging, and on product performance. Detergent powders are often stored in silos and bins, and powder flow issues can cause significant problems with the discharge from the silos. LAS is highly sensitive to temperature, and can vary from a paste-like solid to a much lower viscosity fluid in the temperature range from 20 to 40 °C. This means that in many cases the mechanical strength and behaviour of the granules is strongly influenced by the ambient temperature. The propensity for caking and the ability to store the powders are often inferred by using compression tests [6].

4.4.3.3 Raw Material Selection

A range of materials with complex properties is used in the manufacture of detergent powders. The primary active ingredients are ionic and non-ionic surfactants. In addition, builders are often used, for example zeolite. These aid the surfactant in the wash by reducing the water hardness [5]. Electrolytes are also added, as are sodium carbonate and sodium silicate, which act as pH buffers [6]. In addition, a range of ingredients is added in small quantities, such as enzymes, perfumes, bleaching agents, and other materials. Dusting agents are sometimes used to allow wet mixtures to form more stable agglomerates [6].

There tends to be a very high loading of active ingredients in detergent powders, and this gives little flexibility in choosing or adding extra ingredients.

4.4.4 Foods

4.4.4.1 Overview of the Current State in the Industry

A wide variety of food products is granulated, from cereals and sweets to dairy, beverage, and vegetable powders. The food industry typically uses both fluidised bed and high-shear granulation, employing both batch and continuous processes [78, 82]. A snapshot of various granulated food products is provided in Table 4.6.

Food products are also often formed via spray drying, extrusion, roller compaction, and tabletting [78] (see also Chap. 5). Spray drying is one of the more cost effective formation techniques for foods, producing small, porous, friable particles. Spray-dried foods include beverage powders, infant formulas, instant coffee, powdered flavors, dairy powders, and fruit and vegetable powders. Extruded foods include seasonings, sweet beverages, and instant tea. Roller compacted foods include vitamins, sucrose, dairy powders, encapsulated flavor powders, and soup and seasoning powders. Tabletted foods include stocks, sweets, and vitamins.

Table 4.6 Summary of different granulated food products

Granulated foods	Examples	Main ingredients	Reason for granulation
Dairy powders	Whole and skim milk powders, infant formulas, coffee creamers, buttermilk	Lactose, proteins, fat, minerals, water	Quick dissolution in warm or hot water
Dehydrated convenience foods	Seasonings, stocks, sauces, soups, mashed potatoes, rice and pasta dishes	Flour, starch, sodium chloride, fat, oil, vegetable and yeast extracts, meat powder, flavors, spices, herbs	Improve flowability and dissolution, dosage size, increase color intensity
Beverage powders	Coffee, cocoa, tea	Pure components, milk powder, flavors, sugar	Improve solubility in hot or cold water
Sweets	Sugars	Dextrose, citric acid, flavors, colorings	Dosage size, appealing shape, increase shelf life, improve dispensability, improve compressibility for tabletting
Cereals	Granola, cereal bars	Carbohydrates, grains, dried fruit, often fortified with vitamins and minerals	To combine multiple ingredients

4.4.4.2 Product Specifications and Quality

The primary desired attributes of a food granulated product include flavor, texture, appearance, shape, nutrition, content uniformity, shelf life, flowability, and attrition resistance [78, 82]. Dissolution and compression properties may also be important, depending on the application. Granulation of fine particles can improve their dissolution rate.

The primary processing problems associated with food granulation include sticking and caking [78]. Post-processing issues that may occur during storage include caking and hardening. These problems can sometimes be mitigated by careful control of temperature and humidity.

4.4.4.3 Raw Material Selection

For foods, the raw materials will depend on the desired product. Adding extra ingredients that enhance granulation is not favorable in this industry where taste is the primary objective. Food powders are generally comprised of proteins, lipids, acids, carbohydrates, vitamins, minerals, flavors, additives, and water [78]. Possible food binders include water, lecithin, melted fat, molasses, or water mixed with hydrocolloids or carbohydrates [23]. The binder viscosity is usually higher than that used in other industries. As an example, the creation of granola includes dry mixing of grains and nuts which are granulated with a binder comprising of some combination of honey, water, oil, and molasses, and then baked [82].

Food materials are often sticky, which is a property influenced by their moisture content and temperature [78]. Sweets especially tend to be sticky, and lubricants such as magnesium or calcium stearate are often added to circumvent this.

4.5 Outlook for the Future

Granulation is executed in a wide variety of industries for mixing and agglomerating small ingredient particles into granules having sufficient strength in order to improve the flow properties, to reduce explosion risks, as well as to give other specific performance aspects of the granules. At this moment, improved control of the processes still is a major challenge in view of their complexity and the difficulties encountered in in- and on-line measurement of product parameters.

Granulators are a highly interactive system, where a small change to one variable affects a number of process responses. This is one of the key difficulties with granulation control. For instance, an increase in drum speed in a granulator will increase the powder surface velocity in the spray zone, and reduce the spray flux [60], creating smaller nuclei. The reduction in drum speed will also reduce the energy of inter-particle collisions and reduce the rate of granule growth, the rate of granule consolidation, and the rate of breakage. It is therefore difficult to predict the net effect that reducing drum speed will have on the granule size, even at steady state conditions. It is also possible for the granule size to experience strong transient or even inverse responses [25, 92]. For example, if the drum speed is reduced, the granule size may initially increase due to the reduction in breakage in the short term, but may later decrease as the reduced growth rate works its way through the drum.

The solution to this issue will come from more sophisticated control strategies which control dimensionless groups (i.e., the ratio of several related variables at once) rather than individual parameters, and applying model based control. Examples of dimensionless parameters than can be utilized in control are shown in Table 4.1. Some work on this has begun to be reported: P&G have reported using correlations of dimensionless groups to maintain the key product attributes [71, 73] and an industry-funded drum granulation study has come closest to achieving model based control [25, 84].

The second issue is to measure and control the porosity or density of the granules, which is the best indicator of granule structure currently available. Although there are several lab-based methods available, including bulk density correlations [73]; X-ray tomography [11, 21, 86]; kerosene displacement [44]; and mercury porosimetry [15, 21], these methods are all performed on individual samples or even on individual granules, and are time-consuming and labor intensive measurements. In addition, the granules must be dry prior to analysis. P&G [73] collect detergent granules in a cup of known size, weigh them and use a bulk density correlation to estimate the density of the granules – they then manipulate the liquid binder flowrate at the start of the process. New, fast and efficient granule density sensors are yet to be developed.

4.6 Definitions, Abbreviations and Symbols

FBRM	focused-beam reflectance method
NIR	near-infrared
P&G	Procter & Gamble

h	liquid layer thickness
h_a	height of surface asperities
s	pore saturation level
St_{def}	Stokes' deformation number
St_v	viscous Stokes' number
St_v^*	critical viscous Stokes' number
t_p	drop penetration time
Φ_p	dimensionless particle coating number
Ψ_a	dimensionless spray flux

References

1. Abberger, T., Seo, A., Schæfer, T.: The effect of droplet size and powder particle size on the mechanisms of nucleation and growth in fluid bed melt agglomeration. Int. J. Pharm. **249**, 185–197 (2002)
2. Ash, M., Ash, I.: Handbook of Food Additives. Synapse Information Resources, Endicott (2008)
3. Ash, M., Ash, I.: Handbook of Pharmaceutical Additives. Synapse Information Resources, Endicott (2007)
4. Aulton, M.E., Banks, M.: Fluidised bed granulation – factors influencing the quality of the product. Int. J. Pharm. Technol. Prod. Manuf. **2**(4), 24–29 (1981)
5. Bayly, A.E., et al.: Detergent processing. In: Zoller, U., Sosis, P. (eds.) Handbook of Detergents, Part F: Production. Taylor & Francis, Boca Raton (2008)
6. Boerefijn, R., Dontula, P.-R., Kohlus, R.: Detergent granulation. In: Salman, A.D., Hounslow, M.J., Seville, J.P.K. (eds.) Granulation. Elsevier Science B.V., Amsterdam (2008)
7. Boerefijn, R., Hounslow, M.J.: Studies of fluid bed granulation in an industrial R&D context. Chem. Eng. Sci. **60**(14), 3879–3890 (2005)
8. Butensky, M., Hyman, D.: Rotary drum granulation. An experimental study of the factors affecting granule size. Ind. Eng. Chem. **10**(2), 212–219 (1971)
9. Campanati, M., Fornasari, G., Vaccari, A.: Fundamentals in the preparation of heterogeneous catalysts. Catal. Today **77**(4), 299–314 (2003)
10. Çelik, M.: Pharmaceutical Powder Compaction Technology, 2nd edn. CRC Press, Boca Raton (2011)

11. Dadkhah, M., Peglow, M., Tsotsas, E.: Characterization of the internal morphology of agglomerates produced in a spray fluidized bed by X-ray tomography. Powder Technol. **228**, 349–358 (2012)
12. Degreve, J., et al.: Spray-agglomeration of NPK-fertilizer in a rotating drum granulator. Powder Technol. **163**(3), 188–195 (2006)
13. Deng, S.G., Lin, Y.S.: Granulation of sol-gel-derived nanostructured alumina. AIChE J. **43** (2), 505–514 (1997)
14. Van den Dries, K., et al.: Granule breakage phenomena in a high shear mixer; influence of process and formulation variables and consequences on granule homogeneity. Powder Technol. **113**, 228–236 (2003)
15. Eckhard, S., Nebelung, M.: Investigations of the correlation between granule structure and deformation behavior. Powder Technol. **206**(1–2), 79–87 (2011)
16. Emady, H.N., et al.: Granule formation mechanisms and morphology from single drop impact on powder beds. Powder Technol. **212**(1), 69–79 (2011)
17. Emady, H.N., Kayrak-Talay, D., Litster, J.D.: A regime map for granule formation by drop impact on powder beds. AIChE J. **59**(1), 96–107 (2013)
18. Emady, H.N., Kayrak-Talay, D., Litster, J.D.: Modeling the granule formation mechanism from single drop impact on a powder bed. J. Colloid Interface Sci. **393**, 369–376 (2013)
19. Ennis, B.J., Tardos, G., Pfeffer, R.: A microlevel-based characterization of granulation phenomena. Powder Technol. **65**(1–3), 257–272 (1991)
20. Farber, L., et al.: Unified compaction curve model for tensile strength of tablets made by roller compaction and direct compression. Int. J. Pharm. **346**(1–2), 17–24 (2008)
21. Farber, L., Tardos, G., Michaels, J.N.: Use of X-ray tomography to study the porosity and morphology of granules. Powder Technol. **132**(1), 57–63 (2003)
22. Faure, A., York, P., Rowe, R.C.: Process control and scale-up of pharmaceutical wet granulation processes: a review. Eur. J. Pharm. Biopharm. **52**(3), 269–277 (2001)
23. Franceschinis, E., et al.: High shear mixer granulation using food grade binders with different thickening power. Food Res. Int. **64**, 711–717 (2014)
24. Fries, L., et al.: DEM–CFD modeling of a fluidized bed spray granulator. Chem. Eng. Sci. **66** (11), 2340–2355 (2011)
25. Glaser, T., et al.: Model predictive control of continuous drum granulation. J. Process Control **19**(4), 615–622 (2009)
26. Hapgood, K., Litster, J.D., Smith, R.: Nucleation regime map for liquid bound granules. AIChE J. **49**(2), 350–361 (2003)
27. Hapgood, K.P., et al.: Drop penetration into porous powder beds. J. Colloid Interface Sci. **253** (2), 353–366 (2002)
28. Hapgood, K.P., et al.: Improving liquid distribution by reducing dimensionless spray flux in wet granulation—a pharmaceutical manufacturing case study. Chem. Eng. J. **164**(2–3), 340–349 (2010)
29. Hart, A., Wu, C.Y.: The impact of dry granulation on detergent powder properties. In: Particulate Materials: Synthesis, Characterisation, Processing and Modelling, pp. 102–110. The Royal Society of Chemistry, UK (2012)
30. Hassanpour, A., et al.: Effect of granulation scale-up on the strength of granules. Powder Technol. **189**, 304–312 (2009)
31. Heim, A., Gluba, T., Obraniak, A.: The effect of the wetting droplets size on power consumption during drum granulation. Granul. Matter **6**(2–3), 137–143 (2004)
32. Hemati, M., et al.: Fluidized bed coating and granulation: influence of process-related variables and physicochemical properties on the growth kinetics. Powder Technol. **130**(1–3), 18–34 (2003)
33. Holm, P.: Effect of impeller and chopper design on granulation in a high speed mixer. Drug Dev. Ind. Pharm. **13**(9–11), 1675–1701 (1987)
34. Holm, P., et al.: Granulation in high speed mixers. Part 1. Effects of process variables during kneading. Pharm. Ind. **45**(8), 806–811 (1983)

35. Holt, E.M.: The properties and forming of catalysts and absorbents by granulation. Powder Technol. **140**(3), 194–202 (2004)
36. Hsu, S.-H., Reklaitis, G., Venkatasubramania, V.: Modeling and control of roller compaction for pharmaceutical manufacturing. J. Pharm. Innov. **5**(1–2), 24–36 (2010)
37. Hu, X., Cunningham, J.C., Winstead, D.: Study growth kinetics in fluidized bed granulation with at-line FBRM. Int. J. Pharm. **347**(1–2), 54–61 (2008)
38. Huang, J., et al.: A PAT approach to improve process understanding of high shear wet granulation through in-line particle measurement using FBRM C35. J. Pharm. Sci. **99**(7), 3205–3212 (2010)
39. Ingram, A., et al.: Axial and radial dispersion in rolling mode rotating drums. Powder Technol. **158**(1–3), 76–91 (2005)
40. Islam, A., et al.: Synthesis and characterization of millimetric gamma alumina spherical particles by oil drop granulation method. J. Porous. Mater. **19**(5), 807–817 (2011)
41. Iveson, S.M., Wauters, P.A.L., et al.: Growth regime map for liquid-bound granules: further development and experimental validation. Powder Technol. **117**(1–2), 83–97 (2001)
42. Iveson, S.M., Litster, J.D., et al.: Nucleation, growth and breakage phenomena in agitated wet granulation processes: a review. Powder Technol. **117**(1–2), 3–39 (2001)
43. Iveson, S.M., Beathe, J.A., Page, N.W.: The dynamic strength of partially saturated powder compacts: the effect of liquid properties. Powder Technol. **127**, 149–161 (2002)
44. Iveson, S.M., Litster, J.D.: Fundamental studies of granule consolidation part 2: quantifying the effects of particle and binder properties. Powder Technol. **99**(3), 243–250 (1998)
45. Iveson, S.M., Litster, J.D.: Growth regime map for liquid-bound granules. AIChE J. **44**(7), 1510–1518 (1998)
46. Iveson, S.M., Litster, J.D., Ennis, B.J.: Fundamental studies of granule consolidation Part 1: Effects of binder content and binder viscosity. Powder Technol. **88**(1), 15–20 (1996)
47. Iveson, S.M., Page, N.W.: Brittle to plastic transition in the dynamic mechanical behavior of partially saturated granular materials. J. Appl. Mech. – Trans. ASME **71**, 470–475 (2004)
48. Iveson, S.M., Page, N.W.: Dynamic strength of liquid-bound granular materials: the effect of particle size and shape. Powder Technol. **152**(1–3), 79–89 (2005)
49. Johansen, A., Schæfer, T.: Effects of interactions between powder particle size and binder viscosity on agglomerate growth mechanisms in a high shear mixer. Eur. J. Pharm. Sci. **12**, 297–309 (2001)
50. Kapur, P.C.: Balling and granulation. Adv. Chem. Eng. **10**, 55–123 (1978)
51. Kariuki, W.I.J., et al.: Distribution nucleation: quantifying liquid distribution on the particle surface using the dimensionless particle coating number. Chem. Eng. Sci. **92**, 134–145 (2013)
52. Kayrak-Talay, D., et al.: Quality by design for wet granulation in pharmaceutical processing: assessing models for a priori design and scaling. Powder Technol. **240**, 7–18 (2013)
53. Khadilkar, A., Rozelle, P.L., Pisupati, S.V.: Models of agglomerate growth in fluidized bed reactors: critical review, status and applications. Powder Technol. **264**, 216–228 (2014)
54. Kleinebudde, P.: Roll compaction/dry granulation: pharmaceutical applications. Eur. J. Pharm. BioPharm. **58**(2), 317–326 (2004)
55. Knight, P.C., et al.: An investigation of the effects on agglomeration of changing the speed of a mechanical mixer. Powder Technol. **110**, 204–209 (2000)
56. Kumar, A., et al.: Model-based analysis of high shear wet granulation from batch to continuous processes in pharmaceutical production – a critical review. Eur. J. Pharm. Biopharm. **85**(3 Pt B), 814–832 (2013)
57. Kunii, D., Levenspiel, O.: Ch. 3 – Fluidization and mapping of regimes. In: Levenspiel, D.K. (ed.) Fluidization Engineering, 2nd edn, pp. 61–94. Butterworth-Heinemann, Boston (1991)
58. Le, P.K., et al.: A microscopic study of granulation mechanisms and their effect on granule properties. Powder Technol. **206**(1–2), 18–24 (2011)
59. Li, P., et al.: Effect of granulation on the activity and stability of a Co–Al2O3 aerogel catalyst in a fluidized-bed reactor for CH4–CO2 reforming. RSC Adv. **3**(23), 8939 (2013)

60. Litster, J.D., et al.: Liquid distribution in wet granulation: dimensionless spray flux. Powder Technol. **114**(1–3), 32–39 (2001)
61. Litster, J.D., et al.: Scale-up of mixer granulators for effective liquid distribution. Powder Technol. **124**(3), 272–280 (2002)
62. Litster, J., Ennis, B., Lian, L.: The Science and Engineering of Granulation Processes. Kluwer Academic, Dordrecht (2004)
63. Liu, L.X., et al.: Coalescence of deformable granules in wet granulation processes. AIChE J. **46**(3), 529–539 (2000)
64. Liu, L.X., Smith, R., Litster, J.D.: Wet granule breakage in a breakage only high-shear mixer: effect of formulation properties on breakage behaviour. Powder Technol. **189**(2), 158–164 (2009)
65. Liu, Y., et al.: Granulation processing parameters on the mechanical properties of diatomite-based porous granulates. Powder Technol. **263**, 159–167 (2014)
66. Merkus, H., Meesters, G.H.: In: Merkus, H.G., Meesters, G.M.H. (eds.) Particulate Products: Tailoring Properties for Optimal Performance. Springer International Publishing, Switzerland (2014)
67. Merkus, H.G.: Particle Size Measurements: Fundamentals, Practice, Quality. Springer Science+Business Media B.V., Dordrecht (2009)
68. Michaels, J.N., et al.: Steady states in granulation of pharmaceutical powders with application to scale-up. Powder Technol. **189**(2), 295–303 (2009)
69. Michels, N.-L., et al.: Hierarchically structured zeolite bodies: assembling micro-, meso-, and macroporosity levels in complex materials with enhanced properties. Adv. Funct. Mater. **22** (12), 2509–2518 (2012)
70. Mitchell, S., Michels, N.-L., Pérez-Ramírez, J.: From powder to technical body: the undervalued science of catalyst scale up. Chem. Soc. Rev. **42**(14), 6094–6112 (2013)
71. Mort, P.R.: Scale-up and control of binder agglomeration processes – flow and stress fields. Powder Technol. **189**(2), 313–317 (2009)
72. Mort, P.R.: Scale-up of binder agglomeration processes. Powder Technol. **150**(2), 86–103 (2005)
73. Mort, P.R., Capeci, S.W., Holder, J.W.: Control of agglomerate attributes in a continuous binder-agglomeration process. Powder Technol. **117**(1–2), 173–176 (2001)
74. Newitt, D.M., Conway-Jones, J.M.: A contribution to the theory and practice of granulation. Trans. Inst. Chem. Eng. **36**, 422–442 (1958)
75. Nguyen, T., Shen, W., Hapgood, K.: Drop penetration time in heterogeneous powder beds. Chem. Eng. Sci. **64**(24), 5210–5221 (2009)
76. Nguyen, T.H., Morton, D.A.V., Hapgood, K.P.: Application of the unified compaction curve to link wet granulation and tablet compaction behaviour. Powder Technol. **240**, 103–115 (2013)
77. Niazi, S.K.: Handbook of Pharmaceutical Manufacturing Formulations. Informa Healthcare, Boca Raton (2004)
78. Palzer, S.: Agglomeration of dehydrated consumer foods. In: Salman, A.D., Hounslow, M.J., Seville, J.P.K. (eds.) Granulation, pp. 591–671. Elsevier, Amsterdam (2007)
79. Palzer, S.: Agglomeration of pharmaceutical, detergent, chemical and food powders—similarities and differences of materials and processes. Powder Technol. **206**(1–2), 2–17 (2011)
80. Parikh, D.M. (ed.): Handbook of Pharmaceutical Granulation Technology, 2nd edn. Taylor & Francis, Boca Raton (2005)
81. Parker, D.J., et al.: Positron imaging studies of rotating drums. Can. J. Chem. Eng. **83**(1), 83–87 (2005)
82. Pathare, P.B., Byrne, E.P.: Application of wet granulation processes for granola breakfast cereal production. Food Eng. Rev. **3**, 189–201 (2011)
83. Pearson, J.K.M., Hounslow, M.J., Instone, T.: Tracer studies of high-shear granulation I: experimental results. AIChE **47**(9), 1978–1983 (2001)

84. Poon, J.M.H., et al.: Experimental validation studies on a multi-dimensional and multi-scale population balance model of batch granulation. Chem. Eng. Sci. **64**(4), 775–786 (2009)
85. Van Puyvelde, D.R., et al.: Experimental determination of transverse mixing kinetics in a rolling drum by image analysis. Powder Technol. **106**(3), 183–191 (1999)
86. Rahmanian, N., et al.: Characterisation of granule structure and strength made in a high shear granulator. Powder Technol. **192**(2), 184–194 (2009)
87. Rahmanian, N., Ghadiri, M., Ding, Y.: Effect of scale of operation on granule strength in high shear granulators. Chem. Eng. Sci. **63**(4), 915–923 (2008)
88. Ramaker, J.S., et al.: Scale-down of a high shear pelletisation process: flow profile and growth kinetics. Int. J. Pharm. **166**, 89–97 (1998)
89. Reynolds, G.K., et al.: Direct measurement of surface granular temperature in a high shear granulator. Powder Technol. **182**(2), 211–217 (2008)
90. Rowe, R.C., Sheskey, P.J., Quinn, M.: Handbook of Pharmaceutical Excipients, 6th edn. Pharmaceutical Press, London (2009)
91. Salman, A.D., Hounslow, M.J., Seville, J.P.K.: Granulation. Elsevier Science B.V., Amsterdam (2008)
92. Sanders, C.F.W., Hounslow, M.J., Doyle III, F.J.: Identification of models for control of wet granulation. Powder Technol. **188**(3), 255–263 (2009)
93. Schaafsma, S.H., et al.: Description of agglomerate growth. Powder Technol. **97**(3), 183–190 (1998)
94. Schaafsma, S.H., et al.: Effects and control of humidity and particle mixing in fluid-bed granulation. AIChE J. **45**(6), 1202–1210 (1999)
95. Schaafsma, S.H., Vonk, P., Kossen, N.W.F.: Fluid bed agglomeration with a narrow droplet size distribution. Int. J. Pharm. **193**(2), 175–187 (2000)
96. Schæfer, T., Holm, P., Kristensen, H.G.: Melt pelletization in a high shear mixer I. Effects of process variables and binder. Acta Pharm. Nord. **4**(3), 133–140 (1992)
97. Schæfer, T., Wørts, O.: Control of fluidised bed granulation II: estimation of droplet size of atomised binder solutions. Arch. Pharm. Chem **5**, 178–193 (1977)
98. Schæfer, T., Wørts, O.: Control of fluidised bed granulation III: effects of inlet air temperature and liquid flow rate on granule size and size distribution. Control of moisture content in the drying phase. Arch. Pharm. Chem. **6**, 1–13 (1978)
99. Schmidt, F., et al.: Novel composite spherical granulates with catalytic outer layer and improved conversion efficiency and selectivity. Chem. Eng. Technol. **35**(4), 769–775 (2012)
100. Seo, A., Holm, P., Schaefer, T.: Effects of droplet size and type of binder on the agglomerate growth mechanisms by melt agglomeration in a fluidised bed. Eur. J. Pharm. Sci. **16**(3), 95–105 (2002)
101. Seville, J.P.K.: Chapter 22 Fluidisation of cohesive particles. In: Salman, M.J.H.A.D, Seville, J.P.K. (eds.) Handbook of Powder Technology, pp. 1041–1069. Elsevier Science B.V., Amsterdam (2007)
102. Showell, M.: Handbook of Detergents. In: Showell, M. (ed.) Formulation, p. 128 (2005)
103. Sistare, F., Berry, L.S.P., Mojica, C.A.: Process analytical technology: an investment in process knowledge. Org. Process Res. Dev. **9**(3), 332–336 (2005)
104. Smith, R.M., Litster, J.D.: Examining the failure modes of wet granular materials using dynamic diametrical compression. Powder Technol. **224**, 189–195 (2012)
105. Smith, R.M., Litster, J.D., Howes, T.: Wet granule breakage in high shear mixer granulators. PhD Thesis. Brisbane: The University of Queensland (2008)
106. Smith, R.M., Liu, L.A.X., Litster, J.D.: Breakage of drop nucleated granules in a breakage only high shear mixer. Chem. Eng. Sci. **65**(21), 5651–5657 (2010)
107. Tangboriboon, N., et al.: Ceramic granules forming from calcium sodium aluminosilicate and carboxymethyl cellulose. J. Ceram. Process. Res. **14**(6), 658–666 (2013)
108. Teng, Y., Qiu, Z., Wen, H.: Systematical approach of formulation and process development using roller compaction. Eur. J. Pharm. BioPharm. **73**(2), 219–229 (2009)

109. Tho, I., Bauer-Brandl, A.: Quality by design (QbD) approaches for the compression step of tableting. Expert Opin. Drug Deliv. **8**(12), 1631–1644 (2011)
110. Tok, A., et al.: Monitoring granulation rate processes using three PAT tools in a pilot-scale fluidized bed. AAPS PharmSciTech **9**(4), 1083–1091 (2008)
111. Vonk, P., et al.: Growth mechanisms of high-shear pelletisation. Int. J. Pharm. **157**, 93–102 (1997)
112. Waldie, B.: Growth mechanism and the dependence of granule size on drop size in fluidised bed granulation. Chem. Eng. Sci. **46**(11), 2781–2785 (1991)
113. Walker, G.M., et al.: Drum granulation of NPK fertilizers. Powder Technol. **107**(3), 282–288 (2000)
114. Wang, F.Y., Cameron, I.T.: Review and future directions in the modelling and control of continuous drum granulation. Powder Technol. **124**(3), 238–253 (2002)
115. Watano, S.: Direct control of wet granulation processes by image processing system. Powder Technol. **117**(1–2), 163–172 (2001)
116. Watano, S., et al.: Measurement of moisture content by IR sensor in fluidized bed granulation: effects of operating variables on the relationship between granule moisture content and absorbance of IR spectra. Chem. Pharm. Bull. **44**, 1267–1269 (1996)
117. Watano, S., Miyanami, K.: Image processing for on-line monitoring of granule size distribution and shape in fluidized bed granulation. Powder Technol. **83**(1), 55–60 (1995)
118. Watano, S., Takashima, H., Miyanami, K.: Scale-up of agitation fluidized bed granulation. V. Effect of moisture content on scale-up characteristics. Chem. Pharm. Bull. **45**(4), 710–714 (1997)
119. Watano, S., Yamamoto, A., Miyanami, K.: Effects of operational variables on the properties of granules prepared by moisture control method in tumbling fluidised bed granulation. Chem. Pharm. Bull. **42**(1), 133–137 (1994)
120. Wauters, P.A.L., et al.: Growth and compaction behaviour of copper concentrate granules in a rotating drum. Powder Technol. **124**(3), 230–237 (2002)
121. Yu, W., Erickson, K.: Chord length characterization using focused beam reflectance measurement probe – methodologies and pitfalls. Powder Technol. **185**(1), 24–30 (2008)

Chapter 5
Particulate Flow and Agglomeration in Food Extrusion

Sajid Alavi and R.P. Kingsly Ambrose

Abstract The fundamental principles of extrusion, as applied to food processing, are described. Basic extrusion theory, hardware, various commercial uses, and role of process and design parameters in specific food applications are provided. Innovations in food extrusion technology are enabling its rapid expansion and applicability in diverse areas related to bioprocessing and value addition. Ongoing research spans various cutting edge areas including nutrition and nanotechnology. As an example of ongoing extrusion research that aligns with the main theme of this book, a recent study is described that relates raw material particulate rheology to the granular flow regime in a single screw food extruder. Corn meal and corn flour were used as model particulate systems for the study. Various particulate-scale characteristics and flow parameters of these two materials were determined using a powder rheometer, which is a promising new off-line tool. Properties such as basic flow energy (10.3 mJ/g for corn flour versus 8.7 mJ/g for corn meal), specific energy (7.1 mJ/g versus 4.4 mJ/g), cohesion (1.9 kPa versus 0.8 kPa) and flow function (3.4 versus 10.2) were good indicators of flowability in an extruder. These results were validated in a particulate flow study involving the two model systems in a pilot-scale single screw extruder. Visualization data, obtained using a transparent Plexiglas window, confirmed that corn flour exhibited much higher flow impedance as compared to corn meal during extrusion. It was concluded that powder rheometry is a potentially valuable tool for design of screw metallurgy and geometry, although further studies are needed in this area.

S. Alavi (✉)
Grain Science and Industry, Kansas State University, Manhattan, KS, USA
e-mail: salavi@ksu.edu

R.P. Kingsly Ambrose
Agricultural and Biological Engineering, Purdue University, West Lafayette, IN, USA
e-mail: rambrose@purdue.edu

© Springer International Publishing Switzerland 2016
H.G. Merkus, G.M.H. Meesters (eds.), *Production, Handling and Characterization of Particulate Materials*, Particle Technology Series 25,
DOI 10.1007/978-3-319-20949-4_5

5.1 Introduction

The technology of extrusion is centuries old. In fact, the first known application of the principle of extrusion was more than 2000 years ago, with the invention of Archimedes' screw for pumping water in the third century B.C. It was only in the late eighteenth century that extrusion was applied in the form of a piston press to produce lead pipes, and in form of a single screw extruder in the late nineteenth century for processing rubber. In the 1930s, twin screw extrusion was first used for processing plastics. Thus in the early years of extrusion, industrial applications were predominant, and even now extrusion plays a big role in processing metals, rubber and plastics. The high pressure, temperature and shear based continuous processing capability of extrusion, which has made it an attractive manufacturing technology, was adapted to food applications only over the last few decades. The use of extrusion in food processing, which has continued to evolve and expand rapidly, is the focus of this chapter. The starting materials in food extrusion are usually, though not always, grain-based and particulate in nature (examples: cereal and legume flours, starches, protein fractions and concentrates, fibrous ingredients, vitamin and mineral premixes, etc.) and the final output an agglomerated product (examples: snacks, breakfast cereal, pasta, micronutrient fortified rice, etc.) [7]. An overview of various products is provided in Fig. 5.1. This transformation from particulate to agglomerated form is very complex and involves several overlapping phenomena such as conveying, mixing, kneading, pressurization, cooking, shaping, decompression, expansion and dehydration. The heterogeneous and biopolymeric nature of ingredients only adds to the complexity.

Scientists have employed various fundamental and applied scientific principles and tools to develop some understanding of the food extrusion process, including heat, mass and momentum transfer, rheology, polymer science, and biochemistry [5]. However, there are vast gaps in knowledge and immense potential for refinement and advancement of this technology for use in the fast growing global commercial space for value-addition, nutrition and food security. One such area that needs detailed study is the flow of particulate material in the extruder prior to transformation to a fluidized melt. This chapter describes in detail the characterization of raw materials and their flow in a pilot-scale extrusion system, from this point of view. Also provided are a primer on the basics principles of extrusion processing applied to food systems, a summary of the transformations that biopolymers such as starch and protein undergo, a description of hardware, operations and process parameters employed in modern food extruders, and details of various commercial applications.

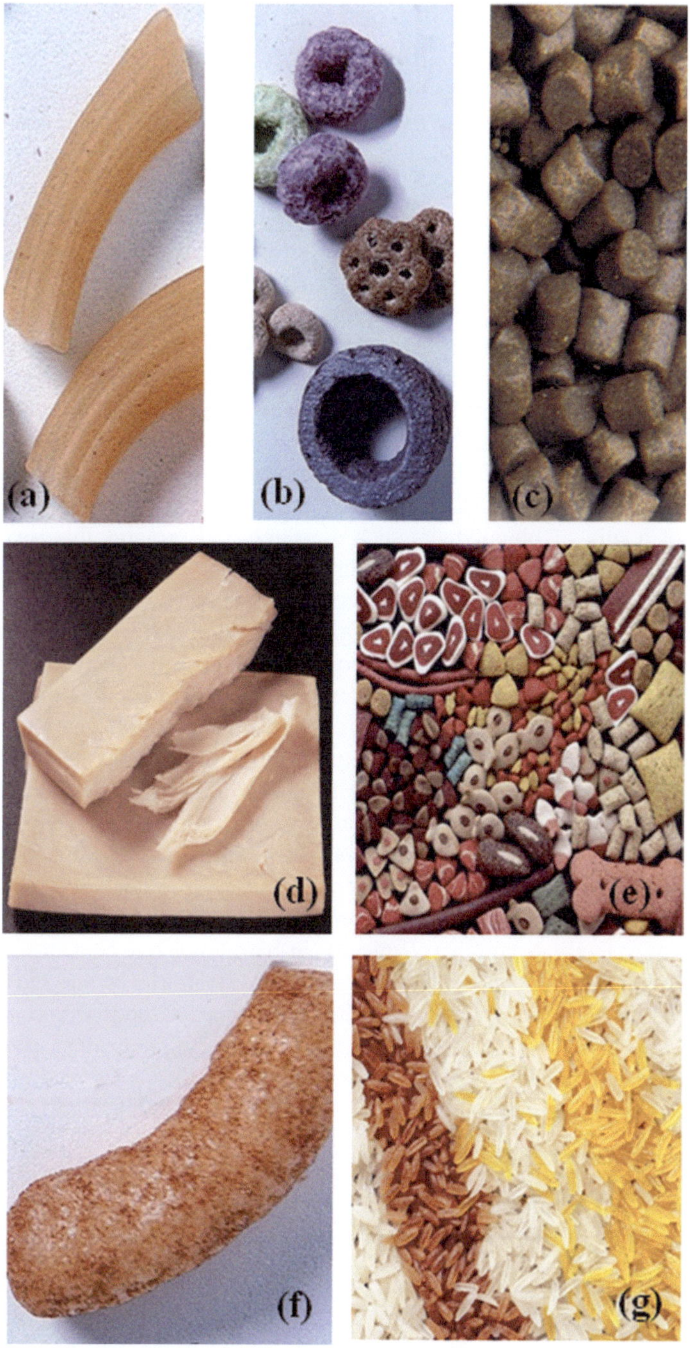

Fig. 5.1 Examples of extruded food products – (**a**) pasta, (**b**) breakfast cereal, (**c**) fish feed, (**d**) texturized vegetable protein, (**e**) pet food, (**f**) snacks and (**g**) fortified rice (Pictures courtesy Wenger Manufacturing, Sabetha, Kansas, USA)

5.2 Basics of Extrusion in Food Applications

The technology of extrusion was first applied to processing of food products in the 1930s in the form of single-screw extrusion processing of pasta (pasta press), ready-to-eat (RTE) cereals, and expanded corn curls. In the 1950s dry expanded pet food was added to the range of products processed using extrusion, and in the 1960s the application of extrusion was extended to a variety of expanded breakfast cereal, texturized plant protein products and even pre-gelatinized starches, leading to its wide acceptance as a major food processing technology.

Extrusion is used for processing various raw materials based on grains ranging from corn, wheat and rice to sorghum, oats, soybeans and other legumes. These raw materials can either be in the form of whole grain flours, de-hulled flours, starches, gluten proteins or bran. Ingredients based on meat, dairy, lipids, sugars and gums are also very common. Vitamins, minerals and other 'minor' additives are often used for micronutrient fortification and functional purposes such as flavor, color, anti-caking action and shelf life. Typically the bulk of any extruded food matrix consists of a mix of biopolymers such as starch and proteins that serve as the continuous phase but can also exist in dispersed phase depending on the product, nature of the ingredients and extent of processing.

From a very basic point-of-view, extrusion (derived from the Latin root 'extrudere' meaning 'to thrust') means to force and push out material or to shape by forcing through a die. However modern-day food extruders not only transport ingredients and shape a product using a combination of screws and dies, but also transform the raw materials from a loosely flowing, granular mass to a well-mixed, kneaded, cooked, fluidized biopolymeric melt that is often expanded or puffed on discharge. Majority of these transformations occur along the length of the extruder barrel in a very short amount of time (less than 2 min) as the screw continuously conveys the raw material forward and mixes it with water. The latter serves multiple and essential roles of plasticizer, puffing agent, and hydration of starch and proteins to facilitate cooking. Typically, in-barrel moisture content ranges from 15 to 30 % wet basis, depending on the application. Steam can be injected as well to provide thermal energy, which can be very important for foods with high protein content. However, most of the energy required for kneading, heating, cooking and expansion is delivered to the material by mechanical means (termed mechanical energy) and generated due to friction between the screw, the barrel and the material itself.

To better understand the physico-chemical transformations taking place inside the extruder barrel, it is often considered to be a series of ***processing zones*** depending on the state of the material inside (Fig. 5.2). These zones are:

1. ***Feeding or metering zone*** at the raw material inlet end, where the pressure is atmospheric, and the material is granular (with individual particle size typically <700 μm) and often loosely agglomerated due to surface moisture;

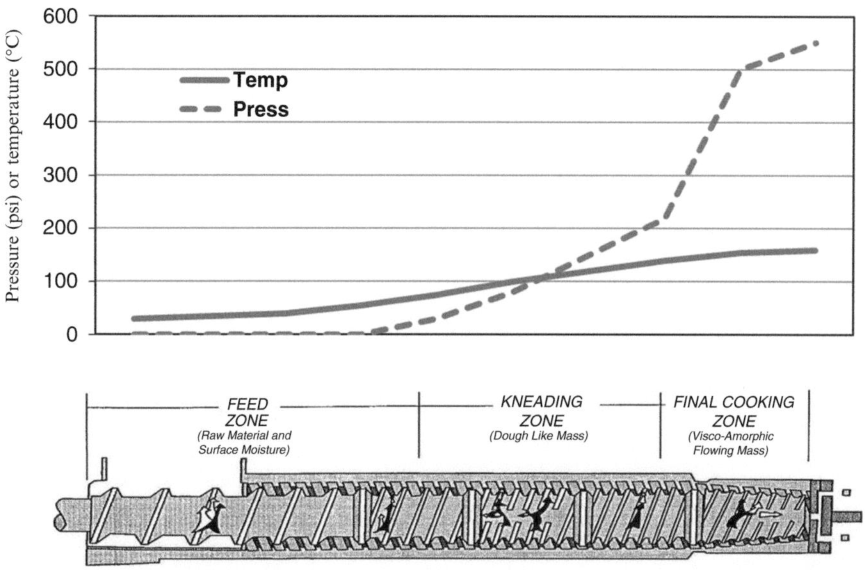

Fig. 5.2 Typical processing zones inside a food extruder [19] (Copyright Taylor & Francis/CRC Press; reproduced with permission)

2. ***Kneading or transition zone***, where compaction starts to occur, pressure and temperature rise, and the loosely flowing material transforms to a cohesive dough; and
3. ***Cooking zone***, typically at the discharge end where the temperature and pressure can be around 180 °C and 500 psi (~30 atm), respectively, or even higher depending on requirements.

The intense mechanical shear, in combination with high temperature and pressure in the cooking zone, allows the extruder to serve as a fast and continuous pressure cooking system. As the cooked, hot and pressurized dough exits the extruder die, the water flashes off to steam and makes the product puff or expand. This is how a highly expanded internal structure is achieved, which is common in various puffed snacks, breakfast cereal and pet foods. A typical puffed snack product might have as much as 85 % of void volume.

As a multi-faceted continuous processing technology, extrusion has several advantages over conventional batch processing methods. Unlike in batch-processing where several pieces of equipment might be needed to make the final product, the same extrusion equipment performs several functions including mixing and unitizing ingredients, cooking, forming of the product to the desired shape, expansion, texture alteration, sterilization and dehydration (due-to steam flash-off). There are literally no effluents in the extrusion process, unlike batch cooking using kettles. The low range of processing moistures implies that extrusion requires much less energy for cooking and also drying of the final product. Moreover, there is better control over the process and product quality, and much greater processing capacity (several tons per hour). These differences make the extrusion process much more cost efficient, and also sustainable from energy and water usage point of view.

5.3 Extrusion Hardware

Extrusion hardware has evolved considerably over the years. Technological advances have led to single and twin-screw extruders with segmented screws and barrels [13], further incremental improvements to hardware and design, highly effective continuous preconditioning methods, and computer-controlled and fully automated control systems. These developments have markedly improved the variety, nutrition, quality and safety of extruded foods.

A conventional food extrusion system has several parts including a ***feeding system***, a ***pre-conditioner***, followed by the 'heart' of the system comprising of a ***barrel*** enclosing one or two ***screws*** (usually rotating in the same direction or co-rotating), and finally a narrow opening or ***die and a knife***. The feeding system, consisting of a feed delivery screw, can be either volumetric or gravimetric. The latter tends to deliver feed with greater accuracy relative to the set-point for the feed rate. The pre-conditioning system adds water and steam to the floury ingredients and serves to pre-moisten and partially cook them. This helps in reducing the motor power required to drive the extruder screws, reduces abrasion of the screw and barrel and thus increases their service-life, and also improves the texture and flavor of the final product. The barrel is usually jacketed for circulation of steam or hot oil to supply thermal energy to the material. Barrels can also be electrically heated. The extruder screws are segmented and can be configured specifically for each application or process by changing the selection and sequence of screw and kneading elements (Fig. 5.2). The die at the discharge end serves to restrict flow and aids in increasing the pressure inside the extruder barrel. A motor-driven knife rotates flush with the die surface to continuously cut the product as it emerges from the extruder. The product is then transferred to a drying oven and if needed to post-processing equipment like drum coaters and enrobers for application of sugar, oil, flavors and/or seasoning as required, and finally packaged.

5.4 Particulate Ingredients to Agglomerated Products: Role of Process Parameters

Extrusion is a very versatile technology for processing food products. By altering the processing conditions, the screw profile or the die, a wide variety of ingredients can be processed via cooking, expansion and further down-stream processing into a range of products using the same equipment. In essence, the technology is a very effective means of transforming raw materials in particulate form into agglomerated products of various sizes, shapes, textures, composition and nutritional content. Extrusion process parameters, in combination with screw and die selection tailored to the application, allow a high degree of flexibility that is very unique in food processing. Some examples in commercial applications are provided below.

Pasta products are usually processed with minimal energy input (<100 kJ/kg) adequate for just kneading and forming, achieved by a combination of relatively high in-barrel moisture (>30 % wet basis), small screw diameter to channel height ratio ($D/H = 3-4$) and low screw speed (50–60 rpm). Intermediate products such as pellets, which can be expanded by frying to make third generation ***3G snacks*** or rolled to make ***flaked breakfast cereal***, require higher energy input (100–200 kJ/kg) to achieve complete cooking in addition to forming. This can be achieved by lowering the in-barrel moisture to ~25 % wet basis, increasing the screw speed (150–200 rpm) and D/H ratio (5–6), using thermal energy in the form of steam, and employing the more versatile twin screw extrusion technology or a combination of two extruders in series for cooking and forming the product. Ready-to-eat ***directly expanded snacks*** and ***puffed breakfast cereal*** require even higher energy input (200–300 kJ/kg), which is achieved by further lowering the in-barrel moisture to 15–18 % wet basis, and increasing the screw D/H (7–9) and speed (300–400 rpm). High protein applications, such as ***dry expanded pet food***, ***floating aquatic feed*** and ***texturized vegetable protein*** require large amount of processing energy input (>300 kJ/kg) and yet adequate moisture (25–30 % wet basis). This requires supplementation of mechanical energy input with equal or higher amounts of steam-based thermal energy in the preconditioner prior to the extruder. The extrusion system can similarly be adapted with relative ease for other applications, including ***infant cereal***, ***confectionery products***, ***animal feed and industrial products*** (example, ethanol from starch and lingo-cellulosic substrates). Due to this versatility and flexibility, today extruded products comprise of a multi-billion dollar market in the U.S. alone. New applications and products based on extrusion are being further researched and developed, and this will be described in Sect. 5.7.

5.5 Characterization of Particulate Rheology

Study of particulate flow and its impact on transport of materials in food delivery systems is of great significance [8]. This is also true for the extruder screw, especially in the feeding and transition zones, as described in the previous section. For this reason, the design and operation of food extruders is at least partly a function of particle-scale characteristics of ingredients and the resultant nature of particulate flow [9, 11, 14, 16, 17, 23]. An understanding of these relationships can aid in design of screw metallurgy and geometry that are tailored for higher operational stability and flexibility of food extruders. This was the overall goal of a preliminary study by our research group. Corn flour and corn meal were selected as model materials representing fine and coarse particulate systems. Particle-scale

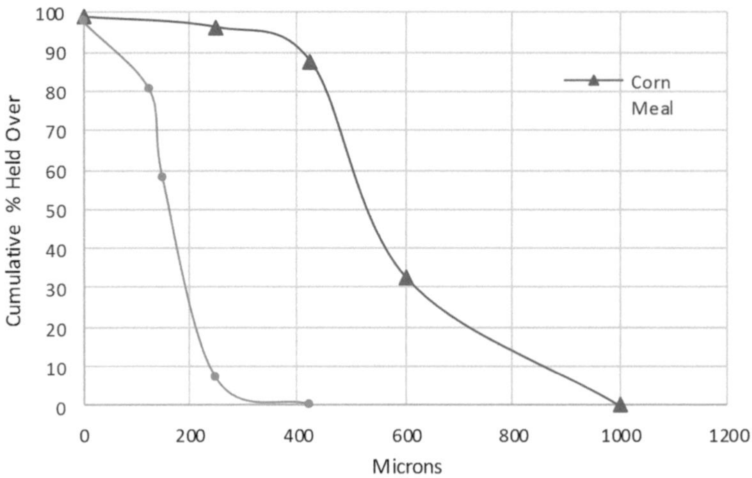

Fig. 5.3 Cumulative oversize particle size distribution of corn flour and corn meal

properties were measured using various techniques including powder rheometry, which characterizes flow parameters of granular materials in contact with a metal surface. These physical properties were then related to particulate flow in a pilot-scale single screw extruder.

The corn flour (de-germed yellow F75) and corn meal (de-germed medium yellow M53) used in the study were obtained from Agricor (Marion, IN). ***Particle size distribution*** was determined using a modified version of AACC 66-20.01 Ro-Tap Sieve Shaker standard method [1] developed by Stoerzinger [20] (Fig. 5.3). Average particle size was 155 μm and 490 μm for individual or primary particles of corn flour and corn meal, respectively.

Initial moisture content of the materials was determined using AOAC standard method 925.10 [4]. Particulate properties were tested using a powder rheometer (FT4, Freeman Technologies, Gloucestershire, UK) after adjusting the moisture to 20 % wet basis. The powder rheometer employs a relative new, unique and reproducible dynamic methodology for measurement of resistance to flow whilst the powder is in motion. It consists of a 48 mm blade or impeller designed for helical movement through the test sample confined in a 50 mm × 160 ml cylindrical vessel (Fig. 5.4a). In addition, the instrument has accessories for measuring shear strength, wall friction and compressibility. The standard ***dynamic test*** consisted of seven test cycles (tests 1–7) at a blade tip speed of 100 mm/s with anticlockwise motion while traversing through the granular material from top to bottom (Fig. 5.4b) and clockwise motion while retracting back to the top (Fig. 5.4c). These identical test cycles were performed to achieve stabilization of flow energy and characterize change in flow behavior due to attrition, agglomeration, segregation, etc. Subsequently four more test cycles were conducted (tests 8–11) with blade tip speed decreasing in gradual steps (100 mm/s, 70 mm/s, 40 mm/s and 10 mm/s, respectively), in order to evaluate sensitivity to different flow rates. Dynamic

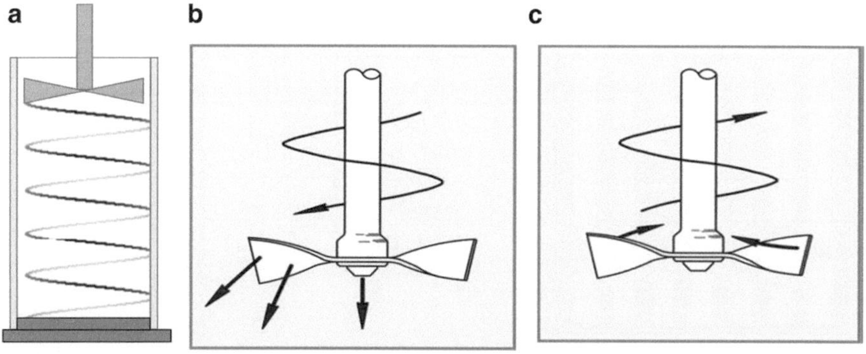

Fig. 5.4 (**a**) Powder rheometer set up showing the helical movement of the blade in the test vessel; (**b**) downward traverse of blade for confined testing; and (**c**) upward traverse of blade during unconfined testing

testing yields several parameters useful for characterizing flow of particulate systems.

The ***Basic Flow Energy*** (***BFE***) is the energy required per unit mass to move the blade through the powder column during confined or forced flow (from top to bottom) after conditioning to achieve stabilization (test 7), as calculated below.

$$BFE(J/g) = \frac{\int_0^{\Delta x} \left(T\dot{\theta} + Fv_x\right)\ v_x^{-1}\ dx}{m} \tag{5.1}$$

where

T = rotation resistance or torque experienced by the blade (N·m)
F = vertical resistance or force experienced by the blade (N)
$\dot{\theta}$ = angular speed of the blade (rad/s)
v_x = vertical speed of the blade (m/s)
Δx = vertical distance traversed by the blade
m = mass of test sample.

Specific Energy (***SE***) is the energy per unit mass needed to displace conditioned powder during unconfined or low stress flow (upwards testing), as described below.

$$SE(J/g) = \frac{(FE_6 + FE_7)/2}{m} \tag{5.2}$$

where

FE_6 = flow energy measured for upward traverse of blade after test 6
FE_7 = flow energy measured for upward traverse of blade after test 7.

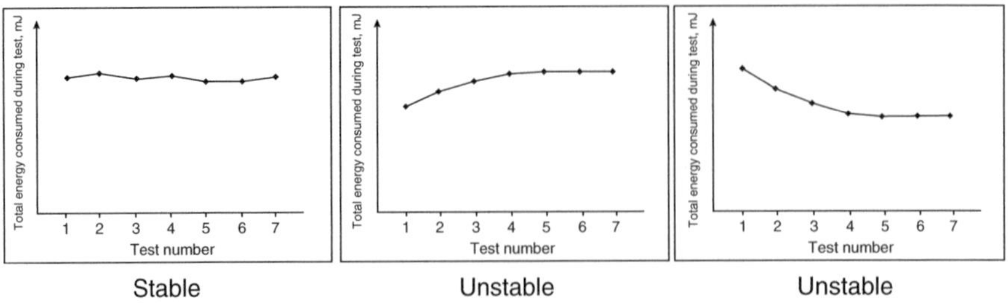

Fig. 5.5 Typical stability curves for powders

Stability Index (***SI***) is a measure of flow energy changes during repeated testing and is an indirect measure of the change in particulate compaction as the blade traverses the samples multiple times, and is calculated as follows.

$$SI = \frac{BFE_7}{BFE_1} \tag{5.3}$$

where

BFE_1 = flow energy during test 1
BFE_7 = flow energy during test 7.

A stable particulate system would have SI ~1.0. whereas unstable systems would have SI $\gg$1.0 or SI $\ll$1.0 (Fig. 5.5).

Flow Rate Index (***FRI***) is the factor by which flow energy is changed when the blade tip speed is reduced by a factor of 10. It evaluates the sensitivity of the powder to different flow rates, and is measured as follows.

$$FRI = \frac{BFE_{11}}{BFE_8} \tag{5.4}$$

where

BFE_8 = flow energy during test 8 with blade tip speed = 100 mm/s and
BFE_{11} = flow energy during test 11 with blade tip speed = 10 mm/s.

Dynamic flow parameters obtained for the two model particulate systems in this study, viz. corn flour and corn meal, are shown in Table 5.1.

The basic flow energy and specific energy data indicate that corn meal has better flow characteristics than corn flour, as the latter requires more energy to flow in both confined and unconfined environments. The lower particle size of corn flour, which allows higher compressibility as well as greater surface to surface electrostatic and other interactions, is a primary reason. Based on these data, corn flour and corn meal can be classified as particulate materials with 'moderate cohesion' and 'low cohesion', respectively. Teunou et al. [22] found that the more cohesive wheat flour had a low flow index of 2.71 while the more granular skim milk with average particle size above 200 μm had a flow index of 11.04 which made it more of a free

Table 5.1 Dynamic flow parameters for corn flour and corn meal

	BFE, mJ/g	SE, mJ/g	SI	FRI
Corn flour	10.28 ± 0.16^a	7.07 ± 0.10^a	1.03 ± 0.03^a	1.29 ± 0.02^a
Corn meal	8.73 ± 0.24^b	4.36 ± 0.05^b	1.02 ± 0.13^a	1.13 ± 0.02^b

N = 3 (replicates); different superscript letters within a column indicate significant difference at $\alpha = 0.05$

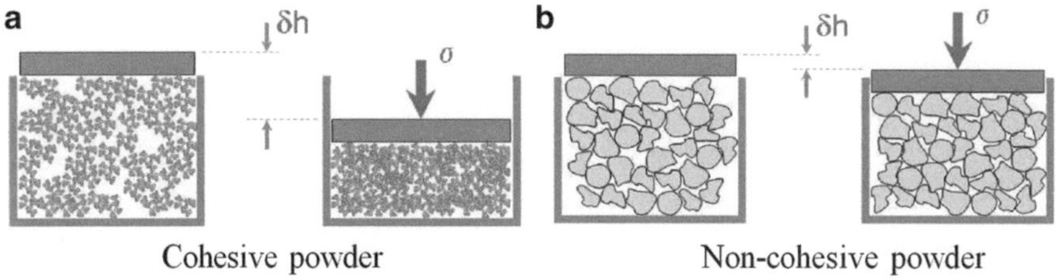

Fig. 5.6 Compressibility tests for powders

flowing powder. Flow index is a parameter derived from shear testing and is similar to flow function as described below in the Mohr circle analysis. Beside particle size other factors that might have contributed to the differences in cohesion and flowability include the surface chemistry and moisture of the wheat flour and skim milk particles. The test results from this study follow a similar trend, that is, the coarser corn meal flows easier than the finer corn flour. BFE is higher than SE for both materials, which is expected as higher stress experienced under confined systems will lead to higher cohesion and greater flow energy requirements. Both corn flour and corn meal have stable behavior, as their stability index is close to unity. This indicates that the energy required for bulk powder flow does not change significantly during repeated testing. FRI data indicate that the sensitivity of corn flour to variable impeller speeds is significantly higher than that of corn meal. This implies that mechanical disruption of surface to surface particulate interactions plays an important role during bulk flow.

After initial conditioning using the standard FT4 cylindrical vessel and blade set up, compression tests were carried out using a vented piston assembly that applied increasing normal stress on the samples (Fig. 5.6). The conditioned ***bulk density*** of corn flour and corn meal at zero stress was 461 and 573 kg/m^3, respectively. Compressibility is a measure of change in bulk density as a function of applied normal stress. Eleven compression steps were performed starting at 0.5 kPa and increasing with each step to a final stress of 22 kPa (0.5, 1, 2, 4, 6, 8, 10, 12, 15, 18 and 22 kPa). Each normal stress was applied for about 25 s to reach equilibrium at the target stress. ***Compressibility*** was calculated as a percentage change in volume, based on the distance travelled by the piston under applied normal stress. The compressibility data for corn flour and corn meal are shown in Fig. 5.7. Corn flour was more easily compacted than corn meal. Higher particle size of corn meal implied lesser surface area and reduced interfacial particulate interactions and less cohesion. On the other hand, cohesive materials, that have poor flow characteristics,

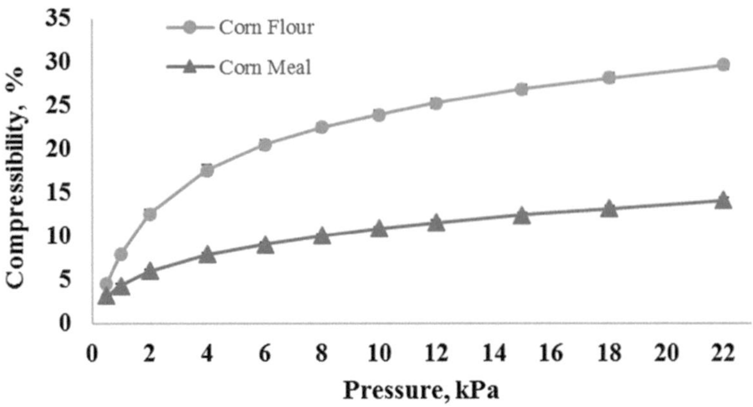

Fig. 5.7 Compressibility data for corn flour and corn meal

Fig. 5.8 Shear testing of powders using a rotational head attachment

are more compressible as can be seen in the case of corn flour. The lower bulk density of corn flour, and thus the presence of higher volume of air voids, is an important factor in its greater compressibility.

During handling and processing, powders can constantly be subjected to consolidation stresses that lead to changes in density and inter-particulate forces. To understand the yield point at which flow of powder initiates, ***shear testing*** of the powder was carried out under consolidation. A rotational shear head, which induced both vertical and rotational stresses, was used in a 50 mm sample holder vessel to conduct the test (Fig. 5.8). As the powder bed resists rotation of the shear head, the shear stress increases until the bed fails or shears, at which time a maximum shear stress was observed. The normal stress is maintained constant throughout the measurement. Mohr circle analysis (Fig. 5.9) was conducted on the yield loci

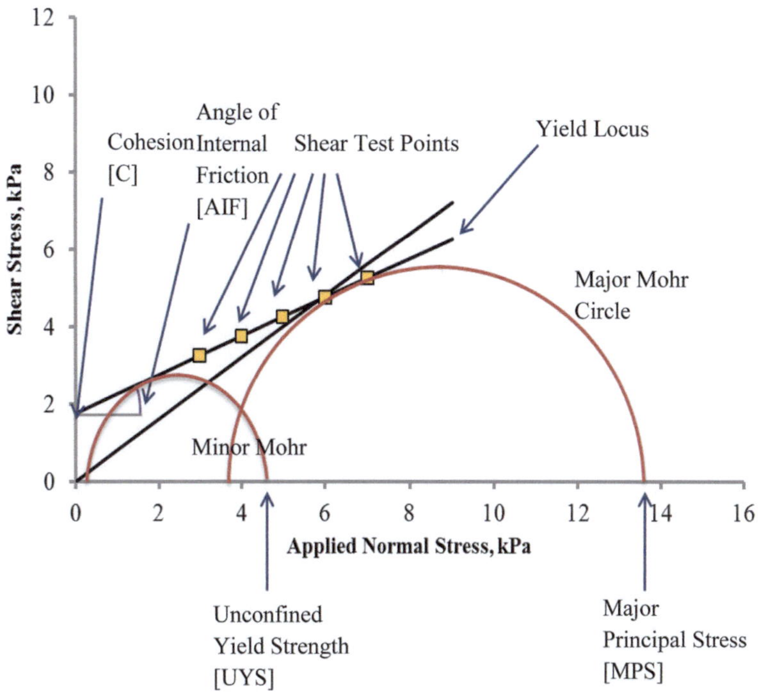

Fig. 5.9 Typical Mohr circle failure plot based on Jenike's theory

(relationship between applied and shear stress) to calculate the ***Cohesion*** (best fit line intercepting y-axis), ***Unconfined Yield Strength or UYS*** (the greater of the two values at which the smaller Mohr circle intercepts the x-axis), ***Flow Function or FF*** (major principal stress/unconfined yield strength), and ***Angle of Internal Friction or AIF*** (angle created by the best fit line with the horizontal axis). For more information on general powder tests, shear tests and Mohr circles see also Chaps. 13 and 15.

Wall friction measures the resistance to flow of powders, previously at rest, in relation to the process equipment surface. The FT4 measures wall friction experienced by particulate material in contact with a stainless steel wall, using a friction head attachment that induces both vertical and rotational stresses on the samples (Fig. 5.10). As the powder bed resists the rotation of the wall friction head, the torque increases until the resistance is eventually overcome. The wall friction head continues to rotate at a fixed velocity for a pre-determined period of time. The torque required to maintain this rotational momentum is measured and used to calculate a 'steady-state' shear stress. The normal stress is maintained constant throughout the measurement. From the relationship between normal stress (σ_w) and shear stress (τ_w), the wall friction angle (ϕ), is calculated using the following relationship:

Fig. 5.10 Wall friction measurement method

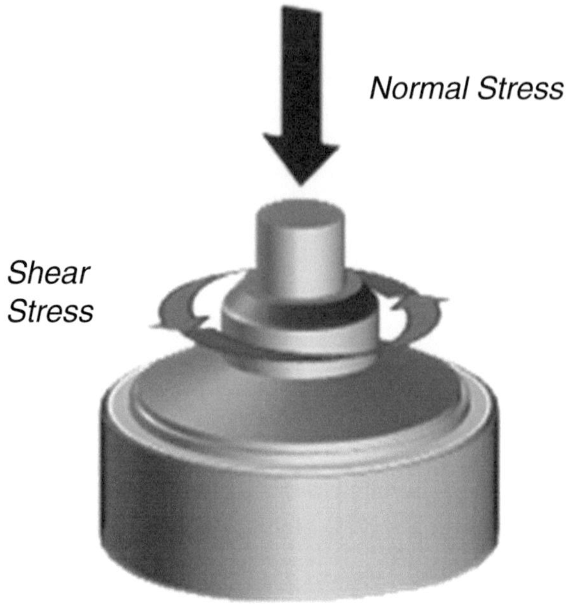

Table 5.2 Shear and wall friction properties of corn flour and corn meal

	Cohesion, kPa	UYS, kPa	FF	AIF, degrees	Φ, degrees
Corn flour	1.92 ± 0.18^{a}	6.07 ± 0.29^{a}	3.43 ± 0.27^{a}	35.09 ± 0.90^{a}	30.24 ± 0.49^{a}
Corn meal	0.75 ± 0.21^{b}	3.54 ± 0.43^{b}	10.16 ± 2.26^{b}	40.58 ± 0.95^{b}	31.30 ± 3.47^{a}

N = 3; different superscript letters within a column indicate significant difference at $\alpha = 0.05$

Table 5.3 Classification of particulate flowability based on flow function value

Type of flow	Flow function value
Easy-flowing	10< FF
Free-flowing	4< FF <10
Cohesive	2< FF <4
Very cohesive and non-flowing	FF <2

$$\phi = \tan^{-1}\left(\frac{\tau_W}{\sigma_W}\right) \tag{5.5}$$

Shear test and wall friction data are shown in Table 5.2. These results supported the dynamic flow property results. Both cohesion and UYS data indicated that corn flour was more cohesive than corn meal. The angle of internal friction and wall friction angle reflected the effect of inter-particle contacts and particle-wall contacts on flow of powders. The higher the angle of internal friction and wall friction angle, the easier is the particulate flow under shear stress in contact with each other and with a wall surface. The flowability function, derived from shear analyses, can be used to classify powder flowability as described in Table 5.3. This will be further discussed in Sect. 5.6.

5.6 Visualization of Particulate Flow During Extrusion

A pilot-scale X-20 single screw extruder (Wenger Manufacturing Inc., Sabetha, KS) was used for experimental study of flow of the two model particulate systems. The extruder screw diameter was 82.1 mm and *L:D* (length:diameter) ratio 8:1. The extruder screw configuration is shown in Fig. 5.11. One half of the extruder barrel was replaced by a clear Plexiglas window so the flow inside the extruder could be visualized. The barrel was not heated for this study. Also the discharge end of the barrel was open and without a die, thus no compression was achieved and the test material was at atmospheric pressure through the length of the barrel. A volumetric feeding system with feeder screw speed control was calibrated to deliver the test materials to the extruder at a feed rate of 103 kg/h after hydration in the preconditioner to 20 % wet basis to match the moisture level in offline tests.

Flow of the two materials was monitored visually at two extruder screw speeds, viz. 200 and 400 rpm through the observation window. Figure 5.12 is a schematic representation of the snapshots of the particulate flows observed at steady state. Fully filled sections of the barrel corresponded to maximum flow impedance. Restrictive elements tended to obstruct flow the most as observed in the form of material accumulation at the steam lock locations. For the same throughput and screw geometry, corn meal experienced much lower flow impedance than corn flour.

Measurements of filled length or the distance from the discharge end to the point where the barrel was completely full, once the extruder was operating at steady state, are shown in Table 5.4 (cf. [12]). At the same screw speed, filled length was 6–11 times higher for corn flour as compared to corn meal. Thus corn flour exhibited substantially lower flowability. The classification of particulate flowability based on flow function FF supported these results. Corn meal (FF = 10.16) is placed in the easy flowing category while corn flour (FF = 3.43) is in the cohesive or poor flow category. Increase in screw speed from 200 to 400 rpm led to 8 % and 50 % decrease in filled length for corn flour and corn meal. Thus, as expected, higher screw speed leads to better flowability in the extruder. Flow rate index FRI-values of greater than 1.0 for both corn meal and corn flour also represented greater flowability at higher impeller speeds during off-line testing of particulate rheology. The temperatures of the material at the inlet and discharge

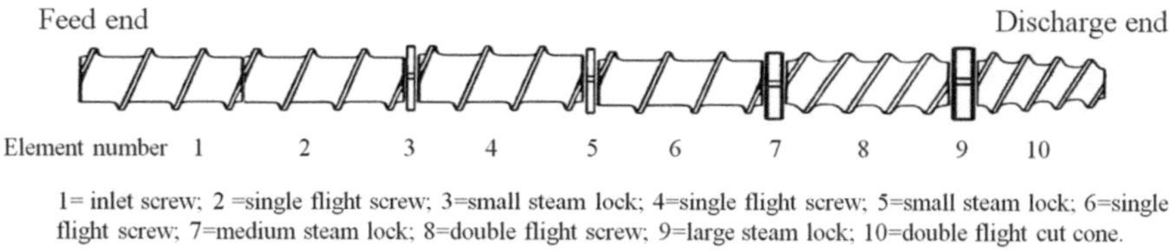

1= inlet screw; 2 =single flight screw; 3=small steam lock; 4=single flight screw; 5=small steam lock; 6=single flight screw; 7=medium steam lock; 8=double flight screw; 9=large steam lock; 10=double flight cut cone.

Fig. 5.11 Screw profile used in pilot-scale extrusion experiments for particulate flow

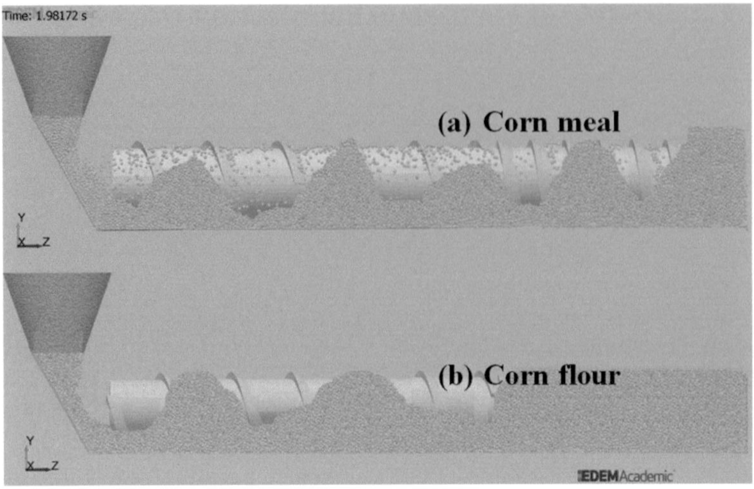

Fig. 5.12 Schematic representing experimental observation of unconfined particulate flow in a single screw pilot-scale extruder (Visualization videos can be accessed from the links http://youtu.be/g_6TWhizjNY for corn meal and http://youtu.be/ifRjF2_4HJw for corn flour)

Table 5.4 Extrusion flow parameters for corn flour and corn meal

	Screw speed (rpm)	Filled length (mm)	Discharge temp.[a] (°C)
Corn flour	200	317.5	26.7
	400	292.1	24.5
Corn meal	200	50.8	25.0
	400	25.4	24.2

[a]Material temperature at inlet was 21.6 °C for all treatments

were also measured. Due to the absence of a die at the discharge end, there was limited energy input and the temperature rise was in the range of 2.6–5.1 °C only (Table 5.4). The particulate rheology is better represented by specific energy SE, which is an unconfined flow parameter. SE for corn flour was higher and corresponded with a higher flow impedance and temperature rise during particulate flow in extruder as compared to corn meal.

5.7 Other Ongoing Research in Extrusion

The inherent economies of scale and process efficiency associated with extrusion, make it a highly viable technology for delivery of enhanced nutrition to huge populations, and addressing diverse health problems such as malnourishment and obesity. Current research is targeting higher nutritional value of food products not just for commercial market but also for addressing macro and micro nutrient deficiencies and food security challenges worldwide [2, 6, 7]. Research is also addressing industrial applications such as pre-treatment of lignocellulosic feedstock for fuel ethanol and development of bio-based nanocomposites for applications such as packaging [21, 24]. Extrusion-based agglomeration is also being investigated for developing high performance pelletized feed for livestock agriculture and aquaculture [3, 15]. Simulations with discrete element modeling (DEM) have also started to yield insight in the behavior of particulate products in extruders [10, 18].

5.8 Conclusions

Extrusion is a widely used process technology involving a multi-step and complex transformation from particulate ingredients to diverse agglomerated end-products. The versatility of the extrusion process makes it easy to adapt for various new food and non-food applications, and there is immense potential to expand research and development to extrusion in industry and academia. One such area is particulate flow characterization, which is important for understanding transport of raw materials in a food extruder. The recently developed powder rheometry method for measuring flow properties of particulate materials is very useful for understanding flow during handling and processing operations including extrusion. Parameters, such as specific energy, compressibility, cohesion and flow function, were found to be good off-line identifiers for flowability in an extruder. This new tool also has the potential for being employed for better design of metallurgy and geometry of extrusion hardware.

5.9 Definitions, Abbreviations and Symbols

Angle of internal friction	angle created by best fit line or yield locus with the horizontal axis (Mohr circle, see Fig. 5.9)
Basic flow energy	energy required per unit mass to move rheometer blade through the powder column during confined flow
Cohesion	intercept of the best fit line or yield locus with the vertical axis (Mohr circle, see Fig. 5.9)
Compressibility	percentage change in volume of powder column based on distance travelled by rheometer piston under normal stress
Flow energy	energy required for upward traverse of rheometer blade through the powder column
Flow function	ratio of major principal stress to unconfined yield strength (Mohr circle, see Fig. 5.9)
Flow rate index	factor by which basic flow energy changes when the rheometer blade tip speed is reduced by a factor of 10
Specific energy	energy required per unit mass to move rheometer blade through the powder column during unconfined flow

Stability index	factor by which basic flow energy changes during repeated testing
Yield locus	best fit line for shear stress versus applied normal stress data (Mohr circle, see Fig. 5.9)

AACCI	American Association of Cereal Chemists International (professional organization)
AIF	angle of internal friction
AOAC	The Association of Analytical Communities (professional organization)
BFE	basic flow energy (see Eq. 5.1)
FE	flow energy
FF	flow function
FRI	flow rate index (see Eq (5.4))
SE	specific energy (see Eq. (5.2))
SI	stability index (see Eq. (5.3))
UYS	unconfined yield strength
D	screw diameter of extruder
F	vertical resistance or force experienced by the blade of the rheometer
H	channel height of extruder
L	length of extruder
m	mass of test sample
T	rotational resistance or torque experienced by the blade of the rheometer
v_x	vertical speed of the blade of the rheometer
δh	distance travelled by rheometer piston under normal stress
Δx	vertical distance traversed by the blade of a rheometer
$\dot{\theta}$	angular speed of the blade of a rheometer
Φ	wall friction angle (see Eq. 5.5)
σ_w	normal stress
τ_w	shear stress

Acknowledgments The authors would like to acknowledge Abhay Patwa, Qi Bian and Andrew Mense for data collection. We would also like to thank Eric Maichel and Pavan Manepalli in the Extrusion Lab at Kansas State University (Manhattan, Kansas, USA) for their assistance in operation of the extruder and flow visualization experiments. Special thanks to Wenger Manufacturing (Sabetha, Kansas, USA) for their long term assistance to extrusion research and teaching at KSU and providing pictures of extruded food products.

References

1. AACC International: Determination of Granularity of Semolina and Farina: Sieving Method. 66-20.01. AACCI, St. Paul (2009)
2. Adedeji, A., Zhou, Y., Fang, X., Davis, D.A., Fahrenholz, A., Alavi, S.: Utilization of sorghum distillers dried grains (sDDGS) in extruded and steam pelleted shrimp diets. Aquacult. Res. (2015). In press.

3. Adedeji, A., Zhu, L., Padmanabhan, N., Alavi, S.: Digestibility and micronutrient retention of micronutrient-fortified extruded sorghum soy blends. IFT Annual Meeting and Expo, July 13–16, Chicago. Book of abstracts #135-06. Institute of Food Technologists, Chicago, IL, USA.
4. AOAC International: Solids (Total) and Loss on Drying (Moisture) in Flour. 925.10. AOAC International, Gaithersburg (2000)
5. Akdogan, H.: High moisture food extrusion. Int. J. Food Technol. **34**, 195–207 (1999)
6. Alavi, S., Giannetta, F., Nanjundaswamy, A., Madl, R., Vadlani, P.: Delivery of antioxidants through fruits and vegetables in extruded foods. Cereal Foods World. **59**(4), 179–185 (2014)
7. Alavi, S., Karkle, E., Adhikari, K., Keller, L.: Extrusion research for addressing the obesity challenge. Cereal Foods World. **56**(2), 56–60 (2011)
8. Fitzpatrick, J.J., Barringer, S.A., Iqbal, T.: Flow property measurement of food powders and sensitivity of Jenike's hopper design methodology to measured values. J. Food Eng. **61**, 399–405 (2004)
9. Garber, B.W., Hsieh, F., Huff, H.E.: Influence of particle size on twin screw extrusion of corn meal. Cereal Chem. **74**, 656–661 (1997)
10. Gonzalez-Montellano, C., Fuentes, J.M., Ayuga-Tellez, E., Ayuga, F.: Determination of mechanical properties of maize grains and olives required for use in DEM simulations. J. Food Eng. **111**, 553–562 (2012)
11. Johanson, J.R.: Modeling of bulk solids. Powder Technol. **5**, 93–99 (1971)
12. Kumar, A., Ganjyal, G.M., Jones, D.D., Hanna, M.A.: Digital image processing for measurement of residence time distribution in a laboratory extruder. J. Food Eng. **75**, 237–244 (2006)
13. Lawal, A., Kalyon, D.M.: Mechanisms of mixing in single and co-rotating twin screw extruders. Polymer. Eng. **35**, 1325–1338 (1995)
14. Leturia, M., Benali, M., Lagarde, S., Ronga, I., Saleh, K.: Characterization of flow properties of cohesive powders: a comparative study of traditional and new testing methods. Powder Technol. **253**, 406–423 (2014)
15. Lundblad, K.K., Hancock, J.D., Behnke, K.C., McKinney, L.J., Alavi, S., Prestløkken, E., Sørensen, M.: Ileal digestibility of crude protein, amino acids, dry matter and phosphorous in pigs fed diets steam conditioned at low and high temperature, expander conditioned or extruder processed. Anim. Feed Sci. Technol. **172**, 237–241 (2012)
16. Onwulata, C.I., Konstance, R.P.: Extruded corn meal and whey protein concentrate: effect on particle size. J. Food Process. Preserv. **30**, 475–487 (2006)
17. Peleg, M., Mannheim, C.H., Passy, N.: Flow properties of some food powders. J. Food Sci. **38**, 959–964 (1973)
18. Quist J.: Cone crusher modeling and simulation. MSci Thesis, Chalmers Univ. Technol. (2012)
19. Rokey, G.: Single screw extruders. In: Riaz, M.N. (ed.) Extruders in Food Applications. CRC Press, Boca Raton (2000)
20. Stoerzinger, K.M.: Effects of postmilling time and temperature on the bread making quality and lipids of whole wheat flour. M.S. Thesis, Kansas State University, Manhattan. Available at http://krex.k-state.edu/dspace/handle/2097/1669 (2009)
21. Tang, X.Z., Kumar, P., Alavi, S., Sandeep, K.P.: Recent advances in biopolymers and biopolymer-based nanocomposites for food packaging materials. Crit. Rev. Food Sci. Nutr. **52**, 426–442 (2012)
22. Teunou, E., Fitzpatrick, J.J., Synnott, E.C.: Characterization of food powder flowability. J. Food Eng. **39**, 31–37 (1999)
23. Van Zuilichem, D.J., Stolp, W.: Aspects of single and twin screw extrusion-cooking of biopolymers. J. Food Eng. **2**, 157–175 (1983)
24. Yoo, J., Alavi, S., Vadlani, P., Behnke, K.: Soybean hulls pretreated using thermo-mechanical extrusion – hydrolysis efficiency, fermentation inhibitors and ethanol yield. Appl. Biochem. Biotechnol. **166**, 576–589 (2012)

Chapter 6
Comminution

Mohsen Yahyaei, Marko Hilden, Fengnian Shi, Lian X. Liu, Grant Ballantyne, and Sam Palaniandy

Abstract Size reduction is an integral part of mineral processing which has the largest contribution to the energy consumption of mineral processing plants. This chapter provides an overview of size reduction theories and describes different mechanisms of breakage which are relevant in the mineral processing context, along with ore characterisation tests which measure the response of ore to different mechanisms of breakage. The most common comminution machines found in mineral processing are briefly described with explanations of how such equipment are typically laid out in comminution circuits. Concepts of novel approaches to designing flexible circuits are presented followed by process operation control strategies. Fine grinding, becoming an integral part of many mineral processing circuits, is also covered in this chapter along with appropriate placements for applications of fine grinding machines and important operating parameters. This chapter covers aspects of comminution which are relevant to mineral processing only and does not extends to other areas such as food, chemical, pharmaceutical, etc.

6.1 Introduction

Comminution is defined as the action of reducing a material, especially a mineral ore, to minute particles or fragments [45]. In the minerals industry, breakage is achieved through blasting, crushing and milling, while in the pharmaceutical and chemical industries size reduction is usually through milling. In the minerals industry, the goal of blasting is to reduce the in situ rock to a size that is readily transportable, whereas crushing and milling aims to liberate the valuable mineral grains from the waste (gangue) in preparation for separation, or to provide particles of a required shape or size. Blasting is achieved via the ignition of explosive

M. Yahyaei (✉) • M. Hilden • F. Shi • L.X. Liu • G. Ballantyne • S. Palaniandy
Julius Kruttschnitt Mineral Research Centre (JKMRC), The University of Queensland, 40 Isles Road, Brisbane 4068, QLD, Australia
e-mail: m.yahyaei@uq.edu.au

© Springer International Publishing Switzerland 2016
H.G. Merkus, G.M.H. Meesters (eds.), *Production, Handling and Characterization of Particulate Materials*, Particle Technology Series 25,
DOI 10.1007/978-3-319-20949-4_6

chemical compounds in the mine, whereas crushing and milling rely on mechanical impacts to comminute the rock or agglomerates. At all levels, the degree of size reduction is governed by the application of energy and the response of the rock in terms of material toughness or strength.

In the pharmaceutical or chemical industries, the main objective of size reduction is usually to improve the properties or the efficient use of a product. Examples are increasing the surface area of the materials so that the adsorption or dissolution rate of the powders is increased significantly and reducing the size of pigments to reach optimum hiding power and gloss of paints. Size reduction usually occurs in stirred media mills.

This chapter covers topics in relation to comminution in the minerals industry. Comminution in the pharmaceutical or chemical industries is not part of scope of this chapter.

6.1.1 Comminution Energy

The process of comminution in mineral processing is dominated by the high energy requirements of breaking rocks to liberate the valuable minerals. Calculating the energy required to reduce the feed ore size to the desired product size is the first step in selecting the appropriate size and installed power of equipment that underpins the design of comminution circuits. According the study of Ballantyne and Powell [1], comminution typically accounts for between 26 and 46 % of the utilised energy on copper and gold mines (Fig. 6.1). Napier-Munn [39] calculated that approximately 1.8 % of global generated electrical energy is consumed in comminution.

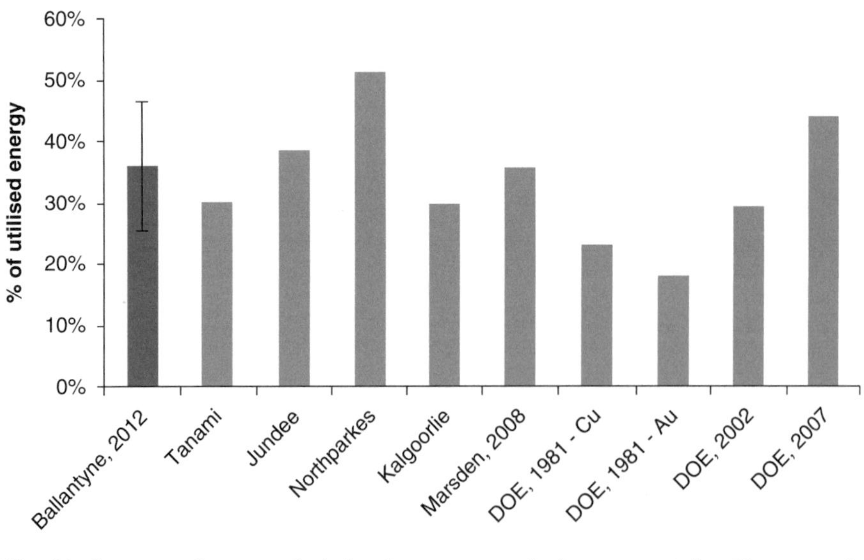

Fig. 6.1 Summary of reports calculating the percentage of mine energy attributable to comminution [1, 12–15, 25, 30, 44]

Curry et al. [10] found that the mill contributed between 43 % and 45 % of ongoing operating costs on a mine.

6.2 Size Reduction Theories

The relationship between energy consumption and size reduction is integral to the science of comminution. Three well-known theories have been proposed to describe how the size reduction can be measured to obtain a linear relationship with energy consumption. The relationship between surface area and energy, commonly attributed to Rittinger [49], was first known and used in the design of explosive charges in the Napoleonic wars [3, in 31]. Lynch and Rowland [27] described the history of the Rittinger theory and how it evolved. Rittinger [49] originally commented that "the throughput (of wet operated stamp mills) is proportional to the 0.4 power of the linear dimension of the openings of the discharge screen". The new surface generated was used as an analogy to help clarify the results. However, the theory that was attributed to him stated that the work done in crushing is proportional to the area of new surface produced. Rittinger's theory has gained significant traction within the scientific community because it is supported thermodynamically by the transfer of strain energy to bond potential energy on the surface of the material. The surface of a material has a higher energy state than the interior due to the balance of long-range chemical attractive forces and short-range Coulomb repulsive forces [18].

The theory of Kick [23] is that the energy required to produce any given reduction ratio in the volume is constant [27]. Tavares and King [67] proved experimentally that Kick's assumption is valid at large sizes where the surface energy is negligible compared to the internal energy required. But this does not exclude the Griffith [21] hypothesis that smaller particles are stronger because they have less structural defects [64].

Bond [4] reasoned that comminution is concerned with both surface and volume; therefore neither Rittinger nor Kick can be completely correct. Strain energy is absorbed by the volume of the particle, but breakage begins with the formation of a crack tip on the surface. Three assumptions from Rittinger [49] or Kick [23] were questioned: neglect of the previous energy imparted on the particles, derivation using cubes, and that the useful work input is equivalent to energy content.

Bond [4] assumed that the energy was equally absorbed by the surface area and volume of a particle (midway between Rittinger [49] and Kick [23]); therefore, the force is proportional not to D^2 or D^3 but $D^{5/2}$. Where D is the size of the particles. Therefore, the energy required to break a unit weight/volume should be proportional to $D^{5/2}/D^3$ or $1/\sqrt{D}$. Bond [4] stated the third 'theory' of comminution in this way: "The total work useful in breakage which has been applied to a stated weight of homogeneous material is inversely proportional to the square root of the diameter of the product particles".

Bond [4] claimed that although the Rittinger theory is concerned with the measurement of surface areas and the Kick theory with the volumes of the product particles, the third theory was proportional with the crack length formed. On the other hand, Stamboliadis [64] stated that both Kick's and Rittinger's statements are not mutually exclusive and Bond is only a partial case of a general expression of Rittinger's statement.

Walker [70] found that neither Rittinger's nor Kick's equation represented all facts properly and gave a general comminution equation that encompassed both theories:

$$dE = -C\frac{dx}{x^n} \tag{6.1}$$

where

E = the work done
x = the particle size
C, n = constants.

Kick's theory results for $n = 1$:

$$E = -Cln\left(\frac{x_2}{x_1}\right) \tag{6.2}$$

For $n = 2$ this becomes Rittinger's equation:

$$E = C\left(\frac{1}{x_2} - \frac{1}{x_1}\right) \tag{6.3}$$

Later Hukki [22] included Bond's third 'law' of comminution ($n = 1.5$):

$$E = 2C\left(\frac{1}{\sqrt{x_2}} - \frac{1}{\sqrt{x_1}}\right) \tag{6.4}$$

There is some evidence that the three theories are not mutually exclusive and each may be accurate within a proper relatively narrow size range. Hukki [22] suggested that the exponent n in the general comminution equation is not in fact a constant, but a variable dependant on the particle size (Fig. 6.2).

There are many experimental results in favour of the theory of Rittinger [49], scant evidence for the theory of Kick [23] and the evidence for the third theory is provided by Bond himself [22]. Generally, the size range covered by these experiments is also relatively narrow [22]. Thus, it is strongly suggested that all the comminution theories are actually incorrect and the increase in energy required to break smaller particles is due to the reduction in internal flaws and weaknesses

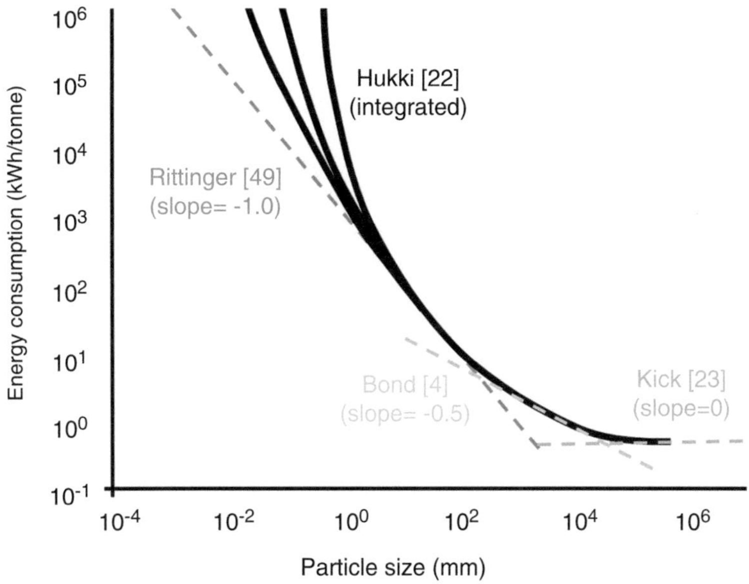

Fig. 6.2 Comminution energy theories (After Hukki [22])

[41, 71]. Schönert [53] hypothesised that three factors combine to increase the apparent strength of smaller particles:

- The presence of flaws is less probable with decreasing size.
- As particles reduce in size, so does their capacity for storing elastic energy.
- Smaller particles fail plastically rather than elastically.

The reduction in competence due to the presence of flaws is often related to the Griffith [21] crack theory. Griffith [21] advanced the scientific concepts of the proportionality between energy and surface area by taking a thermodynamic approach to brittle fracture. He quite logically proposed that if the forces pulling apart molecules in the immediate un-cracked area around the crack rim exceeded the attraction between the molecules, the crack would enlarge until failure was complete [31]. Schönert [53] found that due to plastic deformation at the crack tip, extension of the crack requires at least ten times the energy predicted by the Griffith theory.

The underlying truth of this system may be that the Rittinger [49] hypothesis holds true across all sizes, but that internal free surfaces become external surfaces through breakage. Therefore, for larger particles with a higher availability of internal surfaces, the amount of energy required to generate 'new' external surfaces is lower because of this conversion.

Many other size-reduction energy relationships exist, such as Starkey et al. [65] and Morrell [37]. These are in general equipment specific, often parameterised against a production database. Another specific parameter, the t_{10}, that was developed by Narayanan and Whiten [43] to describe impact breakage, has been applied for the characterization of ore breakage for many decades. t_{10} is the percentage of progeny passing an aperture of one tenth of the original particle size.

Understanding the fundamentals of breakage processes, breakage mechanisms and how the rock responds to different stressing conditions will enable us to study and validate the underlying assumptions of comminution theories.

6.3 Breakage Mechanisms

As with any material, rock particles only break when the applied tensile stress exceeds the tensile strength of the rock. However, in terms of comminution, the main interest is to derive reliable ways of assessing how a particular material breaks in a comminution device such as a crusher or a mill [41]. In general, the breakage of rock particles is dependent on two factors: the "hardness" of the material, and the specific equipment used for particle breakage. In the last few decades, attention has been shifted to separate material property characterization from equipment parameters. This is essential for modelling and optimization of comminution processes in a mineral processing plant [76].

In contrast to standard rock mechanics tests (e.g. uniaxial compressive strength, fracture toughness, etc.) it is usually required in comminution testing to obtain the product size distribution from applying a particular breakage mechanism that relates to specific comminution equipment. Depending on the industrial comminution equipment used, rock particles are broken through three main breakage modes [41]:

- compression
- impact
- shear (abrasion/attrition)

Some comminution equipment only exploit one major breakage mode whereas ball mills, Autogenous Grinding [32] mills, and Semi-Autogenous Grinding (SAG) mills impart both impact and shear breakage. The detailed description of each breakage mechanism is presented in this section.

6.3.1 Compression

One form of load application during a comminution process is compression in which rock particles are compressed slowly (Fig. 6.3) and the particles fail when the uniaxial compressive strength has been exceeded. In terms of measurement response, the main parameters obtained from single particle standard material testing is Uniaxial Compressive Strength (UCS), Young's modulus (ϵ) and Poisson's Ratio (ν) and are often measured by using a cylindrical rock sample

Fig. 6.3 Schematics of different mode of breakage

Table 6.1 Summary of comminution devices and their stressing mode and rate

Comminution equipment	Main breakage mode – stressing method	Applied strain rate (m/s) [66]
High pressure grinding rolls	Compression	Up to 0.1
Jaw crusher	Compression	0.1–1
Gyratory and cone crusher	Compression	0.1–2
Ball, rod, AG and SAG mills	Single impact, double impact and abrasion/ attrition	Up to 15
Stirred media mills	Shear between grinding media	Up to 20
Impact crushers	Single/double impact	20–200
Fluid energy mills	Single impact and abrasion/attrition (rheology also plays a part)	Up to 400

[41]. Measuring the particle size distribution from single particle compression tests is less common due to the difficulties associated with the precise measurement of specific input comminution energy for breakage. Multi-particle bed compression tests are often carried out in a piston-die [16, 17] which relates the specific energy input to size reduction.

Compression is one of the most common loading modes in crushers such as jaw, gyratory and cone crushers as well as high pressure grinding rolls. The typical applied strain rate is between 0.1 and 2 m/s (Table 6.1).

6.3.2 *Impact*

In impact breakage, the particles move perpendicularly to another surface, which may be either a mill wall or another particle surface. Impact breakage has a much higher applied strain rate than compression breakage and the impact speed ranges from tens of meters per second in mills up to 400 m/s in high energy mills. When particles are impacted in a comminution device, they can either have one-sided impact or two-sided impact, as illustrated in Fig. 6.3. In addition, impact breakage of particles can either occur in one single high energy impact or via multiple low energy impacts [66].

6.3.3 Abrasion/Attrition

In abrasion breakage, the surface of particles experience shear while moving parallel to their contact surface and fine particles are rubbed off the surface of the particles. The damage caused by abrasion is not substantial to the body of the particles. As illustrated in Fig. 6.3, attrition breakage has the same mechanism as abrasion, while a small particle is nipped by two larger particles causing substantial damage to the main body of the small particles [41].

With the same input specific energy, different breakage mechanisms lead to different probability of breakage. An example from the work of Schubert [54] is shown in Fig. 6.4. This is because different breakage mechanisms result from application of different stress fields to the particles, even though the same specific energy is applied.

For further information on the different breakage mechanisms, readers are referred to the Powder Technology Handbook chapter by Tavares [66].

To understand response of rocks to different mechanisms of breakage it is useful to develop an energy relationship between the breakage of a given size of particle

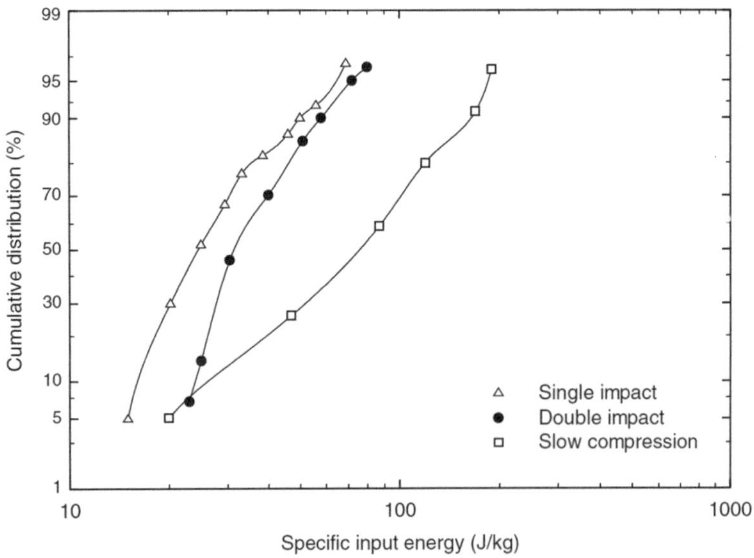

Fig. 6.4 Breakage probability distribution for 12 mm fire clay pellets for different stressing modes. Single and double impact here refer to one-sided and two-sided impact of a particle respectively [54]

and the product size distribution. Different breakage characterisation tests has been developed to measure breakage characteristics of rocks. The next section summarise the breakage characterization tests which are widely used in mineral processing.

6.4 Breakage Characterization

Mining companies have long recognized that the performance of a size reduction machine such as a crusher, an AG or a SAG mill depends not only on the machine operating conditions, but also on the competence of the feed material to be broken. This is very much exemplified at the OK Tedi Mine in Papua New Guinea, where the SAG mill throughput varied from 700 to 3000 t/h due to variations in feed ore competence (Bond work index 16–5 kWh/t) and in feed size distributions [62].

Particle breakage characterization aims to quantify the product size distribution which results from the application of energy to a selected feed size via a specific breakage mechanism. Specifically it aims to establish the relationship between mass specific energy input and resultant product through some type of laboratory test on a given ore. The outcomes from particle breakage characterization may be in the form of a single competence or strength parameter, or a relationship describing the level of size reduction with respect to the applied energy or other test conditions. The result of breakage characterization is useful in assisting in comminution equipment specification, circuit design, machine modelling, and process optimization.

6.4.1 Bond Work Index

One of the most commonly used breakage characterization methods in the mining and mineral industries is the Bond work index method, based on the so-called "Third theory of comminution" [4], which is normally written as Eq. (6.5):

$$W = W_i \left(\frac{10}{\sqrt{P_{80}}} - \frac{10}{\sqrt{F_{80}}} \right) \tag{6.5}$$

where:

W = mass specific energy (kWh/t) calculated as power draw [52] divided by throughput (t/h)
W_i = Bond work index (kWh/t)
P_{80} = aperture size where 80 % of product passing (μm)
F_{80} = aperture size where 80 % of feed passing (μm).

Note that the constant value 10 in the Eq. (6.5) is originally from a reference size of 100 μm in the form $\sqrt{\frac{100}{P_{80}}}$ so that the equation is dimensionally balanced. Bond defined the work index W_i as a comminution parameter which expresses the resistance of the material to crushing and grinding. The work index can be determined in two ways. If plant data (power draw, throughput, F_{80} and P_{80}) are available, Eq. (6.5) can be rearranged to calculate the work index, which is referred to as the operating work index (OW_i). If plant data are not available, laboratory test needs to be conducted on the ore sample of interest. Bond developed crushing, rod mill and ball mill laboratory tests.

In the popular Bond ball mill test [5, 6], the standard mill is 0.305 m diameter by 0.305 m length with rounded corners, running at 70 rpm. The grinding charge consists of 285 balls of various sizes, the total weighing 20.125 kg. The standard feed is prepared by stage crushing to pass a 3.35 mm sieve. A volume of 700 ml is placed in the Bond ball mill and ground dry in a locked cycle mode. After the first grinding period of 100 revolutions, the mill content is dumped, the ball charge is screened out, and the ore is screened at a required closing size, typically 300 μm, 150 μm or 75 μm. The undersize is weighed and removed, and an equal mass of fresh feed is added to mix with the oversize to maintain the original charge mass. The second grinding is performed for a number of mill revolutions calculated to produce a 250 % circulating load. The grinding cycles are continued until the net grams of sieve undersize produced per mill revolution reach equilibrium and just starts decreasing. The net grams of undersize product per mill revolution (Gbp) are defined as the ball mill grindability. Closing sieve size (P_1) is defined in micrometer. The ball mill work index of an ore sample is calculated by Eq. (6.6):

$$W_i = \frac{49.1}{P_1^{0.23}(Gbp)^{0.82}\left(\frac{10}{\sqrt{P_{80}}} - \frac{10}{\sqrt{F_{80}}}\right)} \tag{6.6}$$

The Bond work index can be used in two ways. The first is to differentiate ore competence for a size reduction system. The typical Bond work indices for ores processed in the mineral industry may vary through a range from below 10 kWh/t to over 30 kWh/t; the larger value indicating the more competent ore. The second application is to size a comminution unit. For a given ore F_{80} and a desired product Eq. (6.6) can be used to calculate the specific energy from the measured Bond work index. The multiplication of the specific energy (kWh/t) by the target throughput (t/h) gives the required power draw [5, 6] of the mill. A mill is then selected according to the equipment specifications, which can draw the required power under the chosen operating conditions e.g. ball charge and speed.

The accuracy of the Bond equation has been criticised by various researchers in the literature. Morrell [33] pointed out that the fundamental problem with the Bond equation is in the size exponent. Based on Hukki's suggestion [22], Morrell proposed an energy-size relationship:

$$W = M_i K \left(x_2^{f(x_2)} - x_1^{f(x_1)} \right) \tag{6.7}$$

where

M_i = the comminution index (kWh/t)
K = constant chosen to balance the units of the equation
x_2, x_1 (μm) = equivalent to P_{80} and F_{80}, respectively
f(x_j) takes the following form [35]:

$$f(x_j) = -(0.295 + {}^{x_j}/_{1,000,000}) \tag{6.8}$$

Morrell demonstrates that by taking the different values of K and M_i for coarse grinding and fine grinding respectively, Eq. (6.8) is able to predict the total specific energy requirement in a grinding circuit, with a 95 % confidence limit of ±15 %. Morrell also points out that the predictions of Eq. (6.8) for the individual grinding stages may be very different to the observed values, due to the influence of ore competence and operational conditions [36].

Another limitation of the Bond equation is its underlying requirement for the feed and product size distributions to remain parallel and linear in log–log space, so that one point sizing data (such as the F_{80} or P_{80}) can represent the whole size distribution. Experimental data show that the feed and product in crushing and AG/SAG milling stages often do not meet this requirement. In addition, the breakage mode in AG/SAG mills is predominantly impact, which is different to the Bond ball mill testing conditions.

6.4.2 *Single Particle Breakage Tests*

To characterize particle breakage for crushing and AG/SAG milling, single-particle impact tests are employed. Tavares [66] provides a detailed review on single-particle fracture studies. Many researchers implement the single-particle breakage to study the response of particles in a comminution environment, such as that reported by Gilvarry and Bergstrom [20], Rumpf [50], Narayanan [42], Marktscheffel and Schönert [29], Weichert [73], Tavares and King [67], Bourgeois and Banini [7], Vogel and Peukert [69], Shi et al. [59].

Single-particle breakage can be classified into two groups: single contact point breakage and double contact point breakage. Single contact breakage mimics the breakage mode when a particle lands on a mill shell, hits another particle or is impacted by a rotating shaft – such as in hammer mill or vertical shaft impact mill. Double contact point breakage mimics the breakage mode in jaw/cone crushers and in a mill when a particle is compressed between an impacting balls or rock and the mill shell or particle bed. The single contact point breakage tests are described in Brown and Miles [8] who used a pneumatic gun to propel particles against a hard surface, and in Marktscheffel and Schönert [29] who used a rotor to eject particles against anvils. Double contact point breakage occurs in the drop weight tester [53] and Hopkinson pressure bar test [67].

The Ultrafast Load Cell (UFLC) was developed at the Utah Comminution Centre [74], combining the functions of the simple drop weight tester and the Hopkinson pressure bar. The UFLC is equipped with solid-state strain gauges on the rod, which allows the measurement of three fundamental deformation and fracture characteristics of particulate materials: the particle fracture energy, the particle strength and the particle stiffness [67]. The Short Impact Load Cell (SILC, referring to [7]) is a revised version of the UFLC with a much shorter rod. The typical particle size tested in the above mentioned single-particle breakage devices is in the millimetre range, but has recently been extended to micrometre and nanometre range. Schilde et al. [52] report the use of a nano-indentation device to measure the micromechanical properties of aggregate. The samples are stressed by a "Flat Punch" indenter tip of hardened steel with a diameter of 50 μm. The displacement is controlled by a maximum indentation displacement of 500 nm and a constant indentation velocity of 100 nm/s. One limitation associated with the Hopkinson pressure bar, UFLC, SILC and nano-indenter is their relatively long measurement time and data processing time. They can provide details of fracture characteristics, but at the cost of a representative sample due to the limited number of particles being tested.

Over the past three decades, the Julius Kruttschnitt Mineral Research Centre (JKMRC) has dedicated significant resources to researching and developing techniques to further this cause. These include the Twin pendulum [42], Drop Weight Tester (DWT) [53], Mini-Drop Weight tester [28], Short Impact Load Cell [7] and the Rotary Breakage Tester (JKRBT) [59].

The DWT is now an industry standard breakage characterization test. Figure 6.5 shows a photograph of the DWT installed at the JKMRC. The test apparatus comprises an impactor, an anvil, and guide rails to control the drop path. The mass of the impactor and the height from which it can be dropped are varied to provide a range if impact energies. In a standard drop weight test, 10–30 particles in each size fraction of $\sqrt[2]{2}$ sieve series (i.e. +53–63 mm, +37.5–45 mm, +26.5–31.5 mm, +19–22.4 mm, +13.2–16 mm), are tested with three impact energy levels ranging from 0.1 to 2.5 kWh/t. Each rock specimen is placed on an anvil and is broken by the impactor. The broken pieces for each size fraction and energy level are collected and sized. By varying the drop height and mass as well as the rock size a range of specific impact energies can be generated.

The specific energy used in the drop weight test can be calculated by Eq. (6.9):

Fig. 6.5 JKMRC drop weight tester

$$Ecs = \frac{m_d \cdot g \cdot h}{3600 m} \tag{6.9}$$

where

Ecs = specific energy (kWh/t)
m_d = drop mass (kg)
g = gravitational acceleration (i.e. 9.8 kg/s^2)
h = net drop height (m)
m = particle mass (kg).

As a faster and more cost effective measurement of comparative rock hardness, a reduced version of the drop weight test, the SMC Test was introduced [34]. The SMC Test uses the DWT to break particles in one of the following narrow size fraction: 26.5–31.5 mm or 22.4–19 mm or 16–13.2 mm from drill cores or crushed rocks, with five specific energy levels. The results from conducting the SMC Test are used to determine the so-called drop-weight index (DW_i) which is used to predict mill power requirement and throughput based on empirical equations developed from plant operational databases [38]. The SMC test also provides estimates of full drop weight test results. Over 32,000 SMC tests have been conducted since 2004 [63]. Since particle strength is affected by particle size, row SMC test data requires adjustment to represent the entire sample. The adjustment is based on the impact resistance-particle size relationship measured by a drop weight test or three SMC tests at different size on ore from the same domain. If domain specific data are not available, the average of the SMC testing database is used.

Fig. 6.6 JK rotary breakage tester (JKRBT)

In 2006 the JKMRC developed a Rotary Breakage Tester (JKRBT, Fig. 6.6) for rapid particle breakage characterization tests [59]. The JKRBT uses a rotor-stator impacting system, in which particles gain a controlled kinetic energy while they are spun in the rotor and are then ejected and impacted against the stator, causing particle breakage. The industrialised JKRBT employs a rotor of 450 mm in diameter. The rotor is direct driven by a three phase, 7.5 kW, 5000 rpm electric motor. The motor is controlled by a variable frequency drive (VFD) unit via a Human Machine Interface (HMI) in the control panel. This unit can treat particles in the size range of 5–45 mm. Thus four of the five particle size fractions that are tested in the standard DWT can be treated in the industrialised JKRBT (+37.5–45 mm, +26.5–31.5 mm, +19–22.4 mm and +13.2–16 mm). The JKRBT has a wide range of specific impact energies from 0.001 to 3.8 kWh/t.

The specific energy Ecs in the JKRBT test is determined from Eq. (6.10):

$$Ecs = \frac{0.5\,mV_i^2}{m} = 3.046 \times 10^{-6} C^2 N^2 \left(r + \frac{x_g}{2}\right)^2 \tag{6.10}$$

where

m = feed particle mass (kg)
V_i = impact velocity (m/s) on the anvil
r = rotor radius (m)
x_g = geometric mean size of feed particles (m)
N = rotor speed (rpm)

C = machine design constant that governs the maximum possible impact velocity at a given rotor speed and set of operational conditions.

The constant C is calibrated from the impact velocity measured with a high speed video camera. Note that particle mass has been cancelled in the JKRBT specific energy calculation, hence it no longer affects the specific energy as is the case with the DWT.

Features of the JKRBT include rapid characterization of particle breakage properties; ability to test a large number of particles to offer statistically more valid results; accurately controlled specific energy due to its independency of the particle mass; achieving higher specific energy level than the DWT for coarse particles; good repeatability in product size distribution. The JKRBT has been deployed with some of the world's major mining companies (Anglo American, Barrick, BHP Billiton, Lonmin, Rio Tinto and Teck). JKTech uses it in standard laboratory service and the researchers at the JKMRC use it as a research tool.

The DWT data are processed with Eq. (6.11) [43] to establish the relationship between t_{10} and Ecs for impact breakage characterization:

$$t_{10} = A\left(1 - e^{-b.Ecs}\right) \tag{6.11}$$

where

t_{10} = breakage index defined as the percentage of progeny passing one tenth of the original geometric mean size
A, b = ore impact breakage parameters.

The product of A and b ($A \times b$) is often used as an ore competence index. It is widely accepted by many mineral processing engineers and researchers worldwide to use the A and b parameters in their comminution circuit design, simulation, optimization and operation. In the JKTech database consisting of over 4000 DWT data sets, the $A \times b$ values are typically within the range of 20 and 300, with $A \times b$ close to 20 indicating extremely difficult to break, and any value above 150 very easy to break.

As noted earlier, particle strength is affected by particle size – coarse particles being easier to break than fine particles, possibly due to the difference in the pre-existing crack/micro-crack density in particles. Subjected to the same specific energy in the DWT, larger particles tend to produce larger t_{10} values. When fitting the DWT data with Eq. (6.11), one set of A and b parameters is generated for all sizes. This 'average' set of A and b parameters assumes that particles of different sizes are broken in the same way when subjected to the same impact energy. This is questionable, particularly in an AG/SAG mill where the feed may contain particles from over 200 mm down to minus 1 mm.

A size-dependent breakage model was proposed by Shi and Kojovic [57] to describe the degree of breakage (i.e. the breakage index t_{10}) in relation to input specific energy, particle size and particle breakage properties (Eq. 6.12). This size-dependent breakage model was modified from the Vogel and Peukert [69] breakage

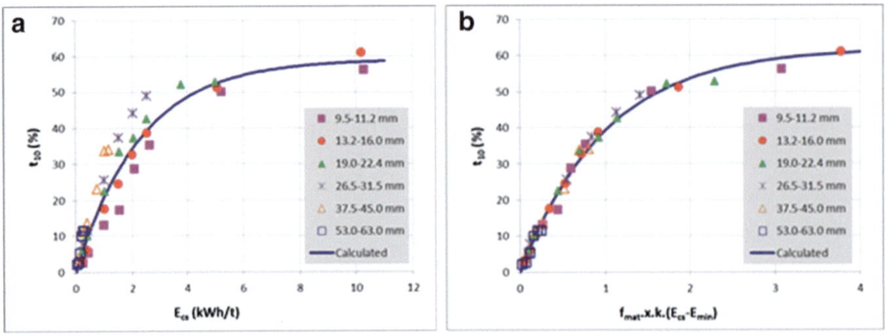

Fig. 6.7 The results using the JKMRC (**a**) the prior-art (Eq. 6.11) and (**b**) the size-dependent breakage models (Eq. 6.12) fitted to the same DWT data of a quarry sample conducted by Banini [2]

probability model, which was developed from a generalised dimensional analysis approach proposed by Rumpf [50], and a fracture mechanical model based on Weibull statistics [72].

$$t_{10} = M\left\{1 - \exp\left[-f_{mat.} \cdot x_g \cdot k(E - E_{\min})\right]\right\} \tag{6.12}$$

where

t_{10} = as defined in Eq. (6.11) (%)
M = maximum t_{10} for a material subject to breakage (%)
f_{mat} = material breakage property described by a size-dependent sub-model (kg/J/m)
x_g = geometric mean size of initial particles
k = successive number of impacts with a single-impact energy
E = mass-specific impact energy (J/kg)
E_{min} = threshold energy below which breakage does not occur (J/kg).

The material parameter f_{mat} and the energy threshold E_{min} can be determined from single particle breakage testing experiments with a laboratory impact device such as the DWT or JKRBT.

Figure 6.7 compares the results of the two equations (Eqs. 6.11 and 6.12) fitted to the same set of DWT data, which demonstrates that the size effect on particle breakage can be well described by the size-dependent model. As the two models have a similar structure, the parameters of the two equations are convertible. The ore competence indicator $A \times b$ can be calculated from the parameters in Eq. (6.12), which allows the mining companies to continue the use of their existing $A \times b$ database.

$$A \times b = 3600 \cdot M \cdot f_{mat.} \cdot x_g \tag{6.13}$$

By using Eq. (6.12) and accounting for differences in loading mechanisms (breakage contact points and strain rate) between the DWT and the JKRBT, it has been shown that the JKRBT can generate similar $A \times b$ values for the same ores as the DWT – established using 54 sets of parallel samples tested by the various mining companies [58].

In addition to the typical use for impact breakage data reduction, Eq. (6.12) has been employed to describe the energy-size reduction relationship for ball mill grinding [60]. Its use has been extended from the mineral industry to the coal industry for processing coal grinding data using a JK Fine-particle Breakage Characteriser (JKFBC) [61], and further expanded as a multi-component breakage model incorporating the coal particle size and density effects [56].

The particle breakage characteristic parameters determined from the DWT, JKRBT or JKFBC tests find two major types of applications. The first type of application is to differentiate ore competence for establishment of a geometallurgy database, for engineering design of comminution circuits and for crushing and grinding operation trouble-shooting and improvement. The second type of application is to provide ore-specific parameters for comminution models. In the past half century, the JKMRC has developed various comminution unit models for crushers, screens, AG/SAG mills, ball mills, HPGR, hydrocyclones, etc., which have been implemented in a comminution process simulation software package, JKSimMet. This package has been widely used by design engineers, plant metallurgists and researchers for plant survey data processing, mass balancing, model parameter fitting, operational trouble-shooting and circuit optimisation. All the models for size reduction machines incorporate ore breakage characteristic parameters in various forms. The models, together with the ore breakage characteristic parameters, are able to accurately predict power requirements, throughput and size distributions of different streams in a circuit for a specific ore.

6.5 Comminution Machines

A wide range of industries and applications demand efficient size reduction equipment. Numerous designs of size reduction equipment are available. Equipment selection depends on specific requirements such as:

- feed size and throughput requirements
- required product size, size distribution and product shape specifications
- material hardness and abrasiveness
- batch or continuous processing
- wet or dry processing
- whether contamination from steel or grinding media is detrimental to product quality
- whether heating or chemical reactions need to be performed simultaneously
- simplicity of maintenance and reliability, and
- capital and operating costs.

Size reduction equipment is characterised here according to the mechanisms employed to break particles.

6.5.1 Crushers

Preparation of materials to a size that can be milled is generally performed by crushers. The breakage mechanism in crushing devices can be both single-particle compression and bed-breakage. Crushers apply high forces to fracture the particles: crushers have been designed to break hard rocks up to a compressive strength of 350 MPa. Crushers can be operated with a reduction ratio of up to approximately 5:1 and sometimes more.

- **Jaw crushers** compress material between two plates, usually one which is static and one that moves with an oscillating motion (Fig. 6.8). Progressive size reduction takes place as the particles work down the crushing chamber. The minimum gap between the plates is referred to as the closed side setting (CSS). Jaw crushers are available from small bench-scale units capable of processing several kg per hour to industrial units able to crush 2000 t/h of run-of-mine ore.
- **Gyratory crushers** work on a similar principle to jaw crushers but have a circular gap (Fig. 6.9). Rock is compressed between a static conical bowl and a concave mantle which oscillates about the central axis. These are generally designed for primary crushing in large-scale rock crushing applications up to 6000 t/h. Typically a mining haul truck will empty its load into the gyratory crusher and reduce the feed with a top-size of up to a few meters down to below around 250 mm.

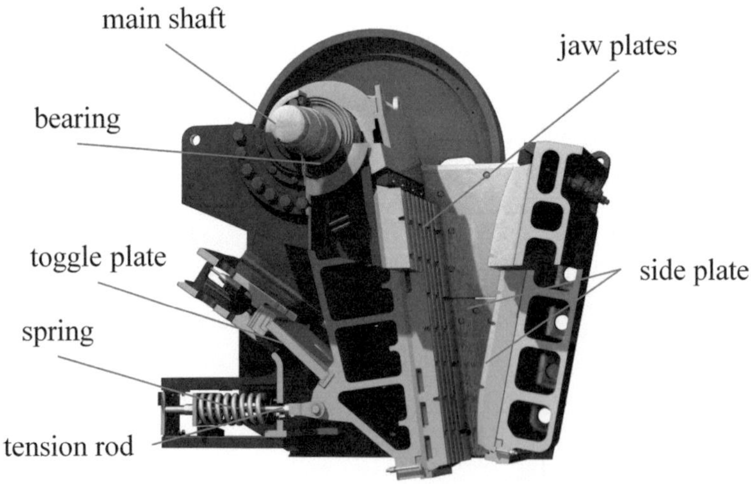

Fig. 6.8 Cross section of a Sandvik CJ Jaw crusher (Courtesy of Sandvik)

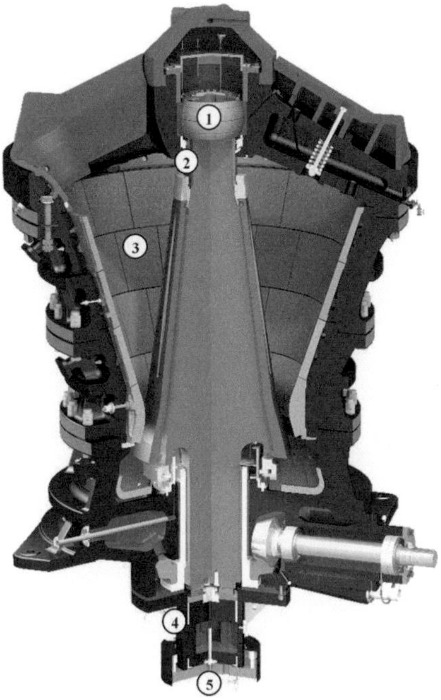

Fig. 6.9 Cross section of a Sandvik gyratory crusher (Courtesy of Sandvik)

- **Cone crushers** have a static bowl and eccentrically driven mantle between which particles are crushed (Fig. 6.10). In minerals applications, these are predominantly used in secondary and tertiary crushing applications producing particles with a top-size of between 50 and 5 mm.
- **Rolls crushers** are manufactured in single-roll and double-roll varieties. Figure 6.11 shows a Sandvik double roll crusher. The rotation draws material into the crushing zone. Toothed rolls crushers have metallic teeth to help draw the material into the gap. In addition to compressive breakage, some shearing can take place. Rolls crushers are more commonly selected for soft to medium hardness rocks and coal applications where they have high capacity in a relatively small plant footprint. These crushers can accept feed size up to 2500 mm.

6.5.2 Compressed-Bed Breakage

Compressed bed breakage devices apply high compressive forces to crush particles that are typically smaller than the crushing gap.

- **High pressure grinding rolls** (HPGR), also known as roller presses are operated at a pre-set crushing pressure through loading a 'floating' roll with pistons

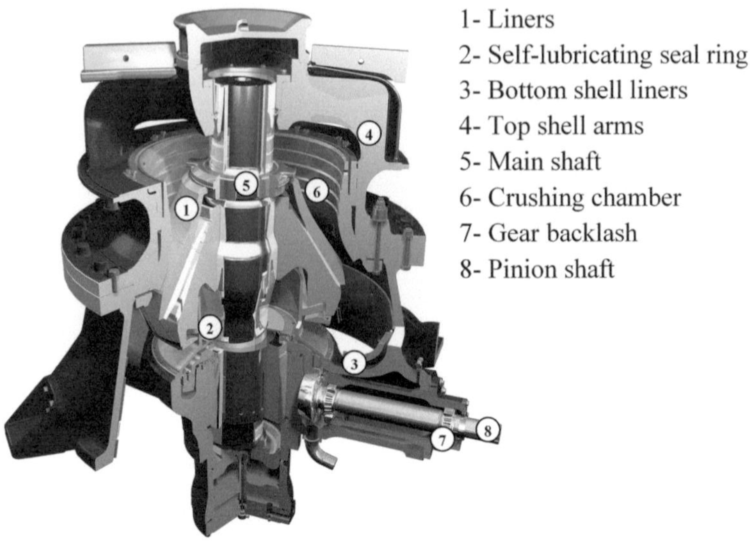

Fig. 6.10 Cross section of a Sandvik cone crusher (Courtesy of Sandvik)

Fig. 6.11 A Sandvik CR810/14-30 double roll crusher (Courtesy of Sandvik)

to provide a compression pressure as high as 200 MPa that compresses the feed material in a bed of particles (Fig. 6.12). A wide distribution of particles in the feed is needed for a continuous bed to form, and the pressure reduces the void fraction between particles to approximately 15–25 % often forming flakes (or cake) which may need to be de-agglomerated. Originally developed for cement applications, advances in the manufacture and design of the roller

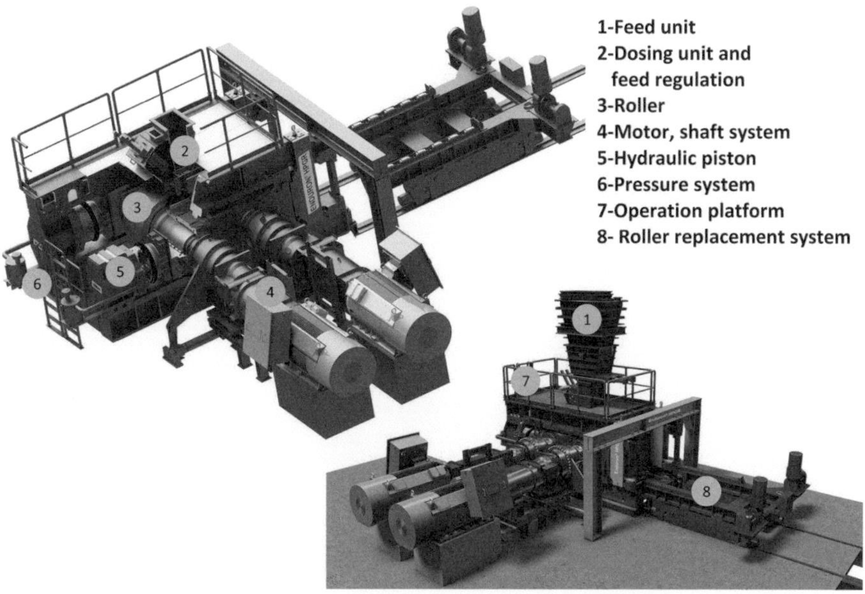

Fig. 6.12 The high pressure grinding rolls (HPGR) (Courtesy of Weir Minerals)

surfaces have allowed the HPGR to be used successfully in hard rock applications. HPGR accept feed sizes up to 80 mm.

- **Vertical roller mills** (VRM) contain a series of rollers that press down on a layer of particles on a circular table, either flat or grooved. Widely used in the cement industry, VRM's usually contain an integrated dry air classification system to draw fines from the grinding zone and return oversize material back to the grinding table (Fig. 6.13). Large size reduction ratios are achieved; however a significant proportion of the energy can be used in drying and by the air classification process.

6.5.3 *Impactors*

Impactors accelerate particles to collide with either a solid steel surface or a bed of particles; or particles are struck by rotating impactors. Impact breakage devices are generally selected for less abrasive and softer materials with compressive strength less than 100–150 MPa. Impactors generally produce a more cubic particle shape, which can be a desirable quality in the product material.

- **Hammer mills** or horizontal shaft impactors have a horizontal shaft with multiple radial strikers which impact against steel surfaces or other particles

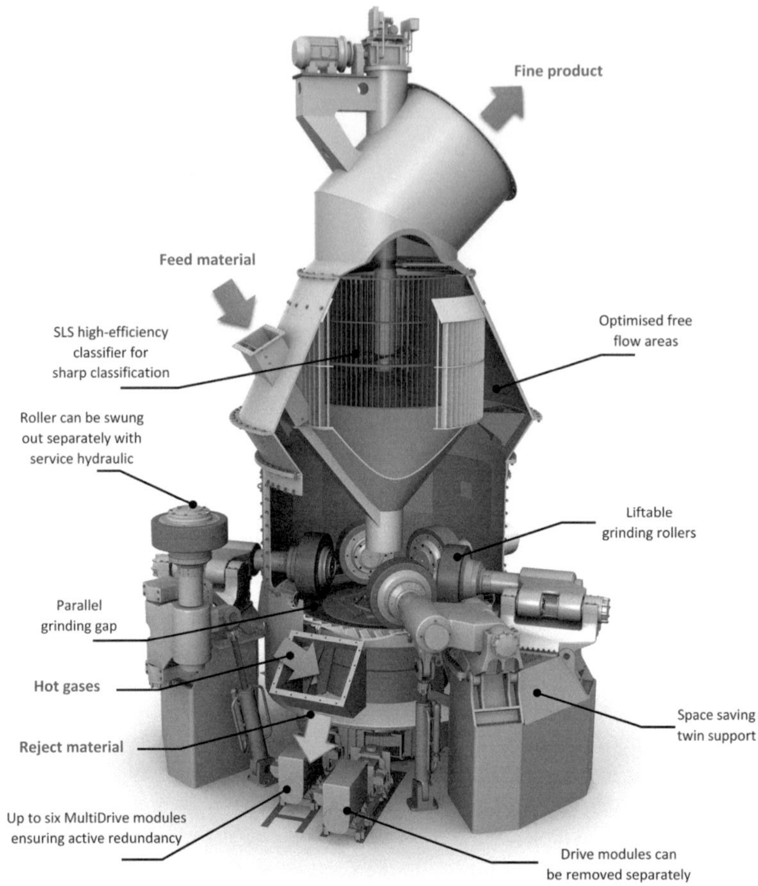

Fig. 6.13 A PFEIFFER VRM vertical roller mill (Courtesy of Gebr. Pfeiffer)

(Fig. 6.14). Hammer mills often contain a grate which retains particles until they are small enough to pass through.

– **Vertical shaft impactors** (VSI) use a vertical spinning shaft to accelerate particles outwards onto the impacting wall. Their designs usually promote rock-on-rock fragmentation. Figure 6.15 illustrates cross section of a VSI from Terex®.
– **Cage mills** consist of one or more concentric rows of pins rotating in opposite directions. Fed from the centre, material is struck by the rotating cage and impacted onto the next row of pins. Cage mills are well suited to sticky and damp materials and achieve a high reduction ratio.

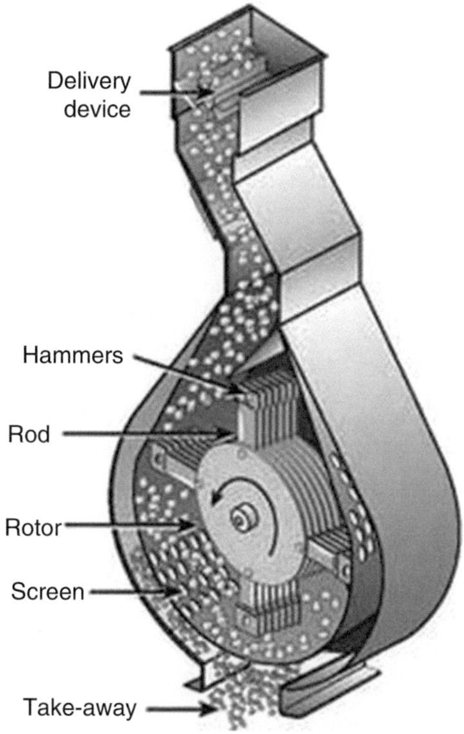

Fig. 6.14 Cross section of a hammer mill

6.5.4 Tumbling Mills

In most size reduction circuits, the greatest proportion of the comminution power is used in tumbling or stirred mills. Milling devices can be categorised in a number of ways depending on whether they are operated with or without grinding media (most commonly spherical steel or ceramic balls), continuously or in batch mode, and wet or dry.

- **Semi-autogenous grinding** (SAG) and **autogenous grinding** [32] mills are cylindrical tumbling mills predominantly operated wet in a continuous mode, however milling rates without media (autogenous grinding) are generally significantly lower. SAG mills can accept particles up to 300 mm and the P_{80} (aperture size that 80 % of product pass that size) of product is in the range of 800–500 μm. SAG mills are usually operated with a charge containing rock and steel balls in an approximate ratio 1:1 by volume, resulting in breakage between rock and steel as well as rock and rock. SAG and AG mills are generally installed in a primary milling duty receiving primary crushed (−250 mm) feed. The largest units in the minerals industry have shells up to 12.6 m in diameter, driven with motors up to 28 MW and able to process up to 6000 t/h. Figure 6.16

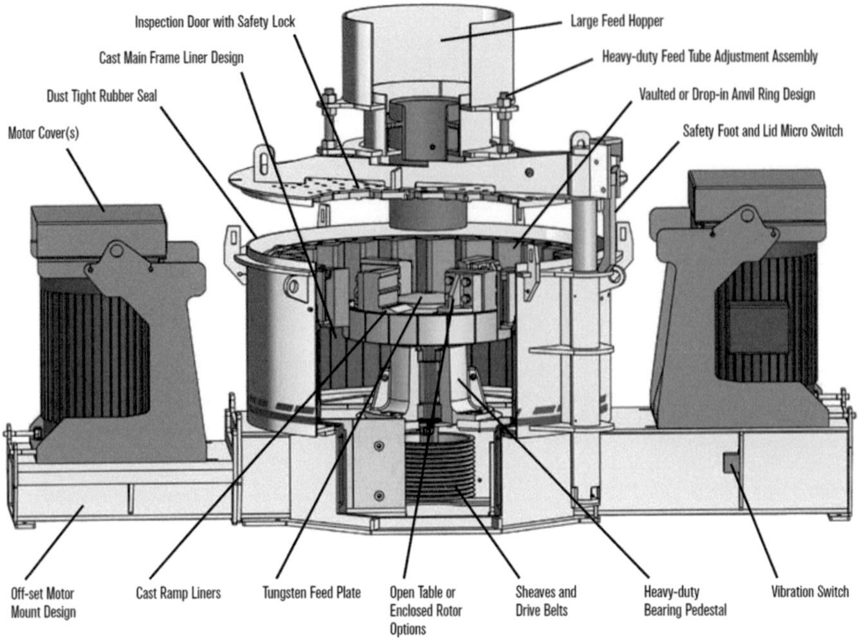

Fig. 6.15 A Terex® Canica VSI (Courtesy of Terex®)

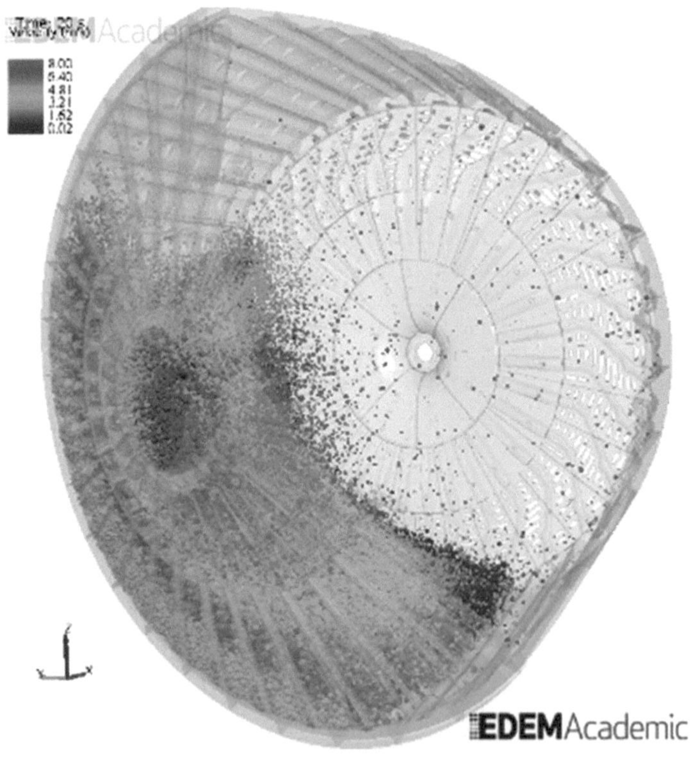

Fig. 6.16 Discrete element model (DEM) simulation of charge motion in a SAG mill (Courtesy of JKMRC (Nirmal Weerasekara))

illustrates Discreet Element Model (DEM(simulation of charge motion in an industrial SAG mill.

- **Ball mills** are cylindrical tumbling mills with steel balls as the grinding media and a small fraction of ore feed material that is far smaller than the balls. They contain a high proportion of grinding media in the charge, often approximately 30 % of the mill volume. Use of smaller grinding media will increase surface area available for grinding, but will have less force for breaking larger particles. Ball mills are used in a wide range of industries. Particle size reduction is efficient down to around 100 μm when operated in closed circuit, but are commonly used to produce product of 60 μm and sometimes even as fine as 20 μm.

6.5.5 *Stirred Mills*

There are many types of fine grinding devices used in industry such as ball mills, stirred mills, jet mills, planetary mill and Kelsey Axial Displacement (KAD) mill. Stirred mills use a static vessel with either a vertical or horizontal agitator to impart motion to the grinding media contained in the vessel. These mills are commonly used for wet grinding duties where the product size is below 100 μm and as small as 20 nm. The stirred mills are well known to produce a steeper particle size distribution that enhances flotation and leaching [9, 46].

Stirred mills can be categorised into two groups i.e. gravity induced and fluidised mills. These mills are grouped based on the stirrer speed. Typically the gravity induce mill has the tip speed of 3 m/s whilst the fluidised mill ranges from 10 to 23 m/s. These mills can be further categorised based on orientation i.e. vertical and horizontal. Table 6.2 shows the grouping of the stirred mills that are commonly used in mining industry.

Stirred mills can be grouped into low intensity and high intensity mills:

- **Low intensity mills** are typically continuous gravity induced **vertical stirred mills**, meaning that the grinding media is contained within the mill by gravity. Examples include the Tower mill, VertiMill® and HIGmills™. Using a lower stirrer speed and lower energy intensity, low cost media can be used. Grind sizes down to approximately 20 μm can be achieved at a high throughput. Figure 6.17 shows a Metso VertiMill®.

Table 6.2 Stirred mill category

Gravity induced mill	Fluidised mill
Vertical orientation	Vertical orientation
VertiMill®	Stirred media detritor
Tower mill	Outotec HIGmills™
CITIC vertical mill	VXP mill
	Horizontal orientation
	IsaMill

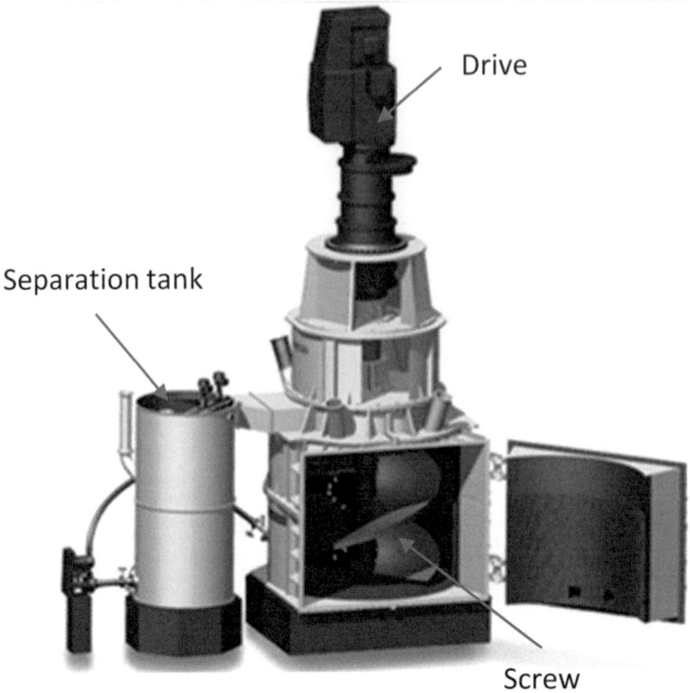

Fig. 6.17 A Metso VertiMill® (Courtesy of Metso)

- **High intensity mills**, known as **agitated bead mills** or agitated media mills, are vertical or horizontal stirred mills with fine grinding media, usually ceramic, from around 6 mm to 50 μm in size. These mills are used in a wide range of industries from pharmaceuticals, food industry and pigments to minerals. Mills can be operated in batch mode at a laboratory through to 10 m^3 continuous industrial units. Sub-micron products can be produced. Continuous units such as the IsaMill™ have an internal centrifugal classifier for removing fine product from the mill while retaining the grinding media (Fig. 6.18).

6.5.6 *Other Mills*

Grinding and size reduction can be accomplished in numerous other ways. These are often used for processing relatively small volumes of material.

- **Planetary ball mills** achieve size reduction by vibrating the entire vessel containing the grinding media and particles. These are usually operated in batch mode (Fig. 6.19).
- **Vibratory disc mills** have a single grinding disc or a series of concentric grinding rings inside a circular mill which is vibrated with a circular motion.

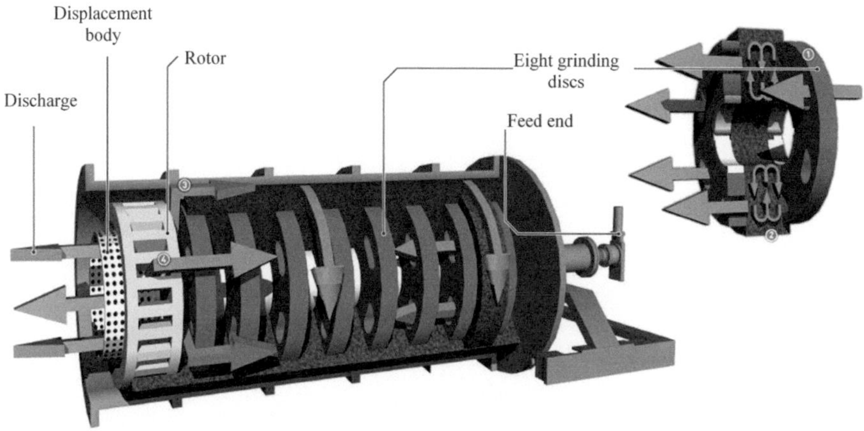

Fig. 6.18 A IsaMill™ (Courtesy of Glencore Xstrata)

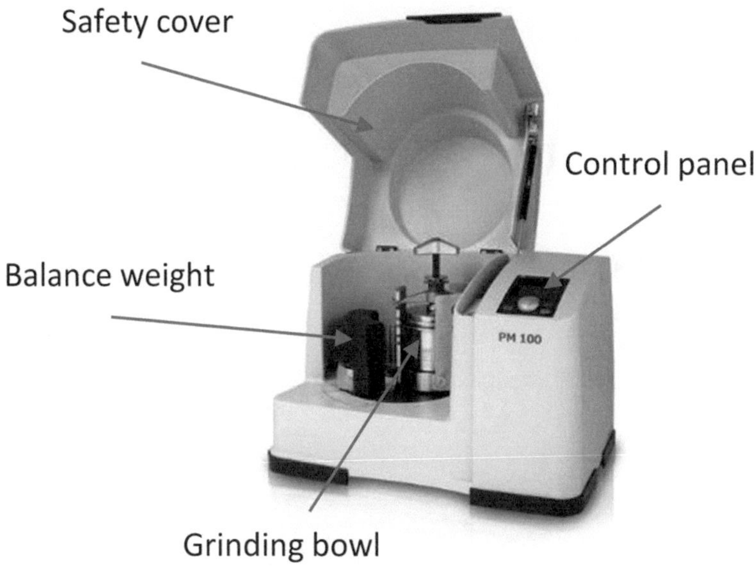

Fig. 6.19 Retsch PM 100 planetary ball mill (Courtesy of Retsch)

These are suited to crushing and pulverising material in a short time. Figure 6.20 shows a laboratory vibratory disc mill.

– **Jet mills** use fine streams of particles driven by high-speed gas flow to break particles by autogenous impacts down to 1–10 μm (Fig. 6.21). These mills have advantages of minimal contamination or heating of particles; while narrow product size distributions can be achieved due to the internal classification.

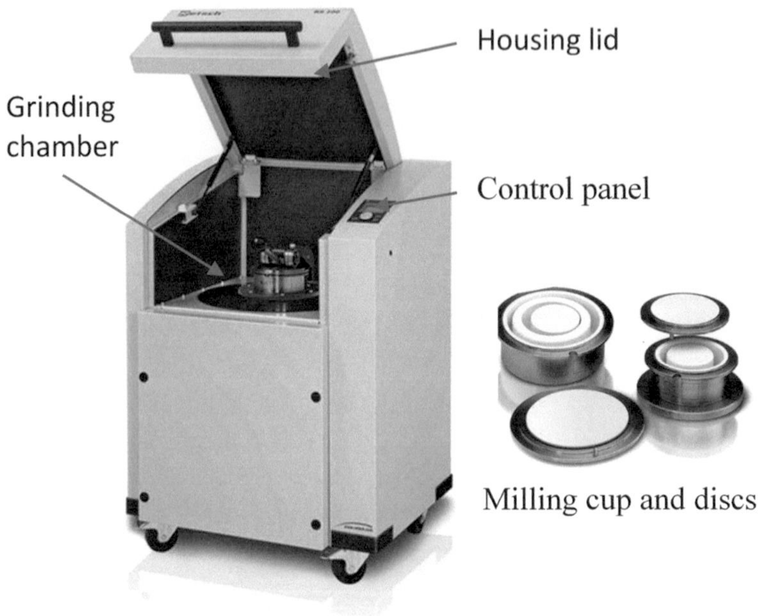

Fig. 6.20 Vibratory disc mill RS 200 (Courtesy of Retsch)

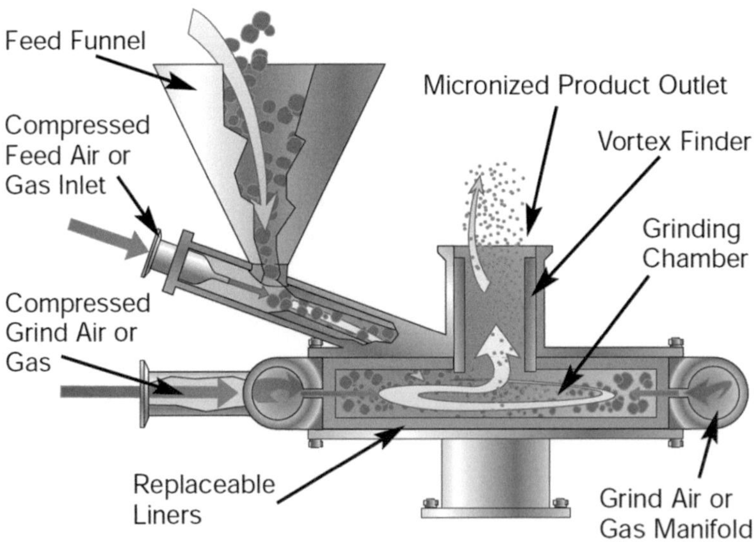

Fig. 6.21 Cross section of a jet pulverizer (Courtesy of Sturtevant)

- **Disc and pin mills** are type of vertical shaft impactor mill, consisting of two rotating disks with pins embedded on one face. The disks are arrayed parallel to each other so that the pins of one disk face those of the other. The substance to be homogenized is fed into the space between the disks and either one or both disks are rotated at high speeds. Pin mills can be used on both dry substances and liquid suspensions [51]. Pin mills are commonly use in the manufacture of pharmaceuticals, as they can achieve particle sizes as fine as a few micrometers [51]. However, heat generated by friction can sometimes be a concern. A manual form of the mill is commonly used to grind marijuana to a useful consistency.

6.5.7 Non-mechanical Breakage

Several techniques have been developed for size reduction using non-mechanical methods. While application of direct mechanical force to particles is often the most effective, non-mechanical methods usually aim to selectively target particular phases within particles. Non-mechanical methods which generate regions of high temperature have the potential to alter the properties of the material being broken.

- It has been known since ancient times that certain rocks are weakened by simple **thermal heating** and quenching. The energy requirements have been shown to be significantly greater than for mechanical methods.
- **Microwave energy** excites certain minerals preferentially while silicate minerals are typically transparent to microwaves. This allows heating energy to be applied selectively. By exploiting differences in heating rate and thermal expansion, inter-granular fracture can be promoted thus causing preferential liberation of target minerals.
- High-voltage **electrical discharge breakage** uses repeated plasma discharges exploiting differences in dielectric properties of the material phases to achieve inter-granular fracturing. The effectiveness of this method is dependent on the material being processed.

6.6 Comminution Circuits

The purpose of comminution circuits may be [26]:

- Liberating valuable minerals for downstream concentration processes such as flotation, gravity separation, magnetic separation, etc.;
- Introduce fractures or increase the surface area available for chemical or bio-chemical reactions such as leaching or bio leaching;
- Produce particles of required size and shape such as in aggregates and cement industry.

Table 6.3 Common comminution unit operations in mining context [24, 40]

Crushers	Tumbling mills	Stirred mills
Jaw crushers	AG and SAG mills	Tower mills
Gyratory crushers	Ball mills	Vertical shaft mills
Cone crushers	Rod mills	Horizontal pin mills
Roll crushers		Vertical roller mills
HPGR		
Impact crushers		
Size classification	**Materials handling**	
Screens	Bins	
Sieve bends	Sumps	
Hydrocyclones	Feeders	
Air classifiers	Conveyor belts	
	Pumps	
	Combiners	
	Separators	

A typical comminution circuit consists of a number of unit operations connected together with material handling units. Materials enter the circuit and pass successively through the size reduction and classification units to reach the desired quality. The crushing section of a comminution circuit treats dry materials while grinding can be conducted either in dry or wet mode. In mining context, process units found in a typical comminution circuits include crushing, milling, size classification, and materials handling and storage [24, 40]. Some common equipment in each operation unit are summarised in Table 6.3.

6.6.1 Typical Comminution Circuits

A comminution circuit generally starts with one or more stages of crushing and screening in dry mode. A typical flow sheet of a crushing circuit to produce feed for ball mills is illustrated in Fig. 6.22.

The grizzly screen prevents fine material entering the Jaw crusher. A double deck screen provides appropriate sizes to be fed to secondary and tertiary cone crushers. Material fine enough to be fed to the ball mill circuit (generally smaller than 12 mm [75]) will leave the circuit as the screen undersize.

AG and SAG mills require coarse particles (i.e. +100 mm) as grinding media, therefore secondary and tertiary crushing stages are not required and the product of the primary crusher can fed directly to the AG/SAG mills. Figure 6.23 illustrates the layout of a primary crushing unit followed by a typical **SA**G mill, **B**all mill grinding circuit closed with **C**yclones – known as an SABC circuit.

After successful implementation of HPGR grinding in the cement industry, application of this equipment in base metal processing circuits has increased

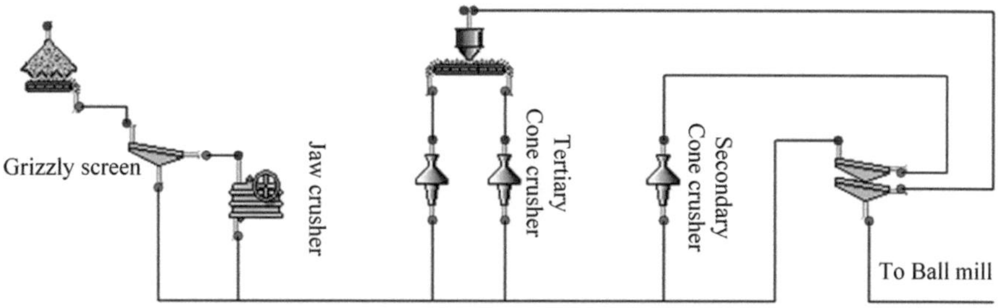

Fig. 6.22 Typical three-stage crushing circuit before ball mill grinding

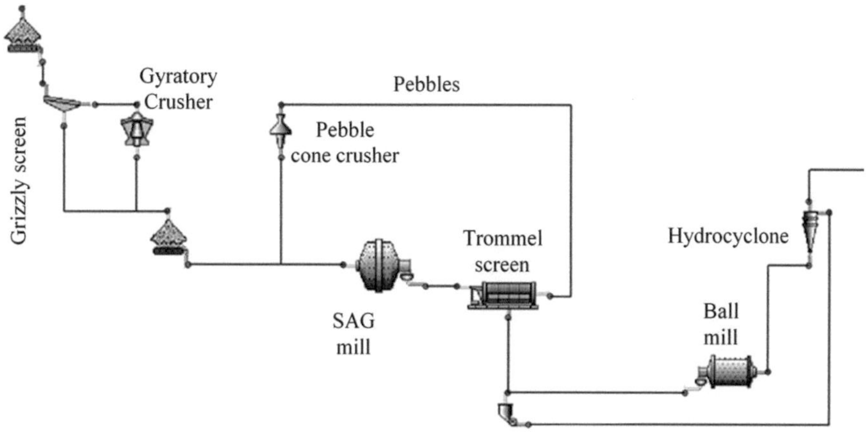

Fig. 6.23 A typical primary crushing unit followed by a SABC comminution circuit

significantly during past years. An HPGR can accept feed size smaller than 90 mm and produce product finer than 4 mm [68]. For a finer top-size, below 40, the product can have a P_{80} of under 1 mm when closed with a screen. Therefore, the HPGR is usually implemented after a secondary crushing stage, or tertiary for a finer feed top size, to produce material in the right size range for ball mills or stirred media mills. Figure 6.24 illustrates a typical crusher, HPGR circuit followed by ball mill grinding. Primary crushed ore is sent to a secondary cone crusher closed with a screen to prevent coarse particles passing to the HPGR.

A feed stock pile ensures fluctuations from the crushing circuit do not pass to the HPGR. It is possible to use a screening stage after the HPGR to recycle coarse material to the HPGR and provide feed with appropriate size distribution for ball mill circuit.

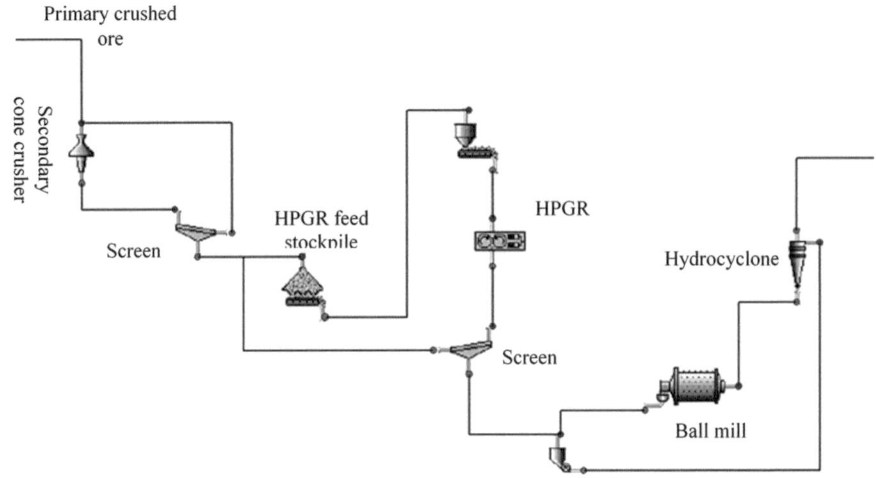

Fig. 6.24 A typical crusher, HPGR circuit followed by ball mill grinding

6.6.2 Flexible Circuit

The increasing demand for natural resources along with the decreasing the grade of ore bodies has driven the plant designers and operators to seek a new approach in designing more efficient mineral processing circuits. Powell and Bye [48] proposed a new approach in circuit design which allows for multi-stream and multi-grade circuits. Flexible circuit design incorporates current technology into the structure of a flexible process which allows the uptake of future advances in processing capabilities. This is unlike the current approach in expansion of the projects through doubling up the existing circuit.

An example of a flexible circuit design is illustrated in Fig. 6.25. The circuit has a standard SABC circuit, but with a screen section added upfront and a HPGR. The screening allows for waste sorting, narrows down the feed sizes for the SAG mill and cone crusher, and incorporates a variable split of finer feed to the SAG mill or HPGR. At different stages of the circuit, waste sorting can allow the rejection of coarse waste. Variable splitters are key control points to deal with variations of ore competence and size. Comminution equipment with different capabilities provides flexibility to handle fluctuations in feed characteristics [48].

Flexible circuits allows balancing the comminution work among comminution units as the feed characteristics fluctuates. Controlling the operation using the concept of flexible circuits provides the ability to maintain the metal production while the ore type changes [19].

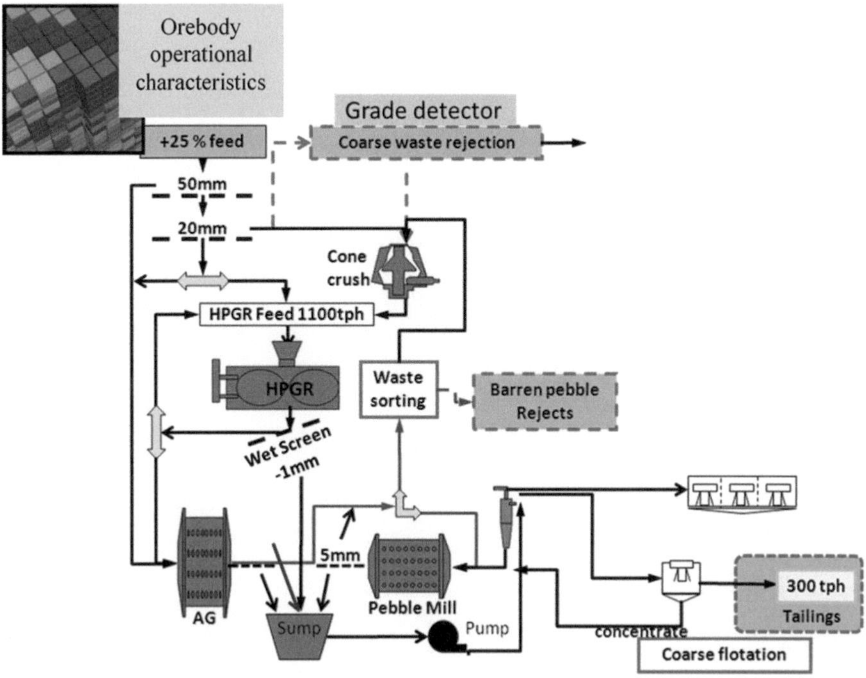

Fig. 6.25 A flexible circuit [48]

6.6.3 *Process Operation Control*

Developing successful process control strategies depends on having access to sensors that are capable of providing accurate and robust measurements. In the mineral processing context a variety of sensors are in use.

In crushing circuits sensors are implemented to measure bins and chute levels, frequency of feeders, lubrication flow rate and temperature, crusher power draw, mass flow rate, particle size distribution (not common), etc.

In grinding circuits the sensors measure parameters such as: pulp density, volumetric flow rate, sump levels, cyclone pressure, mill bearing pressure, mill noise, mill speed, mill power draw, particle size distribution (usually at cyclone overflow), etc.

In the primary crushing circuit, generally the process control objective is protection of the equipment. Sensors of lubrication flow and temperature along with high and low level alarms of bins and crusher chutes are implemented for protection purposes. Secondary and tertiary crushers usually are controlled to maintain the optimum throughput of the circuit while providing product at the required size. Throughput, crusher CSS and feed top size are the three factors affecting crusher performance. An effective method to maximize throughput, is to operate the crusher close to maximum possible power draw. In crushing which is closed with a screen, there is an optimum CSS which maximizes throughput at a certain power draw. Opening the CSS increases the crusher throughput, however due to an increasing recirculating load the throughput only increases up to a certain point before dropping off.

Operating a crusher in choke fed condition and even distribution of feed enhances performance of crushing. This can be controlled by adjusting feeder speed by monitoring crusher power draw and level of material in the crusher bowl, while ensuring that the feed material drops vertically onto the center of the crusher feed chamber.

Controlling the operation of SAG mills depends on available instruments as well as available control elements. In AG/SAG mills with fixed speed, power draw can be controlled by adjusting the mill filling through manipulating feed rate. Factors such as ratio of coarse (+100 mm) to fine particles, ore competence, ball filling and ball size affect the operation. Variable speed mills provide mill speed as another control element in addition to feed rate. Interactions between mill speed, mill filling and power draw are complex and their relationships are different at different operations. In a normal operation, increasing mill speed increases the power draw while producing coarser product and increased throughput. On the other hand, reducing mill speed will result in finer product at a reduced throughput.

Ball mill circuits are usually closed with a classification stage to control the size distribution of product according to the requirements of downstream process. The main objective in closed circuits is to remove final product as quick as possible to prevent overgrinding of final product. Therefore, performance of the classification stage has a significant impact on grinding performance. Ball filling and top size of balls control the energy input and grinding rate for coarse and fine sizes. Reducing the ball size will increase throughput due to providing more surface area at the cost of higher media wear rate. Increasing the ball load improves the power draw and throughput. Water addition to the mill controls the pulp viscosity – which affects grinding efficiency – being the key on-line control parameter along with control of the cut size of the classification circuit that closes the ball mill circuit.

6.7 Fine Grinding Circuits

Fine grinding is an essential process in many industries such as minerals for valuable mineral liberation, pharmaceutical to increase the medicine reactivity and paint for producing ultrafine fillers. This section will focus on stirred mills which are widely used in the mineral industry for fine grinding duty.

Stirred mills were initially installed for regrinding duties (i.e. regrinding rougher concentrate), but lately gravity induced stirred mills have been installed for secondary and tertiary grinding duties [47]. Figure 6.26 shows the application of stirred mills in a typical gold-copper concentration circuit. In order to provide extra grinding power and finer grind size it is possible to install stirred mills in secondary, tertiary and regrind milling duties in a typical SABC circuit [46].

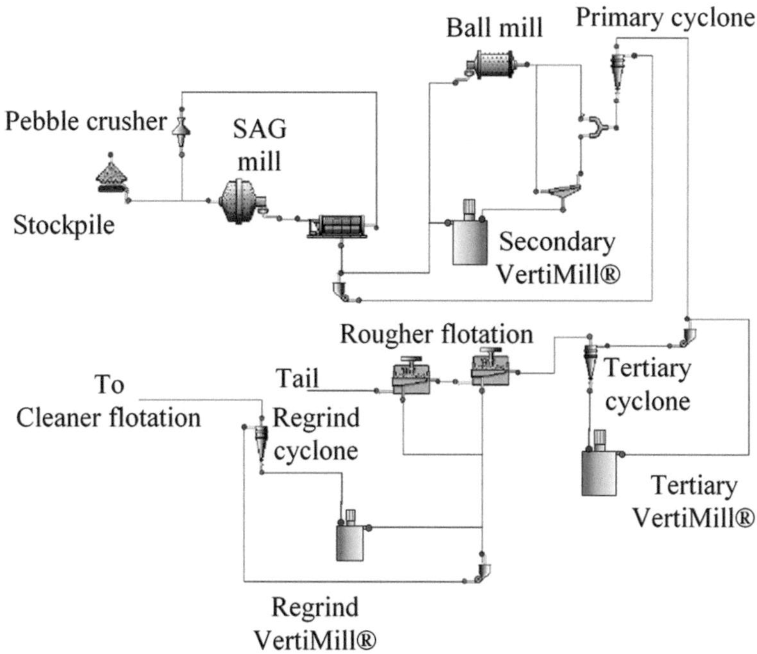

Fig. 6.26 Application of stirred mills as secondary, tertiary and regrind mills in a typical SABC circuit

Table 6.4 Feed and product size of stirred mills

Grinding duty	F_{80} (μm)	P_{80} (μm)
Secondary	450	175
Tertiary	150	90
Regrind	40	7

Gravity induced mills are suitable for all the three grinding duties (i.e. secondary, tertiary and regrind) whilst the fluidised mill is commonly used for ultra-fine grinding of rougher concentrate. Table 6.4 shows the feed and product size distribution of the stirred mill. These results indicate that stirred mills have the capability to grind in a size range overlapping with the ball milling regime. Gravity induced mills have the capability to receive feed up to 6 mm with coarser media (40 mm).

6.7.1 Operational Variables of Stirred Mills

Table 6.5 shows the typical operational variables of a stirred milling circuit. The variables can be categorised into process variables and mill configuration.

Table 6.5 Operational variables of stirred milling circuit

Process	Mill configuration
Particle size and shape distribution	Stirrer design
Slurry density	Tip speed
Feed rate	Media size/type/density
Rheology	

6.7.2 Effect of Operation Variables

Many operational parameters affect the performance of fine grinding circuits.

- **Feed size**
  The feed size is essential as the stirred mills operate in a narrow particle size distribution. For example, gravity induced mills are capable of grinding 6 mm particles whilst the fluidised mill can only accept feed below 500 μm. It is essential to control the top size entering the mill as coarse particles in the feed could build up in the mill which is detrimental to mill performance. Therefore a protection screen is essential to prevent coarse particles entering the circuit.
- **Slurry density**
  Typically the appropriate solid content for gravity induced mills is 65–75 % while for fluidised mills it is 50 %. If the slurry density is too high, the mill losing its grinding efficiency as the kinetic energy of the media will be reduced, whilst low slurry density results in higher media wear.
- **Feed rate**
  There is a critical feed rate for any size of stirred mills. If the feed rate is above that value, due to shorter residence time of particles the mill will start losing its efficiency which leads to a coarse product. Low feed rates will lead to finer particle size distribution as the particles have longer residence time in the grinding zone, but if too low then grinding power is wasted.
- **Rheology of slurry**
  Rheology is mainly controlled by solid content and amount of fines. High slurry viscosity will lead to a reduction in media kinetic energy which consequently reduces grinding efficiency.
- **Stirrer type**
  Figure 6.27 shows different stirrer configurations in stirred mills. The gravity induced mill uses a screw type stirrer whilst the fluidised mill uses pin and disc type stirrers. The stirrer agitates the media for grinding purposes.
- **Stirrer tip speed**
  Stirrer tip speed is a key parameter to control the power draw – increase in tip speed will increase mill power draw, leading to finer grind size.
- **Grinding media**
  Typically grinding media fills approximately 70–80 % of the mill volume. The amount of media in the mill is another of the parameters that control the power draw. The choice of appropriate media is essential to achieve a targeted grind

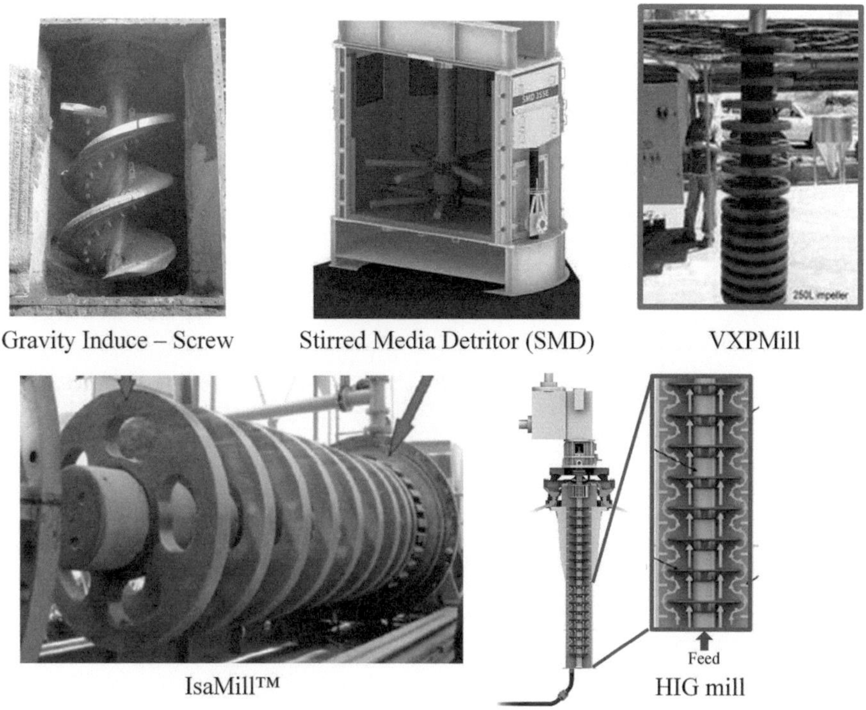

Fig. 6.27 Stirrer types in different stirred mills

size. Type of media (steel or ceramic) is also important in applications where contamination of product is a problem. There are three important media criteria:

1. **Media size**: The choice of media size depends on its ability to break the top particle size. Small media might have capability to produce fine particles but may not be able to break the top size particles.
2. **Media type**: The shape of media has an effect on the product size distribution. Recent studies show cylpebs (i.e. slightly tapered cylindrical grinding media with a ratio of length to diameter of unity) have the ability to break the top size with reduced generation of fines compared to spherical media [55].
3. **Media density**: Gravity induced mills use steel grinding media whilst fluidised mills can use either steel or ceramic media. Media density affects the stressing intensity of collisions between media and particles. Higher media density will provide higher stressing conditions in the mill – leading to finer grind size. The media density ranges from 2.7 to 6.2 g/cm^3.

6.7.3 Configuration of Stirred Mills

Figure 6.28 shows the typical circuit configuration of a stirred mill unit. Stirred mills are typically operated in closed circuit either with hydrocyclones or fine screens.

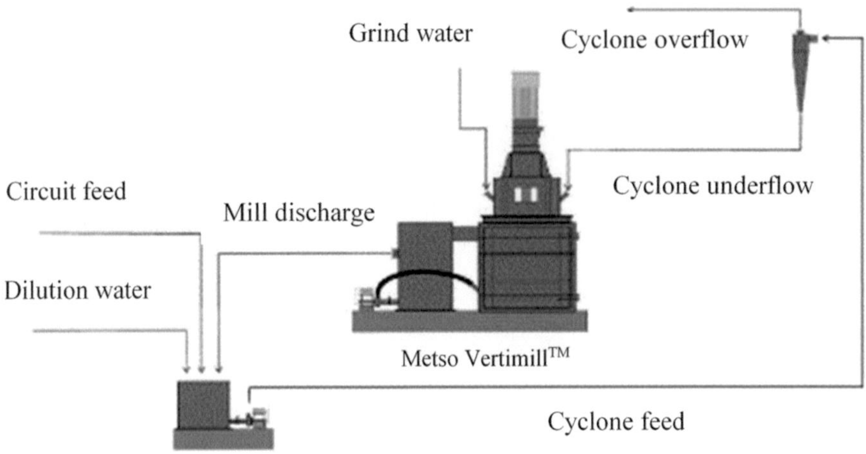

Fig. 6.28 Gravity induced stirred mill circuit configuration [11]

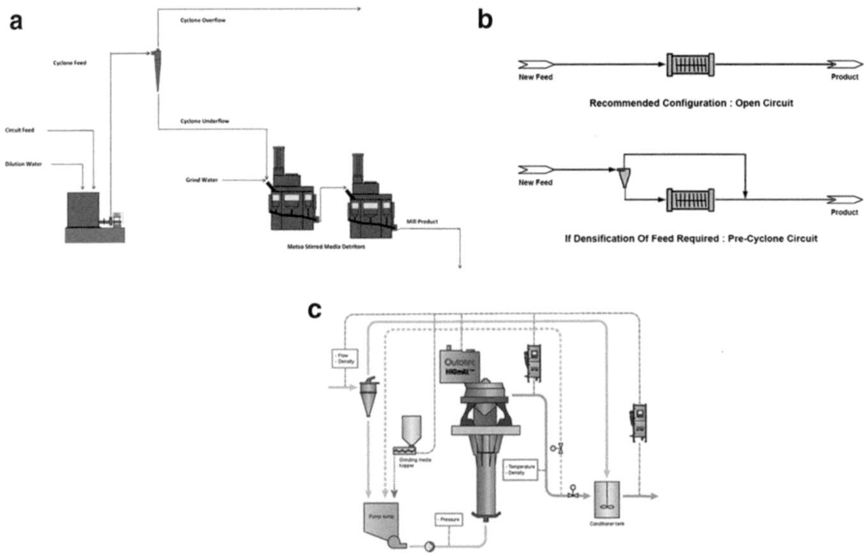

Fig. 6.29 Fluidised mills circuit configuration. (**a**) SMD circuit. (**b**) IsaMill™ circuit. (**c**) HIG mill circuit [9, 11]

Stirred mills can be fed either from top or bottom of the mill. Furthermore, it can be installed with or without a separating tank. A portion of mill discharge can be recycled to the feed of the mill. Cyclone overflow or screen undersize is the unit product.

The fluidised mills are normally operated in an open circuit configuration. Figure 6.29 shows the typical circuit configuration of three types of fluidised mills. There is an option to install scalping cyclones to remove fines before entering the mills. The cyclone overflow combines with mill discharge as the circuit product.

6.8 Definitions, Abbreviations and Symbols

Comminution	the action of reducing size of mineral ore
Gangue	invaluable or waste materials

AG mill	autogenous grinding mill (a mill uses coarse rock particles as grinding media)
CSS	close side setting
Cylpebs	slightly tapered cylindrical grinding media with a ratio of length to diameter of unity
DWT	drop weight tester
HIGmillTM	high intensity grinding mill (Outotec)
HMI	human machine interface
HPGR	high pressure grinding rolls
JKFBC	JK Fine-particle Breakage Characteriser
JKMRC	Julius Kruttschnitt Mineral Research Centre
JKRBT	JK Rotary breakage tester
KAD	Kelsey Axial Displacement
SABC	SAG mill, ball mill grinding closed with cyclone
SAG mill	semi-autogenous grinding mill (a grinding mill uses mixture of coarse rocks and balls as grinding media)
SILC	short impact load cell
SMC test	a reduced version of the drop weight tester
UCS	uniaxial compressive strength
UFLC	ultrafast load cell
VFD	variable frequency drive
VRM	vertical roller mills
VSI	vertical shaft impactor

A, b	ore impact breakage parameters determined from the twin pendulum or DWT results
C	machine design constant that governs the maximum possible impact velocity at a given rotor speed and set of operational conditions
D	particle size
Dw_i	drop-weight index
E	mass-specific impact energy (J/kg)
E_{min}	threshold energy below which breakage does not occur (J/kg)
F_{80}	aperture size, for which 80 % of feed passing (μm)
f_{mat}	material breakage property described by a size-dependent sub-model (kg/J/m)

k	successive number of impacts with a single-impact energy
K	constant in Morrell's energy-size equation (Eq. 6.7)
m	particle mass (kg)
M	maximum t_{10} for a material subject to breakage (%)
M_i	comminution index (kWh/t) in Morrell's energy-size equation (Eq. 6.7)
N	rotor speed (rpm)
Ow_i	operating work index (kWh/t)
P_1	closing sieve size in Bond ball mill work index test (μm)
P_{80}	aperture size, for which 80 % of product is passing (μm)
r	rotor radius (m)
t_{10}	percentage of progeny passing an aperture of on tenth of the original particle size
V_i	impact velocity (m/s)
W	mass specific energy (kWh/t) calculated as power draw divided by throughput (t/h)
W_i	bond work index (kWh/t)
x	aperture size or a particle passing that size
x_1	equivalent to F_{80} size (μm)
x_2	equivalent to P_{80} size (μm)
x_g	geometric mean size of feed particle (m)
ϵ	Young's modulus
ν	Poisson's ratio

References

1. Ballantyne, G.R., Powell, M.S.: Benchmarking comminution energy consumption for the processing of copper and gold ores. Miner. Eng. **65**, 109–114 (2014)
2. Banini, G.A.: An Integrated Description of Rock Breakage in Comminution Machines [Online]. The University of Queensland, St Lucia (2000)
3. Belidor, B.F.: Nouveau cours de mathématique à l'usage d'Artillerie et du Génie. Chez Charles-Antonie Jombert, Paris (1725)
4. Bond, F.C.: The third theory of comminution. Trans. AIME **193**, 484–494 (1952)
5. Bond, F.C.: Crushing and grinding calculations, part I. Br. Chem. Eng. **6**, 378–385 (1961)
6. Bond, F.C.: Crushing and grinding calculations, part II. Br. Chem. Eng. **6**, 543–548 (1961)
7. Bourgeois, F.S., Banini, G.A.: A portable load cell for in-situ ore impact breakage testing. Int. J. Miner. Process. **74S**(65), 31–54 (2002)
8. Brown, G.J., Miles, N.J.: Applied fractal geometry in impact pulverisation. The XIX International Mineral Processing Congress. San Francisco (1995)
9. Burford, B.D., Clark, L.W.: IsaMill™ technology used in efficient grinding circuits. VIII International Conference on Non-Ferrous Ore Processing. Poland (2007)
10. Curry, J.A., Ismay, M.J.L., Jameson, G.J.: Mine operating costs and the potential impacts of energy and grinding. Min. Eng. **56**, 70–80 (2014)
11. Davey, G.: Fine grinding applicants using the Metso VertiMill grinding mill and the Metso Stirred Media Detritor (SMD) in gold processing. 38th Annual Meeting of the Canadian Mineral Processors Ottawa, Canada (2006)

12. DOE: Comminution and Energy Consumption: Report of the Committee on Comminution and Energy Consumption. National Materials Advisory Board, Commission on Sociotechnical Systems, National Research Council, Washington, DC (1981)
13. DOE: Mining industry of the future: energy and environmental profile of the U.S. mining industry. In: U.S. Department of Energy & Office of Energy Efficiency and Renewable Energy (eds.). BCS, Incorporated. U.S. Energy Information Administration (EIA), Washington DC (2002)
14. DOE: Mineral industry energy bandwidth study. U.S. Department of Energy – Industrial Technologies Program (2007)
15. Dorai, S.V.: Energy Analysis of Gold Mining Plants: Case Studies and Technology Analysis – An Australian Case Scenario (MSc). Masters of Science (Industrial Ecology), Chalmers University of Technology, Sweden (2006)
16. Dundar, H., Benzar, H., Aydogan, N.: Application of population balance model to HPGR crushing. Min. Eng. **50–51**, 114–120 (2013)
17. Evertsson, C.M., Bearman, R.A.: Investigation of interparticle breakage as applied to cone crusher. Min. Eng. **10**, 199–214 (1997)
18. Fischer-Cripps, A.C.: Introduction to Contact Mechanics, 2nd edn. Springer, New South Wales (2007)
19. Foggiatto, B., Hilden, M.M., Powell, M.S.: Advances in the simulation of Flexible circuits. 12th Mill Operators' Conference. AusIMM, Townsville (2014)
20. Gilvarry, J.J., Bergstrom, B.H.: Fracture of brittle solids, II. Distribution function for fragment size in single fracture experimental. J. Appl. Phys. **32**, 400–410 (1961)
21. Griffith, A.A.: The phenomena of rupture and flow in solids. Philos. Trans. R. Soc. Lond. A **221**, 163–198 (1921)
22. Hukki, R.T.: Proposal for a solomonic settlement between the theories of von Rittinger, Kick and Bond. Trans. AIME **223**, 403–408 (1962)
23. Kick, F.: Das Gesetz des proportionalen Widerstands und seine Anwendug. In: Leipzig, F. (ed.), Leipzig, Germany (1885)
24. King, R.P.: Modeling and Simulation of Mineral Processing Systems. Butterworth-Heinemann publications, Boston (2001)
25. La Nauze, R.D., Temos, J.: Technologies for sustainable operations. Council for Mining and Metallurgical Institutions Congress. Cairns, (2002)
26. Lynch, A.J.: Mineral Crushing and Grinding Circuits – Their Simulation, Optimisation, Design and Control. Elsevier Scientific Publishing Company, New York (1977)
27. Lynch, A.J., Rowland, C.A.: The History of Grinding. Society for Mining, Metallurgy and Exploration, Inc. (SME), Littleton (2005)
28. Man, Y.T.: A Model-Based Scale-Up Procedure for Wet, Overflow Ball Mills. Ph.D., The University of Queensland, Australia (1999)
29. Marktscheffel, M., Schönert, K.: Liberation of composite particles by single particle compression, shear and impact loading. 6th European Symposium Comminution. Nürnberg (1986)
30. Marsden, J.O.: Energy efficiency and copper hydrometallurgy. 6th International Symposium of Hydrometallurgy. Society for Mining, Metallurgy and Exploration (2008)
31. Mcsaveney, M.J., Davies, T.R.: Surface energy is not one of the energy losses in rock comminution. Eng. Geol. **109**, 109–113 (2009)
32. Morrell, A., Antony, S., Kohlhagen, G., Pommier, Y., Cushman, M.: Synthesis of benz[d] indeno[1,2-b]pyran-5,11-diones: versatile intermediates for the design and synthesis of topoisomerase I inhibitors. Bioorg. Med. Chem. Lett. **16**, 1846–1849 (2006)
33. Morrell, S.: An alternative energy-size relationship to that proposed by Bond for the design and optimisation of grinding circuits. Int. J. Miner. Process. **74**, 133–141 (2004)
34. Morrell, S.: Design of AG/SAG mill circuits using the SMC test. International Autogenous and Semi-Autogenous Grinding Technology. Vancouver, Canada (2006a)
35. Morrell, S.: Rock characterisation for high pressure grinding rolls circuit design. International Autogenous and SemiAutogenous Grinding Technology Vancouver, Canada (2006b)

36. Morrell, S.: A method for predicting the specific energy requirement of comminution circuits and assessing their energy utilisation efficiency. Min. Eng. **21**, 224–233 (2008)
37. Morrell, S.: Predicting the overall specific energy requirement of crushing, high pressure grinding roll and tumbling mill circuits. Min. Eng. **22**, 544–549 (2009)
38. Morrell, S.: Mapping orebody hardness variability for AG/SAG/crushing and HPGR circuit. International Autogenous and Semi-Autogenous Grinding Technology. Vancouver, Canada (2011)
39. Napier-Munn, T.: Is progress in energy-efficient comminution doomed? Comminution'14, 2014 Capetown, South Africa
40. Napier-Munn, T.J., Morrell, S., Morrison, R.D., Kojovic, T.: Mineral Comminution Circuits: Their Operation and Optimisation. Julius Kruttschnitt Mineral Research Centre, Indooroopilly (1996)
41. Napier-Munn, T.J., Morrell, S., Morrison, R.D., Kojovic, T.: Mineral Comminution Circuits – Their Operation and Optimisation. Julius Kruttschnitt Mineral Research Centre, Brisbane (2005)
42. Narayanan, S.S.: Development of a Laboratory Single Particle Breakage Technique and Its Application to Ball Mill Scale-Up. Ph.D., The University of Queensland, Australia (1985)
43. Narayanan, S.S., Whiten, W.J.: Determination of comminution characteristics from single particle breakage tests and its application to ball mill scale-up. Trans. Inst. Min. Metall. Sect. C: Min. Process. Extraction Metall. **97**, C115–C124 (1988)
44. Northparkes: Energy savings action plan: Northparkes mine. In: Bond, K., PTY Ltd (eds.), Australia (2006)
45. Oxford Dictionary: Oxford Dictionary of English. In: Stevenson, A. (ed.) Oxford Dictionary of English, 3rd edn. Oxford University Press. China Translation & Printing Services Ltd, China (2010)
46. Palaniandy, S., Powell, M., Hilden, M., Allen, J., Kermanshahi, K.: VertiMill® – Preparing the feed within floatable regime at lower specific energy. Comminution 14. Cape Town, South Africa (2014)
47. Palaniandy, S., Powell, M., Hilden, M., Kermanshahi, K., Allen, J., Mwansa, S.: VertiMill® – Development of circuit survey and performance evaluation protocols. Metplant 2013. Perth, Australia (2013)
48. Powell, M.S., Bye, A.R.: Beyond mine-to-mill – circuit design for energy efficient resource utilisation. Tenth Mill Operators' Conference. AusIMM, Adelaide (2009)
49. Rittinger, P.R.: Lehrbuch der Aufbereitungskunde. Ernst and Korn, Berlin (1867)
50. Rumpf, H.: Physical aspects of comminution and a new formulation of a law of comminution. Powder Technol. **7**, 145–159 (1973)
51. Sarabakos, G.D., Kosaropoulos, A.E.: Handbook of Food Processing Equipment. Kluwer Academic/Plenum Publishing, New York (2002)
52. Schilde, C., Burmeister, C.F., Kwade, A.: Measurement and simulation of micromechanical properties of nanostructured aggregates via nanoindentation and DEM-simulation. Powder Technol. **259**, 1–13 (2014)
53. Schönert, K.: Role of fracture physics in understanding comminution phenomena. Trans. Soc. Min. Eng. AIME **252**, 21–26 (1972)
54. Schubert, H.: On the microprocesses of comminution. Aufbereitungstechnik Tech. **5**, 237–246 (1987)
55. Shi, F.: Comparison of grinding media—Cylpebs versus balls. Min. Eng. **17**, 1259–1268 (2004)
56. Shi, F.: Coal breakage characterisation part 2: multi-component modelling. Fuel **117**, 1156–1162 (2014)
57. Shi, F., Kojovic, T.: Validation of a model for impact breakage incorporating particle size effect. Int. J. Miner. Process. **82**, 156–163 (2007)
58. Shi, F., Kojovic, T.: Comparison of impact breakage characterisation methods between the JK Rotary Breakage Tester and drop weight tester. International Autogenous and SemiAutogenous Grinding Technology Vancouver, Canada (2011)

59. Shi, F., Kojovic, T., Larbi-Bram, S., Manlapig, E.: Development of a rapid particle breakage characterization device-the JKRBT. Min. Eng. **22**, 602–612 (2009)
60. Shi, F., Xie, W.: A specific energy-based size reduction model for batch grinding ball mill. Mineral. Eng. **70**, 130–140 (2014)
61. Shi, F., Zuo, W.: Coal breakage characterisation part 1: breakage testing with the JKFBC. Fuel **117**, 1148–1155 (2014)
62. Sloan, R., Parker, S., Craven, J., Schaffer, M.: Expert systems on SAG circuits: Three comparative case studies. International Autogenous and SemiAutogenous Grinding Technology Vancouver, Canada (2001)
63. SMC WEBSITE: About the SMC test [Online]. http://www.smctesting.com/about (2015). Accessed 2 Feb 2015
64. Stamboliadis, E.T.: The energy distribution theory of comminution specific surface energy, mill efficiency and distribution mode. Min. Eng. **20**, 140–145 (2007)
65. Starkey, J., Dobby, G., Kosick, G.: A new tool for hardness testing. 26th Annual Meeting of the Canadian Mineral Processors Ottawa Canada (1994)
66. Tavares, L.M.: Breakage of single particles: quasi-static. In: Salman, A.D., Hounslow, M.J., Ghadiri, M. (eds.): Handbook of Powder Technology. Elsevier B.V., Oxford (2007)
67. Tavares, L.M., King, R.P.: Single-particle fracture under impact loading. Int. J. Miner. Process. **54**, 1–28 (1998)
68. Van Der Meer, F.P., Gruendken, A.: Flowsheet considerations for optimal use of high pressure grinding rolls. Min. Eng. **23**, 663–669 (2010)
69. Vogel, L., Peukert, W.: Determination of material properties relevant to grinding by practicable labscale milling tests. Int. J. Miner. Process. **74S**, 329–338 (2004)
70. Walker, W.H.: Principles of Chemical Engineering. McGraw-Hill, New York (1937)
71. Wang, Y., Forssberg, E.: Enhancement of energy efficiency for mechanical production of fine and ultra-fine particles in comminution. China Particuology **5**, 193–201 (2007)
72. Weibull, W.: A statistical distribution function of wide applicability. J. Appl. Mech. **9**, 293–297 (1951)
73. Weichert, R.: Correlation between probability of breakage and fragment size distribution of mineral particles. Int. J. Miner. Process. **22**, 1–8 (1988)
74. Weichert, R., Herbst, J.A.: An ultrafast load cell device for measuring particle breakage. 1st World Congress of Particle Technology. Nürnberg (1986)
75. Wills, B.A., Napier-Munn, T.J.: Will's Mineral Processing Technology. Elsevier Science & Technology Books, London (2006)
76. Yahyaei, M., Weerasekara, N.S., Powell, M.S.: Characterisation of superficial breakage using multi-size pilot mills. Min. Eng. **81**, 71–78 (2015)

Chapter 7
Atomization, Spraying, and Nebulization

Tevfik Gemci and Norman Chigier

Abstract An overview and current state-of-the-art of new developments in atomization, spraying and nebulization for the production of particulate materials are presented. The fundamental mechanisms of atomization physics, Newtonian and non-Newtonian atomization, primary and secondary breakups and disintegration of liquid jets and liquid sheets are briefly introduced, based on the up-to-date literature. The atomizers based on the energy input used to cause atomization are explained with their widely used application areas and their limitations. Spraying processes have a wide variety of industrial applications (automotive, aerospace, combustion, power, agriculture, food, metallurgy, environmental, and others) but in this book we focused only on the explanation of the spraying processes such as spray drying, thermal spraying and nebulization used in the production of particulate materials. Finally we provided a summary of the spray measurement techniques for the spray characterization and visualization.

7.1 Introduction

Liquid atomization is the transformation of Newtonian or Non-Newtonian fluids into liquid spray droplets. Disintegration of liquid or sheets into droplets (atomization process) can be achieved by the kinetic energy of the liquid itself (pressure atomizers), or by exposure to high-velocity air or gas (air-assist or twin-fluid atomizers), or by applying external mechanical energy through a rotating (rotary atomizers) or vibrating device (ultrasonic atomizers), or through the use of electrostatic charge (electrostatic atomizers). The cohesive and disruptive forces acting on the free surface of the jet emerging from the nozzle give rise to oscillations and

T. Gemci (✉)
Consultant Spray and Aerosol Dynamics, Long Beach, CA, USA
e-mail: tgemci@gmail.com

N. Chigier
Em. Prof., Carnegie Mellon University, Pittsburgh, PE, USA
e-mail: chigier@andrew.cmu.edu

© Springer International Publishing Switzerland 2016

H.G. Merkus, G.M.H. Meesters (eds.), *Production, Handling and Characterization of Particulate Materials*, Particle Technology Series 25,
DOI 10.1007/978-3-319-20949-4_7

disturbances (instabilities). Under appropriate conditions, such perturbations grow and the liquid jet disintegrates into droplets. This process is called atomization [11, 24, 27, 59].

The design, testing and analysis of spray combustion for combustors of gas turbine aircraft engines have reached the highest levels in the overall field of spray science and technology to meet the new challenges for (1) high-thrust, low-weight engines for fighter aircraft and (2) low NO_x, high-thrust, low-weight engines for advanced commercial supersonic aircraft.

Agricultural spraying of herbicides, fungicides and insecticides is carried out on a huge scale in all countries using aircraft and tractors. If the drop size and velocity impingement are too large, the crop may be damaged. If the drop size and velocity are too small, the spray is carried by the wind away from the target and may cause damage to a neighboring crop. Foliage is usually dense and there is a requirement to cover both the top and underside of foliage in order to achieve complete coverage of the plants. For insect control, drops must impact directly on insects and penetrate vital organs in order to kill insects.

In the metal processing industry, arc and plasma guns are used to melt metals and other materials and to generate liquid metal and ceramic sprays. To avoid oxidation, liquid metal sprays are introduced into inert gas environments (usually nitrogen). By controlling heat transfer from the gas to the drops, rates of crystal growth during rapid solidification are also controlled. Metal, alloy and ceramic powders are collected and then by sintering and fusing, materials are produced with special material properties which have important applications in the aerospace industry. Liquid metals and ceramic sprays are used to manufacture tools, dyes, gear wheels and a wide variety of objects with complex shapes. Instead of casting and machining materials, material shapes are formed by spraying layer upon layer of materials onto substrates. Robots are used in mass production for forming metal, alloy, ceramic and composite material products. Drop size, velocity, number density, flux and temperature all have critical and determining influences, film formation, adhesion and crystal growth. The material properties, ductility and tensile strengths are all directly related to spray characteristics in the impingement zone.

Spray painting and coating constitute huge industries. Automobiles and other vehicles are painted by robots. Up to 30 % of the spray paint may miss its target. As the spray approaches the coating surface, the air jet is deflected and carries small paint particles away from the surface.

In the carbon-black industries, heavy fuel oil is burned under rich equivalence ratio conditions to promote the formation of soot. There are several hundred different types of soot that are required for applications such as the manufacture of tires and rubber products. In addition to temperature and air/fuel ratio, the spray characteristics play a crucial role in the quantity and quality of soot that is produced. Only recently has this industry begun to use advanced laser diagnostic instrumentation for measurement of spray characteristics. There are very many other chemical processing plants that use spray drying and other spraying processes.

In the food processing industry, spray-drying is used very extensively to remove moisture and produce dry-packaged foods and powders. These foods in liquid form are usually non-Newtonian slurries of solids and liquids and include long-chain polymers. The atomization of non-Newtonian liquid is significantly different from the atomization of Newtonian liquids. Shear and elongational viscosities can be several orders of magnitude larger. Very little is known about the influence of rheological properties on the atomization of non-Newtonian liquids, but fundamental research studies have been initiated.

In the field of medicine, oral and nasal sprays are used very extensively by millions of persons. Meter-dosed and other type inhalers are used by patients suffering from bronchial and pulmonary illnesses. Drops and particles of inhaled medication are expected to pass through the throat and bronchial air passageways and deposit onto the surface of the lungs. If inhaled particles are too large (larger than 5 μm), they will deposit onto the walls of the mouth, throat or bronchial tubes and will not reach the lung surfaces. If particles are too small (less than 0.5 μm), they may be inhaled and exhaled due to the very high drag/momentum ratio. Inhalation velocities of air vary over a very wide range of the spectra of individuals including newborn infants, athletes, the elderly and patients too ill to be able to breathe effectively. Physicians are encouraging patients who require daily dosages of medications (such as diabetics) to employ administration by inhalation instead of by injection into the bloodstream or by swallowing pills. Hospitalized patients suffering from respiratory diseases are housed in tents where they continuously inhale air laden with drops or particles of medication.

Workers in the aluminum and steel industry are researching the production of plates by direct spraying of molten metals. Plates could be produced more efficiently and less expensively in the future by planar sheet spraying instead of the current practice of casting billets, followed by rolling.

Spraying is used extensively in the electronic packaging industry, where layers of material are sprayed onto moving boards. In the nuclear power industry, spray cooling systems are installed for use in case of emergency. Cooling and evaporative systems employ spraying.

Workers in the hair and domestic spray industries have used propellants which contain chlorofluorocarbons, which are banned because of their reaction with the ozone layer in the stratosphere (Montreal Treaty). The quality and characteristics of these hair sprays changed adversely when the banned CFC propellants were replaced by alternative means of atomization.

Fires are usually extinguished by liquid sprays. Some large-scale fires cannot be extinguished, even with large quantities of liquid spray; other fires have been extinguished with relatively small quantities of liquid spray. The production of artificial snow on ski slopes employs compressed air for atomization to simulate natural snowfall. Chemical warfare can be conducted by spraying poisonous chemicals.

Removal of oxides of sulfur, oxides of nitrogen and particulate dusts from flue gases of electric utility, coal power plants, industrial boilers and furnaces are carried out in wet scrubbers, swirl tubes, packed bed separators, cyclones, and wave-plate separators where the combustion exhaust gases are passed through liquid sprays. The encapsulation and microencapsulation of pharmaceuticals and fertilizers are achieved by spraying. In CIP (clean-in-place) and/or COP (clean-out-of-place) techniques, the sanitary cleaning of pharmaceutical and food manufacturing vessels and equipment parts are realized by applying the high pressure water jets with cleaning agents and rinsing with the purified USP water. Pharmaceutical cabinet washers are also used to clean the equipment parts and glassware extensively in the biotechnology and pharmaceutical industries and in testing laboratories.

The list of industrial and domestic applications that use spraying is very long. Most industries use some form of spraying for production, operation, maintenance or cooling processes. Workers in most of these industries have paid little attention, in the past, to atomizer design and control of spray characteristics. For many applications there is no knowledge of optimal or suitable drop size, velocity and number density. Engineers are, however, aware of the influence of changes in atomizer design and flow rates on overall performance of the system. If measurements were to be made of spray characteristics, performance, control and overall cost of production could be improved significantly.

7.1.1 Newtonian and Non-Newtonian Atomization

Atomization of viscous Newtonian fluids has been widely established through many experimental works which have been performed in order to understand the atomization processes such as disintegration, primary and secondary droplet breakups, droplet coalescence and droplet evaporation and condensation, wall impingement and splashes.

Atomization of viscoelastic non-Newtonian fluids is typically complex because the non-Newtonian fluids may contain polymers, surfactants and large concentrations of insoluble solids in suspension. To understand why non-Newtonian fluids differ so dramatically, one must recognize that the hydrodynamics of capillary breakup and atomization are governed by an independent material function – the extensional viscosity – [71]. The jet breakup length increases with increasing extensional viscosity and under these conditions the process of secondary breakup of drop formation is also suppressed [27]. Non-Newtonian fluid atomization has a wide variety of applications, such as in pharmaceutical manufacturing processes, pharmaceutical tablet coating, spray-drying, aerospace propellants, spray painting, fertilizer and pesticide sprays, delivery of airborne drugs, ink-jet printing, metal sprays and many more other applications.

Researches into the atomization of non-Newtonian liquids have recently increased to understand the viscoelastic atomization phenomena [3, 14, 27, 45, 67, 93, 100].

7.1.2 Primary and Secondary Breakup

Basically, atomization can be considered as a disruption of the consolidating influence of surface tension by the action of internal and external forces. In the absence of such disruptive forces, surface tension tends to pull liquid into the form of a sphere, since this configuration has minimum surface energy. The viscosity of liquids exerts a stabilizing influence by opposing any change in system geometry. On the other hand, aerodynamic forces acting on the liquid surface may promote the disruptive process by applying an external distorting force to the bulk liquid. Breakup occurs when the magnitude of the disruptive force just exceeds the consolidating surface tension force [59].

Primary breakup involves the initial formation of drops and other liquid fragments at the surface of a liquid. Primary breakup is important because it controls the initial dispersion of liquid into gas phase and, through the strong effect of drop sizes on interphase transport rates, subsequent mixing properties of sprays. Secondary breakup involves any subsequent breakup of drops or liquid fragments present as dispersed liquids. Secondary breakup is important because drops after primary breakup are intrinsically unstable to secondary breakup, which affects subsequent mixing rates by influencing drop sizes as well [108].

Many theoretical and experimental studies are concluded in the non-dimensional parameters (such as Weber number, Ohnesorge number, Deborah number, and Weissenberg number) to describe the secondary breakup atomization process. Weber number measures the shear forces acting on a drop relative to its surface tension. Depending on the Weber number, a drop will exhibit different breakup modes resulting in different final drop sizes. Newtonian liquids have five distinct breakup modes [84] as illustrated in Fig. 7.1. These are (1) vibrational, (2) bag, (3) multimode, (4) sheet-thinning and (5) catastrophic [93]. Weber number for a liquid drop is defined as $We_L = \rho_G {V_{rel}}^2 d_0/\sigma$. Breakup modes of Newtonian drops can be characterized by Weber numbers with Ohnesorge (Oh) numbers less than

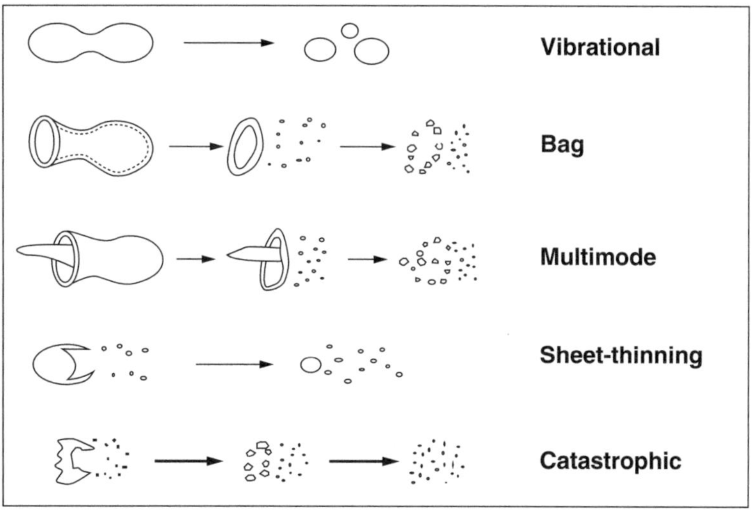

Fig. 7.1 Newtonian liquid drop breakup morphology (Modified from Pilch and Erdman [84]. Copyright Elsevier; reproduced with permission)

Table 7.1 Transition We_L for Newtonian drop breakup modes with Oh <0.1

Vibrational	$0< We_L <11$
Bag	$\sim 11< We_L < \sim 35$
Multimode	$\sim 35< We_L < \sim 80$
Sheet-thinning	$\sim 80< We_L < \sim 350$
Catastrophic	$We_L > \sim 350$

0.1. Oh number is a measure of the viscosity of a liquid relative to its surface tension. The transition Weber number We_L increases from vibrational to catastrophic modes for Newtonian drops with Ohnesorge (Oh) numbers less than 0.1 as shown in Table 7.1 [43].

In addition to the breakup morphology of non-Newtonian fluids, the dimensionless initiation time and total times for secondary atomization processes are also of importance. Pilch and Erdman [84] developed equations for dimensionless initiation times based on Weber and Ohnesorge numbers:

$$T_{ini} = 1.9(We_L - We_c)^{-0.25}\left(1 + 2.2\,Oh^{1.6}\right) \text{ with } We_L < 10^4, Oh < 1.5 \qquad (7.1)$$

Initiation time has been defined as the time elapsed from the last perfectly spherical drop to the point at which the drop has deformed just beyond an oblate spheroidal shape. Total breakup time is that required for all drop fragmentation to have ceased [93]. Snyder et al. [93] studied elastic non-Newtonian drop morphology experimentally and observed that the breakup modes were the same as those for the Newtonian liquids and for inelastic non-Newtonians ones. Breakup mechanics were similar to those for Newtonian liquids, but with more bag growth and presence of ligaments. This behavior has recently been observed for inelastic non-Newtonian drops. In contrast to previous results for Newtonian and inelastic non-Newtonian liquids, the traditional dimensionless groups of Weber and Ohnesorge number fail to characterize suitably the secondary breakup of elastic non-Newtonian fluids [93].

7.1.3 Disintegration of Liquid Jets and Liquid Sheets

G. I. Taylor [101] studied the formation of liquid sheets by the impingement of two opposing jets and by the impingement of a single jet on a plate. The thickness of the liquid sheet decreases due to the radial spreading of the liquid. Viscous shear forces at the plate surface decrease the fluid momentum continuously as the fluid expands radially on the plate and this causes additional reduction in sheet thickness. For the case of Weber numbers (We_{jet} <800), there is a balance between the momentum force of the liquid and the opposing surface tensile force. At the higher Weber numbers (800< We_{jet} <30,000), large amplitude antisymmetric waves are generated, causing premature disintegration of the liquid sheet. These waves are induced by aerodynamic forces acting on the air/liquid interface. As the amplitude of the waves increases, strands are torn from the crests and holes are blown through

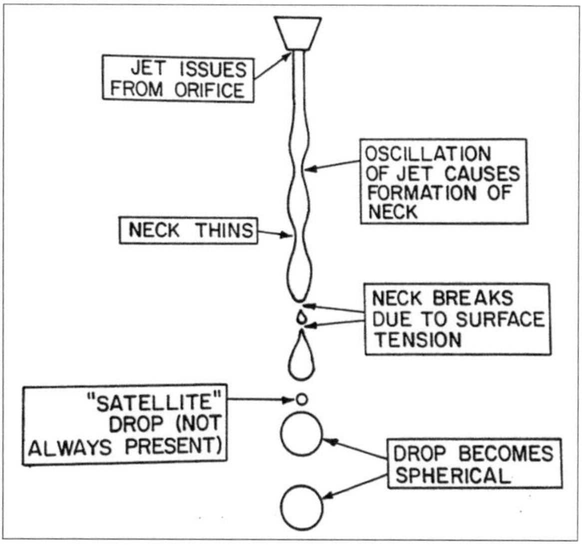

Fig. 7.2 Breakup of a plain circular jet [24]

sheets. Drops are shed from the expanding rims. Fraser et al. [35] described the following idealized wave disintegration mechanism. A single, optimally growing wave increases in amplitude with position downstream. When this wave reaches its critical amplitude, sheet segments are torn off at half-wavelength intervals to form bands of liquid. Surface tension forces rapidly contract these bands into cylindrical strands which ultimately break up into drops.

The air/liquid interface of the liquid sheet can be perturbed by aerodynamic, turbulent, inertial, surface tension, viscous, acoustic or electrical forces. The stability of the sheet and the growth rate of unstable disturbances are determined by the relative magnitude of these forces.

Figure 7.2 shows the breakup of a plain circular jet. First oscillation of the jet causes formation of a neck and then the neck thins and breaks due to the surface tension. Finally, spherical drops form after the neck breaks down and sometimes satellite drops form.

Triggering mechanisms for initiation of liquid surface disturbances can be found in pulsations and disturbances in the liquid and air supply lines to the atomizer, in sharp edges causing separated flows, wall surface roughness, boundary layer growth, turbulence generation and cavitation inside the atomizer and at the orifice exit, and in interaction of the liquid surface with surrounding air flows downstream of the atomizer exit. Once the disturbances have been triggered, further wave growth will largely depend on interactions of the liquid surface with the surrounding air flows. Once the wave growth characteristics have been identified for any particular group of atomizers, means can be found for interfering with wave growth. Aerodynamic, acoustic, and electromagnetic fields could be applied. In-phase fields would result in augmentation of growth rate resulting in more rapid and improved atomization. Out-of-phase fields would be used for damping the growth of

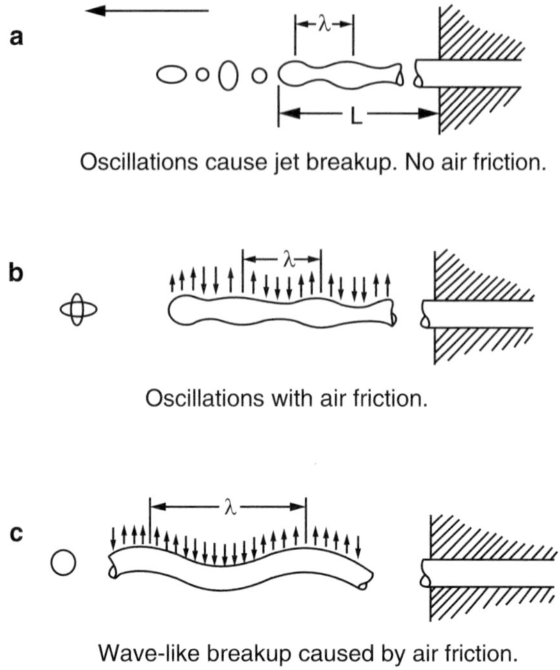

Fig. 7.3 Mechanism of drop formation from a liquid jet [24]

hydrodynamic waves that can resonate with acoustic waves in rocket and space shuttle combustion chambers [23].

Haenlein [44] identified four distinct regimes of breakup in the disintegration of a liquid jet:

1. Drop formation without the influence of air (Fig. 7.3a)
2. Drop formation with air influence (jet with rotationally symmetric disturbance) (Fig. 7.3b)
3. Drop formation due to waviness of the jet (jet disturbance causing sinuous wave formation) (Fig. 7.3c)
4. Complete disintegration of the jet, i.e., atomization.

Perhaps the most commonly quoted criteria for classifying jet disintegration are those proposed by Ohnesorge [78]. From photographic records of jet disintegration, Ohnesorge classified the data according to the relative importance of gravitational, inertial, surface tension, and viscous forces [59]. Another popular categorization of jet disintegration is the jet stability curve. Ohnesorge showed that the various mechanisms of jet breakup could be divided into three regions on a graph of Ohnesorge number $\left[Oh = \mu_L/(\rho_L \sigma\, d_0)^{0.5}\right]$ versus Reynolds number ($Re_L = \rho_L V_L d_0/\mu_L$) according to the rapidity of drop formation:

Region I. At low Reynolds numbers, the jet disintegrates into large drops of fairly uniform size (Rayleigh breakup mechanism),

Region II. At intermediate Reynolds numbers, the jet breakup is by jet oscillations with respect to the jet axis, and

Region III. At high Reynolds numbers, atomization is complete within a short distance from the discharge orifice.

At low Reynolds numbers (region I) the jet structure is predominantly varicose and the mode of breakup follows the Rayleigh mechanism (which is characterized by the lower jet velocities, small surface disturbances, and negligible aerodynamic effects). With increasing Reynolds number, the drop formation mode moves into region II, where the jet oscillates about its axis, having a twisted or sinuous appearance (Fig. 7.3). Passing through the narrow band of region II, region III is reached, in which atomization occurs at the orifice from which the jet emerges.

In a more recent study, Reitz [88] attempted to resolve some of the uncertainties surrounding the Ohnesorge chart. According to Reitz [88], the following four breakup regimes are encountered as the liquid injection velocity is progressively increased:

1. *Rayleigh Jet Breakup Regime* (also called *varicose or dilational regime*) (Fig. 7.4a). It is caused by the growth of axisymmetric oscillations of the jet surface, induced by the surface tension. Drop diameters exceed the jet diameter;
2. *First Wind-Induced Breakup Regime* (also called *sinuous regime*, see in Fig. 7.4b). At the intermediate Reynolds number, the surface tension effect is augmented by the relative velocity between the jet and the ambient gas, which produces a static pressure distribution across the jet. Drop diameters are comparable to the size of the jet diameter;
3. *Second Wind-Induced Breakup Regime* (Fig. 7.4c). Drops are produced by the unstable growth of short-wavelength surface waves on the jet surface caused by the relative motion of the jet and the ambient gas. This wave growth is opposed by the surface tension. Breakup occurs several diameters downstream of the nozzle exit and the resulting average drop diameters are smaller than the jet diameter;

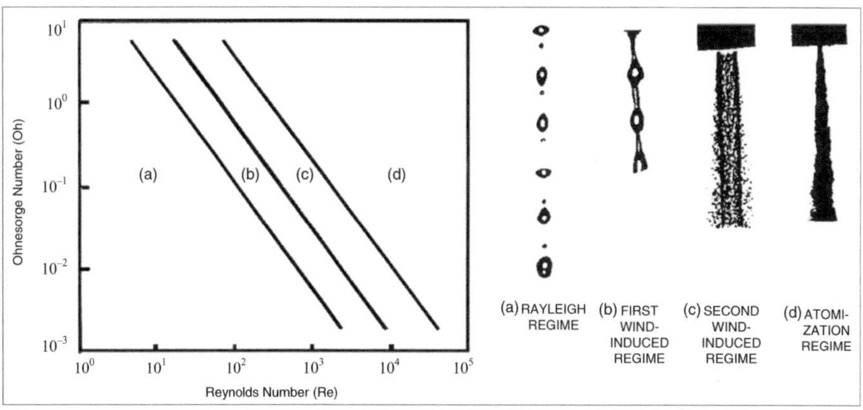

Fig. 7.4 Classification of jet disintegration modes modified from Reitz [88] and jet breakup regime photographs (Modified from Chigier and Reitz [25])

Table 7.2 Ohnesorge's jet breakup classification based on liquid and gas Weber numbers

Jet breakup regime	Weber number range
Rayleigh breakup	$We_L > 8$ and $We_G < 0.4$
First wind-induced breakup	$0.4 < We_G < 13$
Second wind-induced breakup	$13 < We_G < 40$
Atomization breakup	$We_G > 40$

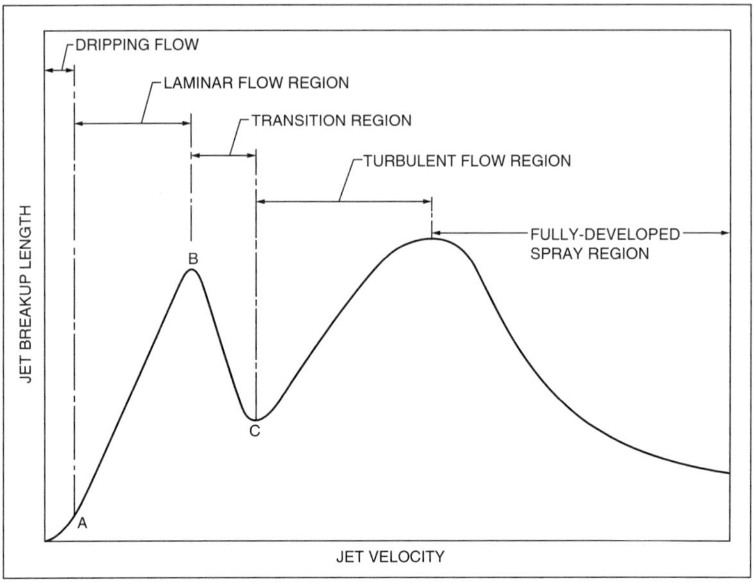

Fig. 7.5 Jet stability curve [24]

4. *Atomization Breakup Regime* (Fig. 7.4d). The jet disrupts completely at the nozzle exit and it is characterized by very high velocities. Average drop diameters are much smaller than the jet diameter [15, 59].

Ohnesorge's jet breakup classification may be alternatively defined by the magnitudes of liquid properties (liquid Weber number We_L) and ambient gas properties (gas Weber number, We_G) as described in Table 7.2 based on the studies of Chigier and Reitz [25], Lin and Reitz [65], and Birouk and Lekic [15].

An alternative way of classifying jet breakup mechanisms is based on the graph of the jet breakup length (L_B) versus jet velocity (V_J), which is known as the jet stability curve. The jet stability curve is divided into the five regions: (1) *dripping region*; (2) *laminar region*; (3) *transition region*; (4) *turbulent region* and (5) *fully developed spray region*. Figure 7.5 shows the jet stability curve indicating change of breakup length with jet velocity.

Birouk and Lekic [15] concluded in their review paper that the breakup of a liquid jet is not completely understood. The typical mechanisms of the breakup associated with various regimes are generally known; however, there is a very limited amount of theoretical framework required to incorporate experimental data. The shortcomings of most of the theories that deal with jet breakup are due to the fact that the influence of the nozzle geometry (and internal flow) on the jet breakup is generally ignored. Furthermore, those theories that incorporate nozzle effects do so empirically [65]. Omission of geometry effects from the theoretical framework is a very important problem because recent studies mentioned by Birouk and Lekic [15] indicate that nozzle geometry (such as contraction ratio, aspect ratio, contraction angle, streamlining related to the cavitation and hydraulic flip and surface roughness) and internal flow have a profound impact on jet characteristics and breakup. The absence of details pertaining to nozzle geometry in the classic literature is a source of confusion and contradiction in many studies. Numerous authors do not report sufficiently detailed descriptions of the nozzles they employ in experiments. Consequently, the reader may encounter two dissimilar observations/conclusions obtained from similar experimental conditions. For example, some classic stability curves experience hysteresis, while others do not [15].

7.2 Atomizers

The atomization of a liquid into a multiplicity of small drops is a process that is fairly easy to accomplish. For most liquids, all that is needed is a high relative velocity between the liquid to be atomized and the surrounding gaseous medium. Some atomizers accomplish this by discharging the liquid at high velocity into a relatively slow-moving stream of air or gas. Notable examples include the various forms of pressure atomizers, and also rotary atomizers which eject the liquid at high velocity from the periphery of a rotating cup or disk. An alternative approach is to expose a relatively slow-moving liquid to a high-velocity airstream. The latter method is generally known as twin-fluid, air-assist, or air-blast atomization. The general classification of atomizers is [11, 13, 24, 56, 59]:

1. Pressure atomizers (jet atomizers, swirl atomizers, jet-swirl atomizers);
2. Rotary atomizers;
3. Air-assist atomizers (twin-fluid atomizers) and air-blast atomizers;
4. Effervescent atomizers;
5. Acoustic atomizers;
6. Electrostatic atomizers;
7. Ultrasonic atomizers; and
8. Whistle atomizers.

An ideal atomizer would possess all the following characteristics:

- Ability to provide good atomization over a wide range of liquid flow rates;
- Rapid response to changes in liquid flow rate;
- Freedom from flow instabilities;

- Low power requirements;
- Capability for scaling, to provide design flexibility;
- Low cost, light weight, ease of manufacture, and ease of removal for servicing;
- Low susceptibility to damage during manufacture and installation;
- Low susceptibility to blockage by contaminants and to carbon buildup on the nozzle face for fuel nozzles;
- Low susceptibility to gum formation by heat soakage for fuel nozzles; and
- Uniform radial and circumferential fuel distribution in fuel nozzles [24, 59].

7.2.1 Pressure Atomizers

As their name suggests, these atomizers rely on the conversion of pressure into kinetic energy to achieve a high relative velocity between the liquid and the surrounding gas. Most of the atomizers in general use are of this type. They include plain orifice and simplex nozzles (pressure-swirl nozzles), as well as various wide-range designs, such as variable geometry, doublex, and dual-orifice injectors [11, 24, 56, 59].

The sprays produced by plain-orifice atomizers have a cone angle that usually lies between 5 and 15°. The cone angle is only slightly affected by the diameter and length/diameter ratio of the orifice and is mainly dependent on the viscosity and surface tension of the liquid and the turbulence of the issuing jet. An increase in turbulence increases the ratio of the radial to the axial component of velocity in the jet and thereby increases the cone angle [59].

The narrow spray cone angles exhibited by plain-orifice atomizers are disadvantageous for most practical applications. Much wider cone angles are achieved in the simplex or pressure-swirl atomizer, in which a swirling motion is imparted to the liquid so that, under the action of centrifugal force, it spreads out in the form of a conical sheet as soon as it leaves the orifice. There are two basic types of simplex (pressure-swirl) nozzles: solid-cone spray and hollow-cone spray. The main drawback of solid-cone spray nozzles is relatively coarse atomization, the drops at the center of the spray being larger than those near the periphery. Hollow-cone spray nozzles provide better atomization and their radial liquid distribution is also preferred for many industrial purposes, especially for combustion applications [24, 47, 59]. Especially Khavkin's book covers the theory and application of swirl atomizers in aspect of original theory developed by Khavkin, novel excellent correlation for drop size and evaporation time calculation, as well as review of swirl atomizer construction [56]. A schematic view of a simplex (pressure-swirl) atomizer and sprays produced by swirl atomizers are shown in Fig. 7.6.

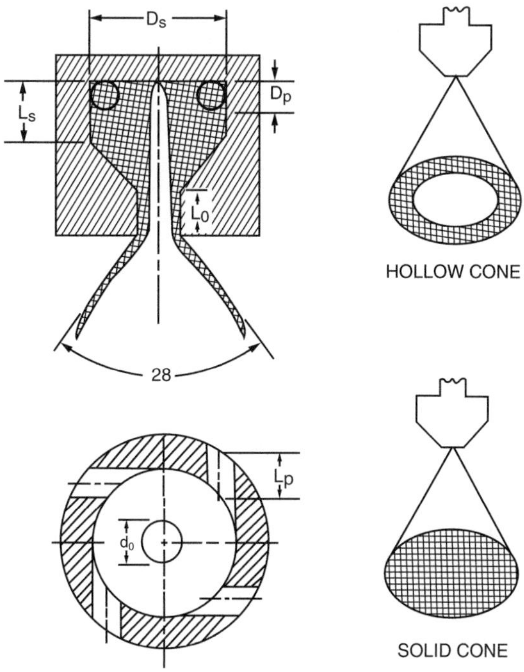

Fig. 7.6 Schematic view of a simplex (pressure-swirl) atomizer (*left*) and sprays produced by swirl atomizers (*right*) [24]

7.2.2 *Rotary Atomizers*

In the rotary atomizer, liquid is fed onto a rotating surface, where it spreads out fairly uniformly under the action of centrifugal force. The rotating surface may take the form of a flat disk, vaned disk, cup, bowl, impeller or slotted wheel. A rotating cup design is shown schematically in Fig. 7.7 [24]. The general advantages of rotary atomizers are as follows:

1. Very viscous and contaminated liquids can be atomized;
2. Liquid can be fed with low pressure;
3. Good atomization can be assured, in some instances close to monodisperse atomization;
4. Possibility of flow rate control without affecting the atomization quality;
5. Rotary atomizers display minimal flow blockage;
6. High flow rates can be used;
7. Rotary atomizers can readily be cleaned;
8. Power consumption is relatively low.

The general disadvantages of rotary atomizers are as follows:

1. Design is complex, including a rotating element, drive with a transmission, bearings, and lubrication system;
2. High rotational speed is required to ensure high atomization quality;
3. High rotational speed reduces the reliability and requires appropriate protections;

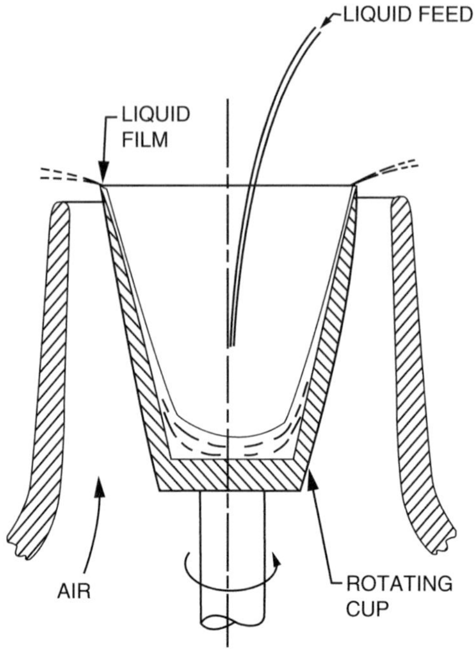

Fig. 7.7 Rotating cup atomizer [24]

4. Rotary atomizers generate large lateral spray cloud dispersion. In many instances this hinders spray utilization;
5. Rotary atomizers have an air pumping effect.

Rotary atomizers reached a high level of design development and found wide industrial application, mainly spray drying, boiler furnaces, air conditioning etc. Rotors with diameters up to 0.44 m or more are built with rotational speed up to 32,000 rpm and flow rate up to 100 m^3/h [11].

7.2.3 Air-Assist (Pneumatic Twin-Fluid) Atomizers and Air-Blast Atomizers

Pneumatic twin-fluid or air-assist atomizers are atomizers in which the energy of the gas, in most cases air or steam, is used for liquid disintegration. There has been intensive development of these atomizers for application to very viscous liquids, oil-in-water fuel mixtures, liquid metals etc. [11]. In all air-assist atomizer designs a high-velocity gas stream impinges on a relatively low-velocity liquid stream, either internally as shown in Fig. 7.8 or externally as shown in Fig. 7.9 [24]. In the internal-mixing type, the spray cone angle is a minimum for a maximum airflow, and the spray widens as the airflow is reduced. This type of atomizer is very suitable for highly viscous liquids, and good atomization can be obtained down to very low liquid flow rates. External-mixing types can be designed to give a constant spray

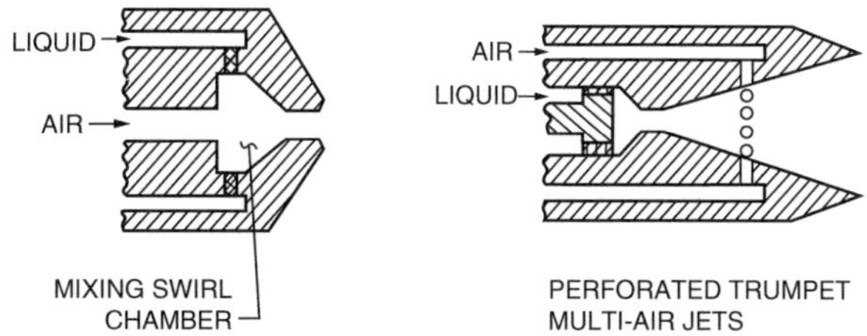

Fig. 7.8 Internal-mixing air-assist atomizer [24]

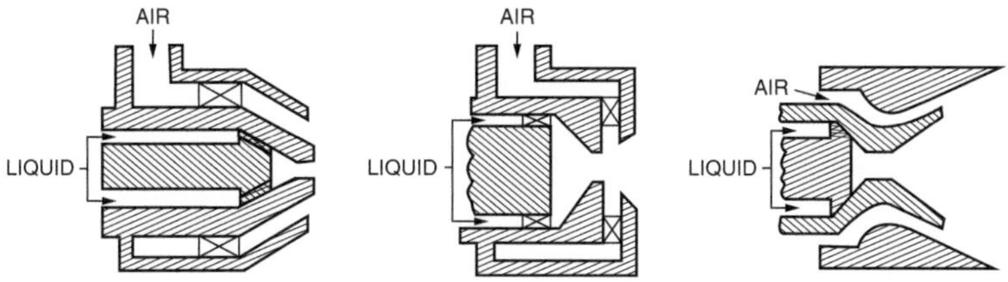

Fig. 7.9 External-mixing air-assist atomizer [24]

angle at all liquid flow rates, and they have the advantage that there is no danger of liquid finding its way into the air line. However, their utilization of air is less efficient, and consequently their power requirements are higher [59].

In principle, the air-blast atomizer functions in exactly the same manner as the air-assist atomizer; both employ the kinetic energy of a flowing airstream to shatter the fuel jet or sheet into ligaments and then drops. The main difference between the two systems lies in the quantity of air employed and its atomization velocity. With the air-assist nozzle, where the air is supplied from a compressor or a high-pressure cylinder, it is important to keep the airflow rate down to a minimum. However, as there is no special restriction on air pressure, the atomizing air velocity can be made very high. Thus air-assist atomizers are characterized by their use of a relatively small quantity of very high velocity air. However, because the air velocity through an air-blast atomizer is limited to a maximum value (usually around 120 m/s), corresponding to a pressure differential across the combustion liner, a larger amount of air is required to achieve good atomization. However, this air is not wasted, because after atomizing the fuel, it conveys the drops into the combustion zone where it meets and mixes with the additional air needed for complete combustion. Air-blast atomizers have many advantages over pressure atomizers, especially in their application to combustion systems operating at high pressures. They require lower fuel pump pressure and produce a finer spray. Moreover, because the air-blast atomization process ensures thorough mixing of air and fuel, the ensuing combustion process is characterized by very low soot formation and a blue flame of low luminosity, resulting in relatively low flame radiation and a minimum of exhaust smoke [24, 59, 73].

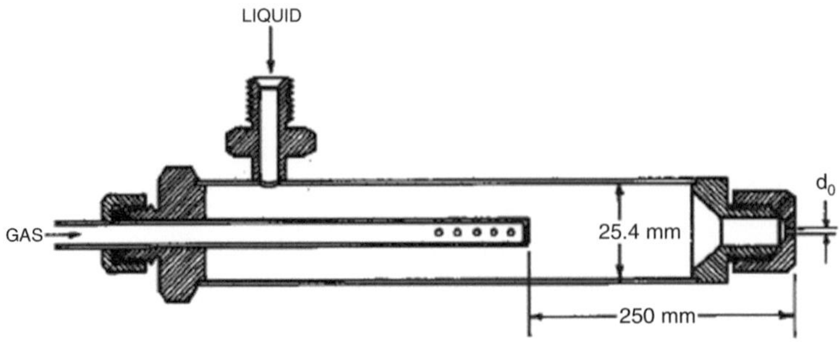

Fig. 7.10 Schematic drawing of effervescent flow atomizer [24]

7.2.4 *Effervescent Atomizers*

All the twin-fluid atomizers described before, in which air is used either to augment atomization or as the primary driving force for atomization, have one important feature in common: the bulk liquid to be atomized is first transformed into a jet or sheet before being exposed to high-velocity air. An alternative approach is to introduce the air or gas directly into the bulk liquid at some point upstream of the nozzle discharge orifice and it is known as effervescent atomizer. One such method is *supercritical* injection, which relies on the *flashing* of dissolved gas in the liquid. The schematic drawing of effervescent flow atomizer is shown in Fig. 7.10 which comprises a plain-orifice atomizer with means for injecting air (or gas) into the bulk liquid at some point upstream of the injector orifice [24, 59].

Flashing dissolved gas systems have been studied by Sher and Elata [92], who developed a correlation of atomization properties in terms of bubble growth rates. Flash atomization of a hydrocarbon solution containing *n*-hexadecane and *n*-butane with nitrogen as the propellant gas was studied by Gemci et al. [40] to obtain a correlation for the Sauter mean diameter with dimensionless degree of superheating and the cavitation number.

7.2.5 *Acoustic Atomizers*

Acoustic atomizers operate on the principle of superposition of the acoustic vibrations generated by a gas flowing with supersonic velocity on the fluid. The vibration frequency in the audible range equals 0.016–20 kHz. Acoustic atomizers require that the gas be used as in pneumatic atomizers, but there are the following differences between those two types:

1. Acoustic atomizers use a different form of gas energy than pneumatic atomizers;
2. Acoustic atomizers ensure better quality and uniformity of atomization for the same conditions;
3. Acoustic atomizers are more economic and have better prospects in the future;
4. Design of acoustic atomizers is more complex [11].

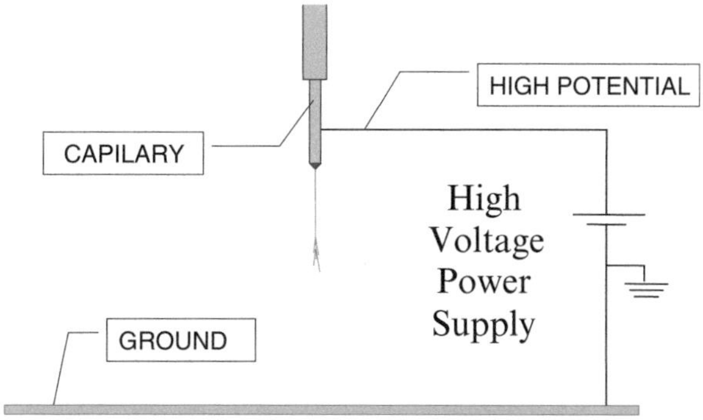

Fig. 7.11 Schematic diagram of an electrospray setup for a capillary orifice (Modified from Gemci et al. [39])

7.2.6 Electrostatic Atomizers

The application of an electrostatic field to liquid emerging from a capillary tube provides a means for generating small droplets without the need for a small orifice and high pressure, as discovered experimentally by Zeleny [111]. Under the optimum conditions, namely with the optimum combination of rheological/electrical properties, voltage, flow rate and system geometry, the fluid forms a conical shape at the exit of the capillary. A fine jet emerges from the tip of this cone, resulting in droplets that are nearly uniform in size and are significantly smaller than the diameter of the capillary itself. This method of atomization shall be referred to as an "electrospray" or "cone-jet" [28]. The microjet breaks up due to varicose wave instabilities into a stream of charged droplets, having diameters roughly twice as large as the jet diameter, similarly to the behavior of electrically neutral jets [38]. The electric field generated in the capillary tube, the charged liquid and the ground electrode cause extensional forces in the liquid, resulting in a liquid cone and a thin jet at the tip of the cone. The potential gradient along the liquid jet generates a tangential electric field, acting on the surface charges. These surface forces stabilize the jet and enable Rayleigh break-up of the jet with generation of nearly monosized, charged droplets, which migrate toward an electrically grounded counter-electrode plate [39]. A simple diagram of the electrospray setup is shown in Fig. 7.11.

The main factors affecting the process are the applied potential-difference, the flow rate and the liquid properties. The use of electric forces on charged droplets provides enhancements in spray atomization that include improved uniformity of deposition, lower energy requirements for atomization, and better control of spray patterns. Applications of electrostatic atomizers in different fields are given in Lefebvre [59], Chigier [24], and Bayvel and Orzechowski [11]. For example production of fine metal powders can be based on electrostatic atomization. A wire is introduced into a region where it is bombarded by electrons generated in a thermoionic emitter. The wire melts and is atomized into particles of the desired sizes. Particles with diameters smaller than 0.01 μm as well as particles with diameters of 10–100 μm can be generated. Use of many such atomizers allows the production of 0.5 ton of powder per day [10]. The very low liquid flow rates that are generally associated with electrostatic atomization have tended to restrict practical applications to electrostatic painting and nonimpact printing. However,

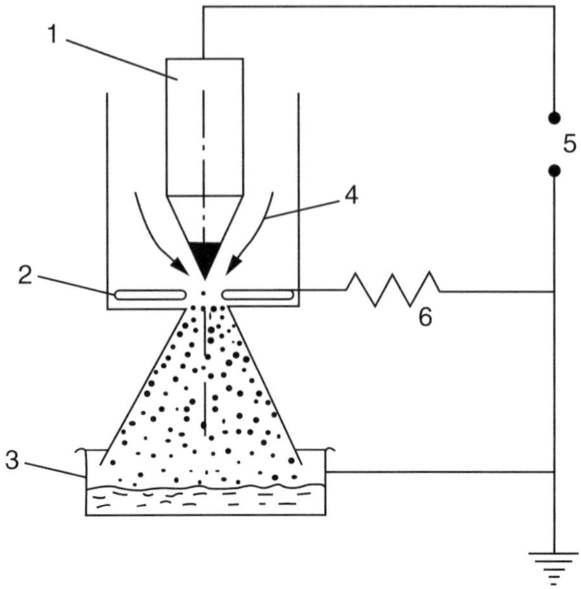

Fig. 7.12 Schematic diagram of electrospray triode atomizer: *1*, emitter electrode; *2*, blunt electrode; *3*, collector electrode; *4*, liquid supply; *5*, voltage source; *6*, resistor [24]

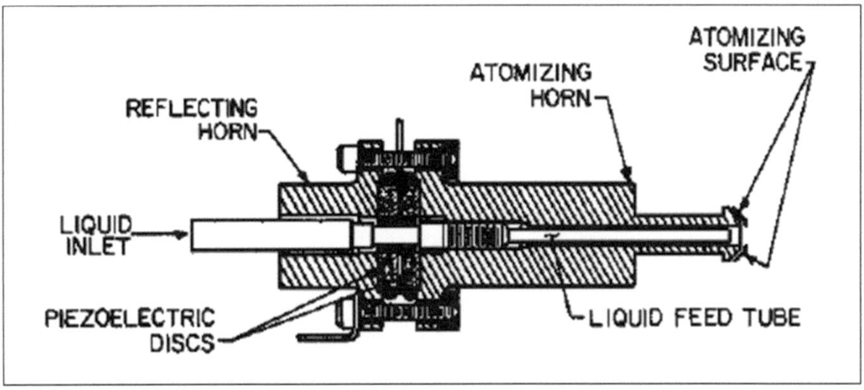

Fig. 7.13 Ultrasonic nozzle assembly [24]

an invention by Kelly [55] called the spray triode, shown schematically in Fig. 7.12, appears to have great promise for the development of electrostatic atomizers capable of handling the high fuel flow rates required by most practical combustion devices [24].

7.2.7 *Ultrasonic Atomizers*

An ultrasonic nozzle is a device designed to generate vibrations of the amplitude required to produce the unstable capillary waves that characterize ultrasonic atomization. One embodiment of an ultrasonic nozzle is shown in Fig. 7.13.

The nozzle is an acoustically resonant device consisting of a pair of piezoelectric transducer elements sandwiched between a pair of titanium horn sections [24]. The transducers are capable of converting high-frequency electrical energy into high-frequency mechanical motion. A pair of tin-plated copper electrodes provides the path through which high-frequency electrical energy is introduced into the device. If the piezo-element is in contact with a liquid in which it generates ultrasonic waves [12, 13].

Ultrasonic nozzle spray technology is used in a wide range of industrial (electronics, medical/biological, combustion, spray drying etc.) and research applications. A brief summary of these applications is as follows [13]:

1. Electronic industry applications
   (a) Applying solder flux to printed circuit assemblies
   (b) Applying solder flux in selective soldering processes such as are used in flip-chip assembly, ball-grid array (BGA) technology, soldering of selected components on a printed circuit assembly, and the soldering of tape-and-reel components
   (c) Coating silicon wafers with photoresist and photoresist developer as part of the photolithography process
   (d) Depositing thin-film coatings using a chemical vapor deposition process on materials intended for use as semiconductors and high-temperature superconductors
   (e) Coating the interior of illumination devices using electrostatic spray principles
2. Medical/biological industry applications
   (a) Coating the interior of blood-collection tubes and related devices with various reagents such as clotting agents, anti-coagulants (such as heparin), and EDTA
   (b) Lubricating syringe barrels with silicones
   (c) Applying cyanoacrylate adhesives to sutures
   (d) Dispensing reagents into well-plates intended for diagnostic testing
   (e) Spraying latex bead suspensions that are used as coatings for test tubes, plates, or paddles in order to attach proteins
   (f) Micro-encapsulating active pharmaceutical ingredients for purposes such as time-release dosing
   (g) Spray-drying pharmaceutical products
   (h) Spraying both soluble and particulate antigens
   (i) Spraying blood and blood plasma in cryo-precipitation processes
   (j) Applying both monoclonal and polyclonal antibodies to surfaces such as membranes
   (k) Spraying a suspension containing bacteria
3. Spray-Drying, Combustion, and other industrial applications
   (a) Web coating such as float glass, paper, and fabrics
   (b) Spray-drying of pharmaceuticals, ceramics, and food products

(c) Slurry and suspension atomization
(d) Fragrance and flavor coatings
(e) Solvent and adhesive bonding
(f) Chemical reaction chambers
(g) Moisturization
(h) Internal combustion engines
(i) Stirling cycle engines
(j) Rankine cycle engines
(k) Fuel cells
(l) Thermoelectric generators
(m) Flame dynamic studies

7.2.8 Whistle Atomizers

As in an ultrasonic atomizer using a transducer, liquid can also be disintegrated into drops by directing high-pressure gas into the center of a liquid jet, as shown in Fig. 7.14. Due to the strong sound waves created inside the nozzle by the focusing airflow, this atomizer is frequently called a whistle or stem-cavity type. It generally operates at a sound frequency of about 10 kHz and produces droplets around 50 μm in diameter at flow rates up to 1.25 l/s [24, 59].

7.3 Spraying Processes

Liquid atomization became popular over 100 years ago and was associated with crude oil refining. At present, it is used in all branches of industry. Power engineering is the largest area of application of devices using atomized liquid. These devices are used in processes such as combustion, steam attemperation, water cooling, water treatment, oil burners, diesel fuel injection, gas turbines, and rocket

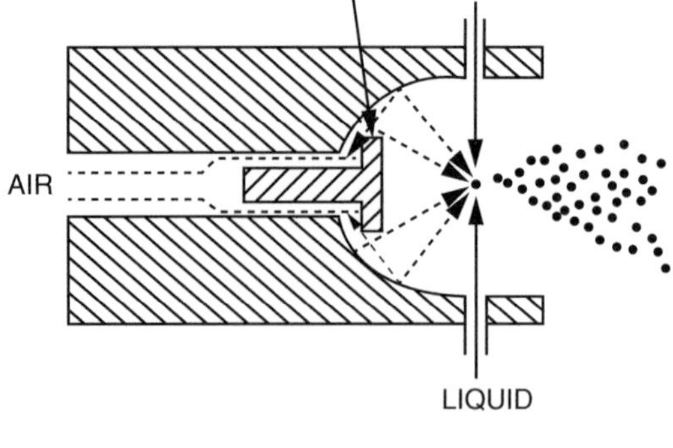

Fig. 7.14 Schematic diagram of whistle-type atomizer [24]

fuel injection etc. Production and process engineering applications are spray drying (dairy products, bakery products, coffee and tea, starch pharmaceuticals, soaps and detergents), distillation, liquid absorption and desorption, injection crystallization, gas elution, spray cooling, powdered metals. Treatment and environmental protection applications are gas scrubbing, fire protection, evaporation and aeration, spray towers, air and gas washing, industrial washing and cleaning, humidification and misting, and air conditioning. Metallurgy and coating industry applications are surface treatment, spray painting, thermal spraying, insulation, fibers, undercoating materials, multicomponent resins, particle coating and encapsulation. Other applications can be listed as medicinal sprays, dispersion of chemical agents, agricultural spraying, foam and fog suppression, printing, and acid etching etc. [11, 24, 59].

The spraying processes such as spray-drying, thermal spraying, and nebulization are explained more in detail because of their specific applications in production and handling of particulate materials.

7.3.1 Spray-Drying Process

Spray-drying is a broadly used manufacturing process which uses the aerosol phase to dry particles. The spray drying technology has been applied in many engineering fields such as the food, ceramic, polymer, pharmaceutical, chemical, and biomedical industries. In food-processing technology, spray-drying methods are widely used in manufacturing of dried eggs, powdered milk, animal feeds, cake mixes, citrus juices, corn syrup, creamers, fish concentrates, infant formulas, potatoes, shortening, starch derivatives, tomato puree, yeast, yogurt, coffee, starch, flavorings, colorings, vitamins, and enzymes. In pharmaceutical industry the spray-drying method is used for the manufacturing of antibiotics and medical ingredients additives. One of the pharmaceutical applications in spray-drying is to manufacture particles that form the basis for dry dosage forms for parental, nasal, or pulmonary delivery, and are administered as suspensions, powders, or aerosols. These particles must be able to stabilize the active pharmaceutical ingredient and provide physical stability for the dosage form on storage. They must have adequate powder flow properties and dispersibility, and, in the case of respiratory delivery, suitable aerodynamic properties [105]. Other pharmaceutical applications of spray-drying are microencapsulation, drying of heat-sensitive materials and the production of amorphous solid dispersions.

Controlled morphology ultrafine particles and nanoparticles produced by spray-drying and spray pyrolysis have great potential for use in applications in the electronic, chemical or mechanical industries, as well as technologies with them, including superconductors, catalyst, sensors, magnetic and optical materials, drug carriers, pigments, and in structural and electronic materials [51, 76, 79, 80]. In catalyst technology, particles having very high porosity (and very large surface area) are required for developing efficient catalysts. The spray-drying of a colloidal mixture of silica nanoparticles and polystyrene latex (PSL) nanoparticles permit the

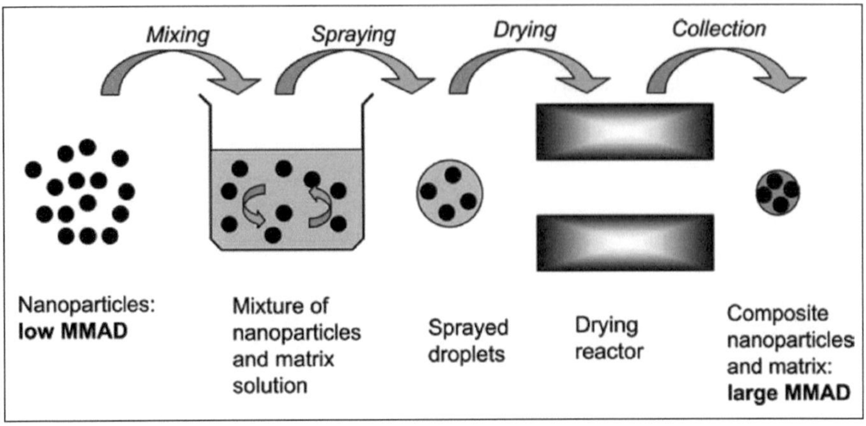

Fig. 7.15 Spray-drying method for producing micrometer-sized particles comprised of nanoparticles for drug delivery where MMAD is mass median aerodynamic diameter [80] (Copyright Elsevier; reproduced with permission)

production of particles with porosities as high as 0.81. For electronic/optoelectronic technologies, this highly porous material exhibits a very low dielectric constant, as low as 1.2, which has great potential for developing ultrahigh memory devices.

One major trend for future medical technology is the use of nanoparticles in treating diseased tissues in the human body such as cancerous tumors, where nanoparticles are introduced into the body through an inhalation process. Their low inertia problem, which results in most of the nanoparticles being exhaled from the lung after inhalation, can be overcome by developing micrometer-sized powders from nanoparticles using the spray-drying method. These could be in the form of large porous particles that microencapsulate nanoparticles [80]. Towards this objective, Sham et al. [91] investigated the feasibility of developing carrier particles to deliver nanoparticles to the lung by incorporating nanoparticles into microparticle carriers. This approach is illustrated in Fig. 7.15. The mass median aerodynamic diameter of such carriers can be adjusted to ensure sufficient lung deposition in the desired upper or lower parts of the lung (either the bronchial region or the alveolar region). After deposition in the lung, the carrier matrix dissolves, releasing the nanoparticles [80].

Another type of carrier particle suitable for drug delivery are large porous particles (LPPs) composed of nanoparticles. This type of particle is characterized by geometrical sizes larger than 5 μm and densities of 0.1 g/cm^3 or less [80]. LPPs have recently become popular for use as carriers for drugs that need to be delivered to the lungs for local and systemic applications [31]. A principal advantage of LPPs over conventional inhaled therapeutic aerosol particles is their aerosolization efficiency [36]. In addition, LPPs possess the potential to avoid alveolar macrophage clearance [97], enabling sustained releases of a drug in the lung [104].

In the conventional aerosol decomposition (CAD) spray-pyrolysis route, each spray-pyrolyzed particle has multiple nanosized crystallites under typical process conditions, but they are virtually inseparable due to the formation of the three-dimensional network. Xia et al. [109] proposed a salt-assisted aerosol decomposition (SAD) spray-pyrolysis method which focused on a strategy for separating these nanocrystallites by introducing some compounds that can be distributed on the nanocrystallite surfaces to prevent them from agglomerating and are then easy to remove, so that a single droplet can produce many finer particles. During the SAD process, in which the particle temperature exceeds the melting point of the salts, the salts melt and act as high-temperature solvents. The material or its components can then dissolve, undergo reactions, and, upon exceeding the solubility limit, precipitate in the solvent. These processes can enhance remarkably the mass transfer due to the liquid-state solvent, in contrast to the very small solid-state diffusion coefficients in the case of the CAD processes. Within an aerosol particle, the dissolution/precipitation cycle can lead to the dissolution of some nanocrystallites and the growth of other crystallites by precipitation. This may break up the three-dimensional network and disintegrate the nanocrystallites, which has been observed in the experiments by Okuyama and Lenggoro [79]. This SAD spray-pyrolysis route can offer good controllability of particle size, chemical composition and material crystallinity, all of which are important to advanced materials. On the other hand, the electrospray-pyrolysis and a low-pressure spray-pyrolysis using the filter expansion aerosol generator are capable of generating nanoparticles and of fragmenting particles to give nanoparticles. Figure 7.16 shows the morphology of particle preparation by a spray method based on the characteristic times of solvent evaporation (τ_{sv}) and solute diffusion (τ_{sl}) from Okuyama and Lenggoro [79].

Properties of precursor, carrier gas flow rate (i.e., residence time of heating or solvent evaporation) and temperature are the main parameters which affect the morphology of particles generated by the spray-pyrolysis. Concerning the evolution of particle morphology during spray-pyrolysis, spherical particles, hollow or fragmented particles, are often formed (Fig. 7.16) in addition to solid (dense) particles. These morphological conditions are undesirable for most applications. Single-crystal or polycrystal particles are also formed, depending on the conditions of operation. A representative system used for the preparation of fine particles by spray-pyrolysis is shown schematically in Fig. 7.17. The main equipment consists of:

1. An atomizer or nebulizer that converts the starting solution into droplets;
2. Carrier gas;
3. Tubular furnace/reactor;
4. Instruments for online measurement of particle size such as a differential mobility analyzer (DMA) and a condensation nucleus/particle counter (CNC/CPC);
5. Sampler or precipitator, and
6. Off-line particle measurements such as X-ray diffraction (XRD), scanning/transmission electron microscopy (SEM/TEM), and luminescence [79].

Figure 7.18 shows the schematic illustration of the experimental apparatus and an illustration of particle formation both for the CAD and the SAD processes, where the solution droplets were generated by an atomizer and were carried by a gas into a

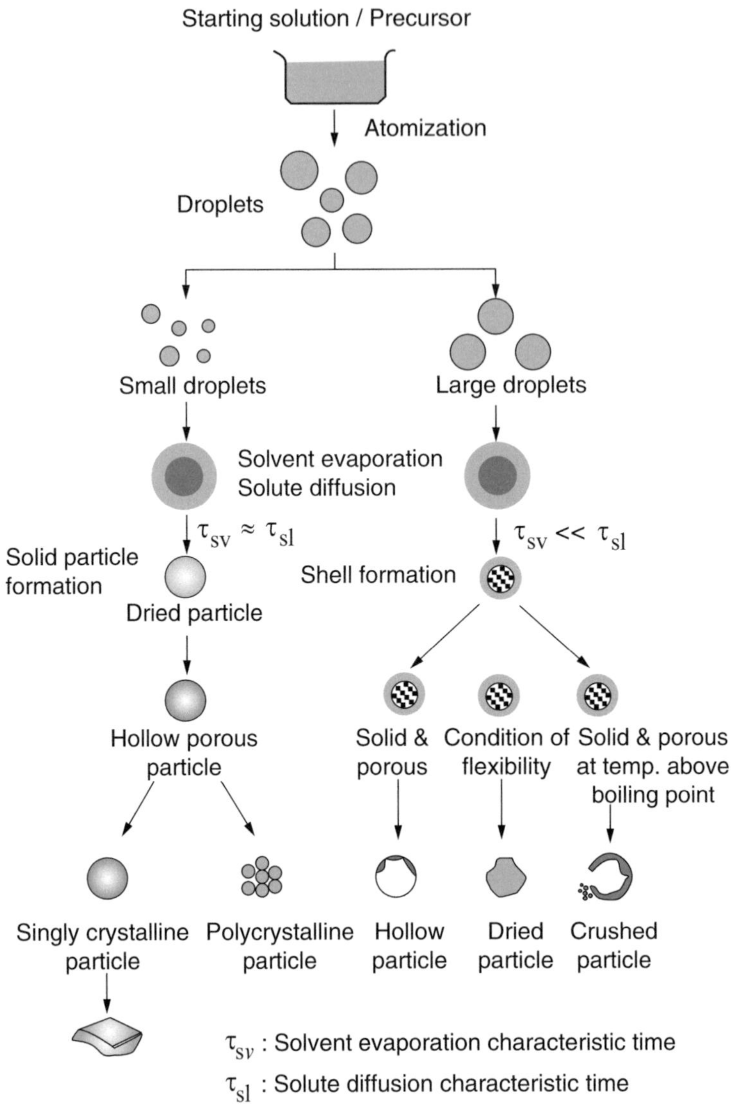

Fig. 7.16 Morphology of particle preparing by spray method [79] (Copyright Elsevier; reproduced with permission)

tube reactor typically heated between 300 and 1,600 °C. The resulting particles were collected in a precipitator [109].

One of the microencapsulation techniques for producing microspheres is spray-drying. The purpose of encapsulation includes achieving controlled/remote drug release, making active materials easier/safer to handle, compartmentalizing multiple component systems, protecting sensitive materials from their environment and

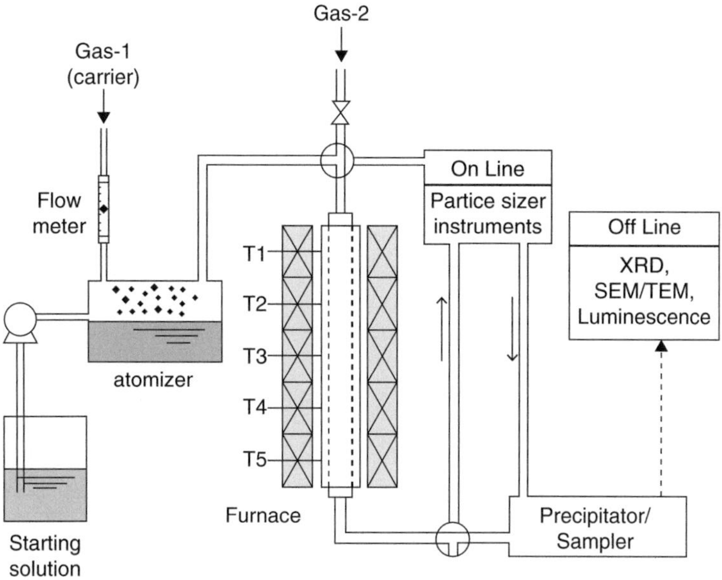

Fig. 7.17 Representative spray-pyrolysis system used for particle preparation [79] (Copyright Elsevier; reproduced with permission)

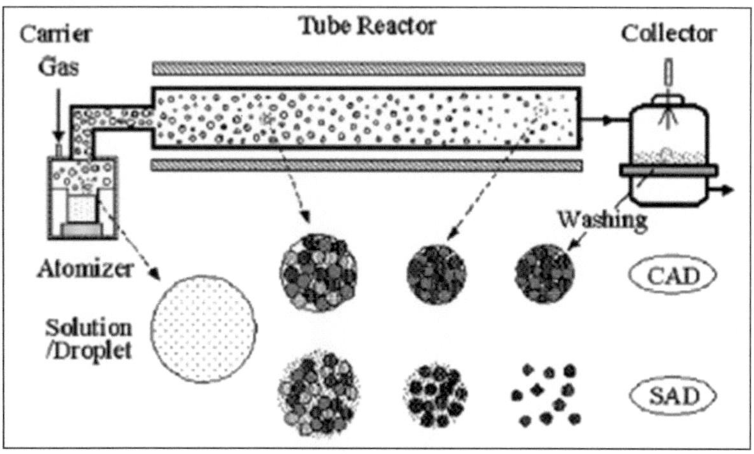

Fig. 7.18 Experimental spray-pyrolysis apparatus and illustration of particle formation for the conventional aerosol decomposition (CAD) and the salt-assisted aerosol decomposition (SAD) processes [79] (Copyright Elsevier; reproduced with permission)

turning liquids into powders or solids. Commercial applications of microcapsules can be found in various formulations such as fertilizers, pesticides, toothpaste, shampoo, bath salts, extended relief pharmaceuticals, carbonless paper, detergents and bleaches, air fresheners, vacuum cleaner bags, fire retardants, and facial tissues

[80]. The spray-drying process has been used to produce drug-loaded microspheres of poly(lactide) and poly(lactide-*co*-glycolide) (PLGA) due to its advantages with respect to the water-in-oil emulsion process [81]. PLGA microspheres are degraded to oligo- and monomers over prolonged periods of time, slowly releasing the entrapped active component [102]. These microspheres also protect the active entities from destructive action by enzymes and the low pH environment of the stomach [32].

In a review article, Nandiyanto and Okuyama [76] defined eight different particle morphologies prepared using the spray-drying method from which the images in Fig. 7.19 were adapted and reprinted from publications by Widiyastuti et al. [106], Okuyama et al. [80], Chang and Okuyama [20], Iskandar et al. [49], Iskandar et al. [50], Nandiyanto et al. [74], and Nandiyanto et al. [75]. These eight controlled morphological particles are named according to their shape or structure as following: (1) dense particle; (2) doughnut particle; (3) raspberry particle; (4) encapsulated particle; (5) well-mixed component particle; (6) hairy particle; (7) porous particle, and (8) hollow particle. Production methods for nano-structured

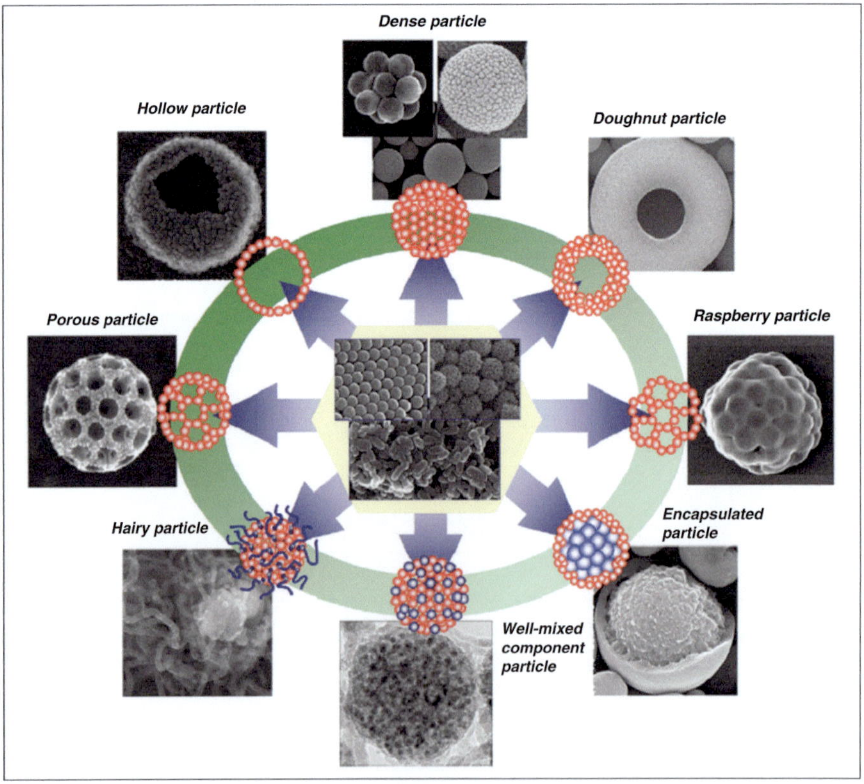

Fig. 7.19 Various particle morphologies prepared using the spray-drying method. Nandiyanto and Okuyama [76] have adapted images from [20, 49, 50, 74, 75, 80, 106] (Copyright Elsevier; reproduced with permission)

particles through using different types of atomizers (ultrasonic nebulizer, two-fluid nozzle, and rotating atomizer), solvent evaporators (electrical furnace, flame evaporator, and diffusion dryer) and particle collectors (bag filter, electrical precipitator, and cyclone collector) are also illustrated schematically in Fig. 7.20.

Vehring et al. [105] conducted an experimental study to generate controlled monodisperse particles (hollow, low-density particles with controlled surface morphology, particles with functional layers, or particles comprising smaller subunits such as nanoparticles or defined voids). A modified vibrating orifice generator head was used as an atomizer in a custom benchtop spray-dryer. The generator was fed with a syringe pump and produced monodisperse droplets in a size range from 10 to 100 μm at production frequencies of 10^5–10^6 Hz. Vehring et al. [105] found that two dimensionless parameters can influence particle formation in a spray-drying process: the Peclet number (ratio of the diffusion coefficient of the solute to the evaporation rate) and the initial saturation of the excipients. Vehring et al. [105] showed that the morphology of dried particles in a spray-drying process can be controlled by varying the air-drying temperature (Fig. 7.21). The gradual increase of the drying gas temperature resulted in a trend of producing more hollow particles. The particles shown consist of a glycoprotein with a molecular weight of 51 kDa.

7.3.2 Thermal Spraying Processes

Thermal spraying is a particulate deposition process whereby solid ceramic particles are injected into a high-temperature gas in which they are heated, eventually melted, and then accelerated and sprayed at relatively high velocity onto a substrate which they impact and on which they form a coating consisting of many layers of overlapping thin lenticular particles, or splats. Fritsching [37] covers in his book the latest development in spray simulation of spray forming metals and spray coatings processes.

Thermal spray processes can be classified into three broad families:

1. The use of combustion heat sources; namely the flame, detonation gun and high velocity oxygen fuel spray (HVOF) processes,
2. By other uses of electrical energy; either in the form of plasma or as an arc, and
3. Use of the energy that evolves from gas expansion and is known as cold spray, kinetic spray or hypersonic spray [4].

Table 7.3 shows the classification of the thermal spray processes according to these energy sources. Figure 7.22 depicts the relationship among the achievable particle velocity, particle temperature and the typical feedstock size [4].

This process map is different from the conventional map of the flame jet temperature and particle velocity (TV relationship) in three respects:

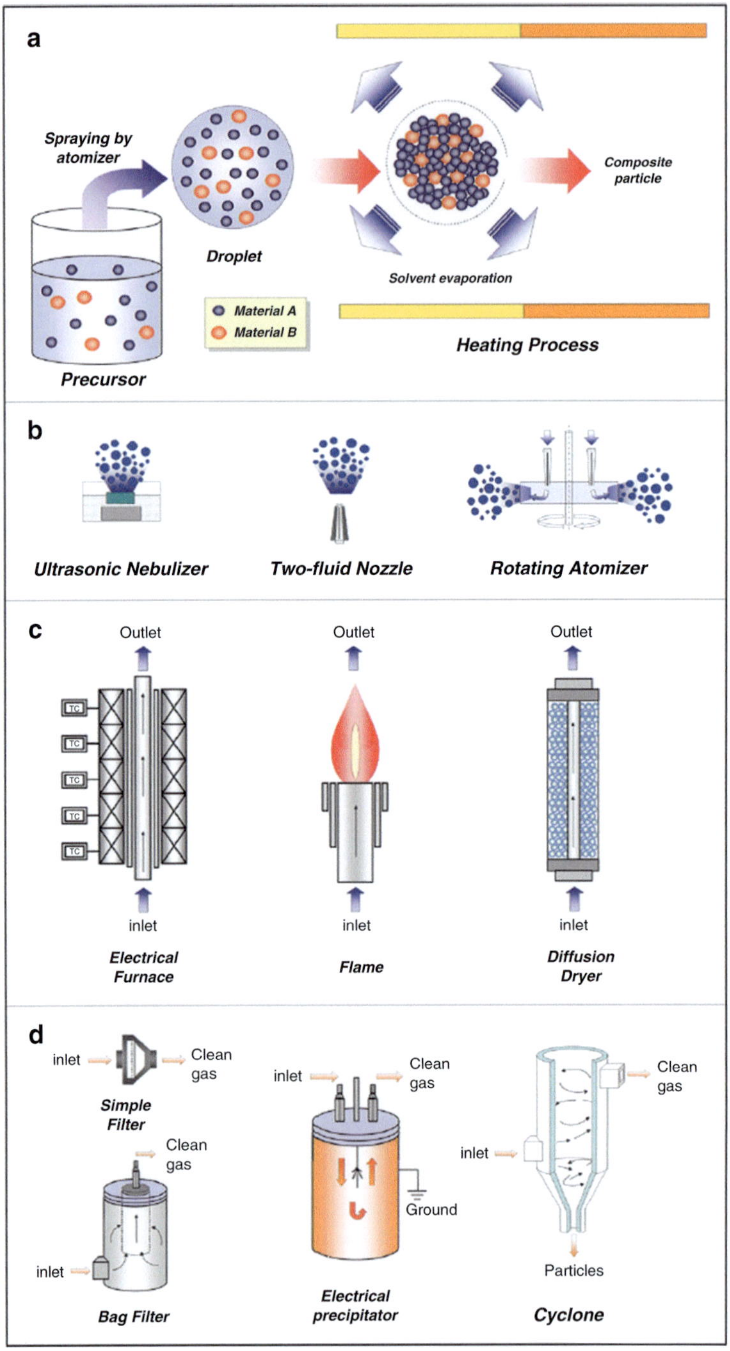

Fig. 7.20 Schematic illustration of nano-structured particle production using the spray-drying method and its apparatus: **a** mechanism; **b** atomizer **c** solvent evaporator and **d** particle collector [76] (Copyright Elsevier; reproduced with permission)

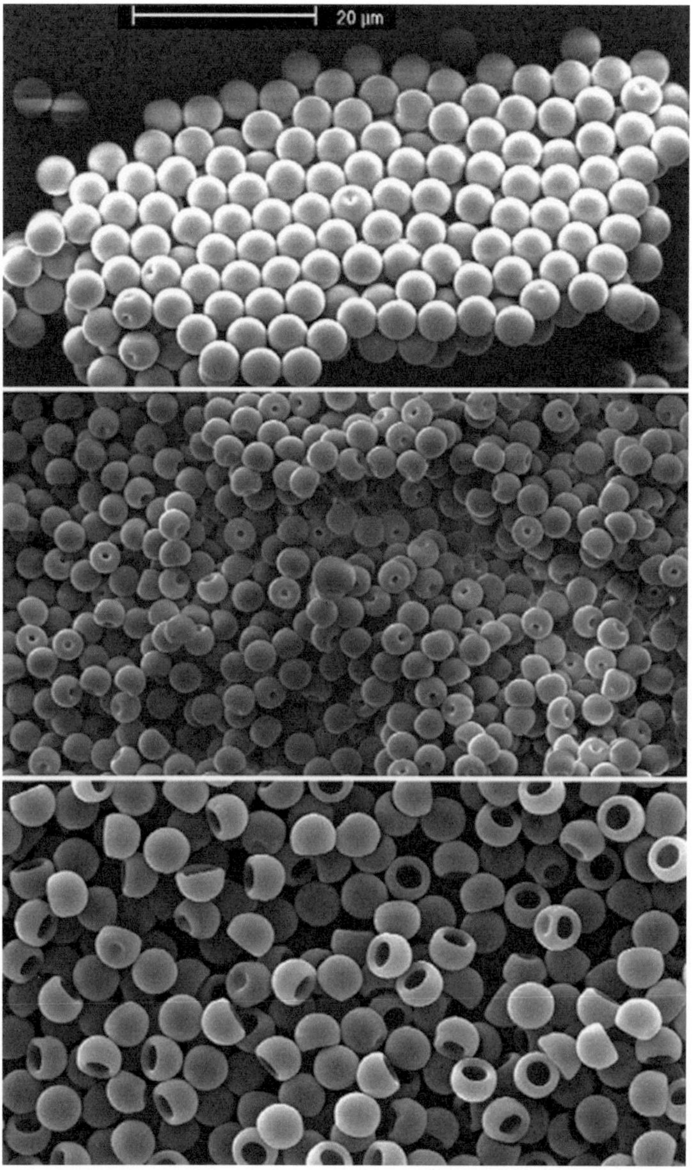

Fig. 7.21 Monodisperse, monomorph glycoprotein particles generated by the spray-drying technique. Gas drying temperatures from *top* to *bottom*: 25 °C, 50 °C, 125 °C [105] (Copyright Elsevier; reproduced with permission)

- The temperature and velocity are represented in terms of the optimized particle properties rather than the effluent environment,
- There is an overlay for material class on the right-hand side, and

Table 7.3 Classification of thermal spray processes according to the energy sources

Electrical energy	Combustion heat energy	Kinetic energy
1. Plasma spray (PS)	1. Flame spray (FS)	1. Cold spray
a. Atmospheric plasma spray (APS)	2. Detonation Gun (D-Gun)	2. Warm spray (kinetic spray or hypersonic spray)
b. Vacuum plasma spray (VPS) or low pressure plasma spray (LPPS)	3. High velocity oxygen fuel (HVOF) spray	
c. Controlled atmospheric plasma spray (CAPS)		
2. Wire arc spray (twin wire arc – TWA)		

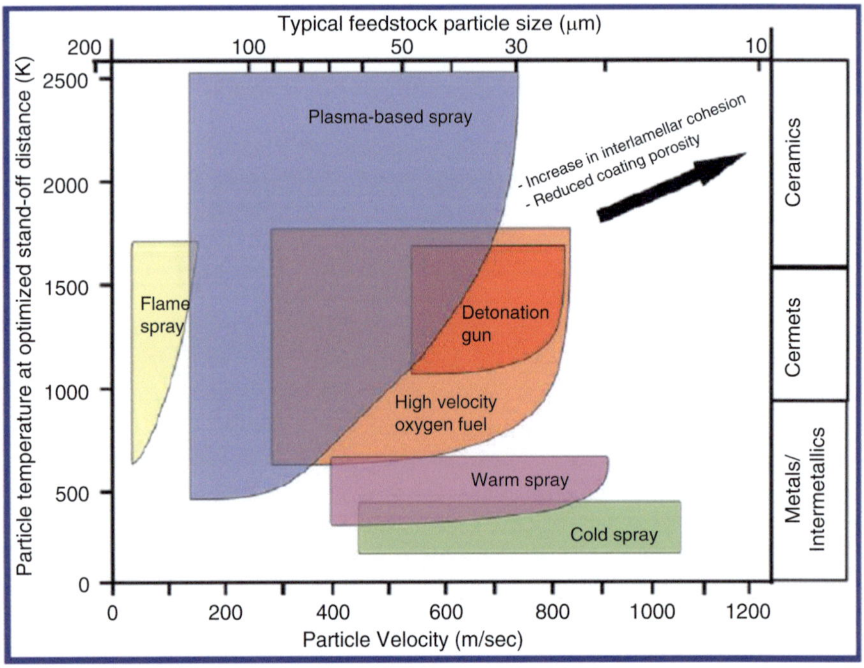

Fig. 7.22 Classification of thermal spray processes in accordance with particle velocity, particle temperature and average feedstock size. The right-hand scale indicates the range of materials that correspond to each spray process [4] (Copyright Maney Publishing; reproduced with permission)

– A logarithmic scale for the feedstock particle size is used.

Pierlot et al. [85] reviewed the methods of designs of experiments applied in thermal spraying and associated processes. The designs enable one to determine a polynomial regression equation which expresses the influence of the process parameters on the response.

Thermally sprayed alumina-based materials, e.g., alumina-titania (Al_2O_3-TiO_2), are commonly applied as wear-resistant coatings in industrial applications. Properties of the coatings depend on the spray process, powder morphology, and chemical composition of the powder. Matikainen et al. [69] have compared experimentally the atmospheric plasma spray (APS) and HVOF spray processes for the wear-resistant coatings from Al_2O_3 and Al_2O_3-13TiO_2 powders. They concluded that the improved coating properties were achieved when agglomerated and sintered nanostructured Al_2O_3-13TiO_2 powder was used in plasma spraying.

Coatings with the highest wear resistance in all tests were produced by HVOF spraying from fused and crushed powders because higher velocity and small particle size resulted in dense and fine microstructure, which provided coatings with high wear resistance comparable to bulk Al_2O_3.

In a review article on testing methods for thermal spray coatings, Ang and Berndt [4] highlighted the extrinsic nature of mechanical property measurements with regard to thermal spray coatings. Thermal spray coatings exhibit anisotropic behavior and microstructural artefacts, such as porosity and the splat structure of coatings. These influence the mechanical characterization methods. Ang and Berndt [4] summarized the mutual relationship between thermal spray process inputs, the anisotropic microstructure and mechanical properties in Fig. 7.23. Thermal spray coatings and their mechanical properties are influenced by the fundamental factors of (1) jet temperature, (2) particle velocity and (3) characteristics of the feedstock.

Pawlowski [82] in his review article discussed the technology, microstructure and properties of nanostructured thermal spray coatings obtained using different feedstocks including: (1) powders composed of agglomerated nanocrystals; (2) solutions and (3) suspensions. The thermal phenomena in flight of these different feedstocks influence the size of crystal. Figure 7.24 gives the summary of major thermal effects at flight of different types of nanophased feedstock in a jet or flame.

Porous structured agglomerated solids at flight have a slower diffusion of heat than do the densely structured particles. Liquid precursors at flight will first have droplets which may experience break-up after being injected into turbulent flow or flame, and can then start to evaporate. Thereafter, the precipitation into a solid phase occurs. Three possible precipitation routes are considered: (1) uniform concentration of solute and volume precipitation leading to solid particles; (2) super-saturation near the droplet surface that leads to the formation of a thick solid shell, which can be entire, fragmented or broken up (the liquid contained inside the broken shells forms small solid spheres); (3) super-saturation near the droplet surface leading to the formation to an elastic shell that initially becomes inflated and then deflates. There is not a lot of specific literature available on the thermal behavior of slurries in flight in jet or flames. However, slurry suspension thermal spraying has the following significant differences with regard to spray-drying: (1) particularly hot heating medium; (2) boiling of the suspension liquid; (3) very short heating time in milliseconds; (4) lack of a binder in the liquid [83].

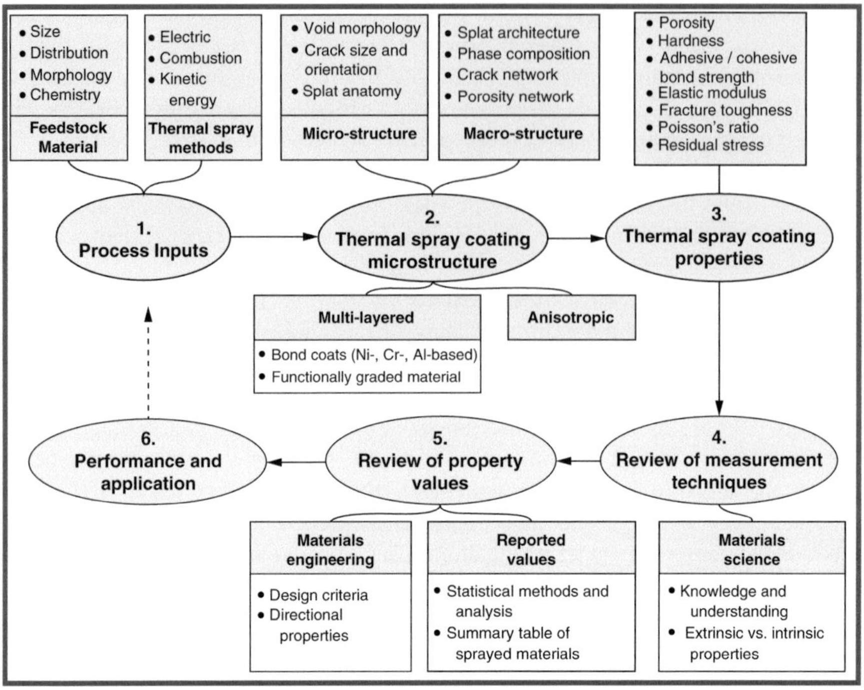

Fig. 7.23 The mutual relationship between thermal spray process inputs, the anisotropic microstructure and mechanical properties [4] (Copyright Maney Publishing; reproduced with permission)

7.3.2.1 Plasma Spraying (PS) Process

PS processing is a well-established and proven technology, which is already in widespread industrial use (aerospace, automotive, medical devices etc.) for a variety of applications (coatings on structural materials for protection against high temperatures such as thermal barrier coatings for exhaust heat management, corrosion, erosion, and wear) due to the low cost and simplicity of the processes.

Energy sources for the plasma spraying are usually the direct current (DC) electric arc or radio-frequency (RF) discharge that create arc energies of 72–720 MJ/h (20–200 kW) [29]. Plasma temperatures in the heating region range from 6,000 to 15,000°K, which are significantly above the melting point of any known materials. The powder of the material to be sprayed is introduced into the plasma where it is melted and projected with a high velocity onto the substrate where it spreads [19]. Coatings can be achieved by different plasma techniques such as atmospheric plasma-spraying (APS), vacuum plasma-spraying (VPS; also called low-pressure plasma spraying, LPPS), and controlled atmospheric plasma-spraying (CAPS) processes [4]. The main parts of the plasma torch are the electrodes and Fig. 7.25 shows a simple schematic diagram of a plasma torch.

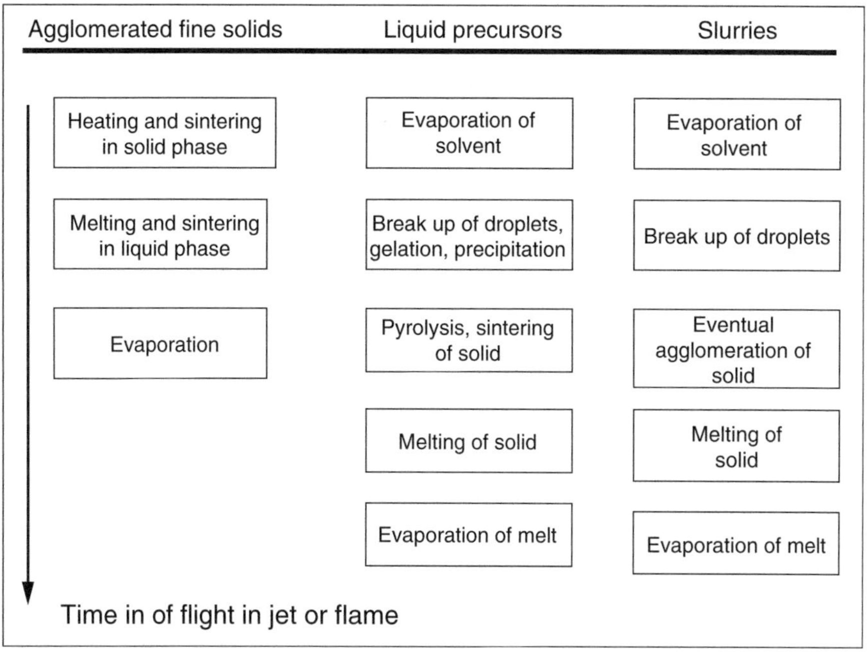

Fig. 7.24 Thermal phenomena occurring at flight of different types of feedstock in jet or flame [82] (Copyright Elsevier; reproduced with permission)

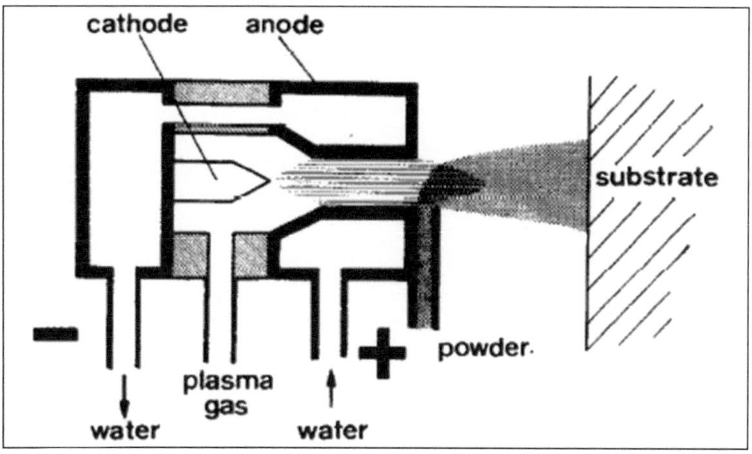

Fig. 7.25 Schematic diagram of a plasma torch [19] (Copyright Elsevier; reproduced with permission)

The cathode is usually a bar of thoriated tungsten and the anode nozzle is usually made of electrolytic copper that is water-cooled. The water-cooling removes the heat dissipated at the anode in order to avoid its destruction. Moreover, the water-cooling stabilizes the arc and the hot gases in the nozzle. When the gas crosses the strong electric field prevailing between the electrodes, it is dissociated and ionized. The atom-atom and electron-ion recombination produce the plasma jet which blows out of the nozzle anode with lengths varying between 30 and 150 mm. In relation to the gas mixture and pressure, the ejection velocity of the gases can vary from a few hundred meters per second to almost 1,000 m/s. The injection velocity of the particles must be perfectly adapted because, as the plasma is very viscous, if the injection velocity is too low, the powders tend to rebound from the jet and pass to its periphery in the cold zone where they are not melted. On the contrary, if the injection velocity is too high, the particles may cross through the jet and again travel into a cold zone of its periphery. This difficulty of introducing the particles into the viscous plasma with its thermophoresis forces (due to the thermal gradients of up to 4,000°K/mm – tending to eject the particles out of the flame) is certainly one of the main problems inherent in the plasma-spraying process in which the injection velocity and thus the pressure of the carrier gas must be adjusted for each type and each granulometry of the sprayed powders [19].

The feedstock particle temperature and velocity during plasma spraying are also sensitive to the parameters of the torch such as input power, arc gas flow, carrier gas, spray distance and type of plasma arc gases used. Gerdeman and Hecht [41] have compiled a summary of the plasma spray-process parameters which is shown in Fig. 7.26. Mash et al. [68] have schematically illustrated the effects of some plasma-spraying process variables on the coating deposition efficiency which has been redrawn by Ang and Berndt [4] and presented in Fig. 7.27.

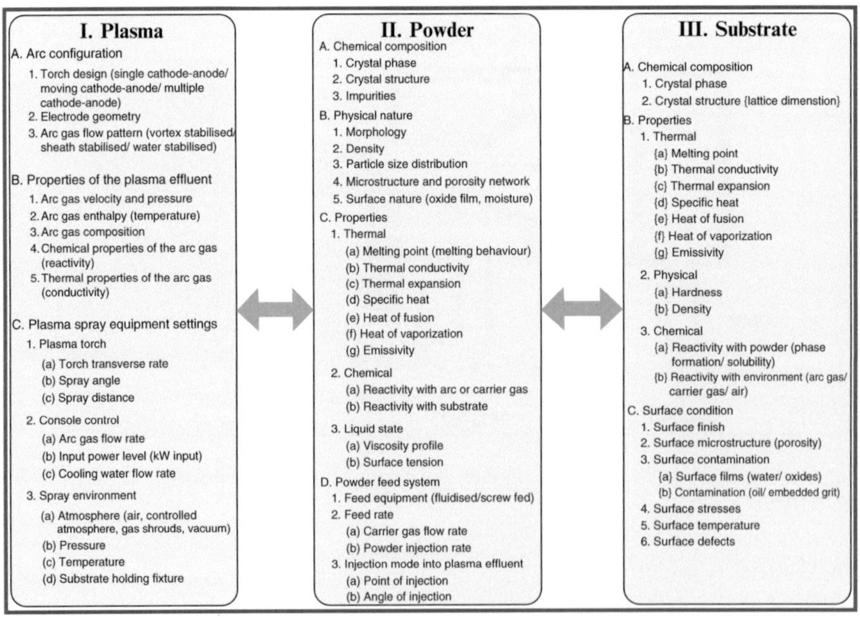

Fig. 7.26 Summary of plasma spray process parameters (Ang and Berndt [4] modified from Gerdeman and Hecht [41]. Copyright Maney Publishing; reproduced with permission)

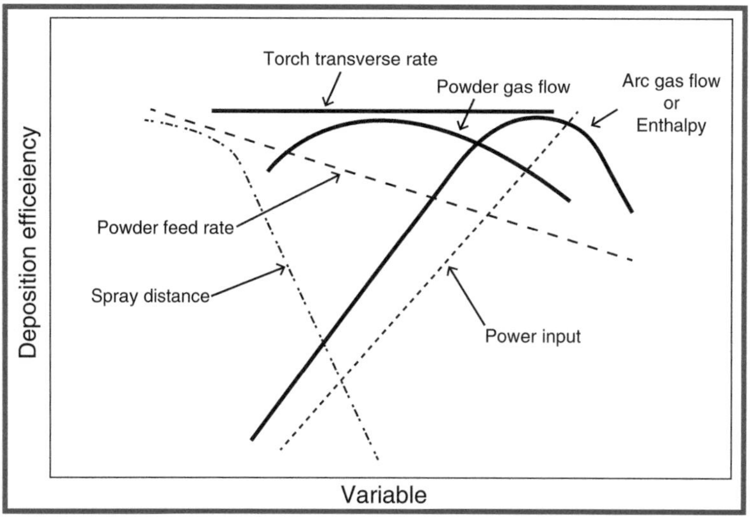

Fig. 7.27 Effect of plasma spraying process parameters on deposition efficiency (Ang and Berndt [4] redrawn from Mash et al. [68]. Copyright Maney Publishing; reproduced with permission)

Rampon et al. [87] studied the influence of the suspension characteristics (i.e., the viscosity and surface tension) on the microstructure of coatings obtained from suspension plasma-spraying. Suspension plasma-spraying consists in injecting a suspension of sub-micrometer particles into a plasma torch which leads to coatings with fine microstructures. When using a conventional DC plasma torch, a suspension can be injected as a straight jet of liquid or as a stream of atomized droplets. The resulting stream of droplets or the jet is commonly injected radially into the DC plasma jet in which the drops are heated. Rampon et al. [87] observed that the nature of the solvent has proven to have great influence on the first breakup: monomodal drop-sized distributions with aqueous suspensions and bimodal drop-size distributions with alcoholic slurries. This means that the aqueous suspensions are more difficult to atomize. In the same way, increased viscosity tends to hinder the atomization, but the effect is less important as it is the surface tension that controls the breakup. It is assumed that above a threshold temperature sufficient to melt the whole particle, the velocity is the key parameter controlling suitable particle spreading and thus the corresponding porosity. Increasing the viscosity of the alcoholic suspension is also a way to decrease the porosity, but it is less effective than the surface tension change. Coatings produced from aqueous suspensions are denser than those elaborated from alcoholic suspensions, whatever the viscosity. So suspension plasma-spraying process is an alternative technique to manufacture the solid oxide fuel cells (SOFCs) in a more effective and less expensive way [87].

Recently also Hui et al. [48] reviewed the applications, potential advantages and challenges of thermal plasma spray (PS) processing for nanopowder production and cell fabrication of SOFCs. Nano-scale powders (1–100 nm in size) are becoming increasingly critical to the innovations in numerous applications, including

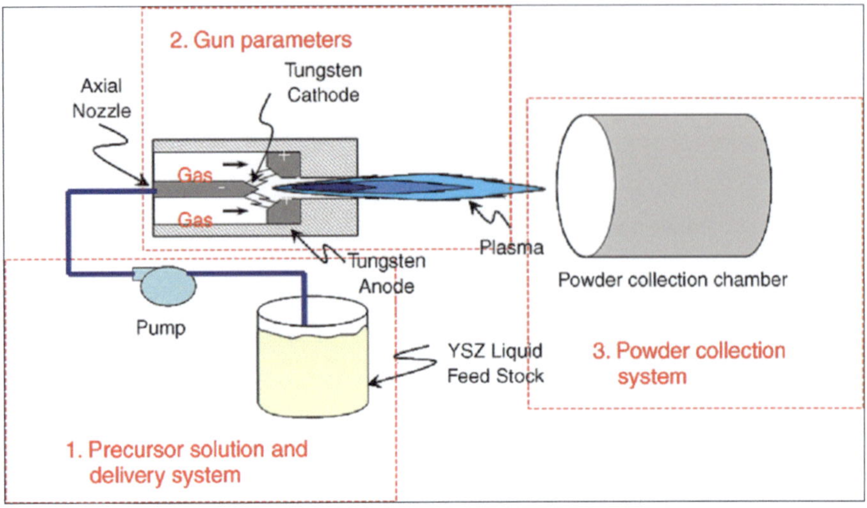

Fig. 7.28 Schematic diagram of a plasma-spray (PS) system for nanopowder synthesis [48] (Copyright Elsevier; reproduced with permission)

catalysis, coatings, cosmetics, electronics, sensors, and drug delivery. Nanopowders offer controlled functionality over coarser materials, such as higher surface area, catalytic activity, and sinterability. Nanomaterials in particular are of considerable interest for use in SOFCs due to their potential to increase the surface area of active sites on which the electrode reactions take place, thereby improving reaction kinetics. SOFC electrode reactions occur mainly at the interfaces between phases that conduct oxygen ions, gases, and electrons, commonly referred to as triple-phase-boundaries (TPBs). Therefore, an extended reaction surface area with an optimized porous microstructure will enhance the electrode performance. Nanostructured materials also possess the enhanced electrical conductivity that is required for SOFC components, either ionic or electronic conductivity. A schematic diagram of a plasma-spray (PS) system for nanopowder synthesis is shown in Fig. 7.28 [48] which consists of (1) precursor feedstock and delivery system, (2) plasma-spraying system (plasma-forming, plasma gas injection nozzle, fuel and oxygen mixtures, plasma reaction, and plasma outlet), and (3) powder collection system (nucleation, particle growth, and filters, cyclones, or cooled surfaces for particle separation from the material flow stream).

7.3.2.2 Wire Arc Spray Process

In the wire arc spray process, a direct current electric arc is formed between two consumable electrode wires and a high-velocity gas jet propels the molten materials towards the substrate. It is a reliable, economical and commonly used method for depositing metal coatings for corrosion protection applications. The temperature within the arch jet (~7,000°K) is sufficient to melt the consumable electrodes, which are atomized into particles, accelerated and then deposited onto the substrate. Thus, the thermal efficiency of the wire arc spray is greater than other thermal spray processes because the particles are in a fully molten state when they enter the spray jet. The molten particles start cooling immediately after leaving the arc zone. Therefore, coating properties such as porosity and adhesion can be effected. Developments of this process, such as high-velocity nozzle caps, have opened up new applications with particle velocities similar to those experienced for HVOF spray processes. Wire arc spraying is not limited to metal wires; for instance, cored wires consist of powdered materials such as carbides or amorphous alloys that are contained within a nickel- or iron-based metal sheath [4].

7.3.2.3 Flame Spray Process

In the flame-spray (FS) process, a stream of fuel and oxygen is combusted externally at the nozzle tip, while powders are introduced axially into the flame via an inert gas such as argon or nitrogen. The feedstock can also be in the form of sticks, wires or rods, for which a carrier gas is not needed [4]. The FS powder systems are similar to Verneuil's burner which is used in the manufacturing of synthetic gems. The commonly used gases are mixtures of O_2-H_2 and O_2-C_2H_2 which give maximum flame temperatures of 2,600 °C and 3,000 °C respectively. Another possibility is to use the O_2-C_2N_2 mixture which generates a flame temperature of 4,700 °C. In the FS stick burner, the combustive-combustible mixture is fed into a circular burner whose apertures are disposed in a crown around the pipe containing the stick as shown in the left-hand frame of Fig. 7.29. The stick, continuously advanced by a variable speed engine, is melted a few millimeters after its exit from the pipe. Then a droplet of liquid material is dragged out as well by the flame as by the compressed air flow which is used to avoid back-flash. The FS stick burner has the best spraying efficiency because the material is sprayed only once it is melted [19].

In the powder burner FS processes, there are three common methods to introduce the coating powder into the flame: (1) a fluidized bed using additional air flow; (2) gravity, and; (3) a fluidized bed using one of the gases of the burner. A FS powder burner with a gravitic powder feeder of a very simple construction is shown in the right frame of Fig. 7.29 [19].

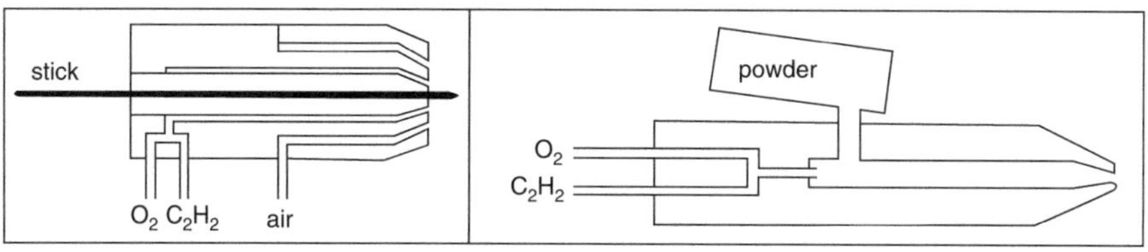

Fig. 7.29 Schematic diagrams of a FS stick burner (*left*) and a FS powder burner (*right*) [19] (Copyright Elsevier; reproduced with permission)

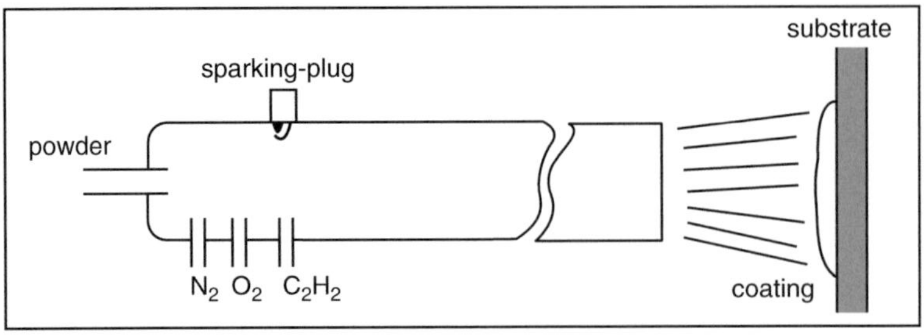

Fig. 7.30 Schematic principle of a detonation gun [19] (Copyright Elsevier; reproduced with permission)

7.3.2.4 Detonation Gun (D-Gun) Process

The working principle of the D-Gun torch is based on generating a high-energy pressure wave through repeated ignition of an explosive mixture, usually acetylene (C_2H_2) and oxygen (O_2), within a long, constricted tube. The thermal output and detonation pressure waves, operating at 3–15 Hz, are intense and represent the TV (flame-jet temperature and particle velocity) source that heats and accelerates the powder particles towards the substrate [4]. Figure 7.30 shows schematically the principle of a D-Gun process. D-Gun coatings have achieved dense coatings with high bond strength [103, 98] and they have been compared to plasma-spray coatings under conditions of abrasion and erosive wear resistance [77]. In general, D-Gun coatings outperform plasma-spray (PS) coatings in niche markets, but they are not as versatile as other thermal spray coatings with regard to materials selection, equipment availability and engineering design of the component [4].

7.3.2.5 High Velocity Oxygen Fuel (HVOF) Spray Process

The operating characteristics of HVOF spray are based on a continuous combustion of the fuel gas mixture supplied at high volumetric flow to the burning chamber of the torch. Figure 7.31 shows a schematic representation of a commonly used HVOF spray process including the combustion chamber, convergent-divergent nozzle, and barrel [89]. In the combustion chamber, continuous combustion of oxygen and fuel occurs under high pressure (0.5–2 MPa). The hot combustion gases are then accelerated to supersonic velocities using a converging-diverging nozzle. The coating powder particles are injected into the gas jet inside the torch and are simultaneously heated and propelled towards the substrate [21].

Process gas selections include propylene, propane, hydrogen or natural gas used in gas-fuelled systems and kerosene in liquid-fuelled systems [53]. Ang and Berndt [4] compiled the summary of typical HVOF fuels and torch combinations in Table 7.4 based on the equipment manufacturers published technical data and typical spray tables provided from feedstock suppliers. Recent developments of

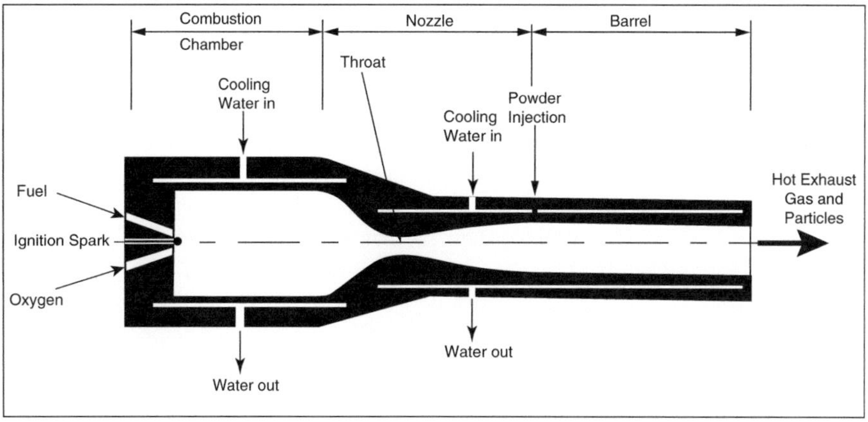

Fig. 7.31 Schematic diagram of a commonly used HVOF spray process [21] redrawn from Richter [89] (© IOP Publishing; reproduced by permission of IOP Publishing. All rights reserved)

Table 7.4 Properties of HVOF fuels and torch combinations* [4] (Copyright Maney Publishing; reproduced with permission)

Torch (year)	Fuel gas option	Fuel flow/L min−1	Oxygen flow/ L min−1	Air flow/L min−1	Oxygen to fuel ratio[a]
JetKote®, Deloro Stellite (1982)	Hydrogen	432	302	–	0.7
	Propane	50	350	–	7.0
	Ethylene	80	337	–	4.2
Top Gun, UTP (1989)	Hydrogen	432	217	–	0.5
	Propane	50	250	–	5.0
	Ethylene	80	240	–	3.0
DJ2600, Sulzer Metco (1989)	Hydrogen	613	214	344	0.47
DJ2700, Sulzer Metco (1994)	Propane	189	278	391	1.90
	Ethylene	111	247	360	2.91
	Propane	68	240	375	4.67
JP-5000®, Praxair (1992)	Kerosene	0.379	876	–	2,314
WokaStar™, Sulzer Metco (2004)	Kerosene	0.379	876	–	2,314
GTV K2, GTV (2005)	Kerosene	0.372	820	–	2,204

*Compiled from (i) equipment manufacturers published technical data and (ii) typical spray tables provided from feedstock suppliers

[a]The values for the liquid feed processes are high since the liquid flow rates are comparatively low

HVOF spray systems have been aimed predominantly at optimizing the combustion, minimizing fuel combustion and increasing the particle acceleration [4].

Depending on the fuel gas and torch design used, the combustion jet temperatures can reach 4,000°K with energies in the order of 288–1,000 MJ/h, 989–278 kW). After passing through the nozzle, gas velocities exceed 1,800 m/s, resulting in the formation of a "shock" diamond pattern within the exiting flame jet. The feedstock, in powdered form, is usually fed axially through the torch using an inert carrier gas such as nitrogen. The ignited gases surround and heat the powdered spray material as it exits the torch. Particle velocities may reach 800 m/s [30, 62]. As a result of the high kinetic energy transferred to the particles through the HVOF process, the coating materials generally do not need to be fully molten [52, 62]. Instead, the powder particles can be in a semi-molten state so that they flatten plastically on impact against the work piece [4].

HVOF spray coatings often exhibit densities of >95 % and adhesion values greater than 10,000 psi (69 MPa) [29, 61]. HVOF spray coatings demonstrate lower residual internal stresses [16, 60] compared to air plasma-spray (PS) coatings because the coating deposition temperatures are significantly lower; and hence reduce the tensile quenching stresses experienced by the splats as they cool and solidify [4]. In addition, peening stresses are generated when particles impact at high velocity. These stresses induce compressive stresses that superimpose on and reduce the previously formed tensile quenching stresses [57]. In this way, dense HVOF coatings of up to 2 mm can be deposited [86, 95]. The feedstock undergoes a short flame residence time, measured in milliseconds [114], that minimizes excessive metal oxidation, decarburization or phase changes [54, 64]. Additional advantages of the HVOF spray process over conventional plasma-spraying (PS) include higher coating bond strength, higher deposition rates, higher hardness, lower oxide content, and improved wear resistance due to a homogeneous distribution of particles [5, 94].

7.3.2.6 Cold Spray Process

Unfortunately, in typical applications of the thermal spray processes, the feedstock materials experience high temperatures that can lead to undesired phase change and oxidation in the coating material. Cold spray is a fast growing coating deposition developed in the last decade due to its ability to deposit coatings without any change in chemical compositions compared with feedstocks and shows excellent flexibility for designing the coating compositions through feedstock [66]. The cold spray deposition process involves the acceleration of solid particles which are maintained at temperatures lower than the melting point of the particle material and at high velocities using a carrier gas which is often supersonic. The subsequent impingement of this high velocity, particle-laden jet onto a substrate is shown in Fig. 7.32 [2].

Intensive investigations revealed that many different types of coatings can be deposited by cold spray. Typical coating materials include ductile materials such as copper, aluminum, zinc, titanium, iron-based alloys and nickel-based superalloys, metallic glass materials, intermetallic materials, ceramic particles dispersed metal matrix composites such as WC-Co, SiC/Al, bronze-diamond, cubic BN/NiCrAl [66]. During cold spray, to build up a deposit successively, upon the impact of a

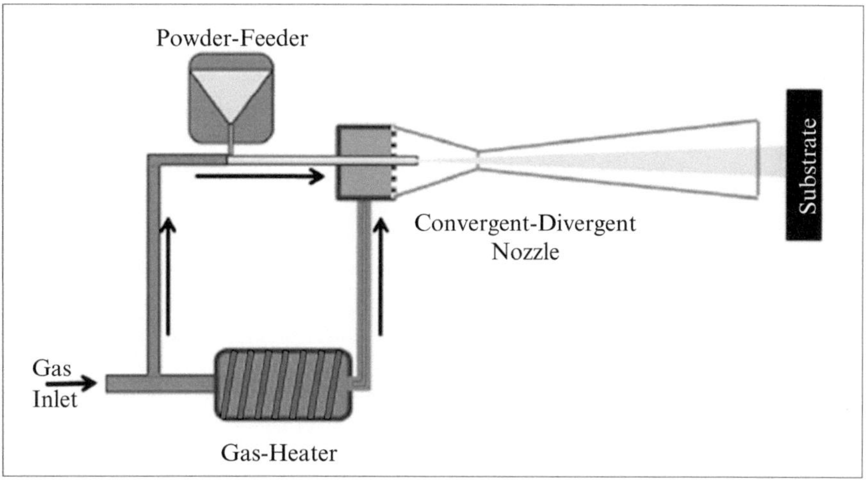

Fig. 7.32 Cold spray process [2] (Copyright Springer Science + Business Media; reproduced with permission)

high velocity solid particle on a substrate plastic deformation is required for both particle and substrate, which creates a bond to make the particle adhere on the substrate. The features of high velocity and low processing temperature significantly limit the reaction, such as oxidation of metallic materials, of spray particles with the atmosphere gas species. The low particle temperature eliminates melting prior to impact. It also benefits by eliminating the melting-induced deteriorating effect on the microstructure of deposits in terms of grain growth and thus possibly retains the microstructure of spray materials in coatings. Therefore, the cold spray process provides the possibility to design the deposit microstructure through feedstock powders [66]. In a cold spray review article, Luo et al. [66] examined the evolution mechanisms of metastable microstructure in terms of in situ densification, residual stress, dislocation multiplication, grain refining, phase transformation and amorphization through typical materials. Proper feedstock microstructure design, in-process control and material selection are essential factors needed to acquire the cold spray coatings with a desired microstructure.

7.3.3 Nebulization Process

Nebulization of liquids or solutions into fine mist, aerosol or fine powders can be achieved by jet nebulizers, ultrasonic (piezoelectric) nebulizers (atomizers), electrostatic atomizers. Nebulizers such as ultrasonic inhalers, dry powder inhalers, metered dose inhalers, and jet inhalers are used in the biomedical field to administer the medications in aerosol, nano or micro size powders or fine mist droplets form into the deep lung region using the upper respiratory tract. Additionally the nebulization process is widely used in thin film powder coatings [42, 90, 99, 107]. Recently, Yeo and Friend [110] provided a summary of the ultrasonic nebulizer techniques used in microfluidic devices.

Selim et al. [90] studied the nebulized spray pyrolysis technique employed to prepare pure and Al-doped ZnO films on single crystalline silicon substrates. They concluded that this nebulization technique generated better-oriented thin films compared to other methods reported in the literature. Gowthami et al. [42] studied nickel oxide thin film prepared by simple nebulizer spray technique. This technique allowed a good quality film growth at different substrate temperatures. The nebulizer technique can be utilized for film preparation with lower volume of solutions under various preparation parameters. In their study, nickel oxide thin films were deposited on glass substrates at temperature 400 °C. The film thickness increased from 1,070 to 2,940 nm as the volume of spraying solution was varied from 1 to 20 ml. The XRD (X-ray diffractometer) results showed that all the prepared films were NiO (cubic phase) of high crystalline quality with the (1 1 1) preferred orientation and the average crystallite size was around 24–12 nm [42].

Suresh et al. [99] studied the deposition of nano crystalline, uniform and crack-free cerium oxide thin films on glass substrates by simple and low cost effective nebulizer spray pyrolysis technique at different substrate temperatures using cerium nitrate as the source material. All films were found to be single phase and well crystalline with most prominent (200) reflection. Crystallinity of the film was found to be increased with increasing substrate temperature up to 400 °C which then decreases. The calculated crystalline size was found to vary in the range of 9–21 nm. Optical analysis revealed that the films were highly transparent (70 %) in the visible region. The high refractive index and band gap were observed as 2.35 and 3.52 eV, respectively. SEM images exhibited the formation of uniform and porous films with golf ball like structure. The conductivity lies in the range $1.65-2.90 \times 10^{-6}$ S/cm. The best film was formed at 400 °C with better crystallinity, high refractive index and smooth morphology. Substrate temperature strongly affected the optical, structural and electrical properties of CeO_2 films due to the decrease of sprayed droplet diameter, presence of unevaporated precursor and densification of the films [99].

Slurry nebulization in plasmas for ICP (inductively coupled plasma) spectrometry has advantages of simplicity, high speed, low cost, minimized analyte loss, and low risk of sample contamination, but it has not been very widely adopted. Wang and Yang [107] surveyed the current state of research on slurry nebulization in plasmas for the analysis of advanced materials. Ceramic samples should be ground to micron-sized or even submicron-sized particles and prepared as slurry with an appropriate amount of a stabilization agent. In theory, addition of the stabilization agent increases the zeta potential between the particles in the chosen medium to improve the slurry stability. A zeta potential of zero means that the slurry particles will agglomerate, and thus the slurry is unstable. The different nebulizers used for the slurry nebulization into plasma are shown in Fig. 7.33. The most popular designs for handling slurries are based on the V-groove Babington-type nebulizer (Fig. 7.33a). The slurry sample flows along a V-groove that has a small orifice through which the carrier gas blows out at high flow rates. This causes the sample to

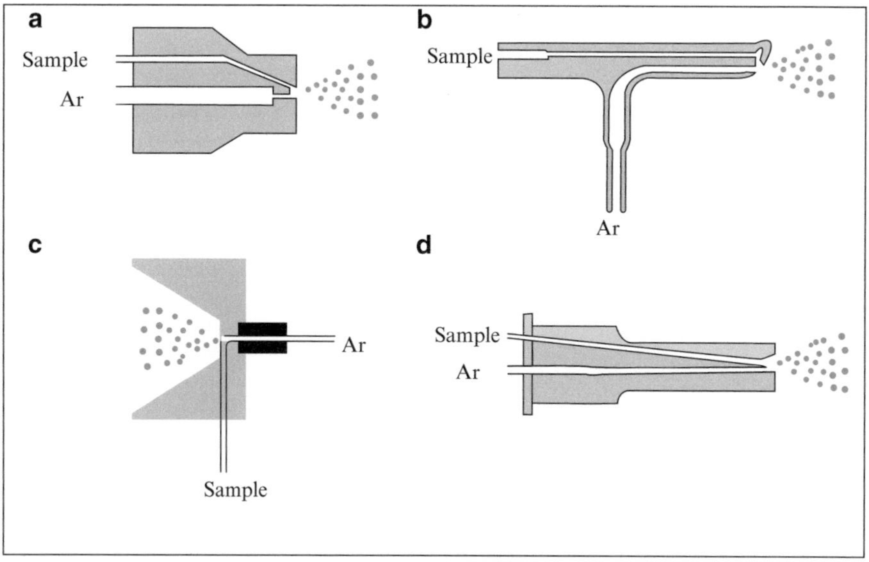

Fig. 7.33 Structural schematics of several nebulization devices: **a** V-groove Babington-type, **b** glass expansion concentric (GEC) type, **c** cross-flow type, and **d** Burgener type nebulizers [107] (Copyright The Royal Society of Chemistry; reproduced with permission)

shatter into an aerosol of small droplets. The high pressure of the argon gas ensures efficient nebulization and prevents blocking of the small orifice. The glass expansion concentric (GEC) nebulizer (Fig. 7.33b) is known to create a very fine mist with small droplets; whereas the Burgener nebulizer (Fig. 7.33d) allows larger droplets to form because of its wider inner diameter. The cross-flow (Fig. 7.33c) and Burgener nebulizers each yielded about the same sensitivity when the same instrument settings were used [107].

The most common medical nebulizer type is the jet nebulizer, which generates aerosols from the liquid medicament using a source of compressed gas. Although relatively inexpensive, treatment with jet nebulizers has long treatment time, the air compressors are bulky and noisy, and expensive medications are wasted in considerable residual volumes. Other types of liquid aerosol generation systems include the Respimat® SoftMist™ Inhaler and AERx® systems, which generate liquid aerosol by mechanically forcing a drug solution through a nozzle array [113]. Ultrasonic nebulizers generally have a higher output rate than jet nebulizers, but a larger average particle size. A metered-dose inhaler (MDI) is a device that delivers a specific amount of medication to the lungs, in the form of a short burst of aerosolized medicine that is usually self-administered by the patient via inhalation. A metered-dose inhaler consists of three major components; the canister which is produced in aluminum or stainless steel by means of deep drawing, where the formulation resides; the metering valve, which allows a metered quantity of the formulation to be dispensed with each actuation; and an actuator (or mouthpiece) which allows the patient to operate the device and directs the aerosol into the patient's lungs [46].

The amount of fine drug particles delivered to the lungs is governed by the balance of varying forces including inter-particulate forces within powders, dispersion forces by airflow through an inhaler device and deposition forces in airways. Fine powders generated through micronization generally have high surface energy and hence are notoriously cohesive. An inhaler device is necessary to disperse such powders into inhalable aerosols. Aerosolization behavior of a dry powder inhaler (DPI) thus depends on both the powder formulations and the inhaler devices. The powder dispersion mechanisms in DPIs are complex, depending on the specific device, and involving air turbulence and impaction. Recent inhaler designs attempt to enhance powder de-agglomeration or drug-carrier detachment by increasing air turbulence and particulate collisions [113].

7.4 Spray Measurements

Sprays represent complex physical processes including the spray formation, the re-distribution of the spray as it enters into the flow field generated by the spray itself due to entrained air flow and the ambient flow, drop collisions and coalescence and evaporation. Thus from the experimental diagnostic viewpoint, the spray formation and distribution constitute a very dynamic and complex process, especially when the spray is injected into a highly turbulent flow wherein the aerodynamics plays an even greater role in the redistribution of the spray [7].

7.4.1 Spray and Particle Characterization Techniques

A wide range of instruments is available for measurement of spray systems, including high-speed motion photography, laser diffraction analyzer, particle image velocimetry (PIV), phase Doppler particle analyzers (PDPA), digital photography, optical interferometry, laser-induced fluorescence, elastic and laser absorption scattering, Mie scattering and shadowgraphy, spontaneous Raman, ultraviolet laser absorption, laser-induced fluorescence and high-energy X-ray radiography [8, 22, 26].

A variety of diagnostic strategies is available to measure particle parameters in the thermal spray processes. The principal objectives of these diagnostic devices are to measure the spray plume characteristics namely, the particle velocities, temperatures, trajectories, and size distributions. These strategies include measurement of individual particle parameters within the spray stream (e.g. Tecnar DVP 2000 and Inflight™ integrated particle sensor) as well as ensemble particle sensors for group measurements of particle velocities and temperatures (e.g. Tecnar Accuraspray, Oseir Spraywatch, Inflight Particle Pyrometer, Stratonics Thermaviz, etc.) [1, 17, 33, 34, 58, 96]. Many of these sensors measure particle temperatures through the use of thermal emission emanating from a molten/semi-molten traveling particle Time-of-flight measurements are used to extract the particle velocities [112].

Mauer et al. [70] summarized the plasma and particle temperature measurement techniques in plasma sprays: (1) the optical emission spectroscopy to measure the optical emission spectrum of plasma allowing the qualitative analysis of the emitting gas species including vaporized particle material and gaseous decomposed constituents of powderous and liquid feedstock and (2) particle diagnostics systems distinguished by their detecting elements of the fiber-optic-based sensors and CCD arrays. The number of available particle diagnostic systems for thermal sprays is small: (2a) DPV-2000 by TECNAR Automation Ltd; (2b) Accuraspray-g3 by TECNAR Automation Ltd.; (2c) Inflight Particle Pyrometer (IPP) sensor by Inflight Ltd.; (2d) ThermaViz by Stratonics Inc.; (2e) SprayWatch by Oseir Ltd.; (2f) Flux Sentinel by Cyber Materials LLC., (2f) NIR Sensor by GTV Verschleissschutz Gmbh [70].

7.4.2 Measurements of Droplet Size and Velocity

The drop-sizing methods may be grouped conveniently into three broad categories [59]:

1. Mechanical
   (a) Drop collection on slides or in cells
   (b) Molten-wax and frozen-drop techniques
2. Electrical
   (a) Charged-wire and hot-wire techniques
3. Optical
   (a) Imaging (photography, holography)
   (b) Nonimaging (single particle counters, light scattering, Malvern particle analyzer)

Optical drop-sizing techniques can be subdivided into three categories: imaging (photography, cinematography, and holography), single drop counting, and ensemble multi drop-sizing. On the other hand, all optical drop-sizing methods can be divided into two classes; one representing spatial averaging (imaging techniques and ensemble drop-sizing) and the second temporal averaging (single drop-counting). The choice of a technique and an instrument for measurement of drop-size distributions that are most appropriate for a particular spray system depends on the application [11, 22].

Whatever measurement technique is applied, a major challenge is to perform the measurements on representative samples or through the whole spray, since both droplet size and velocity often depend upon their location in the spray.

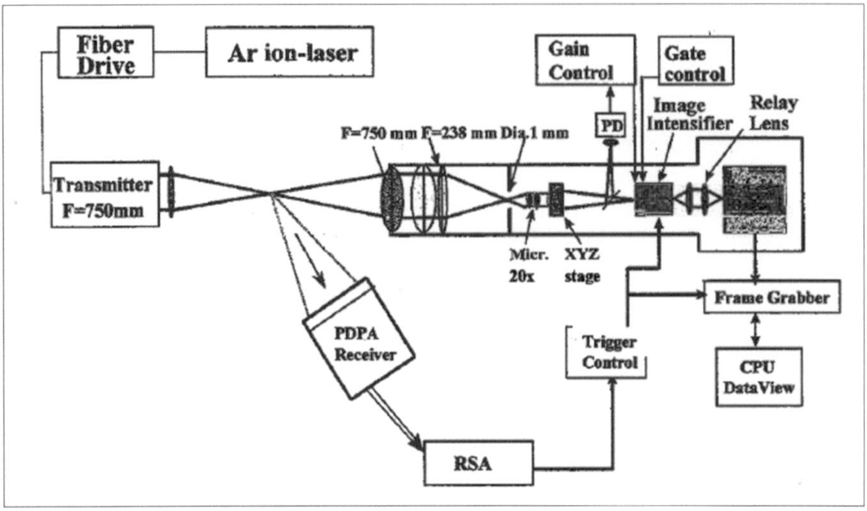

Fig. 7.34 System components for the crossed-beam imaging system [7]

7.4.2.1 Imaging Systems

Digital imaging systems utilize CCD-array detectors to capture the particle images. The instrument is based on the principle of gaining a shadow image for the particles passing through the intersection of two laser beams that form the probe volume. The general set-up of the imaging device is similar to that of the two-component laser Doppler Velocimetry (LDV) or Phase Doppler Particle Analyzer (PDPA) with the imaging components added. With this method, the two beams are transferred to a CCD camera and imaged onto the sensitive area of the CCD camera. Because the image capture must be very fast, a microscope objective is used to magnify the image onto the photocathode of the gated image intensifier. The intensifier amplifies the intensity of the image and acts as a shutter with a 6 ns gate time. The optical set-up of a high-speed imaging system is shown in Fig. 7.34 [7]. As can be seen, an argon ion continuous wave laser is used to produce the laser beams. The standard LDV transmitter optics are used to form a pair of beams which is then caused to focus and intersect using a transmitter lens. Beyond the beam intersection, the beams continue into the imaging receiver where they pass a series of lenses and a spatial filter to remove spurious laser light, and then are focused on the image intensifier. Thus, the lenses continue to transmit the image to the CCD camera. A frame-grabber is used to transfer the image to the computer for image processing. The off-axis PDPA receiver is used as a trigger signal to determine when a particle has passed through the center of the measurement volume and can also be used to measure the particle velocity and the particle size if the particle is spherical. Thus, the imaging system acts like a single particle counter with high spatial resolution. Because crossed beams are used, it eliminates or at least mitigates the depth-of-field problem. The data rate capability of the system is expected to be on the order of 1,000 particles per

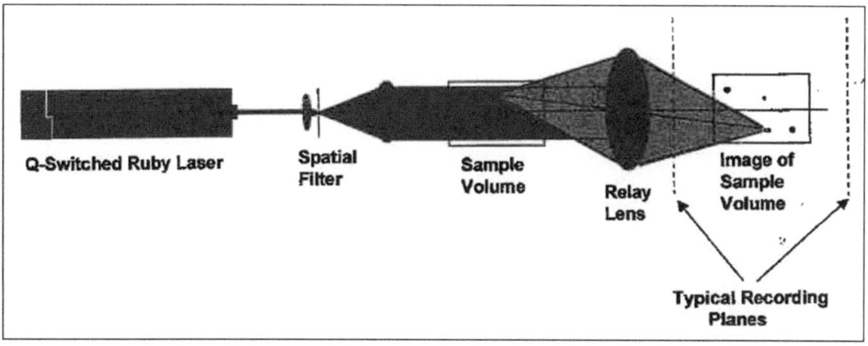

Fig. 7.35 Schematic showing the components of a holographic imaging system [7]

second. This capability depends on the number of pixels used for each frame and on the type of camera used in the system [7].

Holographic imaging of particle fields using pulsed lasers offers an intermediate step which serves to freeze a three-dimensional (volume) of the particle field optically. The hologram essentially serves to extend the depth-of-field of the imaging system greatly. The depth-of-field is limited by the need to record the cones of light from each particle before the particle can be resolved in the reconstructed image. Since the particles are generally moving in the recording space, a very short duration, high power laser pulse is needed. Typically, Q-switched ruby lasers ($\lambda = 0.694$ μm) or frequency-doubled Nd:YAG lasers ($\lambda = 0.53$ μm) are used in holography as portrayed in Fig. 7.35. There are two basic choices in the recording systems: in-line or off-axis [7].

7.4.2.2 Laser Diffraction Analyzer

The laser diffraction method, also known as Fraunhofer diffraction, ensemble light scatter detection, etc. was one of the first optical methods capable of probing sprays with realistic droplet number densities. Single particle-counting systems available at that time were based on forward scatter light detection. This resulted in sample volumes that were too large to handle spray drop number densities, even in relatively dilute sprays. Ensemble methods succeeded since they were able to measure sprays with relatively higher droplet number densities. The method utilizes the scattered light energy produced by particles throughout the exposed laser beam, which is integrated over multiple detectors located in the Fourier transform plane of the receiver lens (Fig. 7.36). Through the use of an appropriate optical light scattering model (based on either Fraunhofer or Mie theory for spherical particles) and an appropriate deconvolution method, the scattered light intensity distribution is converted to a droplet size distribution [7, 26, 72]. Li et al. [63] used a laser diffraction analyzer (Mastersizer 2000, Malvern Instruments) to measure the size distribution of a gas-atomized spherical Cu powder.

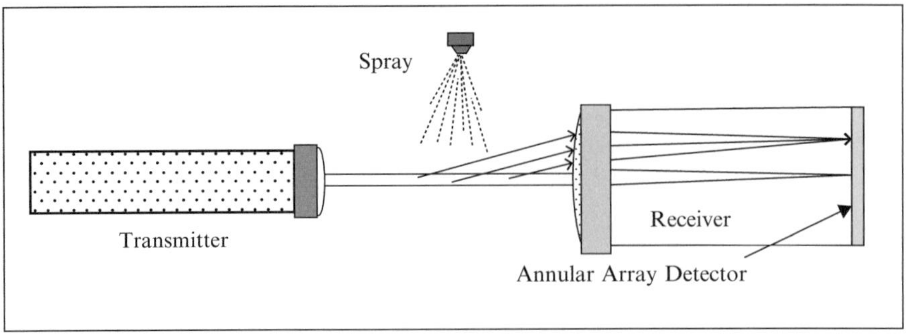

Fig. 7.36 Optical configuration of laser diffraction analyzer [7]

7.4.2.3 Laser Doppler Velocimetry

The laser Doppler velocimetry (LDV), also known as laser Doppler Anemometry (LDA), is the technique of using the Doppler shift in a laser beam to measure the velocity of fluid flows [7, 11, 59]. Although there is a number of optical configurations developed for LDV including the reference beam approach, the simple dual-beam method is most commonly used because the approach is rugged, lends itself to frequency-shifting and the use of fiber optics, and has the least systematic errors associated with it [7]. When using the dual-beam approach, the intersection of two laser beams from a common source defines the region from which measurements can be obtained. The actual measurement region may be a subset of the beam intersection reduced by the field of view of the receiver optics and the detection limits of the signal processor. Particles crossing the measurement region scatter light, which is collected by a receiver probe. The light signal is converted into an electrical "Doppler burst" signal with a frequency related to the particle velocity. In an LDV system, particle velocity and not gas velocity is the actual measured quantity [7]. The most basic one-component system (a single vector component of velocity) LDV system is depicted in Fig. 7.37.

7.4.2.4 Phase Doppler Particle Analyzer

The Phase Doppler Particle Analyzer (PDPA) is an extension of the well-known LDV instrument, which is an advanced laser-based diagnostic instrument that is used for simultaneously measuring the size and velocity of individual spherical particles in polydisperse flow environments [11, 22, 59]. The method was first introduced by [6]. The PDPA makes use of two intersecting beams to measure the particle or drop velocity from the Doppler difference frequency of the scattered light and the drop size interferometrically from the phase shift of the signals produced by pairs of detectors spatially separated by a known distance, as portrayed in Fig. 7.38 [7–9]. Cetegen and Yu [18] used a PDPA instrument to measure the particle size and velocity in DC arc plasma thermal sprays with two different 7 wt% YSZ powders.

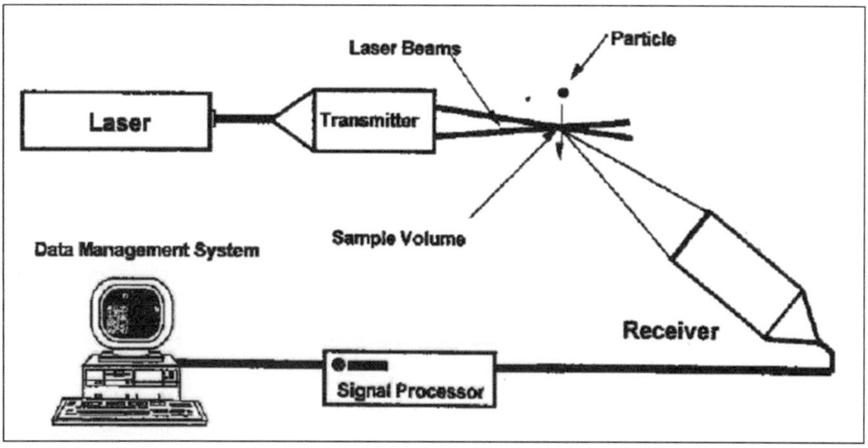

Fig. 7.37 Basic one-component LDV system [7]

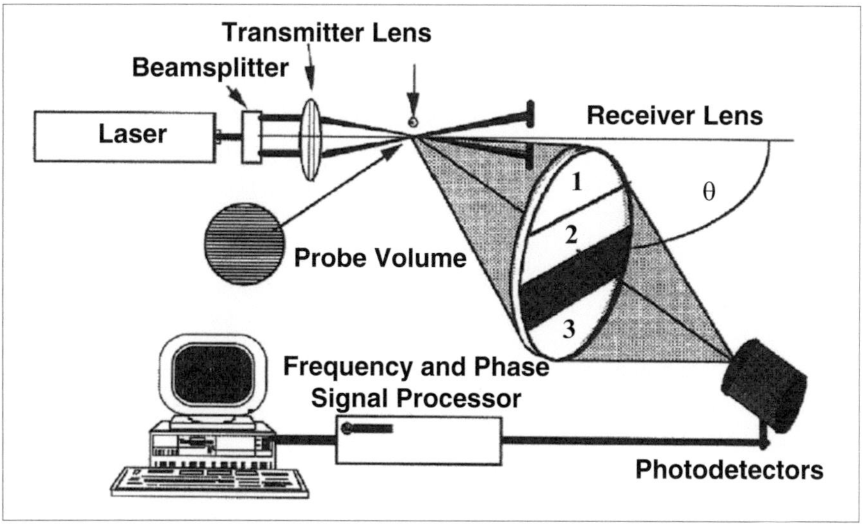

Fig. 7.38 Schematic of the basic PDPA system [7]

7.5 Definitions, Abbreviations and Symbols

Aerodynamic diameter	Diameter of a sphere with density 1,000 kg/m^3 having the same aerodynamic property as the particle of interest
Atomization	Process whereby a volume of liquid is disintegrated into a multiplicity of small drops
Critical Weber number	Represents the criterion for deformations that lead to secondary breakup of a drop where the critical drop diameter is the *maximum drop diameter* (d_{max}) that can exist under given conditions, since all drops bigger than d_{max} should disintegrate (breakup)

Deborah number	Ratio of the relaxation time of a material (characteristic time for material to adjust to applied stresses) to the characteristic time scale of observing the response of the material; it characterizes the fluidity of materials generated by their stress history
Mass median diameter	Median particle diameter of a mass- (or volume-) based size distribution
Nebulization	Process whereby a volume of liquid or solution is disintegrated into fine mist, aerosol or fine powders
Ohnesorge number	Indicator of jet or sheet stability; ratio of viscous forces to the square root of the product of inertia and surface tension; obtained also by dividing the square root of Weber number by Reynolds number, which eliminates velocity from both
Reynolds number	Ratio of inertial forces to viscous forces; for particles, it characterizes the nature of their flow in a liquid (laminar or turbulent)
Sauter mean diameter	Area-weighted mean particle diameter
Spraying	Application of a liquid in the form of small droplets ejected from an atomizer
Weber number	Ratio of disruptive forces to the stabilizing surface tension force; it characterizes the relative importance of a fluid's inertia
Weissenberg number	Ratio of relaxation time of a fluid to a specific process time; it indicates the degree of orientation generated by simple stress

APS	Atmospheric plasma spraying
CAPS	Controlled atmospheric plasma spraying
CAD	Conventional aerosol decomposition
CNC	Condensation nucleus counter
CPC	Condensation particle counter
FC	Flame spraying
DC	Direct current
D-Gun	Detonation gun
DMA	Differential mobility analyzer
DPI	Dry particle inhaler
EDTA	Ethylene-diamine-tetra-acetate
GEC	Glass expansion concentric
HVOF	High-velocity oxygen fuel
ICP	Inductively coupled plasma
LDV	Laser doppler velocimeter
LPP	Large porous particles
LPPS	Low pressure plasma spraying
PDPA	Phase doppler particle analyzer

PIV	Particle image velocimeter
PS	Plasma spraying
SAD	Salt-assisted aerosol decomposition
SEM	Scanning electron microscopy
SOFC	Solid oxide fuel cell
TPB	Triple-phase boundary
TEM	Transmission electron microscopy
TWA	Twin wire arc
VPS	Vacuum plasma spraying
XRD	X-ray diffraction

d_0	Diameter of initial droplet	
d_{jet}	Liquid jet diameter at the jet exit	
d_{max}	Maximum drop diameter for critical Weber number	
$<D_{3,2}>$	Sauter mean diameter	
De	Deborah number	$=\tau_c/\tau_p$
L_B	Jet breakup length	
Oh	Ohnesorge number	$=\mu_L/(\rho_L\cdot\sigma\cdot d_0)^{0.5}=We^{0.5}/Re$
Re_L	Droplet Reynolds number	$=\rho_L\cdot v_L\cdot d_0/\mu_L$
v_j	Liquid jet velocity	
v_L	Liquid droplet velocity	
v_{rel}	Relative velocity of droplet and gas	
We_c	Critical Weber number	$=\rho_G\cdot v_{rel}^2\cdot d_{max}/\sigma$
We_{jet}	Liquid jet Weber number	$=\rho_G\cdot v_{rel}^2\cdot d_{jet}/\sigma$
We_L	Liquid droplet Weber number	$=\rho_G\cdot v_{rel}^2\cdot d_0/\sigma$
μ_L	Liquid viscosity	
ρ_G	Gas density	
ρ_L	Liquid density	
σ	Surface tension	
τ_c	Stress relaxation time	
τ_p	Time scale of observation	
τ_{SV}	Characteristic time for solvent evaporation	
τ_{sl}	Characteristic time for solute diffusion	

References

1. Agapakis, J., Hoffman, T.: Real-time imaging for thermal spray process development and control. J. Therm. Spray Technol. **1**(1), 19–25 (1992)
2. Alhulaifi, A.S., Buck, G.A.: A simplified approach for the determination of critical velocity for cold spray processes. J. Therm. Spray Technol. **23**(8), 1259–1269 (2014)
3. Aliseda, A., Hopfinger, E.J., Lasheras, J.C., Kremer, D.M., Berchielli, A., Connolly, E.K.: Atomization of viscous and non-newtonian liquids by a coaxial, high-speed gas jet. Experiments and droplet size modeling. Int. J. Multiphase Flow **34**, 161–175 (2008)

4. Ang, A.S.M., Berndt, C.C.: A review of testing methods for thermal spray coatings. Int. Mater. Rev. **59**(4), 179–223 (2014). www.maneyonline.com/imr
5. Apelian, D., Wei, D., Farouk, B.: Particle melting in high temperature supersonic low pressure plasma jets. Metall. Trans. B **20B**, 251–262 (1989)
6. Bachalo, W.D., Houser, M.J.: Phase doppler spray analyzer for simultaneous measurements of drop size and velocity distributions. Opt. Eng. **23**(5), 583–590 (1984)
7. Bachalo, W.D.: Experimental methods and spray measurements. In: Spray Technology Short Course Notes, vol. 2, pp. 15213–3890. Published by N. Chigier, Mechanical Engineering Department, Carnegie Melon University, Pittsburgh (1996)
8. Bachalo, W.D.: Spray diagnostics for the twenty-first century. Atomization Sprays **10**(3–5), 439–474 (2000)
9. Bachalo, W.D.: Characterizing particulate and droplet size distributions: exhaust emissions to cloud research. In: Conference on Lasers and Electro-Optics (CLEO) (2009)
10. Bailey, A.G.: The theory and practice of electrostatic spraying. Atomization Spray Technol. **2** (2), 95–134 (1986)
11. Bayvel, L., Orzechowski, Z.: Liquid Atomization. Taylor & Francis, New York (1993)
12. Berger, H.L.: Characterization of a class of widely applicable ultrasonic nozzles. In: Proceedings of the 3rd International Conference on Liquid Atomization and Spray Systems, pp. 1A/2/1–13. London. July 1985 (1985)
13. Berger, H.L.: Ultrasonic Liquid Atomization – Theory and Application. Partridge Hill Publishers, Hyde Park (1998)
14. Bird, R.B., Armstrong, R.C., Hassager, O., Curtiss, C.F.: Dynamics of Polymeric Liquids, Fluid Mechanics (Volume 1) and Kinetic Theory (Volume 2). Wiley-Interscience, Hoboken (1987)
15. Birouk, M., Lekic, N.: Liquid jet breakup in quiescent atmosphere: a review. Atomization Sprays **19**(6), 502–528 (2009)
16. Bolelli, G., Lusvarghi, L., Varis, T., Turunen, E., Leoni, M., Scardi, P., Azanza-Ricardo, C.-L., Berletta, M.: Residual stresses in HVOF-sprayed ceramic coatings. Surf. Coat. Technol. **202**, 4810–4819 (2008)
17. Brogan, J.A., Berndt, C.C., Smith, W.C., Gansert, R.V., Raghu, S., Sampath, S., Herman, H.: Real-time imaging of the plasma spray process. J. Thermal Spray Technol. **4**(4), 374–376 (1995)
18. Cetegen, B.A., Yu, W.: *In situ* particle temperature, velocity, and size measurements in DC arc plasma thermal sprays. J. Thermal Spray Technol. **8**(1), 57–67 (1999)
19. Chagnon, P., Fauchais, P.: Thermal spraying of ceramics. Ceram. Int. **10**(4), 119–131 (1984)
20. Chang, H.W., Okuyama, K.: Optical properties of dense and porous spheroids consisting of primary silica nanoparticles. J. Aerosol Sci. **33**(12), 1701–1720 (2002)
21. Cheng, D., Trapaga, G., McKelliget, J.W., Lavernia, E.J.: Mathematical modelling of high velocity oxygen fuel thermal spraying of nanocrystalline materials: an overview. Model. Simul. Mater. Sci. Eng. **11**, R1–R31 (2003)
22. Chigier, N.: Optical imaging of sprays. Prog. Energy Combust. Sci. **17**(3), 211–262 (1991)
23. Chigier, N.: "Spray Science and Technology," Keynote lecture. In: ASME Winter Annual Meeting, New Orleans (December 1993)
24. Chigier, N.: Spray science and technology. In: Spray Technology Short Course Notes, vol. 1, pp. 15213–3890. Published by Dr. Norman Chigier, Mechanical Engineering Department, Carnegie Melon University, Pittsburgh (1996).
25. Chigier, N., Reitz, R.D.: Regimes of jet breakup and breakup mechanisms (physical aspects). Progr. Astronaut. Aeronaut. **166**, 109–135 (1996)
26. Chigier, N., Bachalo, W., Reitz, R., Bellan, J., Herrmann, M.: Spray control for maximizing energy efficiency and reducing emission in combustion engines. Atomization Sprays **21**(7), 553–574 (2011)
27. Chhabra, R.P.: Bubbles, Drops, and Particles in Non-Newtonian Fluids, 2nd edn. CRC Press, Boca Raton (2006)

28. Cloupeau, M., Prunet-Foch, B.: Electrostatic spraying of liquids in cone jet mode. J. Electrost. **22**, 135–159 (1989)
29. Davis, J.R.: Handbook of thermal spray technology. ASM International, Materials Park (2004)
30. Dongmo, E., Wenzelburger, M., Gadow, R.: Analysis and optimization of the HVOF process by combined experimental and numerical approaches. Surf. Coat. Technol. **202**, 4470–4478 (2008)
31. Edwards, D.A., Haness, J., Caponetti, G., Hrkach, J., BenJebria, A., Eskew, M.L., Mintzes, J., Deaver, D., Lotan, N., Langer, R.: Large porous particles for pulmonary drug delivery. Science **276**, 1868–1871 (1977)
32. Felder, C.B., Vorlaender, B., Gander, B., Mekle, H.P., Bertschinger, H.U.: Microencapsulated enterotoxigenic *Escherichia coli* and detached fimbriae for peroral vaccination of pigs. Vaccine **19**, 706–715 (2001)
33. Fincke, J.R., Swank, W.D., Bewley, R.L., Haggard, D.C., Gevelber, M., Wroblewski, D.: Diagnostics and control in the thermal spray process. Surf. Coat. Technol. **146**, 537–543 (2001)
34. Fincke, J.R., Haggard, D.C., Swank, W.D.: Particle temperature measurement in the thermal spray process. J. Therm. Spray Technol. **10**(2), 255–266 (2001)
35. Fraser, R.D., Eisenklam, P., Dombrowski, N., Hasson, D.: Drop formation from rapidly moving liquid sheets. AICHE J. **8**, 672–680 (1962)
36. French, D.L., Edwards, D.A., Niven, R.W.: The influence of formulation on emission, deaggregation and deposition of dry powders for inhalation. J. Aerosol Sci. **27**, 769–783 (1996)
37. Fritsching, U.: Spray Simulation: Modeling and Numerical Simulation of Sprayforming Metals. Cambridge University Press, Cambridge (2004)
38. Ganan-Calvo, A.M.: On the theory of electrohydrodynamically driven capillary jets. J. Fluid Mech. **335**, 165–188 (1997)
39. Gemci, T., Hitron, R., Chigier, N.: Measuring charge-to-mass ratio of individual droplets using phase doppler interferometery.In: ILASS Americas 15th conference, Madison (2002)
40. Gemci, T., Yakut, K., Chigier, N., Ho, T.C.: Experimental study of flash atomization of binary hydrocarbon liquids. Int. J. Multiphase Flow **30**(4), 395–417 (2004)
41. Gerdeman, D.A., Hecht, N.L.: Arc Plasma Technology in Material Science. Springer-Verlag, New York (1972)
42. Gowthami, V., Perumal, P., Sivakumar, R., Sanjeeviraja, C.: Structural and optical studies on nickel oxide thin film prepared by nebulizer spray technique. Physica B **452**, 1–6 (2014)
43. Guildenbecher, D.R., Lopez-Rivera, C., Sojka, P.E.: Secondary atomization. Exp. Fluids **46** (3), 371–402 (2009)
44. Haenlein, A: Disintegration of liquid jet. NAAC TN 659 (1932)
45. Herrero, E.P., Martín Del Valle, E.M., Galán, M.A.: Development of a new technology for the production of microcapsules based in atomization processes. Chem. Eng. J. **117**, 137–142 (2006)
46. Hickey, A.J.: Pharmaceutical Inhalation Aerosol Technology, 2nd edn. Marcel Dekker, NY (2004)
47. Hiroyasu, H.: Spray breakup mechanism from the hole-type nozzle and its applications. Atomization Sprays **10**, 511–527 (2000)
48. Hui, R., Wang, Z., Kesler, O., Rose, L., Jankovic, J., Yick, S., Maric, R., Ghosh, D.: Thermal plasma spraying for SOFCs: applications, potential advantages, and challenges. J. Power Sources **170**(2), 308–323 (2007)
49. Iskandar, F., Gradon, L., Okuyama, K.: Control of the morphology of nanostructured particles prepared by the spray drying of a nanoparticle sol. J. Colloid Interface Sci. **265**(2), 296–303 (2003)

50. Iskandar, F., Chang, H.W., Okuyama, K.: Preparation of microcapsulated powders by an aerosol spray method and their optical properties. Adv. Powder Technol. **14**(3), 349–367 (2003)
51. Iskandar, F.: Nanoparticle processing for optical applications – a review. Adv. Powder Technol. **20**(4), 283–292 (2009)
52. Joshi, S.V., Sivakumar, R.: Particle behaviour during high velocity oxy—fuel spraying. Surf. Coatings Technol. **50**, 67–74 (1991)
53. Kamnis, S., Gu, S.: 3-D modelling of kerosene-fuelled HVOF thermal spray gun. Chem. Eng. Sci. **61**(16), 5427–5439 (2006)
54. Kear, B.H., Skandan, G., Sadangi, R.K.: Factors controlling decarburization in HVOF sprayed nano-WC/Co hardcoatings. Scr. Mater. **44**, 1703–1707 (2001)
55. Kelly, A.J.: Electrostatic atomization of hydrocarbons. In: Proceedings of the 2nd ICLASS, pp. 57–65. Madison (1982)
56. Khavkin, Y.I.: Theory and Practice of Awirl Atomizers. Taylor & Francis, New York (2004)
57. Kuroda, S., Tashiro, Y., Yumoto, H., Taira, S., Fukanuma, H., Tobe, S.: Peening action and residual stresses in high-velocity oxygen fuel thermal spraying of 316 L stainless steel. J. Therm. Spray Technol. **10**(2), 367–374 (2001)
58. Landes, K.: Diagnostics in plasma spraying techniques. Surf. Coat. Technol. **201**(5), 1948–1954 (2006)
59. Lefebvre, A.H.: Atomization and Sprays. Hemisphere Publishing Corp, New York (1989)
60. Lesage, J., Chicot, D.: Role of residual stresses on interface toughness of thermally sprayed coatings. Thin Solid Films **415**(1-2), 143–150 (2002)
61. Li, C.J., Wang, Y.Y.: Effect of particle state on the adhesive strength of HVOF sprayed metallic coating. J. Therm. Spray Technol. **11**(4), 523–529 (2002)
62. Li, M., Christofides, P.D.: Computational study of particle in-flight behavior in the HVOF thermal spray process. Chem. Eng. Sci. **61**(19), 6540–6552 (2006)
63. Li, W.-Y., Li, C.-J., Wang, H.-T., Li, C.-X., Bang, H.-S.: Measurement and numerical simulation of particle velocity in cold spraying. J. Therm. Spray Technol. **15**(4), 559–562 (2006)
64. Lima, S.R., Marple, B.R.: From APS to HVOF spraying of conventional and nanostructured titania feedstock powders: a study on the enhancement of the mechanical properties. Surf. Coat. Technol. **200**(11), 3428–3437 (2006)
65. Lin, S.P., Reitz, R.D.: Drop and spray formation from a liquid jet. Annu. Rev. Fluid Mech. **30**, 85–105 (1998)
66. Luo, X.-T., Li, C.-X., Shang, F.-L., Yang, G.-J., Wang, Y.-Y., Li, C.-J.: High velocity impact induced microstrucre evolution during deposition of cold spray coatings: a review. Surf. Coat. Technol. **254**, 11–20 (2014)
67. Mansour, A., Chigier, N.: Air-blast atomization of non-Newtonian liquids. J. Non-Newtonian Fluid Mech. **58**, 161–194 (1995)
68. Mash, D.R., Weare, N.E., Walker, D.L.: Process variables in plasma-jet spraying. J. Metals **13**(7), 473–478 (1961)
69. Matikainen, V., Niemi, K., Koivuluoto, H., Vuoristo, P.: Abrasion, erosion and cavitation erosion wear properties of thermally sprayed alumina based coatings. Coatings **4**, 18–36 (2014)
70. Mauer, G., Vaßen, R., Stőver, D.: Plasma and particle temperature measurements in thermal spray: approaches and applications. J. Thermal Spray Technol. **20**(3), 391–406 (2011)
71. McKinley, G.H., Kesharvarz, B., Sharma, V., Serdy, J., Ardekani, A.: Painting by numbers: Spraying and air-assisted atomization of complex fluids, In: Bob Bird's 90th birthday symposium and banquet. University of Wisconsin-Madison, 30–31 Jan 2014
72. Merkus, H.G.: Particle Size Measurements – Fundamentals, Practice, Quality. Springer, Dordrecht (2009)
73. Miesse, C.C.: Correlation of experimental data on the disintegration of liquid jets. Ind. Eng. Chem. **47**(9), 1690–1701 (1955)

74. Nandiyanto, A.B.D., Kaihatsu, Y., Iskandar, F., Okuyama, K.: Rapid synthesis of a BN/CNT composite particle via spray routes using ferrocene/ethanol as a catalyst/carbon source. Mater. Lett. **63**(21), 1847–1850 (2009)
75. Nandiyanto, A.B.D., Kim, S.G., Iskandar, F., Okuyama, K.: Synthesis of spherical mesoporous silica nanoparticles with nanometer-size controllable pores and outer diameters. Microporous Mesoporous Mater. **120**(3), 447–453 (2009)
76. Nandiyanto, A.B.D., Okuyama, K.: Progress in developing spray-drying methods for the production of controlled morphology particles: from the nanometer to submicrometer size ranges. Adv. Powder Technol. **22**(1), 1–19 (2011)
77. Niemi, K., Vuoristo, P., Mänthlä, T.: Properties of alumina-based coatings deposited by plasma spray and detonation gun spray processes. J. Therm. Spray Technol. **3**(2), 199–203 (1994)
78. Ohnesorge, W.: Formation of drops by nozzles and the breakup of liquid jets. Z. Angew. Math. Mech. **16**, 355–358 (1936)
79. Okuyama, K., Lenggoro, I.W.: Preparation of nanoparticles via spray route. Chem. Eng. Sci. **58**, 537–547 (2003)
80. Okuyama, K., Abdullah, M., Lenggoro, I.W., Iskandar, F.: Preparation of functional nanostructured particles by spray drying. Adv. Powder Technol. **17**(6), 587–611 (2006)
81. Pavenetto, F., Genta, I., Giunchedi, P., Conti, B.: Evaluation of spray-drying as a method for polylactide and polylactide-*co*-glycolyde microspheres preparation. J. Microencapsul. **10**, 487–497 (1993)
82. Pawlowski, L.: Finely grained nanometric and submicrometric coatings by thermal spraying: a review. Surf. Coat. Technol. **202**(18), 4318–4328 (2008)
83. Pawlowski, L.: The science and engineering of thermal spray coatings, 2nd edition, ISBN: 978-0-471-49049-4, Wiley, Hoboken (2008b)
84. Pilch, M., Erdman, C.A.: Use of breakup time data and velocity history data to predict the maximum size of stable fragments for acceleration-induced breakup of a liquid drop. Int. J. Multiphase Flow **13**(6), 741–757 (1987)
85. Pierlot, C., Pawlowski, L., Bigan, M., Chagnon, P.: Design of experiments in thermal spraying: a review. Surf. Coat. Technol. **202**, 4483–4490 (2008)
86. Pina, J., Dias, A., Lebrun, J.L.: Study by X-ray diffraction and mechanical analysis of the residual stress generation during thermal spraying. Mater. Sci. Eng. A **347**(1-2), 21–31 (2003)
87. Rampon, R., Marchand, O., Filiatre, C., Bertrand, G.: Influence of suspension characteristics on coatings microstructure obtained by suspension plasma spraying. Surf. Coat. Technol. **202**, 4337–4342 (2008)
88. Reitz, R.D.: Atomization and other breakup regimes of a liquid jet. PhD thesis, Princeton University (1978)
89. Richter, H.J.: In: Tagungsunterlagen Conf. Proc. 4th Colloquium, 13–14 Nov 1997
90. Selim, M.S., Sekhar, M.C., Raju, A.R.: Preparation and characterization of thin films of ZnO: Al by nebulized spray pyrolysis. Appl. Phys. A **78**, 1215–1218 (2004)
91. Sham, J.O.-H., Zhang, Y., Finlay, W.H., Roa, W.H., Löberberg, R.: Formulation and characterization of spray-dried powders containing nanoparticles for aerosol delivery to the lung. Int. J. Pharm. **269**, 457–467 (2004)
92. Sher, E., Elata, D.: Spray formation from pressure cans by flashing. Ind. Eng. Chem. Process. Des. Dev. **16**, 237–242 (1977)
93. Snyder, S., Arockiam, N., Sojka, P.E.: Secondary atomization of elastic non-Newtonian liquid drops. In: 46th AIAA/ASME/SAE/ASEE Joint Propulsion Conference & Exhibit, 25–28 Jul 2010, Nashville (2010)
94. Srivatsan, T.S., Lavernia, E.J.: Use of spray techniques to synthesize particulate-reinforced metal-matrix composites. J. Mater. Sci. **27**(12), 5965–5981 (1992)
95. Stokes, J., Looney, L.: Residual stress in HVOF thermally sprayed thick deposits. Surf. Coat. Technol. **178**, 18–23 (2004)
96. Streibl, T., Vaidya, A., Friis, M., Srinivasan, V., Sampath, A.: Critical assessment of particle temperature distributions during plasma spraying: experimental results for YSZ. Plasma Chem. Plasma Process. **26**(1), 73–102 (2006)

97. Sosnowski, T.R., Gradoń, L., Iskandar, F., Okuyama, K.: Interaction of deposited aerosol particles with the alveolar liquid layer. In: Gradoń, L. and Marijnissen, J. (eds.) Optimization of Aerosol Drug delivery, pp. 205–216. Kluwer, Dordrecht (2003)
98. Sundararajan, G., Prasad, K.U.M., Rao, D.S., Joshi, S.V.: A comparative study of tribological behavior of plasma and D-Gun sprayed coatings under different wear modes. J. Mater. Eng. Perform. **7**(3), 343–351 (1998)
99. Suresh, R., Ponnuswamy, V., Mariappan, R., Kumar, N.S.: Influence of substrate temperature on the properties of CeO2 thin films by simple nebulizer spray pyrolysis technique. Ceram. Int. **40**, 437–445 (2014)
100. Suzzi, D., S. Radl, S., Khinast, J.G.: Local analysis of the tablet coating process: impact of operation conditions on film quality. Chem. Eng. Sci. **01** (2010)
101. Taylor, G.I.: The dynamics of thin sheets of fluid III. Disintegration of fluid sheets. Proc. Roy. Soc. Lond. A **253**, 313–321 (1959)
102. Thomasin, C., Corradin, G., Men, Y., Merkle, H.P., Bertschinger, H.U.: Tetanus toxoid and synthetic malaria antigen containing poly(lactide)/poly(lactide-*co*-glycolide) microspheres: importance of polymer degradation and antigen release for immune response. J. Control Release **41**, 131–145 (1996)
103. Tucker Jr., R.C.: Structure property relationship in deposits produced by plasma spraying and detonation gun techniques. J. Vac. Sci. Technol. **11**(4), 725–734 (1974)
104. Vanbever, R., BenJebria, A., Mitzes, J., Langer, R., Edwards, D.A.: Sustain release of insulin from insoluble inhaled particles. Drug Dev. Res. **48**, 178–185 (1999)
105. Vehring, R., Foss, W.R., Lechuga-Ballesteros, D.: Particle formation in spray drying. J. Aerosol Sci. **38**(7), 728–746 (2007)
106. Widiyastuti, W., Wang, W.N., Lenggoro, L.W., Iskandar, F., Okuyama, K.: Simulation and experimental study of spray pyrolysis of polydispersed droplets. J. Mater. Res. **22**(7), 1888–1898 (2007)
107. Wang, Z., Yang, P.: Slurry nebulization in plasmas for analysis of advanced ceramic materials. J. Anal. At. Spectrom. **29**, 2091–2103 (2014)
108. Wu, P.-K., Hsiang, L.-P., Faeth, G.M.: Chapter 9: aerodynamics effects on primary and secondary spray breakup. In: Yang, V. and Anderson, W.E, (eds.) Liquid Rocket Engine Combustion Instability. Published by AIAA, ISBN: 978-1-56347-183-4 (1995)
109. Xia, B., Lenggoro, I.W., Okuyama, K.: Novel route to nanoparticle synthesis by salt-assisted aerosol decomposition. Adv. Mater. **13**, 1579–1582 (2001)
110. Yeo, L.Y., Friend, J.R.: Surface acoustic wave microfluidics. Annu. Rev. Fluid Mech. **46**, 379–406 (2014)
111. Zeleny, J.: On the conditions of instability of electrified drops, with applications to the electrical discharge from liquid points. Proc. Camb. Philos. Soc. **18**, 71–83 (1914)
112. Zhang, W., Sampath, S.: A universal method for representation of in-flight particle characteristics in thermal spray processes. J. Therm. Spray Technol. **18**(1), 23–34 (2009)
113. Zhou, Q., Tang, P., Leung, S.S.Y., Chan, J.G.Y., Chan, H.-K.: Emerging inhalation aerosol devices and strategies: where are we headed? Adv. Drug Deliv. Rev. **75**, 3–17 (2014)
114. Zhao, L., Maurer, M., Fischer, F., Dicks, R., Lugschneider, E.: Influence of spray parameters on the particle in-flight properties and the properties of HVOF coating of WC-CoCr. Wear **257**(1-2), 41–46 (2004)

Chapter 8
Emulsification: Established and Future Technologies

Karin Schroën and Claire C. Berton-Carabin

Abstract Oil and water don't mix, that is what everyone knows. . . ..but if you are able to convince them; it is very well possible to produce stable emulsions. For this you need the right technology, of which examples will be presented in this chapter, focusing both on established equipment (high pressure homogenization, rotor-stator systems, ultrasound) and technology that is currently developed (microfluidic technology, hybrid systems). Based on the droplet size that is generated and the energy that is required to do so, the technologies will be compared. Besides, attention is given to the emulsion ingredients that stabilize the oil-water interface, and prevent instability of the emulsion through sedimentation, flocculation, and/or coalescence. The chapter concludes with a short outlook on methods that are currently developed to determine emulsion stability, which we expect to become very useful, not only for emulsions but also for derived products.

8.1 Introduction

8.1.1 What Are Emulsions?

Emulsions are dispersions of fluids that are considered immiscible, e.g., oil and water. The basic forms of emulsions are oil-in-water (abbreviated as O/W), and water-in-oil (W/O) emulsions. The droplet phase is called the dispersed phase (prior to emulsification, we will use the term to-be-dispersed phase), the surrounding phase is the continuous phase. Some important examples of emulsions in our

K. Schroën (✉) • C.C. Berton-Carabin
Laboratory of Food Process Engineering, Department of Agrotechnology and Food Process Engineering, Wageningen University, Wageningen, The Netherlands
e-mail: karin.schroen@wur.nl; claire.carabin-berton@wur.nl

© Springer International Publishing Switzerland 2016

H.G. Merkus, G.M.H. Meesters (eds.), *Production, Handling and Characterization of Particulate Materials*, Particle Technology Series 25,
DOI 10.1007/978-3-319-20949-4_8

Fig. 8.1 Examples of food emulsions with different oil/fat content (high concentrations are found on the *top*) and with either the water phase as the continuous phase (*left* part of the image), or the oil phase as the continuous phase (*right* part of the image)

daily life are paints, cosmetic crèmes, pharmaceutical ointments, and many food products (Fig. 8.1) of which we present a few in detail to give you a taste of what an emulsion is, how it is made, and in some cases how you could make them yourselves.

- In raw milk, the fat is present in the form of milk fat globules, which are surrounded by a membrane consisting of a tri-layer of phospholipids, and many proteins derived from the lactating cell. When milk is homogenized in the factory, the globules are broken, and the fat is dispersed into smaller droplets. For stabilization of these newly formed droplets, the material that is initially present at the interface is not sufficient, and components from the plasma such as proteins adsorb and act as emulsifiers. The same processes also occur in cream, and various culinary products, which are concentrated milk fat emulsions; the fat concentration depends on the type of cream.
- Salad dressings are made by whisking vegetable oil in an aqueous mixture that contains acid (e.g. vinegar or lemon) and other taste components such as mustard. When made at home, this emulsion is rather unstable because mustard is not a very efficient emulsifier: the droplets coalesce relatively quickly so one has to prepare just before use. Commercial variants are usually stabilized by other components that do keep the emulsion stable; in some cases also unstable mixtures may be found that need to be shaken before use.

- Mayonnaise is a highly concentrated emulsion of oil droplets in an acidified water phase (lemon or vinegar are mostly used due to the low pH they create), which is stabilized by proteins from egg yolk (amongst others lecithin) that are maximally charged due to the acidic conditions of the water phase. Mayonnaise is very concentrated (70–80 % v/v), and in most recipes it is recommended to whisk in the oil drop by drop until completely dispersed. If the process does not go according to plan, at some stage the emulsion may even convert to a water-in-oil emulsion, and in that case the consistency is irreversibly lost. In a proper mayonnaise, the droplets are squeezed together but do not coalesce due to the charge of the stabilizers, which gives the mayonnaise its nice thick consistency. Note that the previously mentioned egg yolk is an emulsion of its own right; it consists of egg fat (and cholesterol) and protein in an aqueous solution, stabilized by a mixture of phospholipids.
- Margarine is an emulsion of water droplets in fat, stabilized by a packing of needle-like crystals of fat inside the continuous fat phase. The same is also true for butter, although it should be mentioned that this product is obtained through phase inversion.

Various other examples of (food) emulsions have been described in literature, for further reference see e.g. the Encyclopedia of Emulsion Technology [5, 34, 92, 93].

8.1.2 Examples of Complex Emulsions and Other Related Colloidal Systems

Besides single emulsions (O/W, and W/O), it is possible to make water droplets in an oil phase that is subsequently emulsified into another water phase; this is called a water-in-oil-in-water emulsion (W/O/W), and it is also possible to make the reverse (O/W/O) emulsion. Such double or duplex emulsions are used in some medical applications (encapsulation of drugs), and in foods, to enhance the perception of for example fat, but also to mask the taste of for example bitter peptides that are encapsulated in the inner water phase, and to reduce the caloric load. These double or duplex emulsions tend to be very fragile and are very difficult to produce with the technology that is classically used for single emulsions. However, for microfluidic techniques many examples are reported in literature. Good and extensive reviews have been written by Vladisavljevic and Williams [88] and Vladisavljević and co-workers [89].

Foams are closely related to emulsions, and sometimes products contain both to give them a 'lighter' perception. In foams, the dispersed phase is a gas, and one can use similar techniques as applied for emulsion preparation. Well-known examples are e.g. whipped cream containing small air bubbles that are stabilized by small fat crystals which from a structure at the air interface, or ice cream, which consists of cream with a stabilizer (mostly gelatin) that captures the air bubbles incorporated in the liquid ice cream mix, and that is further stabilized by crystals [46, 62].

Finally, liposomes are also colloidal systems that may be used for encapsulation purposes in food or pharmaceutical applications. Liposomes are vesicles consisting of one or more bilayers of phospholipids that encapsulate an inner aqueous compartment. A few applications of liposomes are reported, for example, the encapsulation of water-soluble flavors [15] or the encapsulation of enzymes [24] or antimicrobials [7] in cheese technology.

8.1.3 Main Physical Characteristics of Emulsions

The most important physical characteristics of emulsions are the volume fraction of disperse phase and the droplet size distribution (see also section on stability). The dispersed phase volume fraction (or the concentration of droplets) determines to a large extent the macroscopic properties of emulsions, such as their appearance, texture and rheological behavior. The droplet size is also of great importance, as it affects the emulsion's appearance, and also its flavor, stability, and shelf life. In food emulsions, the droplet size is often expressed as a droplet size distribution, since not all droplets have the same size (i.e., polydisperse emulsions).

Various techniques may be applied to determine particle size distributions; for a detailed reviewed please consult [45]. Briefly, microscopy techniques may be used, often to give a qualitative overview of the emulsion's morphology and range of droplet sizes present. Depending on the emulsion sample and on the type of microscopy applied, none to elaborate sample preparation may be required. Light scattering is probably the most widely used technique to characterize particle size distributions in emulsions. Here we can distinguish between (i) static light scattering (also called laser diffraction), where the angular pattern and extent of light scattering by an emulsion sample can be related to the droplet size distribution, using a mathematical model; and (ii) dynamic light scattering, where the diffusion movement of small particles (typically, submicron emulsion droplets) causes fluctuations in the light scattering pattern in time, from which the droplet size distribution can be calculated using the Stokes-Einstein equation. Both these light scattering techniques generally require some emulsion dilution prior to the measurement, which should be carefully considered as any change in the emulsion droplet environment may alter the physical state and stability of emulsions.

Average droplet diameters can be calculated from the particle size distributions. Several calculations can be performed, that will give more or less weight to the largest or the smallest droplets. Probably most widely used in emulsion science is the area-volume mean diameter (or surface mean diameter, or Sauter diameter, noted d_{32}), and for foams the volume-length diameter (or volume-weighted mean, noted d_{43}) is found of greater importance. These values are calculated as follows:

$$d_{32} = \frac{\sum_{i=1} n_i d_i^3}{\sum_{i=1} n_i d_i^2} \tag{8.1}$$

$$d_{43} = \frac{\sum_{i=1} n_i d_i^4}{\sum_{i=1} n_i d_i^3} \tag{8.2}$$

Where d_i is the diameter of droplets in each size-class, and n_i is the number of particles in each size-class, for a given volume of emulsion. In principle, various average diameters can be calculated, and for emulsions mostly the d_{32} is used, which is also convenient to calculate the required amount of surfactant.

Regarding instability, the largest droplets (or bubbles) will have the greatest effect; see also later sections in which instability in emulsions is discussed in greater detail. Here it is still relevant to mention that in foams unlike in emulsions, instability occurs mostly through Ostwald ripening, and in that process the largest droplets have even greater influence and that is why in that case the d_{43} is preferred. From this it is clear that besides the average value, it is also important to characterize the width of the particle size distribution. For this, for example, the polydispersity value (P) of an emulsion [37] can be used:

$$P = \frac{1}{d_{50,V}} \frac{\sum_i n_i d_i^3 \left| d_{50,V} - d_i \right|}{\sum_i n_i d_i^3} \tag{8.3}$$

Where $d_{50,V}$ is the median diameter, i.e., the diameter for which the cumulative volume fraction is equal to 50 %. Although there is to the best of our knowledge, no fixed convention to set the limit between monodisperse and polydisperse emulsions, it has been considered that emulsions can reasonably be considered monodisperse for a polydispersity value below 25 % [37].

As mentioned above, many food emulsions are highly polydisperse. This can be a cause of emulsion instability, notably when a population of very large droplets is present (the detailed mechanisms of some instability phenomena are explained in the following paragraphs). A high polydispersity also makes it very difficult to understand the connection between the structural characteristics of emulsions (such as the droplet size) and their end-up properties in great detail [43, 44]. Some examples of average and ranges of droplet sizes in common foods are presented in Table 8.1.

A more physical chemical aspect is related to the properties of the interfacial layer that surrounds the emulsion droplets. This interfacial region, typically a few nm thick, is the preferred location for molecules that have dual affinity for oil and water (i.e., amphiphilic, or surface-active molecules). Therefore, the properties and structure of the interface (e.g., electrostatic charge, thickness, and fluidity) are largely governed by the type and concentration of adsorbed amphiphilic molecules. These properties can, in turn, tremendously affect the physical and chemical stability of emulsions as described in the following sections.

Table 8.1 Examples of mean and extreme droplet sizes encountered in some food emulsions

Food product (emulsion type)	Mean droplet size (d_{32}, μm)	Typical droplet size range (μm)	Sources
Raw milk (O/W)	1–2	0.1–10	McClements [43]
Homogenized milk (O/W)	0.25	0.05–0.5	McClements [43]
Salad dressings (O/W)	30–125	A few to >200	Perrechil et al. [56]
Mayonnaise (O/W)	6–20	A few to >50	Langton et al. [36]
Fat spreads (W/O)	2–5	1–10	van Dalen [75]

8.2 Emulsion Stability and Ingredients

Emulsions are, in essence, thermodynamically instable systems, because of the molecular incompatibility of both involved liquid phases. This incompatibility is responsible for the existence of an interfacial tension, denoted as σ (N/m). The total free energy of an emulsion (ΔG, J) thus depends on the total interfacial area (ΔA, m^2), and on the interfacial tension:

$$\Delta G = \sigma \Delta A \quad (8.4)$$

ΔG is almost always positive, meaning that emulsions almost always tend to physically destabilize and go back to their demixed state. However, it is possible to retard destabilization and to obtain metastable emulsions (i.e., stable for a 'reasonable' period of time). This can be achieved, for example, by using emulsifiers. Emulsifiers are surface-active molecules (e.g., amphiphilic biopolymers or surfactants), which can adsorb at the oil-water interface and thereby reduce the interfacial tension.

Yet, emulsions can physically destabilize through a number of phenomena, such as creaming/sedimentation, agglomeration/flocculation and coalescence of droplets. In this section, these three aspects are discussed in some detail, and common methods to determine and quantify the level of destabilization are presented; we are very aware that we cannot be complete in our description of emulsion stability due to the variety of interactions that components may have. Still these selected aspects are the most prominent reasons why emulsions are or may become unstable.

8.2.1 *Creaming/Sedimentation*

The main reason why small droplets are required in emulsions is their creaming/sedimentation behavior, which is caused by the density difference between the two phases that normally is present in emulsion (only very few density matched emulsions have practical value). Obviously, this train of thought only holds for free flowing emulsions, and not for (partially) crystallized emulsions or gelled emulsions that exhibit a high enough yield stress to keep the droplets captured (see later sections). For simplicity reasons, we will use the term sedimentation from now on, but obviously, the same principle holds for creaming. For a droplet in a surrounding liquid the sedimentation velocity can be estimated using Stokes law:

$$v_s = \frac{g\Delta\rho d_d^2}{18\eta_c} \tag{8.5}$$

where:

v_s = sedimentation velocity (m/s)
g = gravitation constant (m/s^2)
$\Delta\rho$ = density difference between the phases (kg/m^3)
d_d = droplet diameter (m)
η_c = viscosity of the continuous phase (Pa s or kg/(m s)).

This equation directly shows that smaller droplets sediment less fast; that highly viscous continuous phases slow down sedimentation, and as mentioned before, that emulsions with a zero density difference between the two phases do not sediment. This also explains why the emulsions that are currently in the market have very small droplets, and mostly have a very viscous continuous phase. Moreover it should be mentioned that small droplets (<0.1 μm) also will move due to Brownian motion. This effect can substantially decrease the creaming/sedimentation rate in emulsions [43].

Besides, it is known that in emulsions with large numbers of droplets sedimentation slows down. This behavior has been described by multiple authors; good summaries can be found in the work of Walstra [91, 92]. Here, we illustrate the effect of multiple droplets on the sedimentation velocity through the so-called Krieger-Dougherty equation:

$$\frac{v}{v_s} \approx \left(1 - \frac{\varphi}{\varphi_{max}}\right)^{k\varphi_{max}} \tag{8.6}$$

where:

v = velocity in a swarm of droplets (m/s)
v_s = Stokes velocity for a single droplet (m/s)
φ = volume fraction of dispersed phase (–)
φ_{max} = maximum volume fraction of dispersed phase, i.e. the volume fraction at which the droplets become closely packed (–) (for spherical, non-deformable droplets, $\varphi_{max} = 0.585$)
k = proportionality constant. For a k-value of 6.5, this equation has been validated for volume fractions up to 40 %, and has shown to be good compared to other models [91].

A basic method to monitor creaming or sedimentation in emulsions consists of placing a certain volume of emulsion in a transparent test tube that is then left immobile for a certain period of time (or even brought under enhanced gravity for very stable emulsions). In the case of creaming, for instance, the droplets will move upwards over time, leading to the formation of a clear, droplet-depleted phase at the bottom of the tube (the serum layer), whereas oil droplets accumulate on the top of the sample, forming a creamed layer. The creaming index (CI, %) can be calculated as follows:

$$\mathrm{CI} = (h_s/h_t) \times 100 \tag{8.7}$$

where:

h_s = height of the serum layer
h_t = total height of emulsion in the tube.

A more sophisticated way of quantifying the creaming (or sedimentation) of emulsions is to measure the amount of transmitted light throughout a transparent tube filled with emulsion, over time (possibly under accelerated gravity conditions), and all along the tube's height. This is possible with automated instruments, such as the LUMiFuge® stability analyzer.

8.2.2 *Bridging and Depletion Flocculation*

To improve the physical stability of emulsions, it is of great importance to cover the available droplet surface that is formed during emulsification, and give it a protective coating. Yet, even if stabilizing components (in particular, emulsifiers) are present in the interface, this does not necessarily mean that the emulsion will be stable; this depends on the 'quality' of the coating. Various short-range interactions as described in the DLVO theory can allow droplets to approach and flocculate in a primary and secondary minimum [39], and these effects may become pronounced if multiple components (that, e.g., carry opposite charges) are used to stabilize an emulsion. Describing these systems is complex, and still part of an on-going debate in literature. Here we will describe the main effects that can be used to establish a good starting point for emulsion preparation; whether this will also work in practice will always have to be tested.

There are two situations that will lead to flocculation, namely one in which the surface is not sufficiently covered, and a surface-active molecule can attach to two or even more droplets (Fig. 8.2, left). The other occurs if a non-adsorbing molecule is overwhelmingly present; this is termed depletion flocculation (Fig. 8.2, middle and right) as described in, e.g., the work of Walstra [91]. The non-adsorbing component cannot come as close to the surface as an adsorbing species would. E.g., a polymer, will mostly remain the distance of its gyration radius away from the interface, and because of that, there will be a lower polymer concentration in this

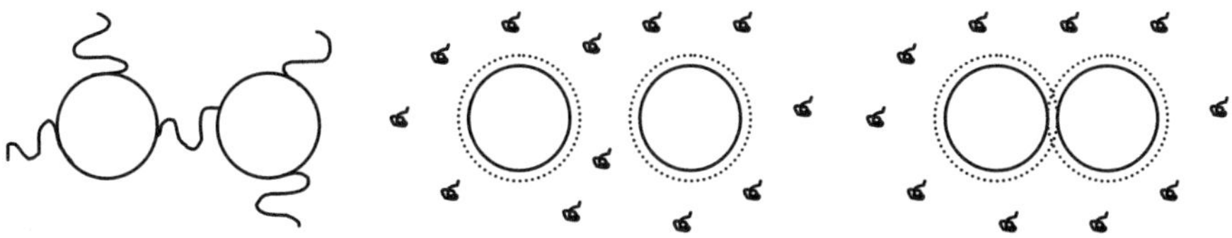

Fig. 8.2 *Left*: Schematic representation of bridging flocculation; one molecule attaches to two or more droplets due to insufficient surface coverage. *Middle and right*: illustration of depletion flocculation by non-adsorbing component. This component will have a non-accessible area near the droplets (indicated by the *dotted line*) that corresponds to the gyration radius of the component, and because of that, there is an osmotic pressure difference that may lead to aggregation (see text for further explanation)

area (depicted by the dotted line in Fig. 8.2). The concentration difference implies that there is also an osmotic pressure difference in the system; the system will strive to minimize this effect and can do so through minimization of the area that is free of polymer, i.e. it does that through aggregation. For spherical droplets this energy gain may be not very big (but still sufficient), but for platelets this effect can be over-dominating.

In the group of Julian McClements (University of Massachusetts, USA) these effects have been very practically translated into so-called stability maps that are very useful in choosing appropriate compositions for emulsions prepared with multiple components, in particular in emulsions stabilized by multiple polyelectrolyte layers. In Fig. 8.3, examples are shown from the field of food, but that we recommend to be taken as a starting point for preparation of any emulsion. The first map (Fig. 8.3a) shows a plot of the critical polyelectrolyte concentration as a function of the dispersed phase fraction (ϕ). It was assumed that the droplets had a radius of 0.3 μm, and the non-adsorbed polyelectrolyte had a molecular weight of 100 kDa and effective radius of 30 nm. The second map (Fig. 8.3b) plots the critical polyelectrolyte concentration as a function of the droplet radius. It was assumed that the droplets had a volume fraction of 0.03 (3 vol.%), and the non-adsorbed polyelectrolyte had a molecular weight of 100 kDa and effective radius of 30 nm. The shaded area in both maps highlights the range of conditions where it should be possible to produce non-flocculated droplets. These plots nicely illustrate that a proper formulation strategy should be taken into account to identify not only the minimum concentration of surface-active molecules needed to cover the entire oil-water interface (C_{Ads}), but also the maximum concentration of such compounds beyond which any more excess induces depletion flocculation (C_{Dep}).

8.2.3 Droplet Coalescence and Emulsion Ingredients

In order for droplets to coalesce, the interfacial film between them needs to drain, and subsequently the film needs to break. Various components that can be present in emulsions will prevent the droplets from approaching this closely that the film can

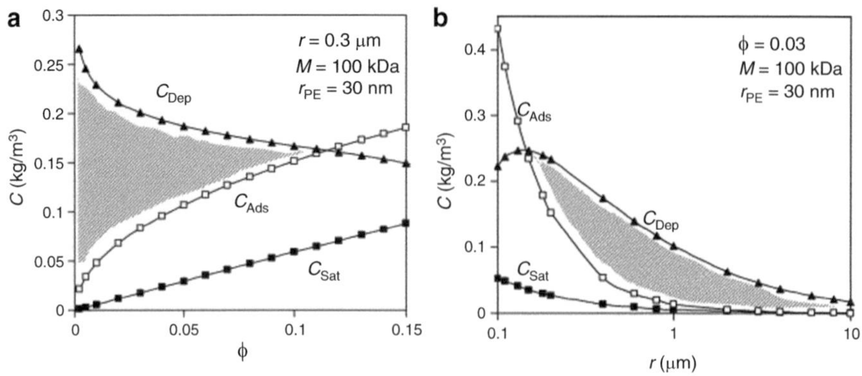

Fig. 8.3 (**a**) Stability map showing the influence of droplet concentration on the critical polyelectrolyte concentrations for saturation, depletion, and adsorption. (**b**) Stability map showing the influence of droplet radius on the critical polyelectrolyte concentrations for saturation, depletion, and adsorption (Reprinted with permission by Elsevier, from Guzey and McClements [18])

break. This can occur through steric repulsion, or charge interactions of similarly charged components.

Various components are used in practice to prepare emulsions, and they may induce both steric and charge effects. Surfactants are mostly low molecular weight components that have both a hydrophilic and hydrophobic part. The special structure of the surfactants causes that most of them cannot dissolve molecularly in one of the two phases of an emulsion. Often, a surfactant will form micelles; for example in water, the hydrophobic parts are clustered in the center and the polar parts are at the outside, in oil reversed micelles are formed. In an emulsion, only a limited amount of surfactant can reside in the interface (typically in the order of 0.1–10 mg/m^2, depending on the surfactant and the available surface area). As soon as the surface is filled, the surplus of surfactant will remain in that phase for which it has most affinity.

In general, a suitable surfactant should be better soluble in the continuous phase of the emulsion. This is the basis of the so-called Bancroft rule. The solubility of a surfactant in oil or in water depends on its molecular structure, and can be quantified through the hydrophilic-lipophilic balance (HLB). This is an empirical value, which takes into account the type and amount of polar and non-polar groups present in the surfactant's molecular structure. Low-HLB surfactants (e.g., polyglycerol polyricinoleate, HLB ~ 3) are more oil-soluble, and suitable for stabilizing W/O emulsions; whereas high-HLB surfactants (e.g., polysorbate 20, HLB ~ 16) are more water-soluble, and suitable for stabilizing O/W emulsions.

An excess of surfactant/stabilizer will be used to fill the interface rapidly when formed; and therewith prevent coalescence. When adsorbed onto an interface, ionic surfactants give mostly rise to electrostatic repulsion, and if they are bulky possibly also to steric repulsion between the droplets. Besides they lower the interfacial tension therewith reducing the energy of the system (Eq. 8.4).

A special class of surface active components is block copolymers, of which one part of the molecule has affinity to the interface, while the other part is extended in the continuous phase. Mostly these molecules form a steric barrier on the interface. Possibly also proteins could qualify as block-copolymers since they contain repetitive units but in this case many different amino acids would be present as units; however, these components are also known to be able to form viscoelastic films on the interface, therewith leading to very stable emulsions. Besides they are able to interact with other surface active components, and form networks that enhance stability. On the other hand, they may also be replaced by surfactants from the interface, and this may induce emulsion instability at longer time scales [93].

Polymers increase the viscosity of the continuous phase, which will reduce the droplet sedimentation velocity (Eq. 8.5), reduce film drainage, and increase through both effects the shelf life of the product. Alternatively, they may also give the product a yield stress, and if the strength of the weak gel is high enough, it can even prevent creaming.

Last but not least, also small particles that are partially wetted by both phases of the emulsion can accumulate in the interface, and give so-called Pickering stabilization, which is mostly a steric effect (recently reviewed by Berton-Carabin and Schroën [9]). Alternatively, in very few cases, it is possible to match the densities of both liquids, and therewith prevent creaming. Still, contact between droplets due to Brownian motion should be prevented in order to prevent coalescence.

More information on surfactants and their behavior can be found, e.g., in the work of Walstra [92], Walstra et al. [93], Lucassen-Reynders [38], and Guzey and McClements [18].

The coalescence of emulsion droplets can be monitored by measuring the particle size distribution (e.g., with a static light scattering instrument) at various time intervals. It should be pointed out that if the measured average particle size increases as compared to an initial value, it may be not only due to coalescence, but also to flocculation (in fact, such an instrument will consider a flock of emulsion droplets as one large particle). To distinguish between flocculation and coalescence, two simple strategies can be used. First, O/W emulsions can be diluted in a concentrated sodium dodecyl sulfate (SDS) solution prior to the measurement. SDS, an anionic surfactant, will rapidly cover the surface of emulsion droplets, providing them with a strong negative surface charge that will disrupt flocs of droplets, if any. Therefore, if a similar average particle size is measured without and with dilution of the emulsion in the SDS solution, most likely, coalescence occurred. Conversely, if the dilution of the emulsion in the SDS solution leads to a smaller particle size compared to the measurement on the non-diluted emulsion, most likely, flocculation occurred. Another strategy can simply be to observe the emulsion morphology under an optical microscope, which is often sufficient to assess whether the size of individual droplets did increase as compared to the initial sample, or if agglomerates (flocs) of small droplets are present. However, a limitation here is the resolution of optical microscopes and the number of droplets that can be viewed, which prevents obtaining accurate details on the structure and morphology of submicron particles.

8.3 Droplet Formation Mechanisms

As can be deduced from Eq. (8.4), making small droplets is hard, because it involves creating a large interfacial area, and a lot of energy is needed to do so. This is also caused by the fact that different time scales are relevant for the various droplet formation mechanisms that occur simultaneously and in relation to the stabilization of the interfaces, as will be discussed in the next section. Now, we focus first on the various mechanisms for droplet formation, both under laminar and turbulent conditions.

In general, when a continuous liquid flows around a discrete droplet, it induces a shear force onto the droplet, which may break the droplet if the exerted force is sufficiently large. That is also why most emulsification methods are designed to generate a very strong flow field that acts on a very small volume through which the emulsion passes.

As stated in Eq. (8.4), the amount of energy needed to create an interface between two phases costs an amount of energy that is proportional to the amount of interface generated, and that in turn is also related to the volume of oil that is present (V, m^3) and with the droplet diameter (d_d, m) this leads to $A/V = 6/d_d$. It is also immediately makes clear that producing small droplets is more energy consuming.

In practice, much more energy than the minimal amount of energy from Eq. (8.4) is needed, because interfaces are not duly stabilized, and because energy will be lost due to heating up of the product. For more information on surface phenomena we recommend, e.g., the work of Hiemenz [19], Hiemenz and Rajagopalan [20] and Lyklema [39].

8.3.1 Interfacial Tension and Shear Forces

In order to break-up a large droplet into smaller droplets, the interfacial tension force that keeps the large droplet together needs to be exceeded by (any) shear force, be it laminar or turbulent. The interfacial tension force is related to the Laplace pressure ($\Delta P_{Laplace}$, Pa) which is defined as:

$$\Delta P_{Laplace} = \sigma\left(\frac{1}{R_1} + \frac{1}{R_2}\right) \tag{8.8}$$

with R_1 and R_2, the respective radii of curvature of an ellipsoid, and σ the interfacial tension (N/m). For a spherical droplet both curvature radii are equal to the droplet radius, R_d, and the resulting Laplace pressure is equal to:

$$\Delta P_{Laplace} = \frac{2\sigma}{R_d} \tag{8.9}$$

For any other oddly shaped droplet the Laplace pressure can be calculated using the actual curvatures.

As mentioned, the interfacial tension force needs to be overcome by shear forces exerted by the emulsification device. Often the Weber number, which is the ratio between the external disruptive stress and the internal coherent stress, is used to characterize droplet formation. Droplet size can be reduced under various conditions. Here we distinguish droplet break-up due to rotation, and elongation both under laminar conditions, or turbulent conditions.

8.3.2 Laminar Plain Shear Flow

In simple laminar flow, a droplet will be subjected to a flow field that leads to rotation of the droplet, and/or to its extension, as indicated in Fig. 8.4. In simple shear flow, the droplet is subjected to the flow field which initially rotates the droplet, which leads to distortion, and eventually the droplet will break up into smaller droplets.

The internal, coherent stress can be estimated with the help of the Laplace pressure in the droplets, equal to $2\sigma/_{Rd}$ as previously discussed (Eq. 8.9). The disruptive stress follows from:

$$\tau_{ext} = \eta_c \frac{dv}{dz} = \eta_c \dot{\gamma} \tag{8.10}$$

where:

$\dot{\gamma}$ = shear rate that is applied (1/s)
η_c = viscosity of the continuous phase (Pa s)
v = velocity of the continuous phase (m/s)
z = position of the droplet (m) perpendicular to the solid surface, which is depicted in the bottom of Fig. 8.4a.

This leads to the following definition of the Weber number:

$$We = \frac{\eta_c \dot{\gamma} R_d}{2\sigma} \tag{8.11}$$

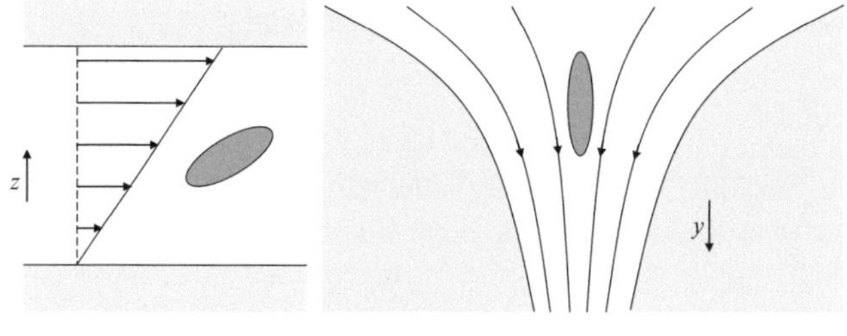

Fig. 8.4 Simple shear flow (**a**) and extensional flow (**b**)

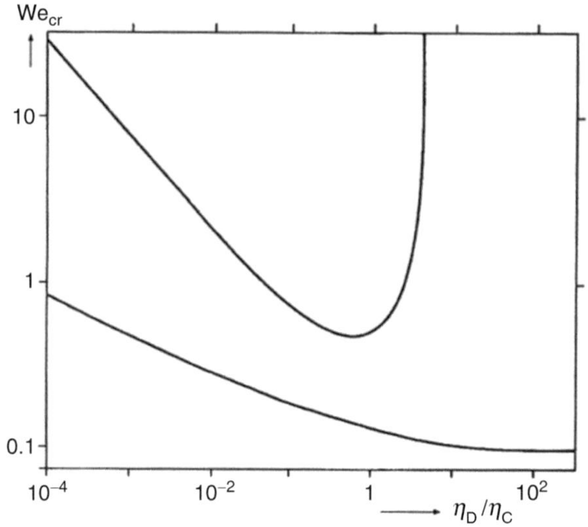

Fig. 8.5 Critical We-numbers (We_{cr}) for laminar flow conditions – plain shear flow (*upper curve*), and for extensional flow (*lower curve*) [17, 92]. η_D/η_C is the ratio of the dispersed phase viscosity over the continuous phase viscosity (Reprinted with permission from Walstra)

In order to achieve droplet break-up, this We-number should exceed a value, called the critical We-number. It was found that the critical Weber number is a function of the viscosity ratio of the water and oil phases. This is because the droplet will deform more when the dispersed phase viscosity is lower, which will give a higher Laplace pressure and a lower external stress, as is also shown in Fig. 8.5 [17, 92]. The figure will be discussed in detail in the section on extensional flow, where both flow patterns are compared.

8.3.3 Laminar Extensional Flow

This type of flow occurs when the liquid is squeezed through a small opening, during which the droplet is extended in the direction of flow, as could be the case in high pressure homogenization (under non-turbulent conditions). In general, during passage through the opening the droplet is extended to form a thread that breaks into many small droplets due to Rayleigh instabilities. In this case, the external force exerted on the droplet is equal to $\eta_c(dv/dy)$, where y is the coordinate in the direction of the extension, and the definition of the Weber number is the same as for simple shear flow.

In simple shear, energy is used to rotate the droplet, and besides the effective viscosity is a factor of two higher in extensional flow [92], leading to more efficient break-up in extensional flow. Therefore, the critical Weber number for extensional flow is lower than for simple shear flow. Figure 8.5 gives an impression of the values for both simple shear flow and for extensional flow; these curves are also known as the 'Grace' curves after the first author to publish them.

For simple shear, the critical Weber number goes through a minimum. At high viscosity ratio it is difficult to make droplets due to the internal viscosity of the droplets that resists deformation, while at low viscosity ratio the droplets may be first deformed into very long threads before they break. Note that the lowest viscosity ratios in Fig. 8.5 correspond to the situation in foams.

In practice, one will always have a mixture of the two types of flow; and the critical We-number will have an intermediate value. Still the diagram is very useful to calculate best- and worst-case scenarios.

8.3.4 Turbulent Flow

As is clear from Figs. 8.7 and 8.8, under turbulent conditions that occur e.g. in high pressure homogenizers, the flow is much more erratic than described for laminar flow (Fig. 8.4). The liquid will start moving in a chaotic way and form swirls and eddies. As will be discussed later, the transition from laminar to turbulent conditions is related to the Reynolds number, and the critical values for transition may be different for different emulsification equipment, as discussed in the conventional emulsification devices section.

Given the chaotic nature of the process, it is more difficult to find a critical Weber-number, since the exact local flow-conditions cannot be determined, because experimentation and simulation on this scale is very difficult. Therefore an average value, the power density (symbol ε, with unit W/m^3 or Pa/s), is taken as a measure for the intensity of the swirls and eddies.

At relatively low turbulence, the surrounding eddies impose shear on the droplets. The external disruptive force can be calculated with the Kolmogorov theory for turbulent flow. This force (τ, Pa) can be estimated as:

$$\tau = \sqrt{\varepsilon \cdot \eta_c} \tag{8.12}$$

where:

ε = power density (Pa/s)
η_c = viscosity of the continuous phase (Pa s).

The Weber number can be defined as before and the critical value now becomes [92]:

$$\mathrm{We_{cr}} = \frac{\tau R_d}{2\sigma} \approx \frac{\sqrt{\varepsilon \eta_c}\, R_d}{2\sigma} \tag{8.13}$$

where:

τ = external stress
R_d = droplet radius
σ = interfacial tension (N/m)

ε = power density (Pa/s)
η_c = viscosity of the continuous phase (Pa s).

Assuming that the critical Weber number is one, the droplet size can be estimated.

When the turbulence becomes very intense, the inertia of the liquid surrounding the droplets becomes the dominant factor. In this case the external disruptive force can be calculated with the aid of the Bernouilli equation as [92]:

$$\tau = \varepsilon^{2/3} R_d^{5/3} \rho_c^{1/3} \tag{8.14}$$

Again this external disruptive force can be substituted in the Weber equation, and by assuming that the critical Weber number will be around unity, one can estimate the droplet radius. The transition from viscosity-dominated break-up to inertia-dominated break-up takes place when the droplets are larger than [92]:

$$R_d > \frac{\eta_c^2}{\sigma \rho_c} \tag{8.15}$$

These flow types relate as follows to the emulsification techniques that are presented in the next section. Flow in a colloid mill can be laminar or turbulent, depending on the viscosity of the product, and the lay-out of the machine (toothed versus regular). Large-scale high-pressure homogenizers operate under turbulent flow and inertial forces dominate, while laboratory scale homogenizers operate in the laminar regime; therefore, translation of results obtained on laboratory scale is not straight forward [69, 94]. Emulsification with ultrasound is always based on inertial forces, created by cavitation of the vapor bubbles.

8.4 Conventional Emulsification Devices

There are many different machines available for emulsification, and they are reviewed in literature by e.g. Walstra (general review [92]), Behrend and Schubert (ultrasound [6]), Schubert and Armbruster (general review [66]), Arbuckle (general review [3]), Brennan (mechanical methods [10]), Karbstein and Schubert (continuous systems [22]). For more detailed information, we would like to refer to these papers. To make an emulsion, mostly all ingredients are mixed (very coarsely), after which a macro-emulsion is obtained that is not stable due to the large size of the droplets (Eq. 8.5). This emulsion needs to be further refined, by passage through an emulsification device. The resulting tiny droplets need rapid stabilization by surfactants in order to prevent coalescence. In industry, high-pressure homogenizers, rotor-stator systems, and ultrasound treatment are mostly used for this purpose (schematic representations are shown in Fig. 8.6), and depending on the formulation, multiple passages may be needed. In the high pressure homogenizer

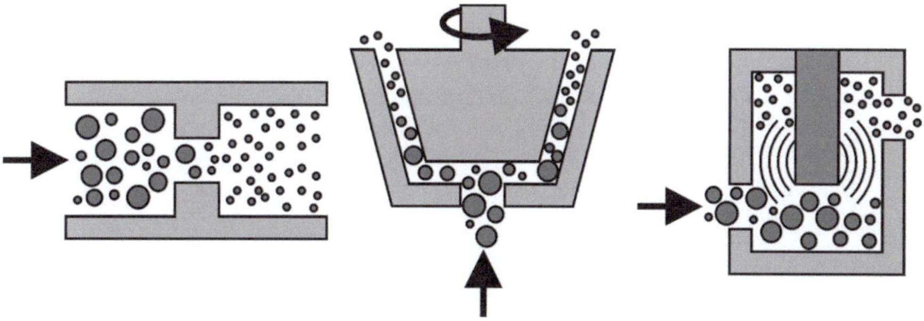

Fig. 8.6 Schematic representation of classic emulsification methods, from *left* to *right*, high pressure homogenizer, rotor-stator system, ultrasound ([80]; reprinted with permission from the author)

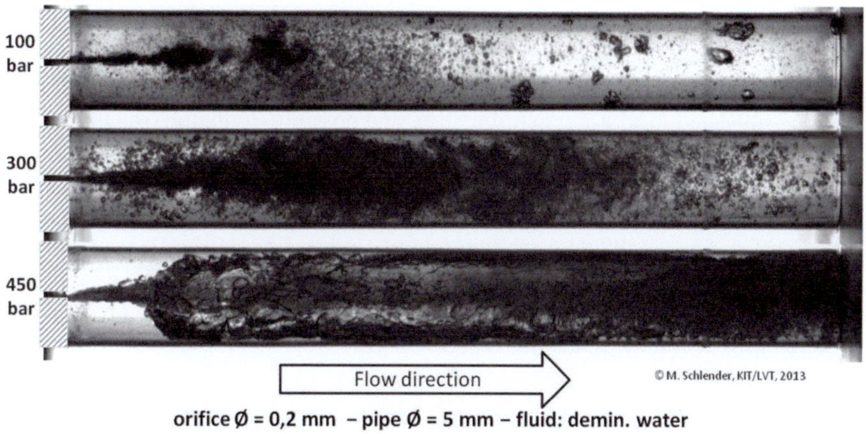

Fig. 8.7 Cavitating fluid leaving a homogenization orifice at different homogenization pressure differences (Images recorded with a VKT Fastcam SAS; record rate, 50,000 fps; shutter speed, 1/216,000 s; total frame, 2001. Reprinted with permission from the author [57])

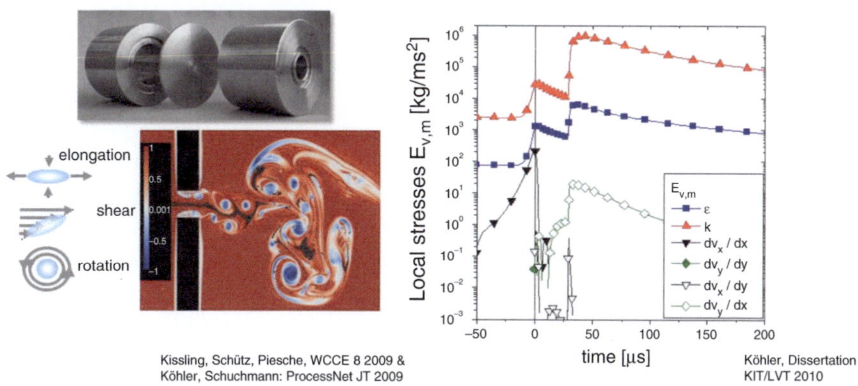

Fig. 8.8 Stresses acting on droplets (*left bottom* image) flowing through a spherical high pressure homogenization orifice (*top left* photograph) [25]. On the *right*, the local stresses on the middle axes as function of time: Stresses resulting from local turbulent energy dissipation ε, *k* as well as shear and elongational stresses in x- and y-direction are shown; the fluid leaves the orifice at time t = 0 µs [28] (Images are reprinted by courtesy of the authors [57])

depicted on the left in Fig. 8.6, the premix emulsion is pushed through a constriction, which results amongst others in shear to break up the droplets. The rotor-stator system in the middle of Fig. 8.6 consists of two concave elements, of which one rotates and one is static. The velocity difference between elements creates the necessary shear for droplet break-up. In the ultrasound equipment on the right in Fig. 8.6, the ultrasound creates cavitation bubbles that collapse, and through which droplets are submitted to shear and break-up. More details are given in the respective sections.

8.4.1 High Pressure Homogenizers

In the high-pressure homogenizer depicted on the left in Fig. 8.6, the pre-emulsion is pressurized and pushed through a tiny orifice. This creates local shear, while the liquid is subjected in addition to turbulence, and possibly cavitation (see Fig. 8.10), which all result in droplet break-up.

Since high pressures are used ranging from 10 to 50 MPa, very intense fields can be reached, that lead to submicron droplets [68]. Where small, lab-scale homogenizers still operate in the laminar flow regime, industrial-scale systems operate in the turbulent regime, which makes it difficult to translate results obtained in the lab to full-scale industrial emulsification. The transition between the two regimes can be derived from the Reynolds number (Re):

$$\mathrm{Re} = \frac{\rho v L}{\eta} \quad \text{Transition from laminar to turbulent is at Re}\tilde{\ }1500 - 3000 \qquad (8.16)$$

where:

ρ = liquid density (kg/m^3)
v = average liquid velocity in the gap (m/s)
L = gap width (m)
H = viscosity (Pa s).

From experiments it was deduced that transition from laminar to turbulent occurs at Re-number 1,500–3,000. The most used designs in industry are valve and nozzle systems. In a valve, a small gap is created by pressurizing the valve with a spring. In lab-scale systems, the gap width will be of the order of 1 μm; for industrial-scale systems, it will be 10–40 μm [69]. In the nozzle system, there is only a small hole through which the pre-mix emulsion is pushed, and in some cases there are a number of openings placed in series, or the liquid is split and recombined in the machine. Also in some cases the emulsion is passed repeatedly through the nozzle. This is all done to create as much shear as possible, and depending on the composition of the emulsion, different lay-outs of the high pressure homogenizers can be applied.

A lot of nice work in this field is carried out in the University of Karlsruhe, in Germany, within the group of prof. Schuchmann, formerly headed by prof. Schubert, and within the group of prof. Windhab at ETH Zurich, in Switzerland. Amongst others, the flow fields in various geometries are visualized, either through experimentation by, e.g., PIV, or through modeling. Both are very challenging, given the obviously extreme conditions in the machine. Besides there are limitations in simulations for which the Navier-Stokes equations need to be solved at a local level, using mathematical grids of nanometer scale [29, 67]. That, combined with the fact that droplets influence each other and local flow in their surroundings, makes these simulations very challenging.

Using local flow velocities, stresses acting on droplets may be calculated. This is illustrated in Fig. 8.8 for single phase flow through a spherical shaped orifice of 200 μm diameter, at 100-bar homogenization pressure [28]. These stresses can be compared to the critical values for the capillary number, which links the exerted shear to the interfacial tension force that resists deformation [17, 90]. The capillary and Weber number are discussed in more detail in the mechanisms section; it is important to remember that the critical capillary number is exceeded by far within microseconds in high pressure homogenizers, and oil-water interface will be readily made. Since interface formation is so rapid, surface active components will not (or hardly) have time to stabilize the formed interface (e.g., [8]), and this may lead to instability in the product unless droplet interactions can be delayed as long as is needed to cover the interface sufficiently. In high pressure homogenizers, this is not likely to be the case given the turbulent nature of the liquid flow (although the turbulent nature will favor rapid transport of surfactants as well). Calculating droplet size distributions that result from a widely distributed field of stresses is therefore one of the main challenges to food (and chemical) process engineers. An indication of the average size can be obtained using the estimates in the droplet formation mechanisms section.

8.4.2 Rotor-Stator Systems

By definition, rotor-systems consist of one part that is immobile and one that rotates; such as stirred vessels but also colloid mills qualify as such. The technology has been extensively reviewed by Urban and co-workers [74], and various examples can be found there by the interested reader.

In a stirred vessel, the liquid movement is rather chaotic and hard to predict, leading to very polydisperse emulsions with droplets typically larger than 10 μm, that are in general not very stable (Eq. 8.5). The energy density of stirrers is low; therefore, making very small droplets is outside the reach of this equipment. In a colloid mill, the central cone is static, while the outside cone rotates at high speed (see Fig. 8.9).

This system can be operated in continuous mode with the pre-emulsion entering from the bottom and the fine emulsion leaving from the top. The distance between

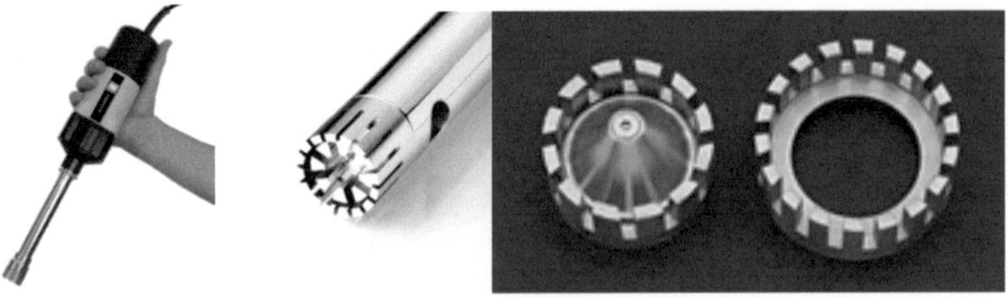

Fig. 8.9 Handheld rotor stator system and its parts. In the middle, the assembled rotor-stator unit, on the right both parts individually (Images taken from the internet)

both parts is small in order to generate high shear rates. A variation to the colloid mill is the so-called toothed mill; in this case, both the rotor and stator have openings, which generate turbulent conditions through which smaller droplets can be made, also at the high throughput that is required at industrial scale.

Since the distance between rotor and stator is very narrow (sometimes much less than a millimeter), the shear forces are very intense. Depending on the viscosity of the emulsion, the flow will be laminar (viscous liquids) or turbulent. The transition towards turbulent flow is again related to the Reynolds number, which characterizes the flow.

$$\mathrm{Re} = \frac{\rho v L}{\eta} \quad \text{Transition from laminar to turbulent is at Re}\tilde{\ }370 \qquad (8.17)$$

where:

L = gap width between rotor and stator (m)
v = tangential speed of the rotor (m/s)
ρ = density of the mixture (kg/m^3)
η = viscosity of the mixture (Pa s).

At Re <370 the flow will be laminar, and at higher values turbulent. This is an experimental finding, as was the case for high pressure homogenizers.

8.4.3 Ultrasound

A third important emulsification method used in industry is ultrasound. At frequencies higher than 20 kHz and up till 100 kHz it can cause physical and chemical changes in matter. The sound is generated by an actuator resulting in pressure fluctuations (standing waves). When the sound is sufficiently intense, the pressure fluctuations become so large that in small regions, the pressure becomes lower than the vapor pressure of water, which induces the formation of small bubbles that implode almost immediately and cause intense, local turbulence (Fig. 8.10). The technology was reviewed by Canselier et al. [11]; the interested reader will find

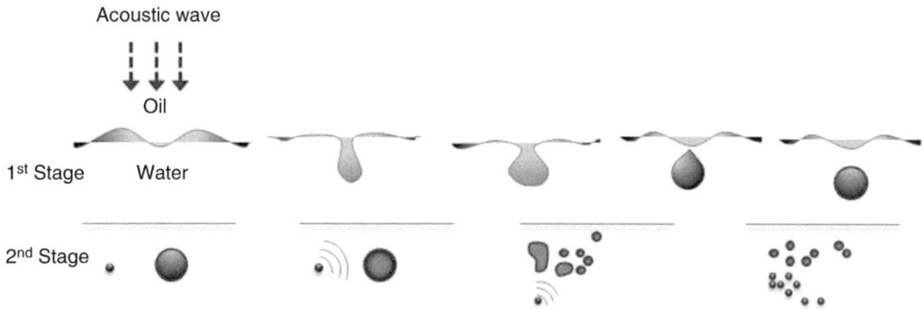

Fig. 8.10 Schematic representation of ultrasound effect. First the oil is dispersed into the water through surface waves, where it undergoes a second stage during which the droplets are further refined due to cavitation (Kendall [23]; image taken from the internet)

many relevant details there. Besides, various demonstrations can be found on the internet, with separated liquids being finely dispersed by ultrasound action.

The effect of ultrasound is very local, so the premix emulsion needs to be brought into close proximity of the actuator where the field is strongest. As soon as the emulsion is led away from the actuator, the effect rapidly becomes less, and this also implies that the treatment chamber should be small in order to be efficient. Besides physical changes also chemical changes can be a result of ultrasound treatment; especially unsaturated fatty acids and oils are notoriously unstable in ultrasound. This technology is rarely used for large scale operation, but if the emulsion components allow, it can generate very small and stable droplets in specialty products.

The energy efficiency of all classic emulsification technologies in regard to the generated droplet size will be compared with those presented for emerging technologies in the following section, and the results can be found at the end of this chapter in the comparison paragraph.

8.5 Emerging Emulsification Technologies

Besides the established technologies that were presented earlier, there is a very lively field of research in which micro-structured devices are used to prepare emulsions, and also derived products, such as double emulsions, bubbles, particles, capsules etc. Mostly these emulsions are fairly monodisperse in droplet size; also the energy efficiency of these methods is rather high. However, these methods are currently not at such a level of development that they can be applied at large scale, although some are promising and closer to large scale application than others [65].

In this section, we will discuss various membrane emulsification techniques that use shear forces to make emulsions, together with microfluidic techniques that may use either shear-based or spontaneous droplet formation mechanisms. At the end of this chapter we will compare all presented methods based on their energy efficiency, but also on their ease of scale-up.

8.5.1 Membrane Emulsification

A membrane is a porous structure mostly used for separation but it is also used to make emulsions and related products. The two membranes that are most frequently applied for emulsification are the Shirazu Porous Glass membrane (SPG; [48]) and the ceramic membrane [63, 64, 86]; see Fig. 8.11, left and middle images. The SPG membrane consists of a matrix of interconnected pores that have similar size all through the membrane, while the ceramic membrane consists of a carrier with large pores onto which a layer with small pores is deposited. The right image in Fig. 8.11 is a microsieve, which has very uniform and also very thin pores. These microsieve membranes are made through photolytic techniques in a clean room environment. The typical pore sizes that can be made are from 0.1 to 100 μm (Aquamarijn microfiltration BV, http://www.aquamarijn.nl/).

The membranes can be used in cross-flow mode [48] or in pre-mix mode (reviewed by Nazir et al. [50]). During cross-flow emulsification (see Fig. 8.12; left image), the to-be-dispersed phase is pressed through the membrane where it forms small droplets on top of the membrane, that are consecutively sheared off by the cross-flowing continuous phase once they have reached a certain size. During pre-mix emulsification, the large droplets of a pre-mix emulsion are broken up into smaller ones as the liquid passes through the membrane, and are sheared off while doing so. The situation is rather similar to what happens with classic emulsification techniques, but now multiple 'nozzles' work in tandem (see Fig. 8.12, scheme on the right).

For both cross-flow and premix emulsification, the membrane needs to be wetted by the continuous phase of the emulsion. This implies using a hydrophilic membrane for O/W emulsions, and a hydrophobic membrane for W/O emulsions; and

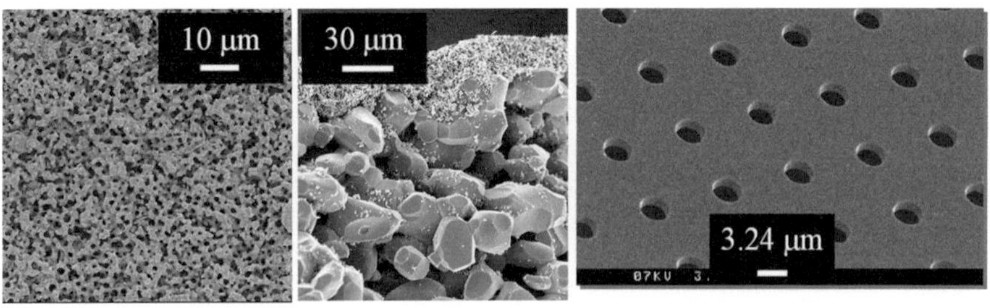

Fig. 8.11 (*Left*) Shirazu Porous Glass (SPG) membrane with interconnected tortuous pores of similar size all through the membrane (Image taken from internet). (*Middle*) Ceramic membrane with an open support structure and much finer top-layer. (*Right*) Microsieve made from silicon with a silicon nitride top layer, with uniform tailor-made pores ([85]; reprinted with permission from Aquamarijn Microfiltration BV)

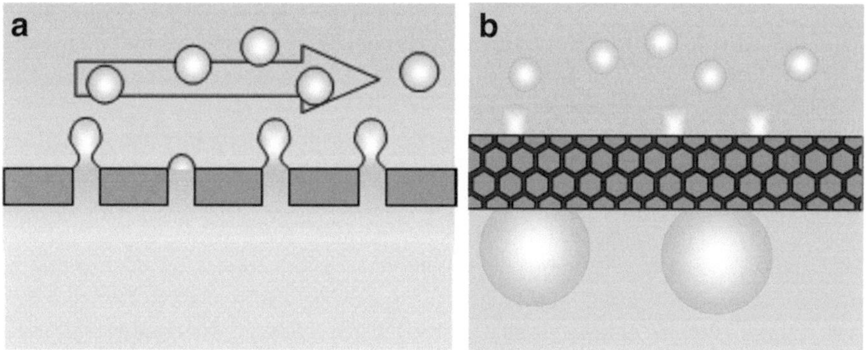

Fig. 8.12 Schematic representations of (**a**) cross-flow membrane emulsification, in which the cross-flowing continuous phase shears-off the droplets that are formed. (**b**) Pre-mix emulsification, in which the droplets of a coarse emulsion are broken up into smaller upon passage through a membrane

most importantly the wettability of the membrane should not change during operation, e.g., because of adsorption of surface-active components that are present in the emulsion.

In very limited cases, it is possible to induce phase inversion during pre-mix emulsification. To achieve this, the membrane should be compatible with the to-be-dispersed phase. Starting from an O/W pre-mix, the membrane needs to be hydrophobic; the oil droplets will wet the membrane, and the continuous water phase will be converted into the dispersed phase during passage through the membrane [73]. If this mode of operation is possible, emulsions with very high dispersed phase fraction may be obtained, but this strongly depends on the components in the emulsion mix, and on their interaction with the membrane.

To generate the required shear for droplet detachment, alternative designs have also been suggested, such as by Stillwell and co-workers [71], who investigated a stirred cell that generated rather polydisperse emulsions due to the differences in shear across the membrane. Eisner [14], Schadler [61], Aryantia and co-workers [4], and Yuan and co-workers [95] used a different approach and rotated (metal) membranes to shear off the droplets. As a result, better control over droplet size is achieved, although it should also be mentioned that compared to regular cross-flow emulsification, the droplets are much larger due to the larger pores in the metal sieves that are used; the mechanical stability required for rotating membranes requires this construction material.

For now, we focus on regular membrane emulsification that was invented in the group of prof. Nakashima in Japan [49]. Various good reviews have been published, and we recommend those by Joscelyne and Trägårdh ([21], general review), Charcosset and co-workers ([13] general review), Vladisavljevic and Williams ([88], general overview with many products), van der Graaf and co-workers ([77], double emulsions), and Charcosset ([12], specific for food). Most information on membrane emulsification is available for O/W emulsions, but also some authors have shown work on W/O emulsions, e.g., Vladisavljevic et al. [87]. On the other hand, pre-mix emulsification is not well documented; only very recently, a review became available by Nazir and co-workers [50], and not much is known about the droplet formation mechanisms.

The droplet formation mechanism during cross-flow membrane emulsification can also be described as a balance between the interfacial tension force that keeps the droplet connected to the pore, and the shear force that tries to remove the droplet, as was the case for the classic emulsification techniques. The first ones to describe this balance were Peng and Williams [55]. Later also more complex droplet formation mechanisms were reported in which two stages were distinguished. During stage one, a certain volume of liquid is pushed into the cross-flowing continuous phase and one reaching a certain value the droplet snap-off process starts. However, during this phase the droplet still grows due to its connection to the pore, and its size is not only determined by the cross-flowing continuous phase, but also by the applied pressure on the to-be-dispersed phase. In general, it can be said that the droplets that are generated are between two and ten times the diameters of the pore.

8.5.2 Microfluidic Techniques

Examples of microfluidic devices. Within the field of microfluidics, both shear based (T-shaped junctions, e.g., [78]; Y-shaped junctions, [70]; flow focusing devices, [1]) and spontaneous droplet formation (e.g., [72]) are used to generate droplets. A very extensive review, covering all these devices, has recently been published by Vladisavljević et al. [89]. In the present chapter we only touch briefly upon these devices since they seem to be still far away from large-scale application, although some may be used for the production of specialty products in the field of pharma.

In the top part of Fig. 8.13, droplet formation in a T-junction is shown together with a simulation result obtained for the same system [76]. At the bottom, an artist's impression is shown of droplet formation in a microchannel due to Laplace pressure differences in the system [79]. As was the case for cross-flow membrane emulsification, the droplet formation mechanisms in microfluidic devices mostly consist of two stages, one formation phase and a snap-off phase during which the droplet can still grow before actually being detached.

When comparing shear-based and spontaneous droplet generation, it is important to notice that the size of the droplets is determined by flow of both phases in the shear-based systems. Both need to be monitored very carefully in order to have monodisperse droplets, which in general are produced at much higher throughput than in the spontaneous systems. In spontaneous droplet formation systems, the continuous phase does not need to flow, and only the dispersed phase should be controlled. Mostly there is a range of disperse phase pressures for which the droplet size is not affected. The droplet formation time in spontaneous systems is in general much longer as in shear-based systems, so the throughput is also accordingly lower.

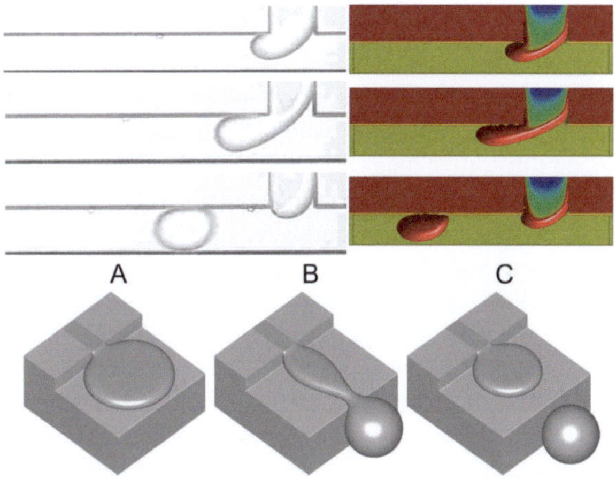

Fig. 8.13 *Top left* row, images taken during cross-flow emulsification in a T-shaped junction. The to-be-dispersed phase is pushed through the top channel into the cross-flowing continuous phase that moves the to-be-dispersed phase into the direction of flow until the shear-force exceeds the interfacial tension force, and a droplet is formed (Reprinted with permission from Van der Graaf et al. [78]. Copyright 2005, American Chemical Society). *Top right row*, simulation results of the system shown on the *left* ([76]; reprinted with permission from Elsevier). *Bottom* image, the three stages of droplet formation in a spontaneous microchannel, stage *A* being intrusion onto the shallow terrace. At stage *B*, the to-be-dispersed liquid leaps from the terrace into a deeper channel, and at stage *C* the size of the droplet is such that the Laplace pressures in the system lead to droplet formation ([79]; reprinted with permission from the author)

Scalability of microfluidic emulsification devices. Although most of the investigations on emulsification in microfluidic devices are focused on single droplet formation units, also some examples are known in which multiple droplet formation units work in tandem as is the case is in straight through microchannels (e.g. [26, 27]), microsieves ([16, 84, 85]; see Fig. 8.11 right image), parallelised flow focusing devices [54] or even work simultaneously, as is the case in so-called EDGE chips [81–83]. These devices have also been prepared partly in metal to comply with industrial demands [37, 38], and it was demonstrated that these systems even have better pressure stability than the original EDGE design. Besides some derived systems have been proposed in which metal sieves as such or in combination with a glass bead bed are used for pre-mix emulsification, which can be operated at very high flux and reasonable monodispersity [51–53, 80].

Currently, these methods can only be applied on rather small scale, but considerable effort is put into surface modification methods that allow for stable wetting of Si-based surfaces that are used in microfluidics (e.g. [2, 58]). Besides, within our own research group we also work on bringing microfluidics toward metal devices, which are the material of choice in industry. Our first attempts were directed towards semi-metal chips, and they could successfully be applied [40–42]. The next step towards a completely metal chip still needs to be made; making micrometer structures at high precision in metal is a great challenge, but the results on semi-metal systems make us optimistic.

8.6 Comparison of Emulsification Techniques

All emulsification methods can be compared based on their energy usage in relation to the droplet size that is generated. Amongst others in the work of Nazir and co-workers [50], an illustrative diagram is shown that summarizes these effects; see Fig. 8.14. When comparing microfluidic emulsification with the more classic techniques, it is clear, that cross-flow emulsification is a much less energy consuming technique than the high pressure homogenizers reported by Lambrich and Schubert [35]. For dilute emulsions, the energy density may be even orders of magnitude smaller. The energy density of cross-flow membrane emulsification is determined by the number of droplets that need to be produced; each volume fraction having its own corresponding line, unlike other emulsification techniques, in which pressurization of the whole volume determines the amount of energy needed. Pre-mix emulsification seems to be more in line with the traditional techniques, but may be useful in the production of relatively large droplets, which could correspond to better defined pre-mixes that can be used in classic emulsification. Regarding microfluidic devices, they are expected to be similar in energy density as cross-flow emulsification. But given the large variety in designs and mode of operation, and also the relatively large droplets that are made compared to classic emulsification devices, it is hard to pinpoint them to the diagram.

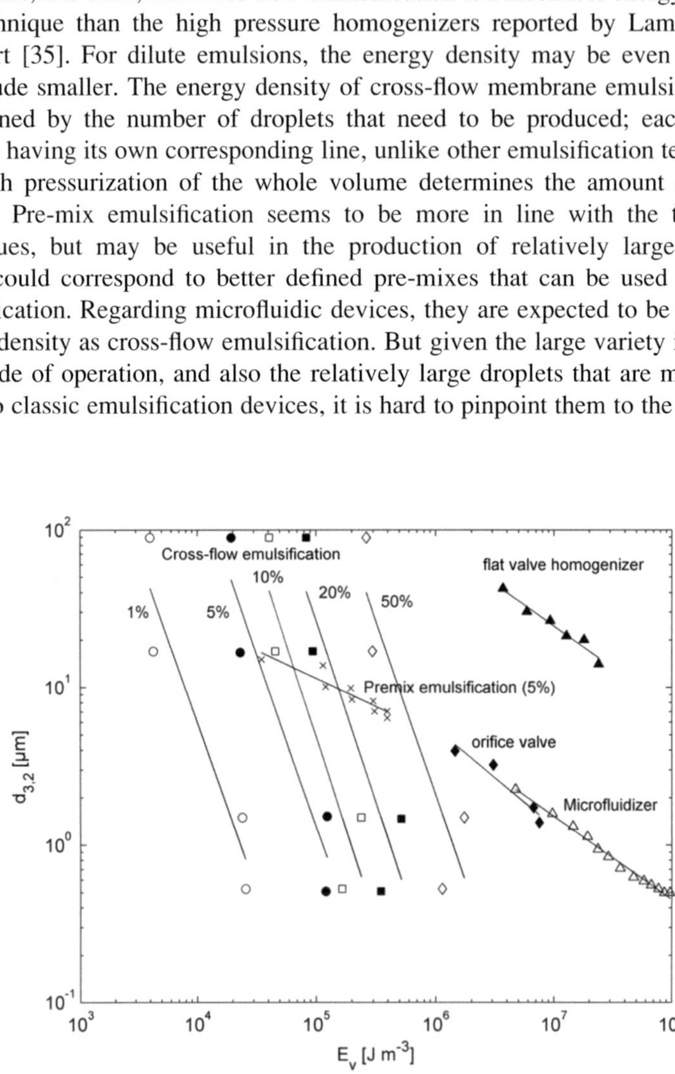

Fig. 8.14 Energy efficiencies of various emulsifying processes: cross-flow emulsification [35], (○) 1, (●) 5, (□) 10, (■) 20 and (◊) 50 vol.%; (×) pre-mix emulsification (5 vol.%) [80]; high pressure homogenization [35], (♦) orifice valve, (▲) flat valve homogenizer and (Δ) Microfluidizer (all 30 vol.%); (Reprinted with permission from Elsevier, from Nazir et al. [50])

8.7 Concluding Remarks

Various emulsification methods are available. Some are established, such as high pressure homogenization, others are still in their early stages of development, such as microfluidic devices. With all methods stable emulsions may be prepared, but only if the time scales of droplet formation are adequately matched with the time scale related to stabilization of the interface. This match can be reached by using stabilizing components that are able to lower the interfacial tension (surfactants), create a steric barrier (e.g. block co-polymers, particles), give rise to repulsive charge interactions (e.g. proteins, depending on the pH), or form a network (mostly through interactions of components).

It is very difficult to predict ab-initio which combination of emulsification device and emulsion composition needs to be chosen. In this chapter, some guidelines are given, such as the stability maps in Fig. 8.3. Still, this is far from perfect. In this light, the new developments in the field of microfluidics that allow emulsion stability testing, both under flow and under enhanced gravity, are very interesting methods that may lead to high-throughput testing of both ingredients and process conditions [31–33]. This is not only important for emulsions, but also for derived products such as double emulsions (a good review is by Muschiolik [47], capsules [59], particles [60], ultrasound contrast agents [30] and many more.

8.8 Definitions, Abbreviations and Symbols

Term	Definition
DLVO theory	Theory describing the various interactions that play a role for colloidal particles
Double emulsion:	Emulsion with three distinct phases, internal-, shell, and continuous phase. These emulsions are O/W/O or W/O/W
Emulsion	Mixture of oil (O) and water (W) in which one phase is finely dispersed as droplets into the other. Emulsions can be oil in water (O/W) or water in oil (W/O)
HLB balance	Indication of hydrophobicity/hydrophilicity of surfactant molecules. Is used in regard to their suitability to make O/W or W/O emulsion
Interfacial tension	The energy related to a liquid/liquid interface
Liposome	Water phase surrounded by a double layer of surfactant that can be used for encapsulation purposes

Abbreviation		
Ca	Capillary number	Dimensionless ratio of viscous effects to interfacial tension effects
Re	Reynolds number	Dimensionless ratio of inertial to viscous effects
		Re_{cr} is the critical value at which transition from laminar to turbulent flow occurs
We	Weber number	Dimensionless ratio of fluid inertia effects to interfacial tension
		We_{cr} is the critical value that has to be overcome to get droplet break-up

Symbol	Meaning	Unit	Remarks
A	Interfacial area	m^2	
ΔG	Gibbs free energy	J	
L	Width	m	
P	Polydispersity		
$\Delta P_{Laplace}$	Laplace pressure difference	Pa	
R_d	Droplet radius	m	
d_d	Droplet diameter	m	
d_i	Diameter of particles in each size-class	m	
d_{32}	Area-volume mean diameter	m	Also called: Sauter diameter, or surface-weighted mean diameter
d_{43}	Volume-length mean diameter	m	Also called volume-weighted mean size
$d_{50,V}$	Median diameter of vol.-based distribution	m	
g	Gravity acceleration	N/kg or m/s^2	
n_i	Number of particles in each size-class	m	
v	Rate/velocity/speed	m/s	
z	Height	m	
ε	Power density	W/m^3	
$\dot{\gamma}$	Shear rate/velocity gradient	1/s	$\dot{\gamma} = dv/dz$
η	Viscosity	Pa s	η_c refers to the viscosity of the continuous phase; η_d refers to the dispersed phase
ρ	Density	kg/m^3	ρ_c refers to the density of the continuous phase; ρ_d refers to the dispersed phase
σ	Interfacial tension	N/m	
τ	Disruptive stress	Pa	

References

1. Anna, S.L., Bontoux, N., Stone, H.A.: Formation of dispersions using "flow focusing" in microchannels. Appl. Phys. Lett. **82**(3), 364–366 (2003)
2. Arafat, A., Giesbers, M., Rosso, M., et al.: Covalent biofunctionalization of silicon nitride surfaces. Langmuir **23**, 6233–6244 (2007)
3. Arbuckle, W.S.: Emulsification. In: Hall, C.W., Farral, A.W., Rippen, A.L. (eds.) Encyclopaedia of Food Engineering, pp. 286–288. Avi Publication Company, Westport (1986)
4. Aryantia, N., Williams, R.A., Houa, R., et al.: Performance of rotating membrane emulsification for o/w production. Desalination **200**, 572–574 (2006)
5. Becher, P. (ed.): Encyclopedia of Emulsion Technology, vol. 1–4. Marcel Dekker, New York (1986)
6. Behrend, O., Schubert, H.: Influence of hydrostatic pressure and gas content on continuous ultrasound emulsification. Ultrason. Sonochem. **8**, 271–276 (2001)
7. Benech, R.O., Kheadr, E.E., Laridi, R., Lacroix, C., Fliss, I.: Inhibition of Listeria innocua in cheddar cheese by addition of nisin Z in liposomes or by in situ production in mixed culture. Appl. Environ. Microbiol. **68**, 3683–3690 (2002)
8. Bentley, B.J., Leal, L.G.: An experimental investigation of drop deformation and breakup in steady, two-dimensional linear flows. J. Fluid Mech. **176**, 241–283 (1986)
9. Berton-Carabin, C.C., Schroën, K.: Pickering emulsions for food applications: background, trends and challenges. Ann. Rev. Food Sci. Technol. **6**, 263–272 (2015). doi:10.1146/annurev-food-081114-110822
10. Brennan, J.G.: Emulsification, mechanical procedures. In: Hall, C.W., Farral, A.W., Rippen, A.L. (eds.) Encyclopaedia of Food Engineering, pp. 288–291. Avi Publication Company, Westport (1986)
11. Canselier, J.P., Delmas, H., Wilhelm, A.M., et al.: Ultrasound emulsification-an overview. J. Dispers. Sci. Technol. **23**(1–3), 333–349 (2002)
12. Charcosset, C.: Preparation of emulsions and particles by membrane emulsification for the food processing industry. J. Food Eng. **92**, 241–249 (2009)
13. Charcosset, C., Limayem, I., Fessi, H.: The membrane emulsification process—a review. J. Chem. Technol. Biotechnol. **79**, 209–218 (2004)
14. Eisner, V.: Emulsion Processing with a Rotating Membrane (ROME). Dissertation ETH Zürich, number 17153 (2007).
15. Gibbs, B.F., Kermasha, S., Alli, I., Mulligan, C.N.: Encapsulation in the food industry. Int. J. Food Sci. Nutr. **50**, 213–224 (1999)
16. Gijsbertsen-Abrahamse, A.J., Van der Padt, A., Boom, R.M.: Status of cross-flow membrane emulsification and outlook for industrial application. J. Membr. Sci. **230**, 149–159 (2004)
17. Grace, H.P.: Dispersion phenomena in high viscosity immiscible fluid systems and application of static mixers as dispersion devices in such systems. Chem. Eng. Commun. **14**, 225–277 (1982)
18. Guzey, D., McClements, D.J.: Formation, stability and properties of multilayer emulsions for application in the food industry. Adv. Colloid. Interface. Sci. **128–130**, 227–248 (2006)
19. Hiemenz, P.C.: Principles of Colloid and Surface Chemistry. Marcel Dekker, New York (1986)
20. Hiemenz, P.C., Rajagopalan, R.: Principles of Colloid and Surface Chemistry. M. Dekker, New York (1997)
21. Joscelyne, S.M., Trägårdh, G.: Membrane emulsification—a literature review. J. Membr. Sci. **169**, 107–117 (2000)
22. Karbstein, H., Schubert, H.: Developments in the continuous mechanical production of oil-in-water macro-emulsions. Chem. Eng. Proc. **34**, 205–211 (1995)
23. Kendall, G.: http://blogs.nottingham.ac.uk/malaysiaknowledgetransfer/2013/06/25/what-is-pharmaceutical-nanoemulsion/. Visited 14 October 2014.

24. Kirby, C.F., Brooker, B.E., Law, B.A.: Accelerated ripening of cheese using liposome-encapsulated enzyme. Int. J. Food Sci. Technol. **22**(4), 355–375 (1987)
25. Kissling, K., Schütz, S., Piesche, M.: Numerical investigation of the flow field and the mechanisms of droplet deformation and break-up in a high-pressure homogenizer. Proceedings of the 8th World Congress Chemical Engeneering, Montreal (2009).
26. Kobayashi, I., Neves, M.A., Uemura, K., et al.: Production characteristics of uniform large soybean oil droplets by microchannel emulsification using asymmetric through-holes. Procedia. Food. Sci. **2011**(1), 123–130 (2011)
27. Kobayashi, I., Nakajima, M., Chun, K., et al.: Silicon array of elongated through-holes for monodisperse emulsion droplets. AICHE J. **48**, 1639–1644 (2002)
28. Köhler, K.: Simultanes Emulgieren und Mischen. Logos Verlag, Berlin (2010). ISBN 978-3-8325-2716-7
29. Köhler, K.: In: Nagel, W.E., Kröner, D.B., Resch, M.M. (eds.) High Performance Computing in Science and Engineering'10. Springer, Heidelberg (2011)
30. Kooiman, K., Böhmer, M.R., Emmer, M.: Oil-filled polymer microcapsules for ultrasound-mediated delivery of lipophilic drugs. J. Control. Release **133**, 109–118 (2009)
31. Krebs, T., Schroën, K., Boom, R.: Coalescence dynamics of surfactant-stabilized emulsions studied with microfluidics. Soft Matter **8**(41), 10650–10657 (2012)
32. Krebs, T., Ershov, D., Schroen, C.G.P.H., et al.: Coalescence and compression in centrifuged emulsions studied with in situ optical microscopy. Soft Matter **9**(15), 4026–4035 (2013)
33. Krebs, T., Schroen, K., Boom, R.: A microfluidic method to study demulsification kinetics. Lab Chip **12**(6), 1060–1070 (2012)
34. Krog, N.J., Riisom, T.H., Larson, K.: Applications in food industry. In: Becher, P. (ed.) Encyclopedia of Emulsion Technology. Applications, vol. 2, pp. 58–127. Marcel Dekker, New York (1985)
35. Lambrich, U., Schubert, H.: Emulsification using microporous systems. J. Membr. Sci. **257**, 76–84 (2005)
36. Langton, M., Jordansson, E., Altskar, A., et al.: Microstructure and image analysis of mayonnaises. Food Hydrocoll. **13**, 113–125 (1999)
37. Leal-Calderon, F., Schmitt, V., Bibette, J.: Emulsion Science – Basic Principles, 2nd edn. Springer, New York (2007)
38. Lucassen-Reynders, E.H.: Dynamic interfacial properties in emulsification. In: Becher, P. (ed.) Encyclopedia of Emulsion Technology, vol. 4, pp. 63–90. Marcel Dekker, New York (1996)
39. Lyklema, J.: Fundamentals of Interface and Colloid Science. Academic, London (1991)
40. Maan, A.A., Schroën, K., Boom, R.: Spontaneous droplet formation techniques for monodisperse emulsions preparation – Perspectives for food applications (Review). J. Food Eng. **107** (3–4), 334–346 (2011)
41. Maan, A.A., Boom, R., Schroën, K.: Preparation of monodispersed oil-in-water emulsions through semi-metal microfluidic EDGE systems. Microfluid. Nanofluid. **14**(5), 775–784 (2013)
42. Maan, A.A., Schroën, K., Boom, R.: Monodispersed water-in-oil emulsions prepared with semi-metal microfluidic EDGE systems. Microfluid. Nanofluid. **14**(1–2), 187–196 (2013)
43. McClements, D.J.: Food Emulsions: Principles, Practices and Techniques. CRC Press, Boca Raton (2005)
44. McClements, D.J., Chanamai, R.: Physicochemical properties of mono disperse oil-in-water emulsions. J. Dispers. Sci. Technol. **23**(1–3), 125–134 (2002)
45. Merkus, H.G.: Particle Size Measurements – Fundamentals, Practice, Quality. Springer, New York (2009)
46. Merkus, H.G., Meesters, G.M.H. (eds.): Particulate Products – Tailoring Properties for Optimal Performance. Springer International Publishing, Switzerland (2014)
47. Muschiolik, G.: Multiple emulsions for food use. Curr. Opin. Colloid. Interface. Sci. **12**, 213–220 (2007)

48. Nakashima, T., Shimizu, M.: Porous glass from calcium alumino boro-silicate glass. Ceram. Jpn. **21**, 408 (1986)
49. Nakashima, T., Shimizu, M., Kukizaki, M.: Membrane emulsification by microporous glass. Key Eng. Mater. **61–62**, 513 (1991)
50. Nazir, A., Schroën, K., Boom, R.: Pre-mix emulsification: a review. J. Membr. Sci. **362**(1–2), 1–11 (2010)
51. Nazir, A., Schroën, K., Boom, R.: High-throughput premix membrane emulsification using nickel sieves having straight-through pores. J. Membr. Sci. **383**(1-2), 116–123 (2011)
52. Nazir, A., Schroën, K., Boom, R.: The effect of pore geometry on premix membrane emulsification using nickel sieves having uniform pores. Chem. Eng. Sci. **93**, 173–180 (2013)
53. Nazir, A., Boom, R., Schroën, K.: Droplet break-up mechanism in premix emulsification using packed beds. Chem. Eng. Sci. **92**, 190–197 (2013)
54. Nisisako, T., Torii, T.: Microfluidic large-scale integration on a chip for mass production of monodisperse droplets and particles. Lab Chip **8**, 287–293 (2008)
55. Peng, S.J., Williams, R.A.: Controlled production of emulsions using a cross-flow membrane. Part I: droplet formation from a single pore. Trans. IChemE. **76**, 894–901 (1998)
56. Perrechil, F., Santana, R., Fasolin, L.H., et al.: Rheological and structural evaluations of commercial Italian salad dressings. Cienc. Tecnol. Aliment. **30**(2), 477–482 (2010)
57. Roos, Y.H., Fryer, P.J., Knorr, D., Schuchmann, H.P., Schroën, K., Schutyser, M.A.I., Trystram, G., Windhab, E.J.: Food engineering at multiple scales: case studies, challenges and the future—a European perspective. Food Eng Rev. doi:10.1007/s12393-015-9125-z (2015, in press)
58. Rosso, M., Giesbers, M., Arafat, A., et al.: Covalently attached organic monolayers on SiC and $SixN_4$ surfaces: formation using UV light at room temperature. Langmuir **25**, 2172–2180 (2009)
59. Sagis, L.M.C., De Ruiter, R., Rossier Miranda, F.J., et al.: Polymer microcapsules with a fiber-reinforced nanocomposite shell. Langmuir **24**, 1608–1612 (2008)
60. Sawalha, H., Purwanti, N., Rinzema, A.: Polylactide microspheres prepared by premix membrane emulsification – effects of solvent removal rate. J. Membr. Sci. **310**, 484–493 (2008)
61. Schadler, V., Windhab, E.J.: Continuous membrane emulsification by using a membrane system with controlled pore distance. Desalination **189**, 130–135 (2006)
62. Scholten, E.: Ice cream (Chapter 9). In: Merkus, H.G, Meesters, G.M.H. (eds.) Particulate Products – Tailoring Properties for Optimal Performance. Springer International Publishing, Switzerland (2014)
63. Schröder, V., Behrend, O., Schubert, H.: Effect of dynamic interfacial tension on the emulsification process using microporous, ceramic membranes. J. Colloid. Interface. Sci. **202**, 334–340 (1998)
64. Schröder, V., Schubert, H.: Production of emulsions using microporous, ceramic membranes. Coll. Surf A Phys. Eng. Asp. **152**, 103–109 (1999)
65. Schroën, K., Blyzniuk, O., Muijlwijk, K. et al.: Microfluidic emulsification devices: from micrometer insights to large-scale food emulsion production. Curr. Trends Food Sci. **3,** 33–40 (2015)
66. Schubert, H., Armbruster, H.: Principles of formation and stability of emulsions. Int. Chem. Eng. **32**, 14 (1992)
67. Schuchmann, H.P.: Food process engineering research and innovation in a fast changing world – paradigms/case studies. In: Advances in Food Process Engineering Research and Applications. Springer (2013).
68. Schuchmann, H.P., Hecht, L.L., Gedrat, M., et al.: High-pressure homogenization for the production of emulsions. In: Eggers, R. (ed.) Industrial High Pressure Applications. Processes, Equipment and Safety, pp. 97–118. Wiley-VCH Verlag, Weinheim (2012)
69. Smulders, P.E.A.: Formation and Stability of Emulsions Made with Proteins and Peptides. PhD thesis. Wageningen University, Wageningen, The Netherlands (2000).

70. Steegmans, M.L.J., Schroën, C.G.P.H., Boom, R.M.: Characterization of emulsification at flat microchannel Y junctions. Langmuir **25**, 3396–3401 (2009)
71. Stillwell, M.T., Holdich, R.G., Kosvintsev, S.R., et al.: Stirred cell membrane emulsification and factors influencing dispersion drop size and uniformity. Ind. Eng. Chem. Res. **46**, 965–972 (2007)
72. Sugiura, S., Nakajima, M., Iwamoto, S., et al.: Interfacial tension driven monodispersed droplet formation from microfabricated channel array. Langmuir **17**, 5562–5566 (2001)
73. Suzuki, K., Hayakawa, K., Hagura, Y.: Preparation of high concentration o/w and w/o emulsions by the membrane phase inversion emulsification using PTFE membranes. Food Sci. Technol. Res. **5**, 234–238 (1999)
74. Urban, K., Wagner, G., Schaffner, D., et al.: Rotor-stator and disc systems for emulsification processes. Chem. Eng. Technol. **29**(1), 1–31 (2006)
75. Van Dalen, G.: Determination of the water droplet size distribution of fat spreads using confocal scanning laser microscopy. J. Microsc. **208**, 116–133 (2002)
76. Van der Graaf, S., Nisisako, T., Schroën, C.G.P.H., et al.: Lattice Boltzmann simulations of droplet formation in a T-shaped micro-channel. Langmuir **22**, 4144–4152 (2006)
77. Van der Graaf, S., Schroën, C.G.P.H., Boom, R.M.: Preparation of double emulsions by membrane emulsification—a review. J. Membr. Sci. **251**, 7–15 (2005)
78. Van der Graaf, S., Steegmans, M.L.J., Van der Sman, R.G.J., et al.: Droplet formation in a T-shaped microchannel junction: a model system for membrane emulsification. Coll. Surf. A Phys. Eng. Asp. **266**, 106–116 (2005)
79. Van der Zwan, E.A.: Emulsification with Microstructured Systems. Process Principles. PhD thesis. Wageningen University, Wageningen, The Netherlands (2008)
80. Van der Zwan, E.A., Schroën, C.G.P.H., Boom, R.M.: Pre-mix membrane emulsification by using a packed layer of glass beads. AICHE J. **54**, 2190–2197 (2008)
81. Van Dijke, K.C., De Ruiter, R., Schroën, K., et al.: The mechanism of droplet formation in microfluidic EDGE systems. Soft Matter **6**, 321–330 (2010)
82. Van Dijke, K.C., Schroën, C.G.P.H., Van der Padt, A., et al.: EDGE emulsification for food-grade dispersions. J. Food Eng. **97**(3), 348–354 (2010)
83. Van Dijke, K.C., Veldhuis, G., Schroën, C.G.P.H., et al.: Parallelized edge-based droplet generation (EDGE) devices. Lab Chip **9**, 2824–2830 (2009)
84. Van Rijn, C.J.M., Nano and Micro Engineered Membrane Technology. Membrane Science and Technology Series 10. Elsevier, Amsterdam. ISBN 0444514899, 9780444514899, 384 p (2004)
85. Van Rijn, C.J.M., Elwenspoek, M.C.: Micro filtration membrane sieve with silicon micro machining for industrial and biomedical applications. Proc. IEEE **29**, 83–87 (1995)
86. Vladisavljevic, G.T., Schubert, H.: Preparation of emulsions with a narrow particle size distribution using microporous -alumina membranes. J. Disp. Sci. Technol. **24**, 811–819 (2003)
87. Vladisavljevic, G.T., Tesch, S., Schubert, H.: Preparation of water-in-oil emulsions using microporous polypropylene hollow fibers: influence of some operating parameters on droplet dize distribution. Chem. Eng. Process. **41**, 231–238 (2002)
88. Vladisavljevic, G.T., Williams, R.A.: Recent developments in manufacturing emulsions and particulate products using membranes. Adv. Colloid Interface. Sci. **113**, 1–20 (2005)
89. Vladisavljević, G.T., Kobayashi, I., Nakajima, M.: Production of uniform droplets using membrane, microchannel, and microfluidic emulsification devices. Microfluid. Nanofluid. **13**, 151–178 (2012)
90. Walstra, P.: Formation of emulsions. In: Becher, P. (ed.) Encyclopedia of Emulsion Technology. Basic aspects, vol. 1, pp. 58–127. Marcel Dekker, New York (1983)
91. Walstra, P.: Emulsion stability. In: Becher, P. (ed.) Encyclopedia of Emulsion Technology, vol. 4. Marcel Dekker, New York (1996)
92. Walstra, P.: Physical Chemistry of Foods. Marcel Dekker, New York (2003)

93. Walstra, P., Wouters, J.T.M., Geurts, T.J.: Dairy Science and Technology. Taylor & Francis, Boca Raton (2006)
94. Walstra, P., Smulders, P.E.A.: Emulsion formation. In: Binks, B.P. (ed.) Modern Aspects of Emulsion Science, pp. 56–99. The Royal Society of Chemistry, Cambridge (1998)
95. Yuan, Q., Houa, R., Aryantia, N., et al.: Manufacture of controlled emulsions and particulates using membrane emulsification. Desalination **224**, 215–220 (2008)

Chapter 9
Mixing of Solid Materials

Ralf Weinekötter

Abstract The chapter focuses on the practical and theoretical aspects of mixing of solids. Dispersion and convection create the process of mixing. The result is endangered by segregating effects if the components of the mixture have very different flow characteristics. The definition of degree of mixing, the measurement methods and statistical evaluation of the data are presented. Classical off-line sampling and analysis of the samples and on-line methods are compared. A classification for the broad range of industrial mixers is established. Design and operational features of batch and continuous mixing processes are highlighted. The efficiency of continuous mixing processes is determined by the interaction of high precision gravimetric feeders with the residence time distribution of the mixer; the concept of the variance reduction ratio is explained. New application of continuous mixing in the pharmaceutical industry is presented.

9.1 Introduction

The mixing of powders, particles, flakes and granules has gained substantial economic importance in a broad range of industries, including for example the mixing of human and animal foodstuffs, pharmaceutical products, detergents, chemicals and plastics. Mixing is achieved while the particles of the ingredients move relative to each other. This particle motion has to be forced by external forces since unlike gas molecules dry particles have no Brownian motion. However, especially for free-flow materials the quality of the mix is endangered by segregation. This segregation is caused by different flow paths of the particles within the mixtures e.g. due to differences in particle size or density. This chapter will present mixing mechanisms and how mixture quality is described with e.g. the variance of

R. Weinekötter (✉)
Gericke AG, Zürich, Switzerland
e-mail: r.weinekoetter@gericke.net

© Springer International Publishing Switzerland 2016

H.G. Merkus, G.M.H. Meesters (eds.), *Production, Handling and Characterization of Particulate Materials*, Particle Technology Series 25,
DOI 10.1007/978-3-319-20949-4_9

concentration. The range of industrial mixers is very large; a classification and checklist will be presented which will help the selection and design of mixing machines. The principle and functionality of batch and continuous mixing processes will be compared.

9.2 Industrial Relevance of Solids Mixing

The mixing of powders, particles, flakes and granules has gained substantial economic importance in a broad range of industries, including for example foods, pharmaceutical products, detergents, chemicals, plastics, etc. As in most cases the mixing process adds significant value to the product, it can be regarded as a key unit operation to the overall process stream.

The term mixing is used for the blending of several ingredients as well as for the homogenization of a batch of one ingredient only. By far the most important use of mixing is the production of a homogenous blend of several ingredients which neutralizes variations in concentration. But if the volume of material consists of one ingredient or compound exhibiting fluctuating properties caused by an upstream production process, or inherent to the raw material itself, the term **homogenization** is used for the neutralization of these fluctuations. By mixing a new product or intermediate is created, for which the quality and price of this product is very often dependent upon the efficiency of the mixing process. This efficiency is determined both by the materials to be mixed, e.g. particle size and particle size distribution, density and surface roughness, and the process and equipment used for performing the mixing. The design and operation of the mixing unit itself has a strong influence on the quality produced, but upstream material handling process steps like feeding, sifting, weighing and transport determine also both quality and capacity of the mixing process. Downstream processing may also destroy the product quality due to segregation (demixing). Continuous mixing is one solution which limits segregation by avoiding storage equipment.

The technical process of mixing is performed by a multitude of equipment available on the market. However, mixing processes are not always designed with the appropriate care. This causes a significant financial loss which arises in two ways:

(a) *The quality of the mix is poor*: In cases where the mixing produces the end product, this will be noticed immediately at the product's quality inspection. Frequently, however, mixing is only one in a series of further processing stages. In this case, the effects of unsatisfactory blending are less apparent, and might possibly be overlooked to the detriment for final product quality.

(b) *The homogeneity is satisfactory but the effort employed is too great (OVERMIXING)*: Overmixing in batch blending is induced by an overlong mixing time or too long residence time in the case of continuous blending. This leads to increased strain on the mixture which can have an adverse effect on the quality of sensitive products. Furthermore, larger or more numerous pieces of equipment must be used than would be necessary in the case of an optimally configured mixing process.

9.3 Mixing Mechanisms: Dispersive and Convective Mixing

The mixing process can be observed in diagrammatic form as an overlap of **dispersion** and **convection** (Fig. 9.1). Movement of the particulate materials is a prerequisite of both mechanisms. **Dispersion** is understood to mean the completely random change of place of the individual particles. The frequency with which the particles of *ingredient A* change place with those of another ingredient is related to the number of particles of the other ingredients in the *direct vicinity* of the particles of *ingredient A*. Dispersion is therefore a local effect (**micromixing**) taking place in the case of premix systems where a number of particles of different ingredients are in proximity, leading to a fine mix localized to very small areas [7]. If the ingredients are spatially separated at the beginning of the process, long times would be required to mix them through dispersion alone, since there is a very low number of *assorted neighbors*. Dispersion corresponds to diffusion in liquid mixtures. However in contrast to diffusion, mixing in the case of dispersion is not

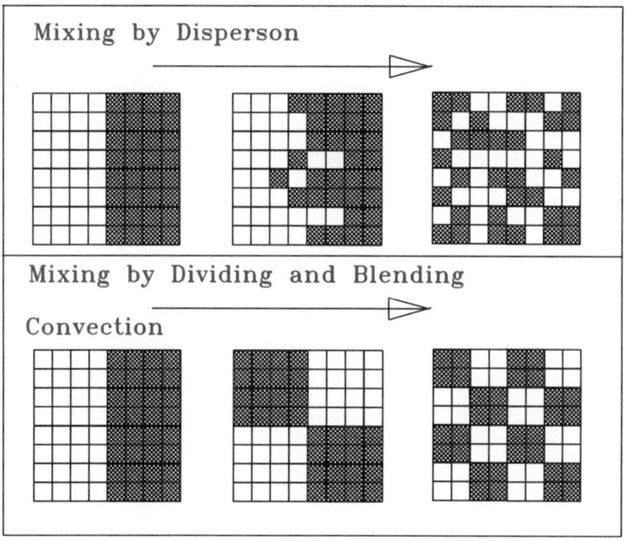

Fig. 9.1 The mixing process can be observed in diagrammatic form as an overlap of dispersion and convection. Mixture is consisting of two components A and B: A symbolized by the *white blocks* and B by the hatched blocks. Dispersion results in a random arrangement of the particles, convection results in a regular pattern

caused by any concentration gradient. The particles have to be in motion for getting dispersed, e.g. by vibration.

Convection causes a movement of large groups of particles relative to each other (**macromixing**). The whole volume of material is continuously divided up and then mixed again after the portions have changed places (Fig. 9.1). This forced convection can be achieved by rotating elements. The dimension of the groups, which are composed of just one unmixed ingredient, is determined by the continuously reduced splitting action of the rotating paddles. Convection increases the number of *assorted neighbors* and thereby promotes the exchange processes of dispersive mixing. A material mass is divided up or convectively mixed through the rearrangement of a solid's layers by rotating devices in the mixer or by the fall of a stream of material in a static gravity mixer, as discussed below.

9.4 Segregation and Demixing in Solids

If the components in a solids mixture possess a selective, individual motional behavior, the mixture's quality can be reduced as a result of segregation. As yet only a partial understanding of such behavior exists, with particle movement behavior being influenced by particle properties such as size, shape, density, surface roughness, forces of attraction and friction. In addition, industrial mixers each possess their own specific flow conditions. Particle size is, however, the dominant influence in segregation [25]. Since there is a divergence of particle sizes in even a single ingredient, nearly all industrial powders can be considered as solid mixtures of particles of different size, and segregation is one of the characteristic problems of solids processing which must be overcome for successful processing. If mixtures are unsuitably stored or transported, they will separate according to particle size and thus segregate Fig. 9.2 illustrates typical mechanisms of segregation.

9.4.1 Agglomeration Segregation

Agglomeration segregation arises through the preferential self-agglomeration of one component in two-ingredient mixtures (Fig. 9.2a). Agglomerates form when there are strong inter-particle forces and for these forces to have an effect the particles must be brought into close contact. In the case of agglomerates the particles stick to each other as a result, for example, of liquid bridges formed in solids, if a small quantity of moisture or other fluid is present. Electrostatic and Van der Waals forces likewise induce cohesion of agglomerates. Van der Waals forces, reciprocally induced and dipolar, operate particularly upon finer grains smaller than 30 μm. High speed impellers or knives are utilized in the mixing chamber to create

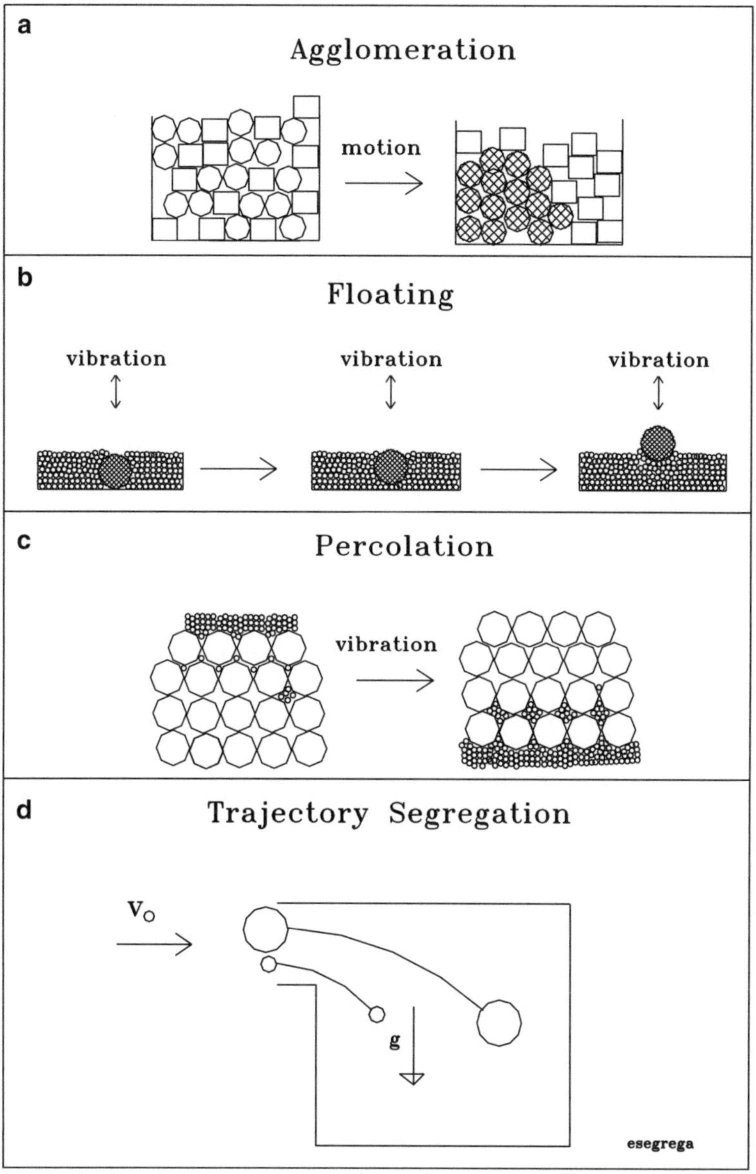

Fig. 9.2 Four mechanism of segregation following [25]

shear forces during mixing to break up these agglomerates. Agglomeration can however have a positive effect on mixing. If a solids mix contains a very fine ingredient with particles in the submicron range (e.g. pigments), these fine particles may adsorb to the coarser ones. An **ordered mixture** occurs which is stabilized by the Van der Waals forces and is thereby protected from segregation.

9.4.2 Flotation Segregation

Flotation segregation can occur if a solids mix is *vibrated*, where the coarser particles *float up* against the gravity force and collect near the top surface, as illustrated in Fig. 9.2b for the case of a large particle in a mix of finer material (the 'Brazil Nut Effect'). During vibration, smaller particles flow into the vacant space created underneath the large particle, thus preventing the larger from reclaiming its original position. If the large particle has a higher density than the fines, it will compact the fines, further reducing their mobility, and the ability of the large particle to sink. Solely because of the blocking effect of the larger particle's geometry there is little probability that this effect will run in reverse and that a bigger particle will take over the place left by a smaller one which has been lifted up. The large particle would in this case have to displace several smaller ones. As a result the probability is higher that coarse particles will climb upwards with vibration.

9.4.3 Percolation Segregation

Percolation segregation is by far the most important segregational effect, which occurs when finer particles trickle down through the gaps between the larger ones (Fig. 9.2c). These gaps act like a sieve. If a solids mixture is moved, gaps between the grains briefly open up, allowing finer particles to selectively pass through the particle bed. Granted a single layer has a low degree of separation, but a bed of powder consists of many layers and interconnecting grades of particles which taken together can produce a significant division between fine and coarse grains, resulting in widespread segregation. Furthermore, percolation occurs even where there is but a small difference in the size of the particles (e.g. 250 and 300 μm particles) [24]. This is most significant in the poured heap appearing, for example, when filling and discharging bunkers or silos. A mobile layer with a high speed gradient forms on the surface of such a cone, which, like a sieve, bars larger particles from passing into the cone's core. Large grains on the cone's mantle obviously slide or roll downwards. However, large, poorly mixed areas occur even inside the cone. Thus, filling a silo or emptying it from a central discharge point is particularly critical. Remixing of such segregated heaps can be achieved through **mass flow** discharge, i.e. the silo's content's move downwards in blocks, slipping at the walls, rather than emptying from the central core (**funnel flow**) (see Chap. 13).

9.4.4 Transport Segregation

Transport segregation encompasses several effects which share the common factor of a gas contributing to the segregation processes. Trajectory and fluidized segregation can be defined, first occurring in cyclones or conveying into a silo where the particles are following the individual trajectories, and second in fluidization. During fluidization particles are exposed to drag and gravity forces which may lead to a segregation.

Williams [25] gives an overview of the literature on the subject and suggests the following measures to counter segregation:

- The addition of a small quantity of water forms water bridges between the particles, reducing their mobility and thus stabilizing the condition of the mixture.
- The tendency to segregate decreases below a grain size of about 30 μm ($\rho_s = 2$–3 kg/l), because of the cohesive behavior of these particles.
- Inclined planes down which the particles can roll should be avoided.
- In general, having ingredients of a uniform grain size is an advantage in blending.

9.5 Mixture Quality: The Statistical Definition of Homogeneity

To judge the efficiency of a solids blender or of a mixing process in general, the status of mixing has to be quantified, thus a *degree of mixing* has to be defined. Here one has to specify what *property* characterizes a mixture, examples being composition, particle size and temperature. The end goal of a mixing process is uniformity of this property throughout the volume of material in the mixer. There are circumstances where a good mix requires uniformity of several properties, e.g. particle size and composition. The mixture's condition is traditionally checked by taking a number of samples, after which these samples are examined for uniformity of the property of interest. The quantity of material sampled or **sample size** and the location of these samples are essential elements in evaluating a solids mixture.

9.5.1 Degree of Mixing and Scale of Scrutiny

Sample size thus represents the resolution by which a mixture can be judged. The smaller the size of sample the more closely the condition of the mixture will be scrutinized (see Fig. 9.3).

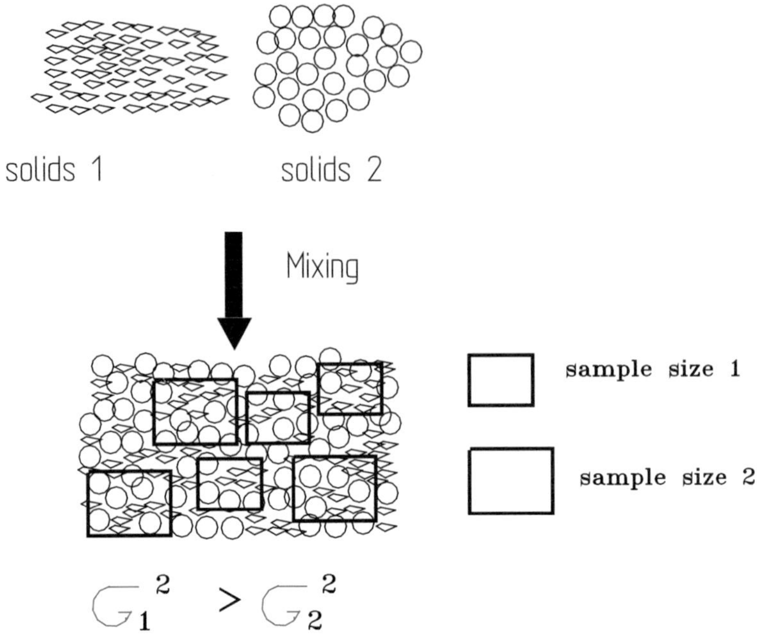

Fig. 9.3 The influence of the size of the sample on the numerical value of the degree of mixing

Dankwerts terms this the **scale of scrutiny** [2]. Specifying the size of the sample is therefore an essential step in analyzing a mixture's quality, since it quantifies the mixing task from the outset. The size of the sample can only be meaningfully specified in connection with the mixture's further application. In pharmaceutical production, active ingredients must be equally distributed e.g. within the individual tablets in a production batch, the sample size for testing the condition of a mixture is one tablet. In less critical industries the sample size can be in the tons. The traditional and general procedure is to take *identically sized* samples of the mixture from various points *at random* and to analyze them in an off-line analysis. Multi-element mixtures can also be described as twin ingredient mixes when a particularly important ingredient, e.g. the active agent in pharmaceutical products, is viewed as a tracer element and all the other constituents are combined into one common ingredient. This is a simplification of the statistical description of solids mixtures. When two-element mixtures are being examined it is sufficient to just trace the concentration path of just one ingredient, the tracer. There will be a complementary concentration of the other ingredients. The description is completely analogous when the property or characteristic feature in which we are interested is not the concentration but e.g. moisture, temperature or the particle's shape. If the tracer's concentration in the mixture is p and that of the other ingredients is q, we have the following relationship: $p + q = 1$. If you take samples of a specified size from the mixture and analyze them for their content of the tracer, the concentration of tracer xi in the samples will fluctuate randomly around that the tracer's concentration p in the whole mixture (the "population"). *Therefore a mixture's quality can only be described using statistical means.* The smaller the fluctuations in the samples' concentration x_i around the mixture's concentration p the better its quality. This can be quantified by the statistical **variance** of sample concentration σ_p^2, which consequently is frequently defined as the **degree of mixing**. There are many more definitions of mix quality in literature on the subject but in most instances these relate to an initial or final variance and are frequently too complicated for industrial application [18]. The theoretical variance for a finite sample number N_g is calculated as follows:

$$\sigma_p^2 = \frac{1}{N_g}\sum_{i=1}^{N_g}(x_i - p)^2 \tag{9.1}$$

where: x_i is the concentration in the sample i.

The relative standard deviation is used as well for judging mixture quality. It is defined by

$$RSD_p = \sqrt{\sigma_p^2} \Big/ p \tag{9.2}$$

The variance is obtained by dividing up the whole mix, the population, into N_g samples of the same size and determining the concentration x_i in each sample. Figure 9.3 illustrates that smaller samples will cause a larger variance or degree of mixing due to the smaller number of particles in the samples (see also pp. 302–304 for size of samples).

If one analyses not the whole mix but a number n of *randomly* distributed samples across the population, one determines instead the ***sample or empirical variance*** S^2. If this procedure is repeated several times a new value for the sample variance will be produced on each occasion resulting in a statistical distribution of the sample variance. Thus, each S^2 represents an estimated value for the unknown variance σ^2. In many cases the concentration p is likewise unknown and the random sample variance is then defined using the **arithmetical average** $<x>$ of the sample's concentration x_i.

$$S^2 = \frac{1}{n-1}\sum\nolimits_{i=1}^{n}(x_i - <x>)^2; \quad <x> = \frac{1}{n}\sum\nolimits_{i=1}^{n} x_i \tag{9.3}$$

Sample variance data is of little utility without knowing how accurately it describes the unknown, true variance σ_p^2. The variance is therefore best stated as a desired **confidence interval** for σ^2 (compare Eq. 9.4). $\Phi(\chi^2)$ is the summation of the χ^2 distribution. χ_1^2 designates lower limit. $\Phi(\chi^2)$ is tabled in statistical text books. The confidence interval used in mixing is mostly a unilateral one, derived by the χ^2 distribution. Interest is focused on the upper confidence limit, which, with a given degree of probability, will not be exceeded by the variance (Eq. 9.4) [15], which is given by:

$$W\left(\sigma_p < (n-1)\frac{S^2}{\chi_1^2}\right) = 1 - \Phi(\chi_1^2) \tag{9.4}$$

Figure 9.4 illustrates how the size of the confidence interval normalized with the sample variance decreases as the number of random samples n increases. The confidence interval depicts the accuracy of the analysis. The smaller the interval the better the mix quality can be estimated from the measured sample variance. If there are few samples the mix quality's confidence interval is very large. An evaluation of the mix quality with a high degree of accuracy (a small confidence interval) requires that a large number of samples be taken and analyzed, which can be expensive and requires large effort. Accuracy and cost of analysis must therefore be balanced for the process at hand.

Example Three tons of a sand (80 % by weight) and cement (20 % by weight) mix have been produced. The quality of this mix has to be checked. Thirty samples at 2 kg of the material mixture have been taken at random and the sand content in these samples established.

The mass fraction of the sand x_i [kg_{sand}/kg_{mix}] in the samples comes to:

3 samples @ 0.75; 7 @ 0.77; 5 @ 0.79; 6 @ 0.81; 7 @ 0.83; 2 @ 0.85

The degree of mixing defined as the variance of the mass fraction of sand in the mix needs to be determined. It has to be compared with the variance for a fully segregated system and the ideal variance of a random mix. First of all the random sample variance S^2 (Eq. 9.3) is calculated and with it an upper limit for the true variance σ^2 can then be laid down. The sand's average concentration p in the whole 3 ton mix is estimated using the random sample average $<x>$:

$$< x > = \frac{1}{n}\sum_{i=1}^{n} x_i = \frac{1}{30}\sum_{i=1}^{30} x_i = 0,797$$

and:

$$\begin{aligned} S^2 &= \frac{1}{n-1}\sum_{i=1}^{n}(x_i - < x >)^2 = \frac{1}{29}\sum_{1}^{30}(x_i - 0,797)^2 \\ &= \frac{1}{29}\left(3\cdot 0,047^2 + 7\cdot 0,027^2 + 5\cdot 0,007^2 + 6\cdot 0,013^2 + 7\cdot 0,033^2 + 2\cdot 0,053^2\right) \\ &= 9,04\cdot 10^{-4} \end{aligned}$$

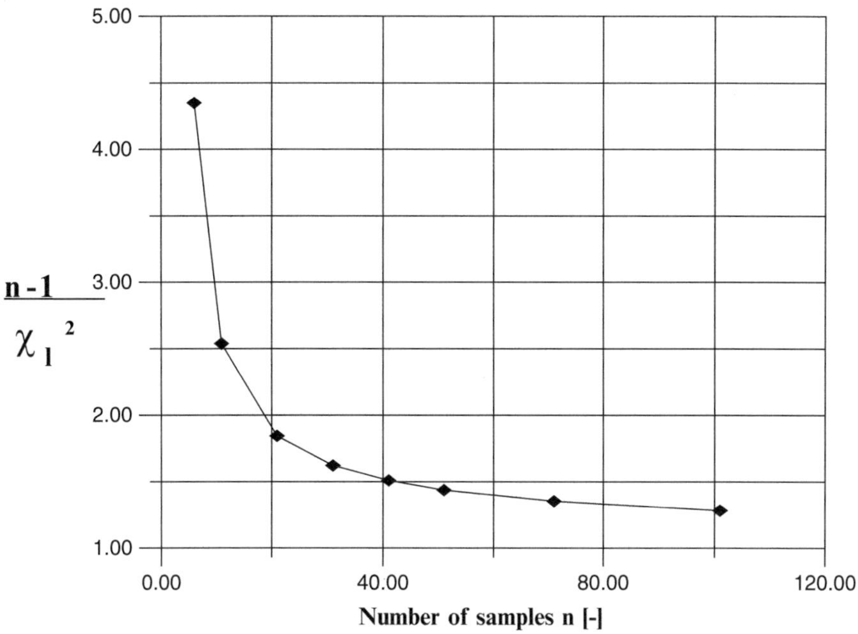

Fig. 9.4 The size of the unilateral confidence interval (95 %) as a function of the number n of samples taken, measured in multiples of S^2 (cf. Eq. 9.5). Example: if 31 samples are taken the upper limit of the variance's confidence interval assumes a value 1.6 times that of the experimental sample variance S^2

Ninety five percent is set as the probability W determining the size of the confidence interval for the variance σ_p^2. An upper limit (unilateral confidence interval) is then calculated for variance σ_p^2:

$$\mathrm{W}\left(\sigma_p^2 < (\mathrm{n}-1)\frac{\mathrm{S}^2}{\chi_l^2}\right) = 0,95 = 1 - \Phi(\chi_1^2) \Rightarrow \Phi(\chi_1^2) = 0,05$$

From the table of the χ^2 distribution summation function (in statistical books), $\Phi(\chi_l^2; \mathrm{n}-1)$, the value 17.7 is derived for a degree of freedom of 29 ($= n-1$). Figure 9.4, showing values for $(n-1)/\chi_1^2$ for different number of samples n, allows a fast judgment of these values.

$$\sigma^2 < (\mathrm{n}-1)\frac{\mathrm{S}^2}{\chi_1^2} = 29 \cdot \frac{9,04 \cdot 10^{-4}}{17,7} = 14,8 \cdot 10^{-4} \tag{9.5}$$

It can therefore be conclusively stated with a probability of 95 % that the mix quality σ_p^2 is smaller than $14.8 \cdot 10^{-4}$. Thus, $\sigma = \sqrt{(14.810^{-4})} = 3.810^{-2}$ and $\mathrm{RSD} = 3.8 \cdot 10^{-2}/0.797 = 4.8 \cdot 10^{-2} = 4.8\%$.

9.5.2 *Ideal Mixtures*

A **perfect mixture** is when the concentration at any randomly selected point in the mix in a sample of any size is the same as that of the overall concentration. The variance of a perfect mixture has a value of zero. This is only possible with gases and liquids which can be mixed molecularly and where sample volumes of the mixture are many times larger than its ingredients, i.e. molecules. In the case of solids mixtures, perfect mixtures do not exist and variance must be considered in comparison to both sample size and sensor area in relation to particle size. σ_p^2 thus depends on the size of the sample (Fig. 9.5).

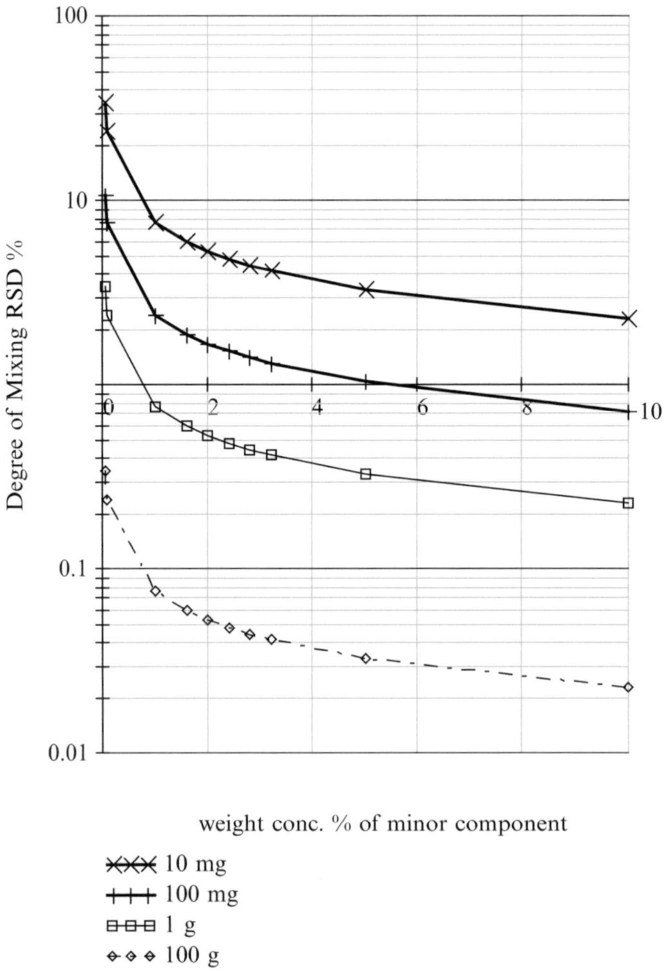

Fig. 9.5 Degree of Mixing expressed as RSD $= \sqrt{\sigma^2}/_p$ for a random mixture calculated following Sommer. The two components have the same particle size distribution, $d_p(50) = 50$ µm, $d_{max} = 130$ µm, $m = 0.7$ (exponent of the power density distribution of the particle size) parameter: sample size ranging from 10 to 100 g [23]

There are two limiting conditions of optimum homogeneity which are the equivalent of a minimum variance: an **ordered** and a **random mixture**.

9.5.3 Ordered Mixtures

The components align themselves according to a defined pattern. Whether this ever happens in practice is debatable. There exists the notion that because of inter-particle processes of attraction, this mix condition can be achieved. The inter-particle forces find themselves in an interplay with those of gravity and other dispersive forces, which would prevent this type of ordered mix in the case of coarser particles. Inter-particle forces predominate in the case of finer particles, i.e. cohesive powders. Ordered agglomerates or layered particles can arise. Sometimes not only the mix condition but also the mixing of powders in which these forces of attraction are significant is termed **ordered mixing** [6]. However, Egermann [3] points to the fact that one should only use ordered mixing to describe the condition and not the mixing of fine particles using powerful inter-particle forces. Note that both the number of adsorbed small particles to the surface of larger ones and the number of particles in agglomerates will in practice show some variation.

9.5.4 Random Mixtures

A random mixture also represents an optimum condition. It is defined as follows: A uniform random mix occurs when the probability of coming across an ingredient of the mix in any subsection of the area being examined is equal to any other point in time for all subsections of the same size, provided that the condition exists that the particles can move freely.

The variance of a random mixture σ_z^2 is calculated as follows for a (quasi-) two-ingredient blend [9, 12, 17]:

$$\sigma_z^2 = \frac{p^*q}{n_p} \tag{9.6}$$

p is the concentration of one of the ingredients in the mix, q is the other ($q = 1-p$) and n_p is total number of particles in the sample. It should be noted that the variance of the random mix grows if the sample size, i.e. the number of particles n_p in the sample, decreases. The variance for a **completely segregated system** is eq. given by:

$$\sigma^2_{segregated} = p \cdot q \tag{9.7}$$

For application of Eq. (9.6) it is a practical disadvantage that the number of particles in the sample has to be known in order to calculate the variance, rather than the usually specified, sample mass. Stange [19] calculated the variance of a random mix on the basis of the mass of both fractions. His approach is based on the fact that an ingredient possessing a distribution in particle size by necessity also has a distribution in particle mass. He made an assumption for the average mass m_p and m_q of the particles in each component and the particle mass's standard deviation σ_p and σ_q. He designated the **variability** c as the quotient of the standard deviation and average particle mass, or:

$$c_p = \frac{\sigma_p}{m_p}; \quad c_q = \frac{\sigma_q}{m_q} \tag{9.8}$$

Variability is a measure for the width of the particle size distribution. The higher the value of c the broader the particle size distribution.

The size of the sample is now specified in practice by its mass M and no longer by the number of particles n_p as shown in Eq. (9.6). The variance in random mixture for the case of (quasi-) two-component mixes can be given by:

$$\sigma^2 = \frac{p^*q}{M_s} * \left[p * m_q\left(1 + c_q^2\right) + q * m_p\left(1 + c_p^2\right)\right] \tag{9.9}$$

Equation 9.9 estimates the variance of a random mixture, even if the components have different particle size distributions. If the components have a small size (i.e. small mean particle mass) or a narrow particle size distribution, i.e. c_q and c_p are low, the random mix's variance falls. Sommer has presented mathematical models for calculating the variance of random mixtures for particulate systems with a particle size distribution [17]. This model has been used for deriving Fig. 9.5.

9.6 Measuring the Degree of Mixing and the Mixing Time

The mixing process aims at uniformly distributing one or more properties within a quantity of material. These can be physically recordable properties such as e.g. size, shape, moisture, temperature, or color. Frequently it is the mixing of **chemically** differing components which forms the subject under examination. Off-line and on-line procedures are used to for this examination. In off-line procedures, specified portions are (randomly or systematically) taken from the volume of material. These samples are often too large for a subsequent analysis and must then be split. Many

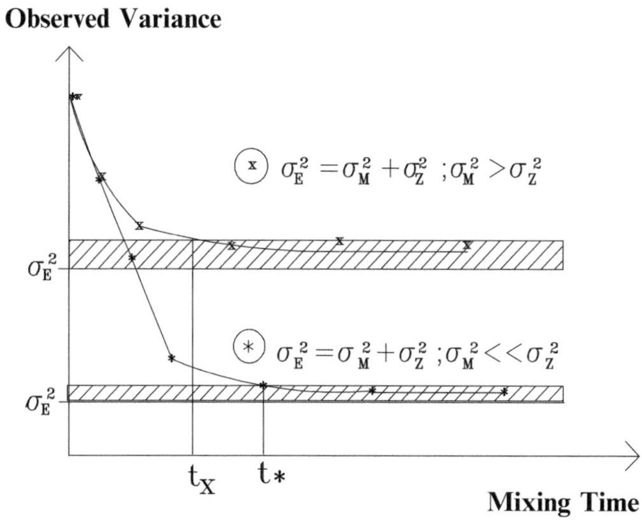

Fig. 9.6 Illustration of the influence of the measurements' accuracy on the variance as a function of the mixing time following Sommer [16]. A set of samples have been taken at different mixing times for computing the sample variance. Special attention has to be paid whether the experimental sample variance monitors the errors of the analysis procedure (x) or detects really the mixing process (*). Confidence intervals for the final status σ_E^2 are shown as hatched sections

analytical processes, e.g. the chemical analysis of solids using infrared spectroscopy, require the samples to be prepared beforehand. At all of these stages there exists the danger that the mix status within the samples will be changed. As a consequence, when examining a mixing process, whose efficiency can be characterized by the variance expression $\sigma^2_{\text{process}}$, all off- and on-line procedures only give this variance indirectly:

$$\sigma^2_{\text{observed}} = \sigma^2_{\text{process}} + \sigma^2_{\text{measurement}} \tag{9.10}$$

The observed variance $\sigma^2_{\text{observed}}$ also contains the variance $\sigma^2_{\text{measurement}}$ resulting from the test procedure and which arises from errors in the systematic or random taking, splitting and preparation of the samples and from the actual analysis. A lot of attention is often paid to the accuracy of an analyzer when it is being bought. However, the preceding steps of sampling and preparation also have to fulfill exacting requirements so that the following can apply:

$$\sigma^2_{\text{process}} \gg \sigma^2_{\text{measurement}} \Rightarrow \sigma^2_{\text{process}} = \sigma^2_{\text{observed}} \tag{9.11}$$

Figure 9.6 illustrates the impact of precision of the determination of mixing time for batch mixers. It is not yet possible to theoretically forecast mixing times for solids and therefore these have to be ascertained by experiments. The traditional method of determining mixing times is once again sampling followed by off-line analysis.

The mixer is loaded and started. After the mixer has been loaded with the ingredients in accordance with a defined procedure it is run and samples are taken at set time intervals. To do this the mixer usually has to be halted. The concentration of the tracer in the samples is established and the random sample variance S^2 ascertained. This random sample variance serves as an estimated value for the variance σ_p^2 which defines the mixture's condition. All analyses are burdened by errors and this is expressed in a variance σ_M^2 derived from the sampling itself and from the analysis procedure. Initially there is a sharp fall in the random sample variance and it runs asymptotically towards a final value of σ_E^2 as the mixing time increases. This stationary end value σ_E^2 is set by the variance of the mix in the stationary condition σ_Z^2, for which the minimum would be the variance of an ideal random mix, and the variance S_M^2 caused by errors in the analyzing process. The **mixing time** denotes that period in which the experimental random sample variance S^2 falls within the confidence interval of the stationary final condition σ_E^2. Two cases can be considered [16]. In the first case with large measurement errors, σ_E^2 is determined by the analyzing process itself since for sufficient mixing time the mixing process fluctuations in its stationary condition are much smaller than those arising out of the analysis or $\sigma_p^2 \ll \sigma_M^2$. In this case, the mixing process can only be tracked at its commencement, where $\sigma_p^2 > \sigma_M^2$. The "mixing time" t_x (compare Fig. 9.6) obtained under these conditions does not characterize the process. In the second case where the measurement errors are small, or $\sigma_p^2 \gg \sigma_M^2$, the analyzing process is sufficiently accurate for the mixing process to be followed through to its stationary condition. This allows an accurate determination of the true mixing time t^*. The "mixing time" t_x obtained on the basis of an unsatisfactory analysis is always deceptively shorter than the true time t_*.

9.6.1 On-Line Procedures

Advances in sensor technology and data processing are enabling an increased number of process parameters to be completely monitored using on-line procedures. The great leap forward from off-line to on-line procedures lies in the fact that the whole process of preparing and analyzing samples has been automated. As a result of this automation the amount of collectable test data has risen considerably, thereby enabling a more comprehensive statistical analysis and, in ideal cases, even regulation of the process. On-line procedures must in most cases be precisely matched to the process and the expense in terms of equipment and investment is disparately higher. The accuracy of laboratory analyses in the case off-line procedures cannot be produced using on-line processes. There are as yet few on-line procedures for chemically analyzing solids. Near-infrared (NIR) spectrometers fitted with fiber optic sensors are used solely in the field of foodstuffs and for identifying raw materials in the pharmaceuticals industry and have also been applied to mixtures [26; 20]. For pharmaceutical mixes the NIR method has been proposed for the control of mixing efficiency [14]. This method records the specific NIR absorption of functional chemical groups. If these spectrometers are based on modern diode array technology a spectrum covering the whole wavelength range is obtained in a fraction of a second. The exact sample size of a fiber optic probe which detects the reflected light can only be estimated [8, 22]. In addition it is very small, thus few particles will be detected. However, in recent years bigger probes have been developed e.g. by Bruker instruments.

9.6.2 Sampling Procedures

The purpose of taking samples is to record the properties of the whole volume of material from a small, analyzed portion of it. This is difficult to achieve with solids since industrial mixes in particular always present a distribution of grain sizes, shape or density and specific grains may be selectively sampled on account of the ingredients' specific motional behavior. Two basic 'golden' rules should be handled for sampling (see further Chap. 15):

1. A powder should be sampled when in motion.
2. The whole of the process stream should be taken for short increments of time in preference of the stream being taken for a prolonged period.

9.7 Equipment for Mixing of Solids

A wide variety of equipment is commercially available to suit a multiplicity of mixing tasks. In this overview mixers and devices for mixing solids are divided into four groups as follows: (1) mixed stockpiles; (2) bunker mixers, (3) rotating mixers or mixers with rotating tools, and (4) direct mixing of feeding streams.

9.7.1 Mixing Stockpiles

Many bulk goods, which are often stored in very large stockpiles, do not possess uniform material properties within these stockpiles. In the case of raw materials this may be caused by natural variations in deposits or in the case of primary material by variations between different production batches; e.g. in the iron and steel industry there are fluctuations in the ore and carbon content of the finished material. If these stockpiles are emptied in the "first-in-first-out" principle, material with a variance in its properties will find its way into the subsequent process and reduce its efficiency. To provide a uniform finished material a mix is obtained by following

Fig. 9.7 Recovery of the fine homogenized coal by system Chevron (Central Coking Plant Saar GmbH, Germany); width of the bridge scraper: 57.5 m; capacity 1200 t/h; (Courtesy PWH – Krupp Engineering)

a defined scheme for building up and emptying large stockpiles (Fig. 9.7). Such mixing processes are also called *homogenization*. As in any mixing process the volume of material is homogenized by moving portions of it relative to each other. A long stockpile is built up by a movable conveyor belt or other corresponding device travelling *lengthwise*. During loading the belt continuously travels up and down the whole length. In the strata thereby created a temporal record, is stored of the material's delivery. If the material is now systematically removed *crosswise* to these layers, each portion removed from the stockpile (Fig. 9.7) will contain material from all the strata and therefore from the times it was supplied. Since such bins are built up over days or weeks, mixed stockpiles reduce the degree of long term fluctuations in the material's properties.

9.7.2 Bunker and Silo Mixers

Bunker and silo mixers (Fig. 9.8) are sealed vessels, the biggest of which may likewise serve to homogenize large quantities of solids. They are operated batchwise, continuously or with partial recirculation of the mixture. Their sealed construction also enables material to be conditioned, e.g. humidified, granulated, dried or rendered inert, as well as mixed. In gravity mixers, granular material is simultaneously drawn off by a system of tubes at various heights and radial locations, brought together and mixed. Other types of construction use a central take-off tube into which the solids travel through openings arranged at various

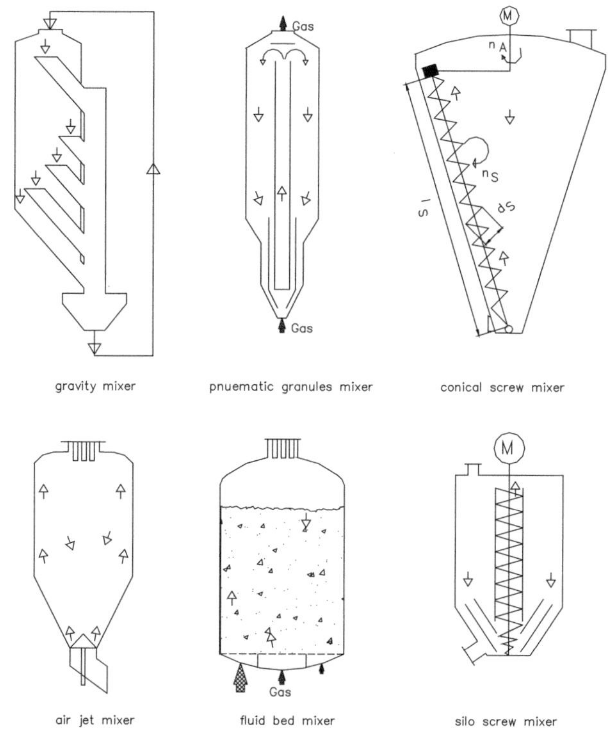

Fig. 9.8 Classification of bunker or silo mixers following [13]

heights up this pipe. If the quality of the mix does not meet requirements, the withdrawn material is fed back into the bunker [4]. In this fashion the bunker's entire contents are recirculated several times and thus homogenized. The material drawn off is in most cases carried to the top of the bunker by air pressure (using an external circulation system). Gravity mixers are designed for free-flowing powders and are offered in volumes ranging between 5 and 200 m^3. The specific energy consumption, i.e. the energy input per product mass, is very low at under 1–3 kWh/t. Silo screw mixers are silos with a special funnel mixer at their outlet and are grouped with the gravity mixers. A concentric double cone gives a different residence time period for the material in the inner and outer cones, inducing remixing. Such mixers are available for quantities of material between 3 and 100 m^3. In the case of granulate mixers, material from various areas of the vessel is brought together in its lower section and then carried upwards by air pressure in a central pipe (using an internal circulating system) where the solids are separated from the gas and at the same time distributed on the surface. Design sizes reach up to 600 m^3 and the specific energy input, like that of gravity mixers, is low. The rotating screw of a **conical screw mixer** transports the material upwards from the bottom. This screw is at the same time driven along the wall of the vessel by a swiveling arm. This type of mixer processes also both pastes and cohesive powders. The solids at the container wall are continuously replaced by the action of the screw so that the mix can be indirectly heated or cooled through the container's outer wall. It is also used for granulation and drying. Mixers of this design are offered in capacities of between 25 l and 60 m^3.

In **blast air** or **air jet mixers**, air is blown in through jets arranged around the circumference of a mixing head placed in the bottom of the vessel. The specific air consumption is 10–30 Nm^3/t and the largest mixers have a capacity of 100 m^3. If a fluid flowing through a bed of particles against the force of gravity reaches a critical speed (minimum fluidization velocity), the particles become suspended or fluidized by the fluid.

Through increased particle mobility, **fluidized beds** possess excellent mix properties for solids in both a vertical and radial axis. In **circulating fluidized beds** often used in reaction processes, this is combined with elevated heat transfer and material circulation as a result of the high relative velocities of the gas and solids. Lower fluidizing speeds to limit air consumption are generally used if the fluidized bed only serves the purpose of mixing. Furthermore, differing volumes of air are fed to the air permeable segments installed in the container's floor which serve to distribute air. The largest fluidized bed mixers as used in cement making reach a capacity of 10^4 m^3. The material must be fluidizable, i.e. free flowing (with a particle size greater than 50 μm) and dry. The specific power input lies between 1 and 2 kWh/t, but air consumption rises sharply in the case of particle sizes above 500 μm. **Fluidized bed granulators** utilize the mixing properties of fluidization for granulation, atomized fluid distribution, and drying.

9.7.3 Rotating Mixers or Mixers with Rotating Component

Figure 9.9 shows four categories of mixers where the mix is agitated by rotating the whole unit or where movement in the mix is produced by rotating components built into the apparatus. These mixers are classified according to their Froude number (Fr):

$$Fr = \frac{r\omega^2}{g} = \frac{r{n_{rev}}^2 4\pi^2}{g} \tag{9.12}$$

Here r denotes the mixer's radius or that of the mixer's agitators, g the gravitational acceleration and ω the angular velocity. The Froude number therefore represents a dimensionless rotating frequency. The Froude number is the relationship between centrifugal and gravitational acceleration. No material properties are accounted for in the Froude number: Subject to this limitation, a distinction is drawn in Fig. 9.9 between Fr <1, Fr >1 and Fr $\gg 1$.

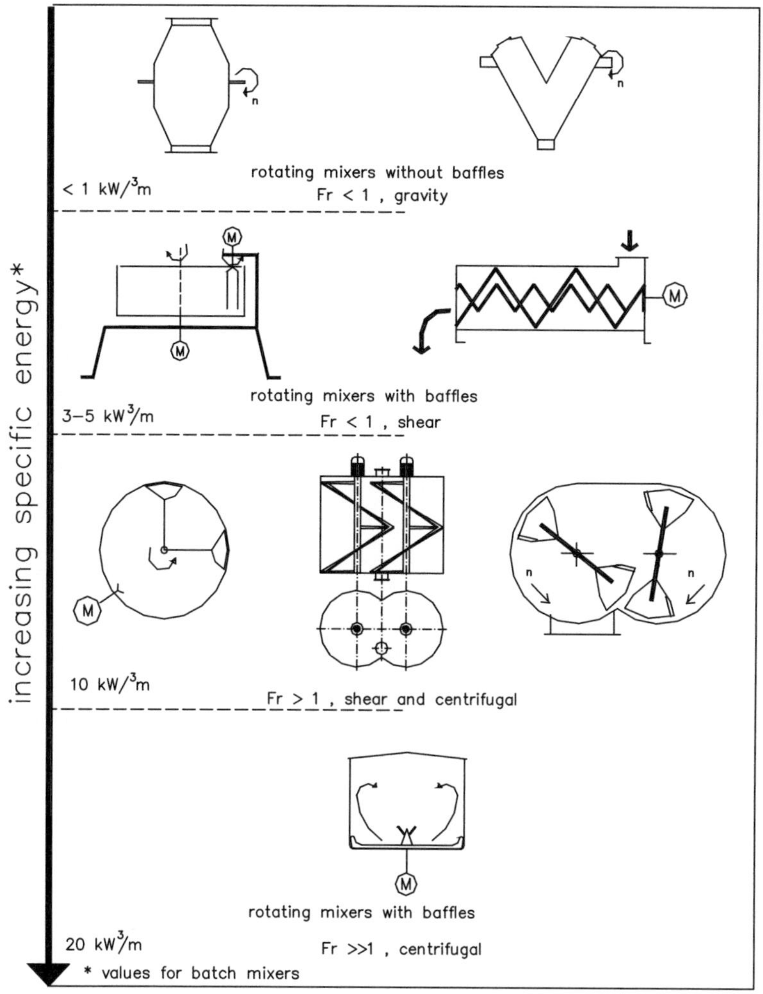

Fig. 9.9 Classification of mixers – movement of material by rotating agitators or revolving containers. * Specific output data applies to batch mixers [13]

Free fall mixers are only suitable for free-flowing solids. Familiar examples of free fall units are drum mixers and V-blenders. However, as the solids are free-flowing, demixing and segregation may also occur, leading to separation of the ingredients. Since drums are also used in related processes such as rotary tubular kilns or granulating drums for solids, these processes may also be prone to size segregation. In some cases, this may even be intentional, as for example with rotating disc granulators common in iron ore processing. Despite these risks of segregation, mixers without built in agitators are particularly widely used in the pharmaceuticals and foodstuffs industries since they can be cleaned very thoroughly.

Asymmetrically moved mixers in which for example a cylinder is tilted obliquely to the main axis, turning over the mix, also belong in the free fall category, e.g. in fertilizer drum granulation processes. Mixing is done gently. Because of the material's distance from the central axis, high torques have to be applied by the drive motor and these moments have to be supported by the mixer's bearings and bed. Units with a capacity of 5000 l are offered. There are also mixers operating range Fr <1, where the work of moving the mix is undertaken by rotating agitators. The particles of solids are displaced relative to each other by agitators inside the mixer. This design is suitable for both cohesive, moist products and those which are free flowing.

Examples of **displacement mixers** are ribbon blenders or paddle mixers. Because of their low rpm the load on the machine is slight but the mixing process is relatively slow. The specific energy input is low and lies under 5 kW/m^3.

Ploughshear and centrifugal mixers operate in a range with Fr >1. The consequence is that at least in the vicinity of the outer edge of the agitator the centrifugal forces exceed that of gravity and the particles are spun off. Thus, instead of a pushing motion there is a flying one. This accelerates the mixing process both radially and axially. If the ingredients still need to be de-agglomerated, high speed cutters are brought into the mixing space to de-agglomerate the mix by impact. At very high Froude number ranges (Fr >7) there is a sharp increase in the shear forces acting on the mix. The impact load is large and sufficient to heats the product as a result of dissipated energy. The heat is caused by friction between the mixer's tools and the solids as well as by friction amongst the solids' particles. As well as simple mixing, here the mixer's task is often de-agglomeration, agglomeration, moistening and sintering. Such mixers are especially used for producing plastics and in pharmaceutical industry for granulation.

9.7.4 Mixing by Feeding

Direct mixing of feed streams represents a continuous mixing process (Fig. 9.10). The solids are blended by metering in each ingredient and bringing these streams of solids together locally. There is no axial mixing (transverse or back mixing) or as such it is very low, with the result that the quality of the metering determines the mix's homogeneity. Metered feeder units should therefore ideally be used, preferably operated **gravimetrically** with appropriate feedback control of weight loss. According to the requirements of the case in question, mixing is also required obliquely to the direction of travel. If the ingredients are brought together in a perpendicular fall, this is achieved by their merging together. If this oblique mixing is not sufficient, **static mixers** can be used for free flowing powders or granules where, for example, the stream of solids is repeatedly divided up and brought back together by baffles as it drops down a tube. The energy input into the mixer is very low but such systems need sufficient height to achieve mix quality.

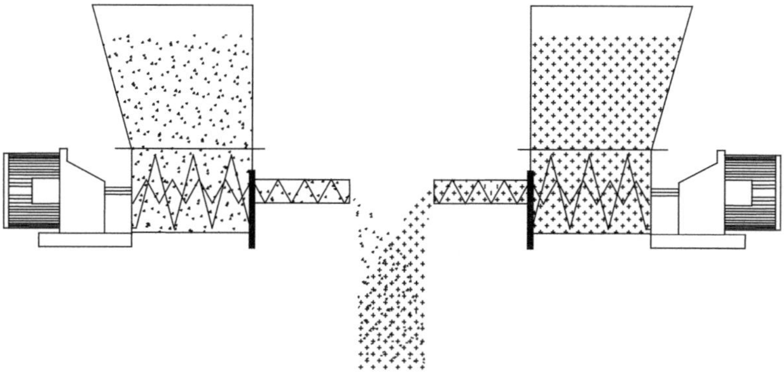

Fig. 9.10 Direct mixing by merging of feeder streams

It was shown that the efficiency for radial mixing depends on the gas phase as well [5]. At best they operate with low volume concentration and for particles in the size range between 20 and 200 μm. Static mixers have been used for very abrasive free-flow materials like Silicon Carbide. Since any rotating equipment is avoided inside static mixers, abrasion is limited. As it will be shown below mixture quality is dependent upon feed consistency and residence time within the static mixer. Since the latter is very short in static mixers (seconds or fractions of a second), short time feeding precision has to be very high for achieving high quality mix.

9.8 Designing Solids Mixing Processes

9.8.1 Goal and Task Formulation

An essential prerequisite for the efficient design of a mixing process is a clear, exact and comprehensive formulation of the task and objective. Applying Table 9.1 as a check list guarantees a systematic formulation of the *mixing task* along with the major *formative conditions*. Priority *objectives* covering the economic requirements, quality targets and operating conditions have to be met when engineering a mixing system. Besides a definition of the stipulated quality of the mix and an average production throughput (minimum or maximum) the *quality target* can also include additional physical (moisture, grain size, temperature) and chemical properties required of the mixed product. Furthermore, the general principles of quality assurance frequently demand production documentation. This means that material batches must be coded, mixture recipes recorded and the flow of materials in and out balanced out against their inventories and consumption. Clearly *formative economic conditions* such as investment, maintenance requirements and utilization of existing space often determine the actual technical features of a design when it is put into practice. Specifications arising from the mixing system's operation are

Table 9.1 Check list for formulating a mixing task

A	**Mix recipes (mixture composition)**
	Number and designation of the recipes
	The preparation's composition (the ingredients' percentages and margins of accuracy to be observed, particularly in the case of low dosage ingredients)
	The percentage of each recipe as part of the total production output
	The frequency with which the recipe is changed and any desired sequence
	Cleaning operations when a recipe is changed (dry, wet, cleaning in place CIP)
	Sampling and analyses
B	**Ingredients**
	Designation
	Origin, supplier, packaging
	Bulk density, solids density
	Grain size (grain size distribution) and shape
	Flow properties, gradient
	Abrasiveness
	Moistness (damp, hygroscopic, dry)
	Temperature, sensitivity to thermal stress
	Sensitivity to mechanical stress (crushing, abrasion, fracture)
C	**Product (mixture)**
	Mix quality
	Bulk density
	Fluidizability (air take-up during mixing)
	Tendency to segregation
	The mix's flow properties
	Agglomeration, de-agglomeration required
D	**The mixer performance**
	Mix performance: production volume per unit of production (average, minimum, maximum)
For batch mixers:	
	Batch mix size (final volume after mixing)
	Start-up filling level
	The filled mixer's idle time
For continuous mixers:	
	The production volume with an unchanged recipe
	Feed/mix output tolerance range
E	**Integrating the mixers into the system**
	Material flow diagram (average, maximum and minimum figures)
	The ingredients' in and outflow
	Spatial requirements, height, layout
	The mixture's usage
	Storing, feeding and weighing devices
	The type of process inspection, process control, storage and data exchange
	Safety requirements

(continued)

Table 9.1 (continued)

F	**Mixer design**
	Raw material, surfaces and the inflow and outflow configuration
	Heating, cooling, inertizing, pressurization, vacuum
	The addition of liquid into the mixer
	De-agglomeration
	Current, steam and water connections, adjutants, types of protection, protection against explosion
G	**Formative economic conditions**
	Investment costs
	Maintenance, running and staff costs
	Profitability

grouped together under *formative operating conditions*. These set the requirements on, for example:

- Staff numbers and training
- Process monitoring, process management system design and the degree of automation
- Operating, cleaning, maintenance
- Safety, dust, explosion, emission protection and the alarm system.

Sometimes, raw material costs exceed the processing cost by far; or manufacturing contributes a negligible part of the overall cost, if e.g. the marketing and R&D determine the manufacturing cost of a newly patented pharmaceutical product.

9.8.2 The Choice: Mixing with Batch or Continuous Mixers

Mixing processes can be designed as batch or continuous process. Table 9.2 gives a detailed comparison of discontinuous and continuous mixing processes, to help guide the selection of the mixing method.

9.8.3 Batch Mixing

Batch or discontinuous mixing is characterized by the fact that the mixer is filled with the ingredients and after a certain **mixing time**, the mixture is discharged. The feeding (or filling), mixing and discharging operations are performed *one after the other*. Batch processing presents advantages for small quantities of material because of its lower investment costs and greater flexibility. Batch mixers are used even when very large volumes of material are being homogenized since continuous mixers are limited by their lower volume. However, in the batch mixer's very flexibility lies the danger that it is not being optimally utilized. For example,

Table 9.2 Comparison of discontinuous and continuous mixing processes

Implementation data	Discontinuous	Continuous
Number of ingredients	As many as wanted	2–10. Any more ingredients are usually combined in a premix.
Frequency with which the recipe is changed	Several times per hour	A recipe must remain unchanged for several hours
Cleaning frequency or idle time	Several times a day	Once a day or less
Production output, throughput	Any rate	More than 100 kg/h. Exception: feeding laboratory extrusions
Risk of separation	Present, therefore there must be short transportation paths, few intermediate silos	Low risk when the material is taken directly to the next processing stage or directly drawn off
Spatial requirement	Large amount of space and intermediate silos required for machines with a throughput greater than 5000 kg/h	Low spatial requirement even for machines with a high throughput
Requirements placed on the equipment	Simple feeding but high demands on the mixer	Accurate continuous feeding (feeding scales necessary) but low demands on the mixer
Safety	Steps have to be taken in the case of materials with a risk of explosion	The small quantities of material present during processing have a low potential risk, which simplifies safety design
Automation	Variable degree of automation	Contained in the processing

overmixing can occur, whereby the product could be damaged and the process's effectiveness suffers.

9.8.4 Feeding and Weighing Equipment for a Batch Mixing Process

The number of mix cycles multiplied by the usable mixer capacity gives the set mixture output per hour. The mix cycle consists of the filling, mixing, discharge and idle times (Fig. 9.11). To this is added in special cases the time taken for sampling and analysis and that for associated processes such as de-agglomeration and granulation. The capacity (throughput rate) of a batch mixing process having a mixture charge with a mass M_{batch} gives the following as shown in Eq. (9.13):

Fig. 9.11 Classical automated batch mixing installation. The components are stored in small silos shown at the top of picture. The materials are extracted from these hoppers into a downstream weighing hopper according to the recipe. Once all component are fed into this weighing hopper, a valve is opened and the exact batch falls into the downstream batch mixer (Courtesy Gericke AG)

$$\dot{m} = \frac{M_{batch}}{t_f + t_m + t_d + t_i} \quad \left[\frac{kg}{s}\right] \tag{9.13}$$

The mixing time t_m depends on the selected mixer design and size, the filling time t_f on the system's configuration whilst the discharge time t_d depends on both the mixer's design and the system's layout. The idle time is t_i. The choice of feed and weighing devices is determined by the number of ingredients, their mass and proportions, the throughput volume, the stocking and mode of delivery, the spatial circumstances, degree of automation etc. In the simplest case the ingredients are manually weighed into the mixer. In some cases, *sandwiching* of specific ingredients may be desirable, i.e. staged delivery of multiple layers of key ingredients between other excipients. Where there are higher requirements in respect of accuracy, safety and recording, a hopper scale represents a simple device for weighing and releasing the components into the mixing equipment (Fig. 9.12).

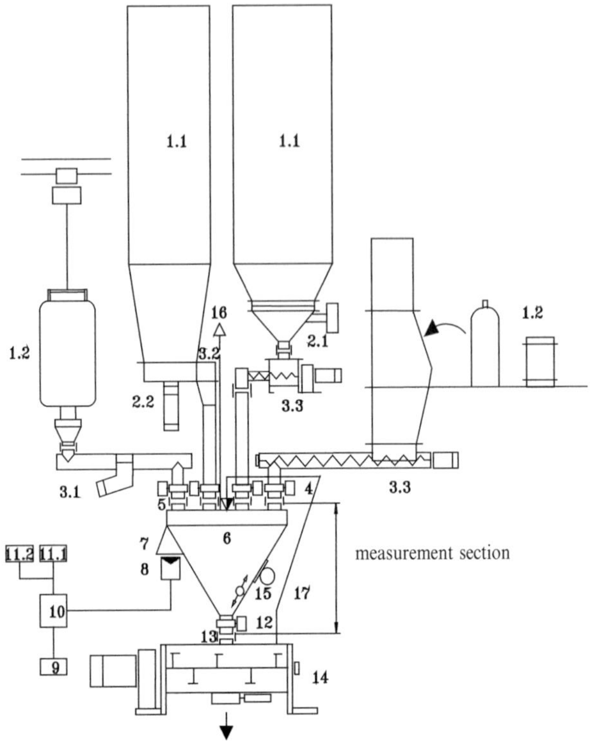

Fig. 9.12 Weighing hopper with additive weighing for feeding a batch mixer; *1.1* storage silos; *1.2* big bag, bag, drum; *2.1–2.2* dischargers; *3.1–3.3* feeder units; *4* cut-off; *5* flexible connections; *6* weighing hopper; *7* support for gravity force; *8* gravity operated sensor (load cell); *9* set point; *10* weighing analysis and regulation; *11.1* measured value indicator or output; *11.2* recorder (printer); *12* cut-off; *13* flexible connection; *14* mixer; *15* discharger; *16* dust extraction and weighing hopper ventilation; *17* mixer ventilation

Besides speed and precision of batch mixing installation, adequate hygienic design (Fig. 9.13) is required in food and infant formula installations.

9.8.5 Continuous Mixing

In a **continuous mixing** process (compare Figs. 9.14 and 9.15) the ingredients are continuously fed into the mixer, then mixed and prepared for the next processing stage. The operations of feeding, mixing and discharging follow each other locally but occur simultaneously. In continuous mixing the weighing and filling of a batch mixer are replaced by the ingredients' controlled continuous addition. The blending time in a continuous mixer is in fact the material's **residence time**, which is determined by the feed rate to the mixer. Losses of product during start up or

Fig. 9.13 Batch mixing installation for nutritional products. Hygienic design principles have been applied to minimize the risk of cross contamination between batches (Courtesy Gericke AG)

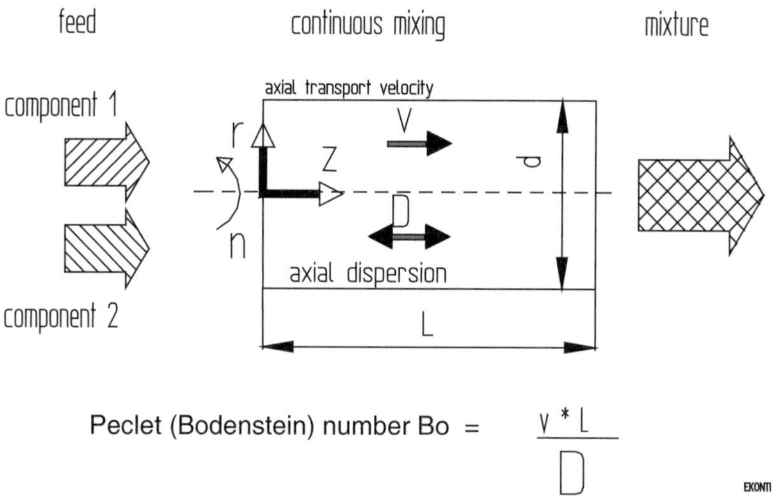

Fig. 9.14 Continuous mixing of two ingredients. Axial mixing or dispersion shows up as well as residence time distribution of the product inside the mixer

shut down added to this lower degree of flexibility come as further disadvantages of the continuous process. Yet it possesses considerable advances over batch processing both in financial terms and in respect of process control. Even high-throughput continuous mixers are compact. A smaller volume scale provides short mixing paths and ease of mixing. When integrated into a continuous production system, a continuous mixing process saves on reservoirs or silos and automating the

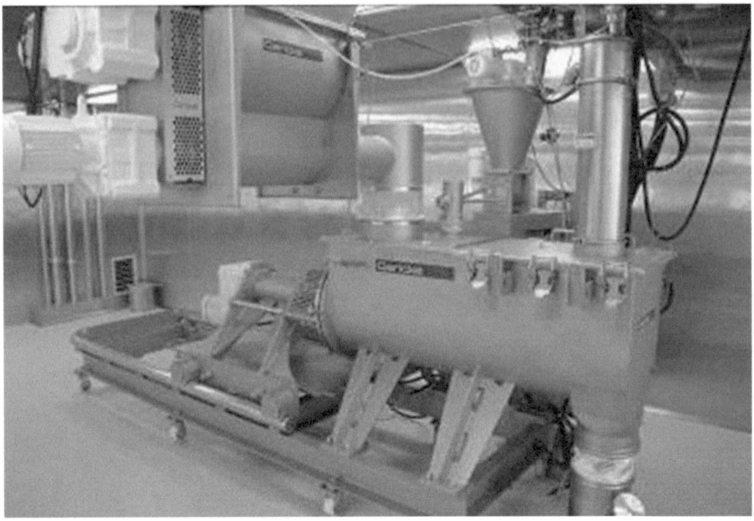

Fig. 9.15 Continuous mixing for the production of milk powder mixers: Continuous gravimetric solids feeder (loss-in-weight feeding) supply at constant rate the components (base powder and micro ingredients) directly to the continuous mixer (Courtesy Gericke AG)

course of the process is simplified. In the case of dangerous products or base materials there is less potential risk with a continuous process since only a small quantity of material accumulates in the mixer. Segregation can be limited in a continuous mixer by its smaller required scale. A continuous mixer, which on account of its compact construction can be positioned before the next station in the processing chain, guarantees that a mix of a higher quality will in fact be made available to that next stage of the process, with smaller material handling distances. Due to axial dispersion D the product is exposed to a distributed mixing time, the residence time distribution. However, due to the much smaller dimension, the stress pattern of the product is far more uniform than in large batch blender.

9.8.6 Axial Dispersion, Residence Time Distribution and Variance Reduction Ratio

The continuous mixer has principally two tasks (Fig. 9.14): (1) The ingredients, which in an extreme case arrive in the mixer side by side, have to be **radially mixed** (r). In this case radial means lateral to the direction of the material's conveyance into the mixer. (2) If in addition there are large feed rate fluctuations or the ingredients are themselves unhomogenized, the mixer must also minimize any differences in concentration in an axial direction (z), i.e. in the direction of the material's conveyance, or the mixture must be **axially mixed** as well.

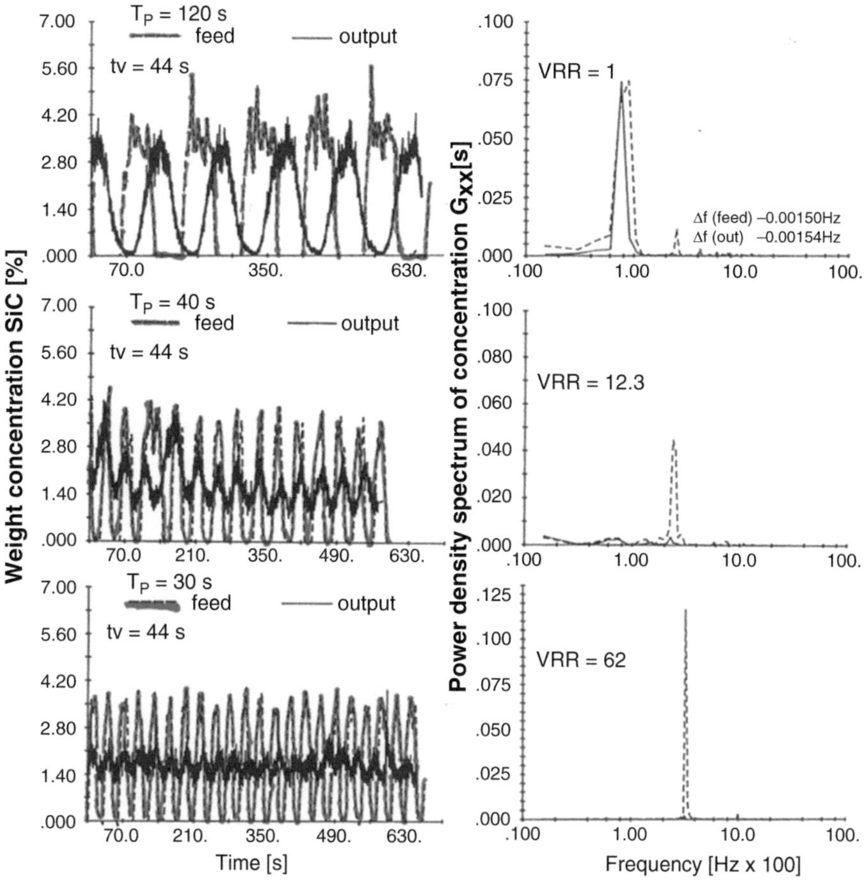

Fig. 9.16 Dampening of feed fluctuation in a continuous mixer – the Variance Reduction Ratio (*VRR*) describes the efficiency of continuous mixing processes; it compares the variances in concentration of inlet and outlet. Tracer-feed oscillating with different periods T_p, main component feed at constant rate (20 g/s), mean residence time in the continuous mixer $t_v = 44$ s. (**a**) Variation in time of SiC concentration: ***dotted line*** at the entrance of the continuous mixer, ***bold line*** at the outlet of the continuous mixer. (**b**) Power density spectrum of SiC concentration

If a mixer only has to perform its task radially, it can have a very compact structure, since slim line mixers with a high rpm very quickly equalize concentrations radially over short mixing paths. Feed fluctuations (see Fig. 9.16) are damped by the residence time distribution of the material inside the mixer [20]. The residence time distribution describes the degree of axial dispersion occurring in the mixer. The Bodenstein (= Péclet) number Bo (Fig. 9.14) characterizes the ratio of axial transport velocity and axial dispersion coefficient *D*. The capability to reduce incoming fluctuations (thus variance) inside continuous mixers depends on the ratio of period-of-entrance fluctuation to the mean residence time as well as the residence time distribution (Fig. 9.16). Besides the quality of a continuously produced mix, it is also of great interest to what extent by axial mixing a continuous mixer can remove inconsistencies in the ingredients being fed in or fluctuations in the feeding. Dankwerts [2] put forward the term "variance reduction ratio" or VRR to describe this extent. It is defined as the ratio of the variances at the mixer's inlet and outlet:

$$VRR = \frac{\sigma_{in}^2}{\sigma_{out}^2} \tag{9.14}$$

This concept developed for liquid mixer was extended to solid mixers [21]. Thus the variance reduction ratio compares the mix quality at the machine's inlet and outlet and thereby distinguishes the efficiency of a continuous mixer.

Besides the number of ingredients in the mix, a decisive feature in selecting the process is the individual component's flow rates. Since the feed's constancy can only be maintained with a limited degree of accuracy at continuous feeding rates below 300 g/h, ingredients with low flow rates necessitate a premixing operation. There is an increasing trend towards continuous mixing installations. Widely used are continuous processes in plastics industry, detergents and foodstuffs. Although less common, pharmaceutical processes utilizing continuous mixing are growing in appeal due to the small volume of the apparatus. The US Food and Drug Administration for example has promoted a Process Analytical Technology (PAT) Initiative with the objective of facilitating continuous processing to improve efficiency and manage variability [1, 11, 27].

High Variance Reduction Ratios are achieved if the oscillation period of the tracer feed T_p is small compared to the mean residence time in the mixer (Fig. 9.17).

9.9 Protective Measures for Avoiding Dust Explosions in Mixing Installations

Organic and other powders that are sensitive to oxidation and have a particle size of <300 μm can be explosive if they are distributed in the atmosphere within a specified concentration range (between lower and upper explosion limit). Then, protective measures against the occurrence of explosions are necessary [10]. The can be divided into primary and secondary measures.

Primary Measures

It has to be assured that no sources of ignition are present in the equipment. Ignition is caused by sparks produced mechanically, by heat produced from friction, by static charges, by electrical sparks, by too high a temperature in the material or in parts of the equipment and by chemical reactions. Therefore mixers are designed in a safe manner to exclude contact of e.g. paddles with the housing, by limiting the circumferential velocity of rotating components in general to less than 1 m/s, or are equipped with temperatures sensors in the area of drive or shaft sealing.

Fig. 9.17 Continuous feeding and mixing installation for pharmaceutical ingredients. These elements are used for continuous manufacturing production upstream direct compact, dry or wet granulation and hot melt extrusion (Courtesy Gericke AG)

Inertization serves to avoid a potentially explosive atmosphere. Atmospheric oxygen can be replaced by a protective inert gas, mainly nitrogen, so that the reduced O_2 content no longer enables any combustion. With many dusts a reduction of the oxygen concentration to below 10 % by volume is sufficient. Monitoring the oxygen level in the equipment is necessary.

Secondary Measures

An explosion resistant mixer, silo and container design is a possible secondary measure. There are two types of construction, pressure proof or pressure shock proof.

9.10 Definitions, Abbreviations and Symbols

Bodenstein number	dimensionless ratio of convective to diffusive transport forces
Froude number	dimensionless ratio of inertial to centrifugal (or gravitational) forces
Péclet number	see Bodenstein number
Variance Reduction Ratio	dimensionless ratio of the variances in inlet and outlet concentration of a mixer

Bo	Bodenstein number (compare Fig. 9.14)
Fr	Froude number (compare Eq. 9.12)
Pe	Péclet number (compare Fig. 9.14)

RSD	Relative standard deviation (compare Eq. 9.2)
VRR	Variance reduction ratio (compare Fig. 9.14)

d	mixer diameter
d_{p50}	median particle diameter, represents the 50 % point in the cumulative undersized particle size distribution
d_{max}	maximal particle diameter
D	axial coefficient of dispersion
E_{Mix}	mixing energy
g	gravitational acceleration
H	height of the fluidized bed
L	mixer length
m	the exponent of the power distribution used for characterizing the cumulative screen size distribution
m_p, m_q	average particle weight of the two components p and q in the mixture
M_{batch}	mass of a batch
M_s	mass of a sample
n	number of samples randomly taken in the population
n_p	number of particles in a sample
n_{rev}	rotational frequency
N_g	number of samples of the total population
p	fractional concentration tracer component in the population
p_g	fractional mass of the coarse ingredient
P	power
q	1-p
r	mixer radius
S	empirical standard deviation
S^2	random sample variance
t, t'	time
t_v	mean residence time
t_f,t_m,t_e, t_i	filling, mixing, discharging and idle time
t^*	mixing time
T_p	feed fluctuation period
v	axial velocity
$W\{\}$	probability
x	concentration of the tracer component
x_i	concentration in the 'i'-th sample
$<x>$	mean concentration
ρ_{bulk}	bulk density
ρ_s	density of the solids
σ_p, σ_q	standard deviation of the particle weight for the two ingredients in the mix
σ^2	variance

σ^2_z variance of a random mix
$\Phi(\chi^2)$ cumulative function of Chi-squared distribution
χ^2 chi-squared distribution
$\chi^2 l;\chi^2 u$ lower (l) and upper (u) limit of a two-sided confidence interval in a chi-squared distribution
ω angular velocity

References

1. Berthiaux, H. et al.: Continuous Mixing of Pharmaceutical Powder Mixtures, 5th World Congress on Particle Technology. http://www.fda.gov/cder/ops/Pat.htm (2006)
2. Dankwerts, P.V.: The definition and measurement of some characteristics of mixtures. Appl. Sci. Res. **3**, 279 (1952)
3. Egermann, H., Orr, N.A.: Comments on the paper "Recent developments in solids mixing" by Fan, L.T: et al. Powder Technol. **68**, 195–196 (1991)
4. Eichler, P., Dau, G.: Geometry and mixing of gravity discharge silo mixers, First European Congress on Chemical Engineering, Florence, Italy, Proceedings, vol. 2, pp. 971–974 (1997)
5. Eichstädt, O.: Continuous Mixing of Fine Particles within Fluid Dynamic Vertical Tube Mixers, Dissertation, in German, ETH-Zurich (1997)
6. Fan, L.T., Chen, Y., Lai, F.S.: Recent developments in solids mixing. Powder Technol. **61**, 255–287 (1990)
7. Harnby, N., Edwards, M.F., Nienow, A.W. (eds.): Mixing in the Process Industries, 2nd edn. Butterworth- Heinemann, ISBN 0 7506 1110 3 (1992).
8. Kaye, B.H.: Powder Mixing, Chapman & Hall, London, ISBN 0-412-40340-4 (1997).
9. Lacey, P.M.C.: The mixing of solid particles. Trans. Instn. Chem. Engrs. **21**, 53–59 (1943)
10. Lemkowitz, S.M., Pasman, H.J.: Chapter 3 'assessment and control of fire and explosion hazards and risks of particulates particle technology series. In: Merkus, H.G., Meesters, G.J.M. (eds.) Particulate Products – Tailoring Properties for Optimal Performance, vol. 19. Springer, New York (2014)
11. Llusa, M., Muzzio, F.: The effect of shear mixing on the blending of cohesive lubricants and drugs. Pharm. Technol. **29**, 26 (2005)
12. Merkus, H.G.: Particle size measurements. Particle Technology Series, vol. 19. Springer, New York (2009). ISBN 978-1-4020-9015-8
13. Müller, W.: Methoden und derzeitiger Kenntnisstand für Auslegungen beim Mischen von Feststoffen (Methods and the current state of the art in solids mixing configurations). Chem. Eng. **53**, 831–844 (1981)
14. Niemöller, A.: Conformity test for evaluation of near infrared data. Proceedings of the Internationa Meeting on Pharmaceutics, Biopharmaceutics and Pharmaceutical Technology, Nuremberg, 15–18 March 2004
15. Raasch, J., Sommer, K.: The application of statistical test procedures in the field of mixing technology, in German. Chem. Eng. **62**(1), 17–22 (1990)
16. Sommer, K.: How to compare the mixing properties of solids mixers (in German). Prep. Technol. **5**, 266–269 (1982)
17. Sommer, K.: Sampling of Powders and Bulk Materials, e.g. p. 164. Springer, Berlin (1986)
18. Sommer, K.: Mixing of solids. In: Ulmann's Encyclopaedia of Industrial Chemistry, vol. B4, Chapter 27, VCH Publishers, Weinheim (1992)

19. Stange, K.: Die Mischgüte einer Zufallmischung als Grundlage zur Beurteilung von Mischversuchen (The mix quality of a random mix as the basis for evaluating mixing trials). Chem. Eng. **26**(6), 331–337 (1954)
20. Weinekötter, R., Davies R., Steichen, J.C.: Determination of the degree of mixing and the degree of dispersion in concentrated suspensions. Proceedings of the Second World Congress Particle Technology, pp. 239–247, Kyoto, Japan (1990)
21. Weinekötter, R., Reh, L.: Continuous mixing of fine particles. Part. Part. Syst. Charact. **12**, 46–53 (1995)
22. Weinekötter, R., Gericke, H.: Mixing of Solids. Particle Technology Series, vol. 12. Kluwer Academic Publishers, Dordrecht (2000). ISBN 978-94-015-9580-3
23. Weinekötter, R.: Degree of Mixing and Precision for Continuous Mixing Processes Proceedings Partec, Nuremberg (2007)
24. Williams, J.C.: Fuel Soc. J. Univ. Sheffield **14**, 29 (1963)
25. Williams, J.C.: In: Uhl, V.W., Gray, J.B. (eds.) Mixing, Theory and Practice, vol. 3. Academic, Orlando (1986)
26. Williams, P., Norris, K. (eds.): Near-Infrared Technology in the Agricultural and Food Industries. American Association of Cereal Chemists, St. Paul (1987)
27. www.fda.gov/cder/ops/Pat.htm

Chapter 10
Particle Separations by Filtration and Sedimentation

Steve Tarleton and Richard Wakeman

Abstract This chapter presents an overview of methods for separating particles (and the like) from fluids by filtration and sedimentation. Focussing primarily on separations from liquids involving recovery of the dispersed 'solids' phase and thickening, features of the types of equipment available are provided together with their advantages, limitations etc., and indicative process models for predicting or characterising their performance. A section on membranes, which are frequently used for the separation of sub-micron (micrometre) particles and nano-sized materials, is also included.

10.1 Introduction

The separation of solids (particles) from fluids by filtration and sedimentation is practised in almost every industry sector including chemical, oil and gas, pharmaceutical, food, water, aeronautical and automotive as well as pollution control. The product may be the particles themselves, the liquid phase, a solute dissolved within the liquid, and sometimes all three components. Cake filtration and sedimentation are common methods for the handling and recovery of particles and this chapter concentrates on descriptions of these two processes.

S. Tarleton (✉)
Department of Chemical Engineering, Loughborough University, Loughborough, UK
e-mail: e.s.tarleton@lboro.ac.uk; http://www-staff.lboro.ac.uk/~cgest

R. Wakeman
The Well House, West Hill Road, West Hill, Ottery St Mary, Devon, UK
e-mail: richard@richardwakeman.co.uk

© Springer International Publishing Switzerland 2016
H.G. Merkus, G.M.H. Meesters (eds.), *Production, Handling and Characterization of Particulate Materials*, Particle Technology Series 25,
DOI 10.1007/978-3-319-20949-4_10

Fig. 10.1 A 92 m^2 roller discharge rotary vacuum filter (*left*, Courtesy of Dorr-Oliver Eimco) and a vertical leaf pressure vessel filter (*middle*, Courtesy of LFC). Micrograph of a typical woven filter medium (*right*, Courtesy of Sefar)

10.1.1 Cake Filtration Processes

Industrial filters are available in forms ranging from units capable of handling different filtration applications to those restricted to use with specific fluids and process conditions, from very small to very large scale, and for either continuous or batch operation. Continuous filters are essentially capable of carrying out filter cake formation, deliquoring, washing and discharge, sequentially without interrupting flow of the process feed. Conversely, batch filters can perform the same range of functions but need to be stopped in order to discharge the cake. The surfaces on which separation takes place, the filter media, may be orientated horizontally or vertically, and be either planar or cylindrical. Woven media are frequently used in cake filters, but there are many alternatives and the range of variants is considerable [see 18]. The driving force may be pressure, either a positive pressure or a vacuum, centrifugal or in a few cases just gravity. Photographs of some industrial filters and a typical filter medium are shown in Fig. 10.1.

The above and other factors have led to a bewildering choice in filter design, but even so the underlying principles of all cake filters are similar. In cake filtration a particulate deposit, the cake, accumulates on the surface of the semi-permeable filter medium whilst liquid passes through. Ideally, particles bridge over the pores in the medium and thus limit so called 'blinding' where pores can be blocked by particle ingress. Particle bridging is promoted by a raised feed concentration (see Fig. 10.2). After an initial period of deposition the cake itself starts to act as the filter medium whilst further particles are deposited. This accumulation continues until the pressure drop across the cake exceeds the maximum permitted by economic or technical considerations, or until the filtrate (liquid) flow rate falls to an unacceptable level. Provided that cakes are of sufficient thickness to discharge efficiently and any washing/gas deliquoring can be carried out to the required level without cracking, it is generally more productive to operate a filter with thinner cakes.

The most important factor in cake filtration is permeability of the filter cake. Fundamental understanding, as well as the development of useable process models, is conventionally based upon Darcy's Law which describes the flow of fluid through a porous medium such that

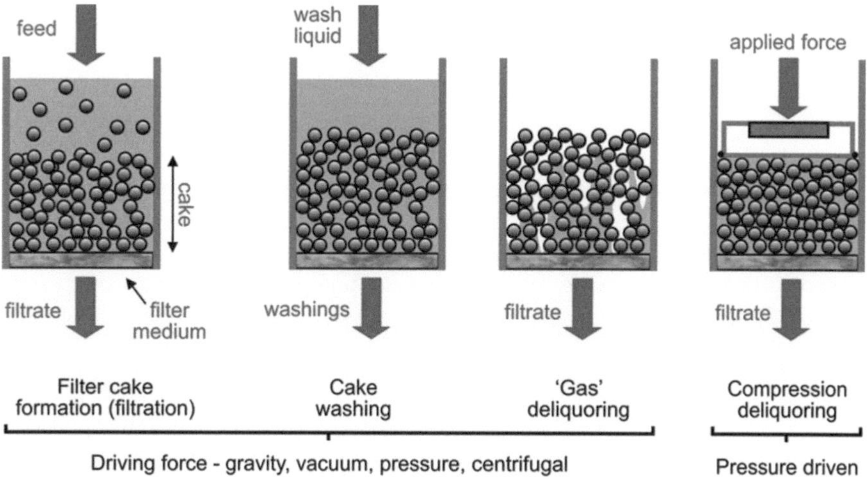

Fig. 10.2 The basic processes involved in deadend cake filters. As water is often the process liquid, the term 'dewatering' is often used interchangeably with deliquoring (Courtesy of Filtration Solutions, UK)

$$u = -\frac{k}{\mu}\frac{dp}{dz} \tag{10.1}$$

where dp is the dynamic (hydraulic) pressure difference across thickness dz of porous medium of permeability k, and u the filtrate flux (superficial velocity) of fluid with a viscosity μ flowing through the bed. Those factors which determine the permeability, the porosity of the filter cake (ε) and size of particles (x) in the cake, together with the particle size distribution and state of aggregation of the particles, are fundamental parameters. They dictate the ease with which any filtration process, and post-treatment process such as compression deliquoring (where pressure is directly applied to the cake to compress it and thereby improve its homogeneity), displacement washing (where clean liquid is passed through the cake to remove solute(s) and, for instance, 'clean' the constituent particles) and gas deliquoring (where gas displaces the liquid in the pores of the cake to 'dry' it), will be accomplished. The combination of sequences shown in Fig. 10.2, together with cake discharge, is known as the filter cycle.

In cake filtration, the cake permeability is most often interpreted via specific resistance (α)

$$k = \frac{1}{\rho_s(1-\varepsilon)\alpha} \tag{10.2}$$

where ρ_s is the true solids density. Important terms such as α, ε and x can be related through, for example, Eq. (10.2) and the Kozeny-Carman equation which is expressed in terms of specific surface (S_0) as

$$k = \frac{\varepsilon^3}{5S_0^2(1-\varepsilon)^2} \Rightarrow \alpha = \frac{180}{\rho_s x^2}\frac{1-\varepsilon}{\varepsilon^3} = \frac{180}{\rho_s x^2}\frac{C}{(1-C)^3} \tag{10.3}$$

where C is the solids concentration in the cake ($=1-\varepsilon$) and spherical particles are assumed ($S_0 = 6/x$). In reality the relationship between variables is complex and made more difficult because of filter cake compressibility which makes a separation more difficult. Here, the cake structure, and hence α and ε, are variant with pressure and generally variant with spatial position in the cake and with time. The consequence is that average values of cake resistance and porosity (or solids concentration) often need to be used in modelling. Despite the inherent complexities, the following general observations can be made:

- *Particle size and size distribution*. Greater rates of filtration can be expected with larger particles in the feed and narrower size distributions are usually favoured. Particles may vary in size from very fine or colloidal matter or molecular aggregates to coarse granular solids.
- *Solution environment and particle surface charge*. Particles less than about 50 μm have a tendency to aggregate and form more compressible filter cakes. Aggregation may arise as a result of the chemical composition of the suspension, or it may be induced/changed by the addition of chemical coagulants (e.g. acids, alkalis, metal salts) or flocculants (e.g. polyelectrolytes) prior to filtration; with chemical addition the particle charge is generally altered. The resulting aggregates are referred to as flocs which can range from being relatively compact and strong assemblies to fragile and 'stringy' collections of loosely bound particles depending on the chosen conditions, particle/liquid interactions etc. The effects of particle charge become more pronounced below 5–10 μm and tend to dominate over hydrodynamic effects below (say) 1 μm. Around the isoelectric point (IEP) of the suspension, i.e. the point of net zero surface charge, more rapid filter cake formation (lower cake resistance) and faster settling rates can be expected.
- *Particle shape*. The majority of particles are of irregular shape, but it is a property that is often difficult to control, and characterise/account for in filtration (and sedimentation) processes. Particle shape affects the volume and surface area of a particle, and hence the specific surface. In general, extremes of particle shape have undesirable effects in filtration.

10.1.2 Sedimentation Processes

Sedimentation is a term that is used to describe particle settling phenomena in suspensions, where particles or aggregates are suspended by hydrodynamic or particle-particle interaction forces with compression being absent. Initially, the particles or aggregates, which are most often more dense than the suspending liquid, settle with return (upward) flow of liquid between the downward moving particles. Eventually the aggregates come into close proximity of one another and

Fig. 10.3 A view of two thickeners showing the central driveheads which drive submerged rotating arms (*left*, Courtesy of Dorr-Oliver Eimco) and a vane decanter centrifuge with the motor drive and gearbox being visible at the right hand end of the unit (*right*, Courtesy of Mitsubishi Kakoki Kaisha)

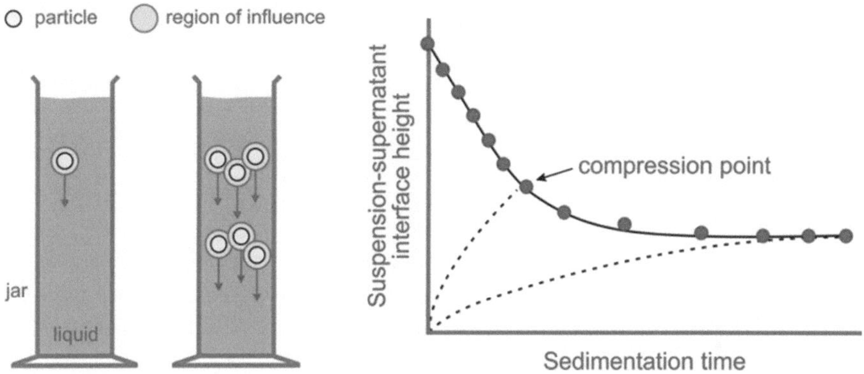

Fig. 10.4 Single particle and hindered settling (*left*) and the typical form of a batch settling curve (*right*) (Courtesy of Filtration Solutions, UK)

those at lower levels feel compressive forces due to the presence of those higher in the mixture; the settled material is known as sediment. Gravity driven sedimentation processes are frequently carried out industrially on a continuous basis at large scale in order to ideally separate the feed suspension into an overflow of clear liquid, the supernatant which is removed toward the top of the unit, and an underflow of more concentrated solids, the particles, that are removed in the form of a sludge/paste toward the bottom (see Figs. 10.3 and 10.4).

In order to increase the settling rate of smaller particles, the force applied to the particles must be increased. One way to do this is to use a centrifugal force. Whereas in a gravity force field particle motion is upwards or downwards, depending on whether the particle is less or more dense than its suspending liquid, in a centrifugal field the motion is radial through the liquid either inwards or outwards, again depending on relative densities. Centrifugal sedimentation can be performed industrially either batchwise or on a continuous basis and units generally occupy a smaller footprint than gravity driven variants.

Particle size, particle density and fluid viscosity are the primary factors to be considered in a sedimentation process, but suspension concentration and particle shape can also have a significant influence. Suspensions with particle diameters of the order of microns settle too slowly for most practical operations. So, to increase their settling rate, the particles are aggregated or flocculated into relatively larger particles known as flocs. Fundamental considerations of sedimentation are normally based upon Stokes Law which relates the drag force on a sphere (F_D) for a low Reynolds number (Re) to liquid properties and particle size such that

$$F_D = 3\pi\mu u x \tag{10.4}$$

where μ is the viscosity of the liquid and u the particle-liquid relative velocity. A single (discrete) spherical particle settling in a gravity field at low concentration is subjected primarily to drag and gravity forces and buoyancy, and a force balance gives:

$$\begin{array}{ccccccccc} m\dfrac{du}{dt} & = & mg & - & m\dfrac{\rho}{\rho_s}g & - & F_D \\ \left\{\begin{array}{l}\text{inertial}\\ \text{force}\end{array}\right\} & = & \left\{\begin{array}{l}\text{gravity}\\ \text{force}\end{array}\right\} & - & \{\text{buoyancy}\} & - & \left\{\begin{array}{l}\text{drag}\\ \text{force}\end{array}\right\} \end{array} \tag{10.5}$$

where m is the particle mass, g the acceleration due to gravity and t the time. Incorporating Stokes' law for F_D, integrating and considering longer times gives the terminal (or Stokes) settling velocity (u_t) which is achieved rapidly for small particles:

$$u_t = \frac{x^2(\rho_s - \rho_l)g}{18\mu} \tag{10.6}$$

where ρ_l is the liquid density. If the settling occurs in a centrifugal force field, such as in the spinning basket of a centrifuge, then the particle never reaches a terminal velocity (because the value of r continually increases) and its equation of motion can be written as

$$\frac{dr}{dt} = \frac{x^2(\rho_s - \rho_l)r\omega^2}{18\mu} = u_t\frac{r\omega^2}{g} \tag{10.7}$$

where r is the radius relative to the basket centreline and ω the angular velocity of the basket. That is, the instantaneous velocity of the particle is equal to the terminal velocity in a gravitational field increased by $r\omega^2/g$. The term $r\omega^2/g$ is known as the g-factor and is a basic measure of the separating power in an industrial centrifuge.

When the concentration of the feed suspension increases (say >1 % by volume), the particles are closer together and the motion of any single particle is usually affected by the motion of neighbouring particles. For most practical suspensions the settling rate declines with increasing concentration and the process is referred to as hindered settling. Although there are alternative approaches, for non-flocculated systems Richardson and Zaki [20] showed that the hindered settling velocity (u_h) can be equated to the product of u_t and ε^n, where ε is the voidage or porosity of the suspension and n can take a range of values dependent upon x, D (the diameter of the vessel in which the sedimentation is taking place) and Re_t (the Reynolds number based on u_t); by way of example, for $\mathrm{Re}_t < 0.2$, $n = 4.6 + 20x/D$.

10.1.3 Overview of the Chapter

The many forms of solid/liquid separator developed over the years encompass a wide range of variants which manufacturers claim to give productivity and/or cost benefits. Whilst it is not possible to describe all variants or aspects, Sects. 10.2, 10.3, 10.4, 10.5, 10.6, and 10.7 give a descriptive overview of exemplar equipment types with advantages and limitations highlighted. Typical values for filter sizes, operating parameters etc. are provided, however, the reader should be aware that exceptions are likely to exist. Capital costing equations are given where available and whilst these should be used with some caution, they do provide some basis for comparison; no attempt has been made to correct costings for inflation.

Tables are used to show a range of information, including the solids concentration and particle size found in a typical feed; n/a is used to indicate where information is not available. Process indices give relative values between 0 and 9 for cake dryness (and state), washing performance, liquid product clarity and particle breakage where 9 represents the best performance currently available; a '–' indicates that either a rating is not applicable or the equipment is not capable of performing the operation. For instance, the '1 S, 2, 5, 9' ratings shown in Table 10.6 for the circular basin thickener signify a wet solids discharge in the form of a slurry ('C' designates a cake and 'S' a slurry), poor washing performance, near average liquid product clarity and minimal breakage of particles/aggregates. These indices, and knowledge of typical solids concentration and particle size, can be used as an aid to equipment selection [8].

Figure 10.5 shows the broad classifications of filtration and sedimentation equipment.

In Sects. 10.2, 10.3, 10.4, and 10.5 the principal equations for process models are also presented. Whilst these models can facilitate equipment sizing and predictions of equipment performance in terms of the solids, liquid and solute throughputs, it should be recognised that their successful implementation often requires the experimental measurement of characterising parameters. Such empirical relationships are necessary as it is not currently possible to routinely predict the behaviour of suspension, sediment and cake properties from a knowledge of fundamental particle (e.g. size, shape, charge) and liquid (e.g. viscosity, pH) properties. The interested reader is directed to texts such as Tarleton and Wakeman [27] where full

Gravity filtration

Batch:
Single leaf Nutsche

Semi-continuous:
Sand bed

Continuous:
Gravity belt
Stationary screen
Vibrating screen

Vacuum filtration

Batch:	*Continuous:*
Multi-element leaf (Moore's)	Horizontal belt
Single leaf Nutsche	Horizontal rotary table
Single leaf tilting pan	Horizontal tilting pan
	Precoat rotary drum
	Rotary disc
	Rotary drum (bottom fed)
	Rotary drum (top fed)
	Rotary drum (internal fed)

Centrifugal filtration

Baffle ring centrifuge
Basket centrifuges (pendulum and peeler)
Cone screen centrifuges (slip discharge, vibratory/oscillatory, tumbling, worm screen)
Inverting bag centrifuge
Screen baffle centrifuge
Single and multi- stage pusher

Pressure filtration

Batch:	*Semi-continuous:*	*Continuous:*	*Variable volume:*
Multi-element candle	Bag	Belt press	Diaphragm filter press
Multi-element leaf	Cartridge	Duplex strainer	Expression (screw) press
Plate & frame press	Dead-end membrane	High shear crossflow	Horizontal element tube press
Precoat Nutsche and multi-element leaf	Fibre bed	Rotary disc	Vertical diaphragm filter press
Precoat plate & frame press	Low shear crossflow	Rotary drum	Vertical element tube press
Recessed plate filter press	Sand bed	Sand bed	
Sheet filter	Simplex strainer	Tower press	
Single leaf Nutsche			

Gravity sedimentation

Circular basin thickener	Blanket clarifier
Circular high capacity thickener	Circular clarifier
Deep cone thickener	Inclined plate clarifier
Lagoon thickener	Rectangular clarifier
Lamella separator	Vertical flow clarifier
Settling tank thickener	Helical screw classifier
	Hydraulic classifier
Flotation	Rake classifier

Centrifugal sedimentation

Disc stack centrifuge
Scroll decanter centrifuge
Single and multi- bowl basket centrifuges
Tubular bowl centrifuge

Circulating bed hydrocyclone
Reverse flow hydrocyclone

Fig. 10.5 Broad classifications that highlight some of the many forms of solid/liquid separator (not all are described in this chapter)

descriptions and worked examples of model implementations in filters are provided. See also Wakeman and Tarleton [32] for more details about filtration, washing, deliquoring and sedimentation fundamentals, and Wakeman and Tarleton [33] for details of scale-up procedures for filters, centrifuges and membrane separators.

10.2 Pressure Filters and Presses

Pressure filters/presses most often operate in a batchwise manner and use positive pressure above the semi-permeable separating surface(s) to remove liquid and retain solids in the form of cakes. They are used in a wide range of chemical and process industries for the separation of suspensions which contain finer particles that settle slowly and exhibit poor filterability, and/or suspensions that contain higher solids contents. Filtration pressures are typically in the range from 0 to 800 kPa, but can be higher, and usually provided by centrifugal or positive displacement pumps. Smaller units employ compressed gas as the driving medium. Many types can be fully automated to sequence cake formation, washing and deliquoring operations. Some filter processes allow for cake consolidation through the inclusion of flexible diaphragms and several pressure filters have been designed for semi-continuous and continuous operation.

Table 10.1 shows some typical characteristics whilst capital equipment costs for some filters can be estimated over the specified filter area ranges using equations adapted from Couper et al. [6]:

$$\begin{array}{ll}\text{Leaf filter}\,(A = 3 - 230\,\text{m}^2): & 4600A^{0.7}\ \$ \\ \text{Plate and frame filter press}\,(A = 1 - 100\,\text{m}^2): & 2070A^{0.55}\ \$\end{array} \tag{10.8}$$

10.2.1 Single Leaf (Nutsche)

Typical uses: Fully enclosed batch processing of a wide range of feeds requiring good solids washing.

The Nutsche is a versatile batch filter comprising of a cylindrical pressure vessel with a single planar filter medium (leaf) at the bottom (see Fig. 10.6). The feed suspension, which may be toxic, volatile or flammable, is introduced to the fully enclosed vessel and a constant pressure (typically up to 800 kPa but can be higher) is applied within the vessel to initiate downward filtration. Cake formation may be followed by deliquoring (by gas-blowing) and cake washing (by displacement or reslurry, where the latter involves a repeated cycle of clean wash liquid addition, cake redispersion and subsequent filtration to reform the cake). Smaller Nutsche filters utilise a manual cake discharge and are generally able to operate at higher pressures, whilst larger machines employ mechanical ploughs or rakes to discharge the cake either centrally or through a side port. Fully automated versions of the pressure Nutsche filter are readily available and most include cake smoothing devices to help reduce the problems of cake cracking. A range of filter media can be accommodated with filter areas up to a maximum of 30 m^2, although up to 15 m^2 is more normal.

10.2.2 Multi-element Leaf or Candle

Typical uses: Batch operations with solids forming lower compressibility or incompressible cakes.

Multi-element pressure filters find widespread use and comprise of a cylindrical vessel inside which many horizontal or vertical permeable elements covered by filter cloths are placed (see Figs. 10.1, 10.7, and 10.8). The process suspension, which may be toxic, volatile or flammable is pumped into the vessel at pressures

Table 10.1 Typical characteristics of some industrial pressure filters and filter presses

	Single leaf[a]	Multi- leaf or candle	Filter press	Diaphragm filter press	Tube press	Vertical diaphragm filter press	Belt press	Rotary pressure drum/disc
Operation mode	Batch	Batch	Batch	Batch	Batch	Batch[c]	Continuous	Continuous
Max. pressure (kPa)	800	600	1000	1600	10,000	1600	<500	700
Max. cake thickness (m)	0.5	<0.1	<0.05[b]	0.08	0.03	0.08	Varies	0.18/varies
Filter area (m^2)	0.1–15	2–300	5–2000	5–2000	1–3	2–234	1–75	Up to 40/170
Solids dryness index	6 C	5 C	6 C	8 C	8 C	8 C	8 C	5–6 C
Washing index	8	6–8	8	8	4	8	7	6/–
Liquid clarity index	8	8	8	8	7	8	7	6
Particle breakage index	8	8	8	7	7	7	7	8
Particle size in feed (μm)	1–200	0.5–100	1–100	1–200	1–200	1–200	1–200	1–100
Feed conc. (%w/w)	<1–20+	<1–20	<1–30+	0.3–30+	0.3–30+	0.2–30+	0.2–30+	5–30+
Installed power (kW)[d]	<20	n/a	<25	<35	10–30	20–140	<10	<30

[a]Vacuum and gravity driven variants also exist
[b]Can be up to 0.2 m with additional spacers
[c]Semi-continuous as cake discharge time is relatively short
[d]Can be higher (typically +15 kW) depending on compressed air/gas requirements.

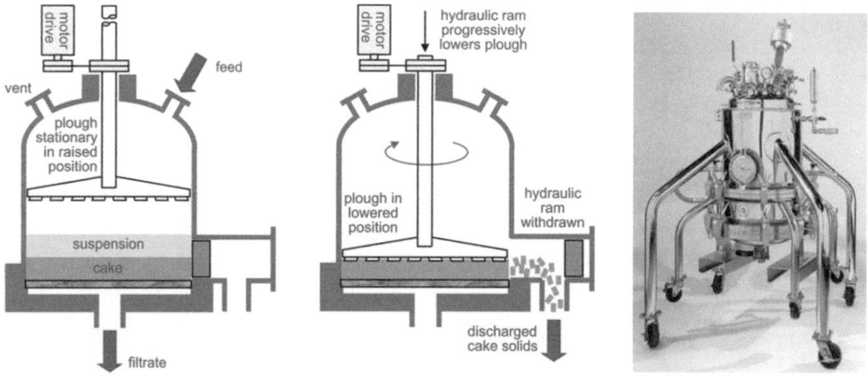

Fig. 10.6 Filtration (*left*) and side port cake discharge (*middle*) in a single leaf (Nutsche) filter (Courtesy of Filtration Solutions, UK). Photograph (*right*) shows a fully assembled 0.15 m^2 Nutsche filter (Courtesy of Pope Scientific). Units can be pressure, vacuum or gravity driven

typically up to 600 kPa, although this can be higher in special units. The positive pressure induces cakes to grow on the outer surfaces of the cloths and filtrate is transported away through the elements via a suitable manifold system. Either flat elements, in the form of square, circular or rectangular leaves, or tubular candles are used. These are spaced sufficiently far apart (up to 20 cm) to avoid the possibility of cakes touching on adjacent elements.

The pressure vessel can be jacketed for operation at elevated (or lowered) temperatures. Up to 150 °C is a typical limit, but this very much depends upon the integrity of the filter medium at the raised temperature. Filter leaves can be automatically extracted for cake discharge if adequate floor/height provisions are made. Relatively frequent cake discharge is normally required and solids are usually removed with the filter leaves in-situ either by vibration, rotating blades, centrifugal force (horizontal elements only) or liquid sluicing to give a wet discharge. Element precoating can be used to, for instance, help prevent particle ingress into the filter medium, and for more extreme duties metallic or ceramic filter elements can be employed.

Vertically mounted vessels: These filters contain either horizontal or vertical leaves with a normal maximum filter area of 120 m^2 (see Fig. 10.7); a specialist filter that is used in alumina processing has a significantly larger filtration area of 440 m^2. They utilise floor area economically but can require excessive height allowance, particularly when leaves need to be withdrawn vertically for cake discharge or cloth cleaning. Horizontal leaves are preferred when either washing is required, rapidly settling feeds are processed or intermittent operation is envisaged, however, the installed cost can be higher as filtration takes place only on the upper surfaces. Multi-element filters having vertical rectangular leaves are best suited to the processing of feeds with a particle settling velocity less than 3 cm s^{-1} but give relatively poor washing performance as cakes tend to prematurely fall off the filter leaves. With tubular candles, cakes form on the outer surfaces of the candle elements and this arrangement is most frequently used when washing is not required.

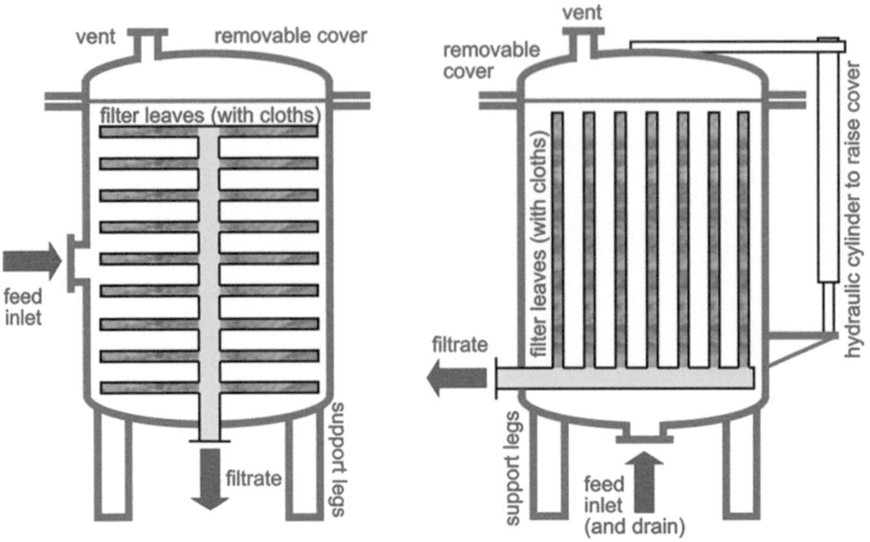

Fig. 10.7 Cross-sectional views of vertical vessel, multi-element pressure filters with horizontal (*left*) and vertical (*right*) leaves (Courtesy of Filtration Solutions, UK)

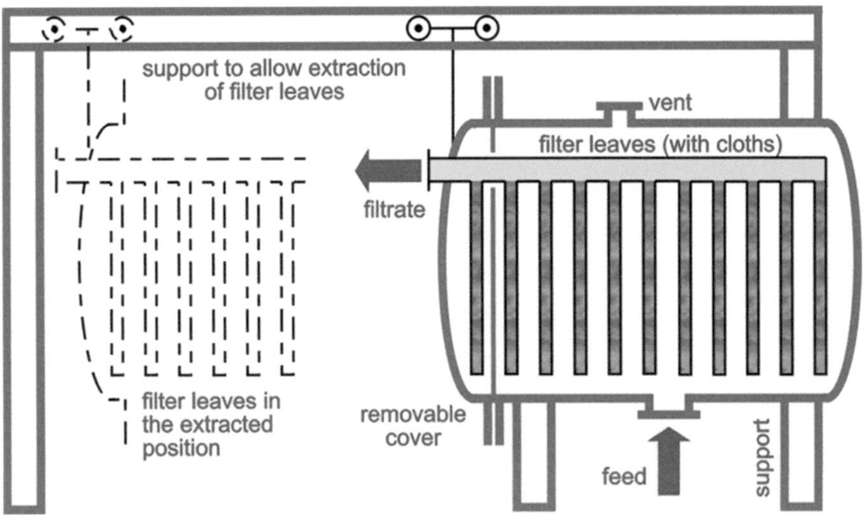

Fig. 10.8 Cross-sectional view of a horizontal vessel, multi-element pressure filter with vertical leaves showing the in-situ and extracted positions of the filter leaves (Courtesy of Filtration Solutions, UK)

Horizontally mounted vessels: These filters contain vertically mounted flat elements with filtration areas up to 300 m^2 (see Figs. 10.1 and 10.8). Whilst needing little height, a large floor space can be required, particularly where filter elements are withdrawn for cake discharge or cloth cleaning. Washing, although possible, can be troublesome if cakes fall off the filter leaves prematurely.

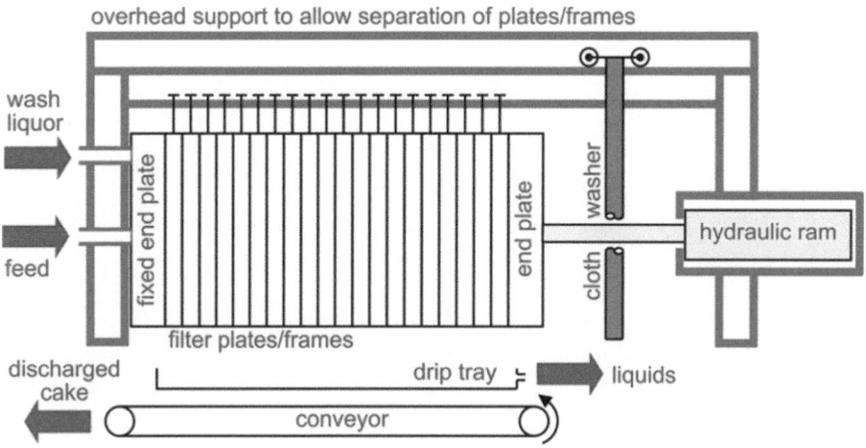

Fig. 10.9 Schematic of a typical overhead bar horizontal filter press showing aspects of filter cycle operations (Courtesy of Filtration Solutions, UK)

10.2.3 Filter Presses

Typical uses: Batch processing of solids forming incompressible and moderately compressible filter cakes.

Although variants exist, the basic unit shown in Fig. 10.9 comprises of many narrow vertical chambers lined on both sides by filter cloths. The chambers, formed between hollow frames and flat filter plates or between adjacent recessed plates, allow for filter cake formations as well as washing and gas-blown deliquoring operations. Suspension is fed to the square, rectangular or circular chambers through a variety of plate porting arrangements and a suitable positive displacement or centrifugal pump. The cakes grow inside each chamber until they meet. Pressures, typically limited to a maximum of 800–1000 kPa, are sufficient to allow centre ported plates to deliver higher solids content feeds whilst bottom and top fed plates generally facilitate more even cake formations and the processing of faster settling suspensions, respectively. Some specially reinforced presses utilise filtration pressures up to 7000 kPa.

Although filter cakes can be removed by reslurrying, they are usually discharged by releasing the mechanical/hydraulic clamping pressure on the press and manually or automatically separating the plates and/or frames. Good cake/cloth release properties are thus preferable, particularly in an automated press. Cloth washing using sprays can also be performed when the filter plates are separated. Modern filter plates are made from either polymers or steel with polymer coatings and formed to provide good drainage surfaces for the covering filter cloths. Plates with dimensions up to 4×5 m are used and filter cloth areas can be as large as 2000 m^2, however, plates up to 2×2 m and cloth areas in the range 50–1000 m^2 are far more typical.

Filter presses are available in two basic forms as shown schematically in Fig. 10.10:

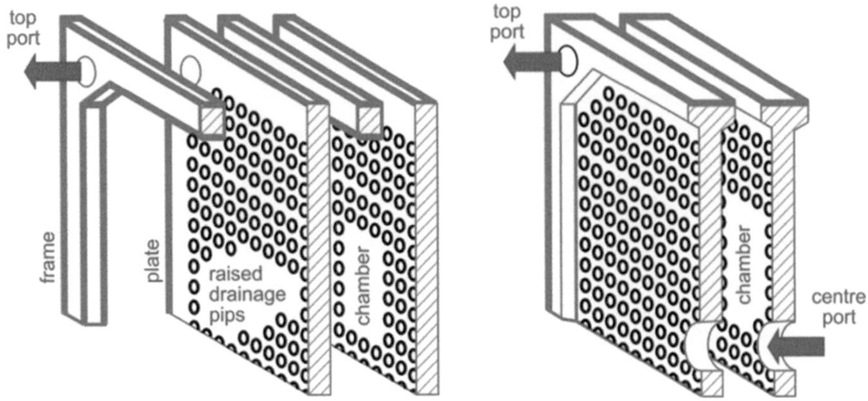

Fig. 10.10 Examples of a top ported plate and frame (*left*) and a centrally ported recessed filter plate (*right*); only one corner of each square/rectangular plate is shown (Courtesy of Filtration Solutions, UK)

Plate and frame press: The basic arrangement comprises an alternating sequence of flat filter plates and frames to facilitate cake formations, normally up to 50 mm thick. The feed suspension and wash liquor enter through the same ports to allow filtration and 'simple' washing, respectively. A more sophisticated arrangement incorporates flat wash plates where suspension and wash liquor enter through separate ports to facilitate improved 'through' washing. In the context of filter presses, plate and frame units offer the advantages of longer cloth life, easily replaced cloths, more uniform cakes and an ability to accommodate alternative filter media such as paper. Their disadvantages include higher capital cost, inlet ports which are prone to blockage at higher feed concentrations and a tendency towards leakage.

Recessed plate press: Here, the functions of the plate and frame are combined such that cake is formed within a recess on each plate. Unlike the plate and frame press, cake thickness is restricted to 32 mm unless additional frames are used as spacers. Feed suspension usually enters through centrally ported plates. The inherent advantages of recessed plate presses include lower initial costs, less tendency towards leakage, an ability to process higher concentration feeds and ease of automation. Their disadvantages include shorter cloth life, longer cloth change times, a tendency to form uneven cakes and an inability to accommodate filter papers.

10.2.4 Variable Volume Filters and Presses

These comprise a family of filters devised to handle suspensions of finer solids which are difficult to pump and/or filter. Typical feeds include suspensions of gelatinous and fibrous materials and those particulates containing occluded liquid within an inherently porous structure.

Fig. 10.11 A sidebar diaphragm press installation (Copyright © Outotec)

10.2.4.1 Horizontal Diaphragm Filter Press

Typical uses: Batch processing of suspensions forming compressible filter cakes where dry cakes and/or efficient post-treatment are required.

These machines are similar in form and general operation to filter presses (see Sect. 10.2.3 and Fig. 10.11). However, the plate surfaces are modified by the addition of flexible diaphragms to form 'membrane plates'. Although different processing conditions are employed, feed pumping can be stopped after ~80 % of the required volume of filtrate has been produced. In this state the chambers in the press are partially filled with cake and residual unfiltered suspension. The diaphragms on each membrane plate are then inflated from behind to induce pressures up to 1600 kPa which filters the remaining suspension and squeezes the now joined cakes in each chamber. The combination of compression by the diaphragms and subsequent gas deliquoring reduces cake moisture content by up to ~25 % more than that achieved in a conventional filter press. The compression process also tends to produce more uniform cake with improved washing characteristics and release properties; the latter also being aided by the correct choice of filter cloth. Although diaphragm presses are significantly more expensive than conventional filter presses the additional capital and operating costs are often justified by shorter cycle times and the beneficial properties of the final cake.

10.2.4.2 Vertical Diaphragm Filter Press

Typical uses: Semi-continuous processing of solids forming compressible filter cakes that require efficient post-treatment.

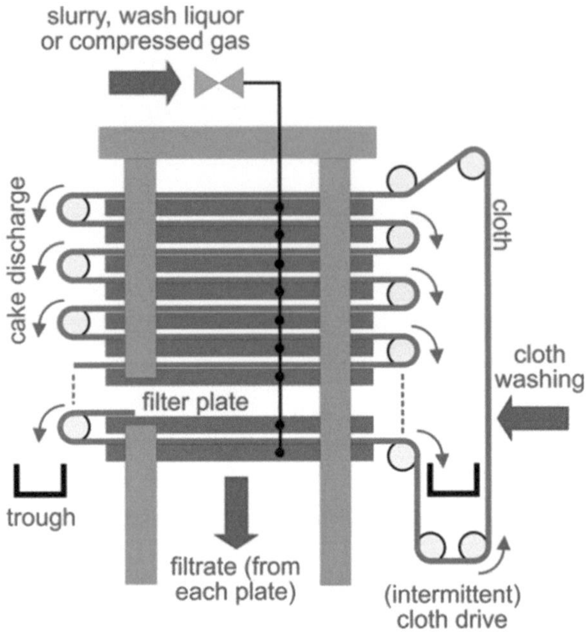

Fig. 10.12 Schematic of the vertical diaphragm filter press (Courtesy of Filtration Solutions, UK)

The vertical diaphragm filter press (or tower press, see Fig. 10.12), may be thought of as a conventional horizontal diaphragm press mounted on its end. In place of the fixed filter cloths, a continuous cloth zigzags through the plate pack and is supported on grids within the horizontal chambers. After hydraulically closing and sealing the plate pack, pressure driven cake filtration takes place in the downward direction via a pump. Compression with elastomer diaphragms at up to 1600 kPa, cake washing, and gas-blown deliquoring can then performed in sequence within each chamber. At the end of the cycle the plate pack opens and the cloth is driven forward to discharge the cakes without manual assistance. At the same time, the filter cloth leaving the plate pack is washed by high pressure water sprays to help maintain its permeability.

The largest available units can produce in excess of 200 te h^{-1} (metric tonnes per hour) of dry solids and offer significantly better washing characteristics due to the preferable orientation of the cakes. A filtration area up to 234 m^2 is technically feasible whilst individual plate areas are in the range 0.4–9 m^2. Although filtration usually takes place only on the upward facing part of the filter cloth, this disadvantage is reduced by complete automation and short down-times. Some more complex machines have a filter cloth on both sides of each chamber to form so called 'double sided presses', however, particle sizes must be toward the lower end of the allowable range to prevent excessive sedimentation.

10.2.4.3 Tube Press

Typical uses: Batch processing of compressible materials where drier cakes are required.

A tube press comprises of two concentric cylinders where a permeable tube covered with a filter cloth is positioned centrally within a solid outer tube lined by an elastomer diaphragm. The filter cycle is initiated by pumping the feed suspension into the annular space so formed and with sufficient suspension in the press, pressure is applied to induce filtration in the radial direction. This process is most often performed at constant pressure via the diaphragm in two stages where a lower pressure is used initially to promote more even cake formation. When filtration is complete the elastomer diaphragm is further inflated (hydraulically) to compression deliquor the cake.

The most widely used version, the vertical axis tube press, is usually used in parallel groups to give the desired filter area as individual units are limited to ~3 m^2. Squeeze pressures are high, up to 10,000 kPa, and thus very low moisture content cakes can be obtained. Cake discharge is achieved by automatically opening the bottom end cover, lowering the central element by ~0.3 m and applying a reverse back pulse of compressed air to dislodge the cake. Although capital and running costs can be higher, tube presses offer the potential for short cycle times, near optimum cake thickness and reduced thermal drying requirements for the discharged cake.

10.2.4.4 Expression Press

Typical uses: Batch or continuous deliquoring of finer particle suspensions forming compressible filter cakes.

Batch units are characterised by series of cylindrical or square boxes containing semi-permeable cloths at their closed ends. The feed is introduced to the boxes and a moving piston expresses liquid through the cloth at pressures up to 40,000 kPa to leave the deliquored solids. Although many variants exist, the continuous expression press, which may also be classed as a continuous pressure filter, is typified by the most widely used screw press shown schematically in Fig. 10.13. This comprises a variable pitch helical screw rotating at up to 2 rpm inside a perforated cylindrical or conical screen surround. As the (usually flocculated) feed moves through the unit, pressure is progressively increased to continuously express the liquid phase and discharge cake through a variable orifice nozzle. Larger units can have a length and screen diameter in excess of 8 m and 1 m, respectively, and are capable of dry solids throughputs greater than 1 te h^{-1}.

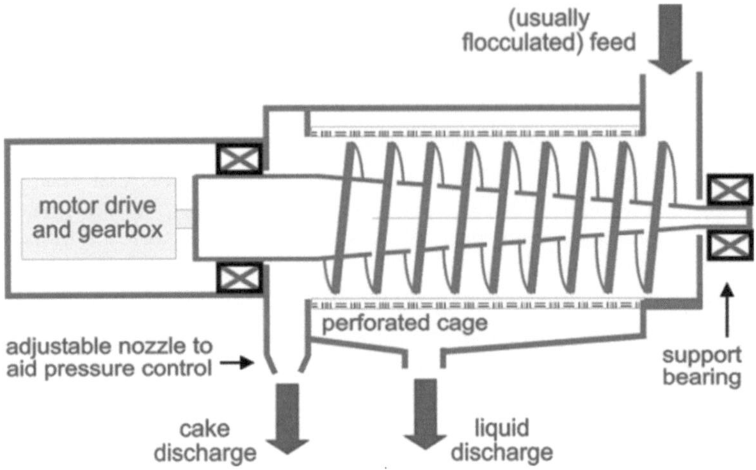

Fig. 10.13 Schematic of a typical expression press (Courtesy of Filtration Solutions, UK)

10.2.5 Continuous Pressure Filters

Continuous pressure filters are often inherently complex and some are based on vacuum driven filters with the addition of an enclosing pressurised shell.

10.2.5.1 Belt Press

Typical uses: Flocculated sludge deliquoring, wastewater treatment.

The belt press shown in Fig. 10.14 was originally conceived to deliquor highly flocculated materials but may be used to process a range of feeds. Although several variants exist, belt presses are characterised by two continuous, tensioned belts. Flocculated material is introduced onto the lower semi-permeable cloth (belt) and then progressively squeezed under pressure as the belts move over a sequence of successively smaller diameter rollers. Liquor is removed through the cloth by a combination of gravity drainage and mechanical squeezing to (ideally) produce a dry, crumbly cake. Filter cloth washing, to help recover initial permeability, may be performed using water sprays at a convenient place after cake discharge. Whilst power consumptions are relatively low, these complex machines have several inherent disadvantages including high flocculant use (dependent on the nature of the feed), relatively low squeeze pressures and a need to use long, strong filter cloths which can be expensive to replace. A guideline (maximum) feed flow rate is 10–15 $m^3\ h^{-1}$ per metre width of filter cloth which typically equates to *ca.* 850 kg $h^{-1}\ m^{-1}$ of dry solids loading.

Fig. 10.14 Photograph of a belt filter press (Courtesy of Sernagiotto)

10.2.5.2 Rotary Pressure Drum

Typical uses: Continuous separation of finer particle suspensions where cakes require post treatment.

The rotary pressure drum filter is similar in principle and basic form to the rotary vacuum drum filter (see Sect. 10.3.3). It comprises a rotating, bottom fed drum of area up to 40 m^2 enclosed within a sealed housing (see Fig. 10.15). Rather than applying a vacuum inside the compartments of the drum, the pressure inside the housing is raised by compressed gas up to 700 kPa and this facilitates constant pressure filtration at the outer drum surfaces. Raised temperatures can be accommodated as can volatile and toxic feeds. Both cake deliquoring by gas blowing and displacement washing can be performed reasonably effectively at different pressures through a multi-compartment arrangement within the housing. Cake discharge usually occurs at atmospheric pressure.

10.2.5.3 Rotary Pressure Disc

Typical uses: Continuous, generally larger scale, separation of finer particle suspensions where cake washing is not required.

The rotary pressure disc filter is again similar in form and general operation to its vacuum driven counterpart (see Sect. 10.3.4) with the addition of an enclosing housing. Rotating cloth covered discs, having a total filtration area of between 2 and 170 m^2, are pressurised externally up to 700 kPa to promote cake formations and

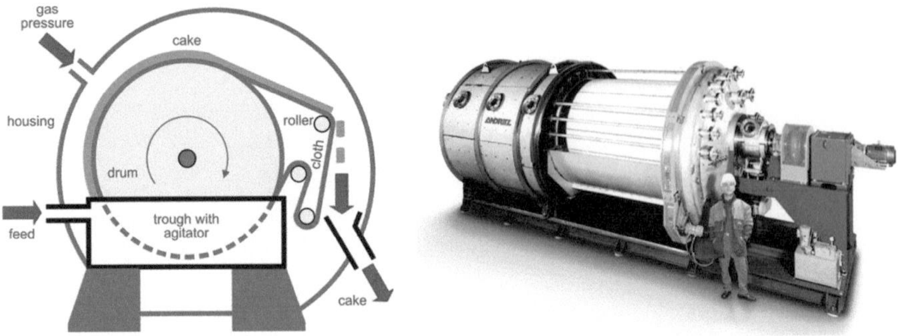

Fig. 10.15 Schematic of a continuous rotary pressure drum filter with belt discharge (*left*, Courtesy of Filtration Solutions, UK) and photograph of a multi-cell pressure drum filter (*right*, Courtesy of Andritz)

generally higher throughputs than equivalent size vacuum units. Although able to handle volatile liquids more readily, the pressure disc filter's inherent advantages are offset by increased costs and the difficulties which can be experienced in cake discharge; the latter is usually achieved by either reslurrying or the use of helical screw conveyors.

10.2.6 Modelling of Filtration

The most widely used process design models for pressure (and vacuum) cake filtration are based on the general filtration equation which is derived from Darcy's Law, Eq. (10.1). The general filtration equation relates the cumulative volume of filtrate (V_f), i.e. the amount of liquid removed, to the filtration time (t_f) and is usually stated as:

$$\frac{dV_f}{dt_f} = \frac{A_f^2 \Delta p_f}{\mu_l \left(\alpha_{av} c V_f + A_f R \right)} \tag{10.9}$$

where A_f is the filter medium area devoted to filtration, Δp_f the filtration pressure, μ_l the viscosity of liquid, c the effective feed concentration and R the filter medium resistance. As most filter cakes are compressible the cake properties, which are characterised by average values of specific cake resistance (α_{av}) and cake solids concentration (C_{av}), are invariably taken to be functions of filtration pressure alone such that:

$$\alpha_{av} = \alpha_0 (1 - n) \Delta p_f^n \tag{10.10}$$

$$C_{av} = C_0 \Delta p_f^{\beta} \tag{10.11}$$

$$c = \frac{s\rho_l}{1 - m_{av}s} = \frac{s\rho_l}{1 - \left(1 + \frac{\rho_l}{\rho_s}\left(\frac{1 - C_{av}}{C_{av}}\right)\right)s} \tag{10.12}$$

where m_{av} is the ratio of mass wet/dry cake and s the mass fraction of solids in the feed. The empirical scale-up constants α_0, n, C_0 and β are most often derived from sequences of constant pressure or vacuum experiments. A value of $\alpha_{av} \leq 10^9$ m kg^{-1} is representative of a very easy to filter material whereas $\alpha_{av} \geq 10^{13}$ m kg^{-1} represents a very difficult to filter material. The value of n is often taken to be the characteristic measure of cake compressibility, i.e. the extent to which a filter cake will compress when it is subjected to a compressive force. When $n \approx 1$ a cake is regarded as very compressible and $n < 0.2$ indicates essentially incompressible.

The cake thickness (L_f) can be related to the cumulative volume of filtrate and cake/particle properties through a mass balance. For example, in the single leaf (Nutsche) filter and filter presses

$$L_f = \frac{V_f}{A_f}\frac{s(\rho_s(m_{av} - 1) + \rho_l)}{\rho_s(1 - m_{av}s)} \tag{10.13}$$

A general feature of pressure driven batch filters is that cake formation is not generally limited by time, but rather by the dimensions and capacity of the filter itself.

10.2.6.1 Cake Formation at Constant Pressure

For a single leaf (Nutsche) filter the filtration pressure is fixed such that α_{av}, C_{av}, m_{av} and c remain constant throughout cake formation as given by Eqs. (10.10), (10.11), and (10.12). Equation (10.9) is then integrated at constant pressure to give:

$$t_f = \frac{\alpha_{av} c \mu_l}{2A_f^2 \Delta p_f} V_f^2 + \frac{\mu_l R}{A_f \Delta p_f} V_f \tag{10.14}$$

With curved elements, such as in the candle filter and the tube press, a modified version of Eq. (10.9) gives:

$$t_f = V_f\left(\frac{\alpha_{av} c \mu_l}{2A_f^2 \Delta p_f} V_f + \frac{\mu_l R}{A_f \Delta p_f} + \frac{4c}{d\rho_c A_f}\left(\frac{\alpha_{av} c \mu_l}{3A_f^2 \Delta p_f} V_f^2 + \frac{\mu_l R}{2A_f \Delta p_f} V_f\right)\right) \tag{10.15}$$

where d is the element diameter and ρ_c the bulk density of the filter cake. In both cases the filtrate flux reduces with time as the cake builds. Cake formation on rotary pressure disc and drum filters is modelled in a similar manner to that described for the corresponding vacuum filter in Sect. 10.3.5.

10.2.6.2 Cake Formation at Variable Pressure

In the case of, for instance, filter and diaphragm presses and multi-element leaf filters, the pressure generally increases throughout cake formation according to a pump curve. That is, the Δp_f vs. q relationship for a positive displacement pump (i.e. constant flow filtration where the flow rate of filtrate is constant with time) or a centrifugal pump (i.e. variable pressure-variable flow filtration where both pressure and flow rate are variant). Substituting $q = dV_f/dt_f$ into Eq. (10.9) and rearranging gives an expression for the volume of filtrate:

$$V_f = \frac{A_f}{\alpha_{av}\mu_l c}\left(A_f\frac{\Delta p_f}{q} - \mu_l R\right) \tag{10.16}$$

As the pressure varies several parameters are time variant whereby α_{av}, C_{av}, m_{av} and c change throughout filtration to an extent dictated by Eqs. (10.10), (10.11), and (10.12). The pressure is specified by a given position on the pump curve and the filtration time is obtained from:

$$t_f = \int_0^{V_f} (1/q)dV_f \tag{10.17}$$

It is generally necessary to solve Eqs. (10.16) and (10.17) using numerical techniques in conjunction with a knowledge of the pump characteristics in order to find sequential values of V_f and t_f, and consequently the amount of solids processed.

10.2.7 *Modelling of Compression Deliquoring*

In the tube press and all variants of the diaphragm press, flexible diaphragms can be inflated at a constant pressure to facilitate compression deliquoring of the filter cake (s). This consolidation is most frequently performed immediately after cake formation to remove unwanted liquor and/or improve the distribution of cake solids to aid any subsequent cake washing and gas deliquoring. As a rule-of-thumb, the duration of cake compression equals the total filtration time.

The cake properties during compression deliquoring are related to the compression pressure (Δp_c) by:

$$C_e = C_{e0}\Delta p_c^{\gamma} \tag{10.18}$$

$$(C_{av})_\infty = C_0\Delta p_c^{\beta} \tag{10.19}$$

where C_e is the modified consolidation coefficient which characterises the process, $(C_{av})_\infty$ the equilibrium cake solids volume fraction at infinite consolidation time, and C_{e0}, γ, C_0 and β are empirical scale-up constants derived from sequences of constant pressure consolidation experiments. Process design calculations for compression deliquoring are primarily based on the theories of Shirato et al. [22, 23] where the cake thickness (L_c) at a given time (t_c) is related to a dimensionless consolidation time (T_c) by:

$$L_c = L_0 - (L_0 - L_\infty)\frac{\sqrt{\frac{4(j_{II})^2 T_c}{\pi}}}{\left(1 + \left(\sqrt{\frac{4(j_{II})^2 T_c}{\pi}}\right)^{2v}\right)^{1/2v}} \tag{10.20}$$

$$T_c = \frac{i^2 C_e t_c}{\omega_0^2} = i^2 C_e t_c \left(\frac{A_c \rho_s}{M_s}\right)^2 \tag{10.21}$$

$$L_\infty = \frac{M_s}{A_c \rho_s (C_{av})_\infty} \tag{10.22}$$

where L_∞ is the cake height at infinite consolidation time, j_{II} is a factor that accounts for any filter element curvature, i the number of drainage surfaces (normally = 1), M_s the mass of solids in the cake, ω_0 the volume of solids per unit filter area, A_c the active filter area during consolidation and v is an empirical scale-up constant. Values for i, A_c and j_{II} vary according to the type of press, but use of Eq. (10.20) allows the change in cake thickness with time to be evaluated.

10.2.8 Modelling of Cake Washing

Design equations for the most frequently used displacement washing process are usually based on the dispersion model [see 30]. The model requires the determination of a dispersion number (D_n) that characterises the washing process and use of a design chart (Fig. 10.16) which allows the amount of wash liquid used (specified as the number of wash ratios, W) and the amount of solute removed from the cake (e.g. fractional solute recovery, F) to be found. The pressure (Δp_w) is fixed throughout washing and cake properties such as specific resistance, solids concentration and thickness (L_w) are normally assumed to remain constant.

In the dispersion model the superficial velocity (u) and pore velocity (v) of wash liquor are related to the intrinsic properties of the cake through a version of Darcy's Law

$$u = \frac{\Delta p_w}{\mu_w(\alpha_{av}\rho_s L_w C_{av} + R)} \tag{10.23}$$

$$v = \frac{u}{\varepsilon_{av}} = \frac{u}{1 - C_{av}} \tag{10.24}$$

where μ_w is the viscosity of the wash liquor. For the solute, the ratio of the molecular diffusion coefficient (D) to the axial dispersion coefficient (D_L) is

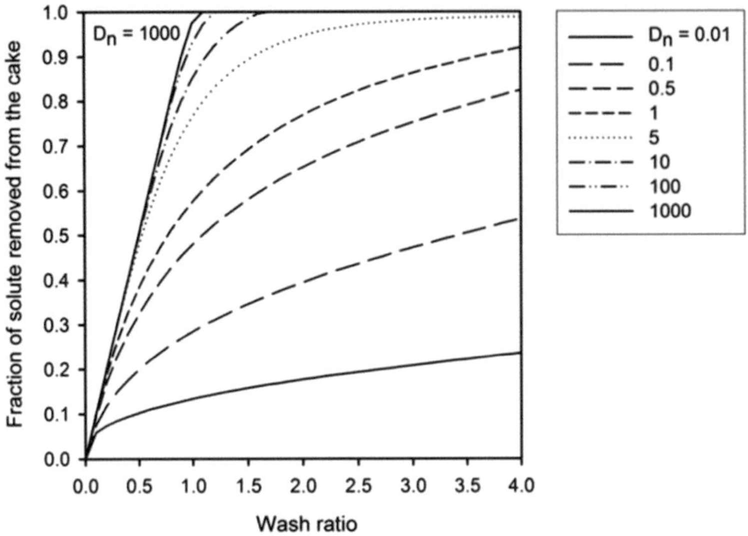

Fig. 10.16 Variation of the fraction of solute removed (F) from a saturated filter cake with wash ratio (W) and dispersion number (D_n)

dependent on the product of the Reynolds number (Re) and Schmidt number (Sc) as well as the cake thickness. For instance, with ReSc ($= vx/D$) >1 and L_w <10 cm

$$\frac{D_L}{D} = 0.707 + 55.5(\text{ReSc})^{0.96} \tag{10.25}$$

and hence

$$D_n = \text{ReSc}\frac{L_w\,D}{x\,D_L} \tag{10.26}$$

Depending on whether t_w, W or F are specified the other values can be found with Eqs. (10.23,) (10.24), (10.25), and (10.26) and use of Fig. 10.16.

10.2.9 *Modelling of Cake Gas Deliquoring*

Process design calculations for the gas deliquoring of filter cakes require the specification of a threshold pressure/vacuum (p_b), which is the pressure or vacuum required to start deliquoring, and an irreducible saturation (S_∞), which is the cake saturation where deliquoring due to liquid displacement stops [see 31]. These can be either measured in a capillary pressure experiment or the former can be calculated with some confidence using:

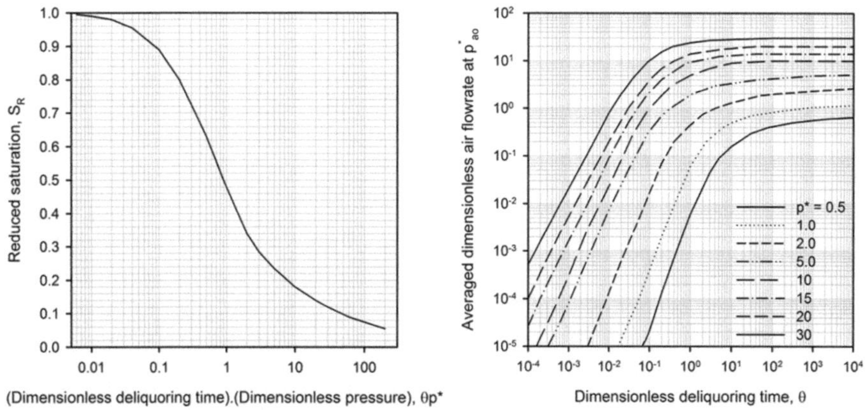

Fig. 10.17 Reduced saturation of a filter cake as a function of dimensionless deliquoring time (*left*) and the dimensionless air/gas flow rate through a filter cake as a function of dimensionless pressure p^* (*right*). In both cases deliquoring is by or pressure (or vacuum) applied in the gas phase

$$p_b = \frac{4.6C_{av}\sigma}{x(1 - C_{av})} \tag{10.27}$$

where σ is the cake liquid surface tension. Knowledge of p_b and S_∞, together with the intrinsic properties of the cake and two design charts (Fig. 10.17), allows the evaluation of cake moisture content (M) or deliquoring time (t_d) as well as the flux of gas (or gas/air rate) required to deliquor the cake. The applied gas pressure (Δp_d) is fixed and properties of the cake such as thickness (L_d), solids concentration and specific resistance are assumed to remain constant.

The basis of the approach requires the calculation of a dimensionless time (θ) or reduced saturation (S_R) depending on whether time (t_d) or the required saturation (S) is known, respectively, whereby

$$\theta p^* = \left(\frac{t_d p_b}{\alpha_{av}\rho_s C_{av}(1 - C_{av})\mu_l (L_d)^2 (1 - S_\infty)}\right)\left(\frac{\Delta p_d}{p_b}\right) \tag{10.28}$$

$$S_R = \frac{S - S_\infty}{1 - S_\infty} \tag{10.29}$$

In this way the unknown value (either t_d or S) can be determined with the aid of Fig. 10.17 (*right*) and the cake moisture content (M) can be calculated using

$$M = \frac{100}{1 + \frac{\rho_s}{S\rho_l}\left(\frac{C_{av}}{1 - C_{av}}\right)} \tag{10.30}$$

Calculations for the amount of gas/air required to drive the gas deliquoring are more complex and best described by worked example [see 27]. In general, calculations for air rate tend to over-estimate requirements which leads to an over-specification of compressor (or vacuum pump) requirements.

10.3 Vacuum Filters

A category of filter that uses vacuum induced driving forces and semi-permeable media to facilitate a separation. Whilst pressure differences across a filter are limited to less than 85 kPa (usually <75 kPa), most units are capable of processing a wide range of (coarser particle size) feed materials in a continuous manner. Many types employ a rotary valve arrangement to set different vacuum levels over sequential phases in a filter cycle thus facilitating more control over cake formation, deliquoring and washing; any given phase is operated at a constant vacuum level. Several vacuum filters have counter-current washing capability where the washings from downstream are used to wash filter cakes further upstream in order to reduce overall wash liquor consumption. Although it is possible to enclose some types to conserve heat and/or vapours the processing of more volatile constituents at higher altitudes can cause significant problems. Most variants have a minimum cake thickness requirement to help ensure adequate cake discharge. Woven filter cloths or specially developed coated media are used almost exclusively.

Table 10.2 shows typical characteristics of the most common continuous vacuum filters; some batch variants are listed in Fig. 10.5. Capital costs for continuous vacuum filters can be estimated over the specified filter area ranges using equations adapted from Couper et al. [6]:

$$\begin{array}{ll} \text{Horizontal belt}(A = 1 - 110\,\text{m}^2): & 1.1 \times 10^5 A^{0.5}\ \$ \\ \text{Rotary disc}(A = 10 - 300\,\text{m}^2): & 3.2 \times 10^4 A^{0.43}\ \$ \\ \text{Rotary drum, knife discharge}(A = 1 - 140\,\text{m}^2): & -2.9A^2 + 2470A + 65100\ \$ \\ \text{Rotary drum, belt discharge}(A = 1 - 75\,\text{m}^2): & -14A^2 + 3710A + 72800\ \$ \end{array} \tag{10.31}$$

10.3.1 *Horizontal Belt*

Typical uses: Separation of relatively free filtering solids where good post treatment is required.

The horizontal belt is a continuous filter with an endless cloth supported on a perforated belt (see Fig. 10.18). The belt and cloth are driven around two rollers and across a sequence of evacuated suction boxes at linear speeds up to 0.5 m s^{-1}. The feed suspension is introduced at one end and filtered to produce a cake. The length of the filter, which can be in excess of 60 m, is arranged to allow adequate cake formation as well as the chosen number of deliquoring and washing operations. Due to the ease with which wash liquors can be segregated, it is relatively simple to perform counter-current washing to exacting requirements provided wash liquor

Table 10.2 Typical characteristics of some industrial continuous vacuum filters

	Rotary drum	Horizontal belt	Rotary table	Rotary tilting pan	Rotary disc
Max. submergence (% of cycle)	30[a]	As req'd	As req'd	As req'd	28[a]
Area under active vacuum (% of cycle)	75–80[e,b]	As req'd	80	75	75[b]
Max. for washing (% of cycle)	29[c]	As req'd	As req'd	As req'd	None
Max. for deliquoring only (% of cycle)[(d)]	45–60[e]	As req'd	As req'd	As req'd	45–50
Cake discharge (% of cycle)	10–25[e]	0	20	25	25
Min. cake discharge thickness (mm)	1–6[e]	3–5	20	20–25	10–13
Filter area (m^2)	0.1–180	3–170	5–320	5–200	1–300
Solids dryness index	6 C	7 C	7 C	7 C	4 C
Washing index	7	9	8	9	–
Liquid clarity index	7	7	7	7	6
Particle breakage index	8	8	8	8	8
Particle size in feed (μm)	1–200[e]	20–80,000	20–80,000	20–80,000	1–700
Feed conc. (%w/w)	1–20[e]	5–30+	10–30+	5–30+	5–20
Motor drive power (kW)	0.5–20[f]	1–90	n/a	3–45	5–60[f]
Power to generate vacuum (kW)	≈ (filter area in m^2)$^{1.25}$				

[a]Higher submergence (cake formation period) is available, consult manufacturer
[b]Assumes no trunnion stuffing boxes
[c]Washing starts at the horizontal centreline on the rising side and extends up to 15° past top dead centre
[d]Deliquoring means drainage of liquor from cake formed during submergence
[e]Depends on variant
[f]Including agitator

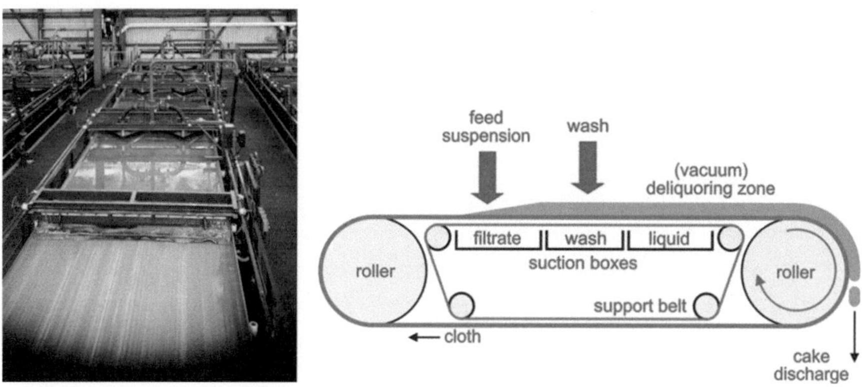

Fig. 10.18 Photograph of a vacuum driven horizontal belt filter (*left*, Courtesy of Clear Edge) and a typical filter cycle (*right*, Courtesy of Filtration Solutions, UK)

carry-over into the next suction box is avoided. The final cake is naturally discharged as it passes over the second roller and separation of the belt and cloth beneath the filter allows the cloth to be cleaned by sprays as it returns.

Horizontal belt filters are best suited to the larger scale filtration of medium and faster settling slurries. Although they occupy a large floor space and the cost of installation per unit filter area is relatively high, these disadvantages are generally offset by full automation, flexibility, high capacity and relatively high speeds of operation. Filters may be sealed to prevent the escape of heat and/or vapours, however, should the belt or cloth be damaged then replacement of either component can be expensive. Some units are programmed to operate semi-continuously via intermittent motion of the belt.

10.3.2 Horizontal Rotary Filters

Typical uses: Processing of fast settling slurries where good washing is required.

The two forms of horizontal rotary filter differ primarily in the manner in which the filter cloth is arranged around the periphery of the circular separation surface (see Fig. 10.19).

Table: A filter comprising of a rotating horizontal table with an annular filter cloth. Vacuum is applied over individual segments of the table to initiate filtration and the formed cake is subsequently deliquored and/or washed by sprays according to requirements. The final cake is continuously discharged via a screw conveyor to typically leave a 3–4 mm residual heel of cake on the cloth. Dependent on the properties of the solids, the presence of the heel can have undesirable consequences for future cycles and necessitates the use of more open filter media with the potential for cloudier filtrates. As the cloth surface is not physically divided into individual sectors some short-circuiting of the feed may occur as well as

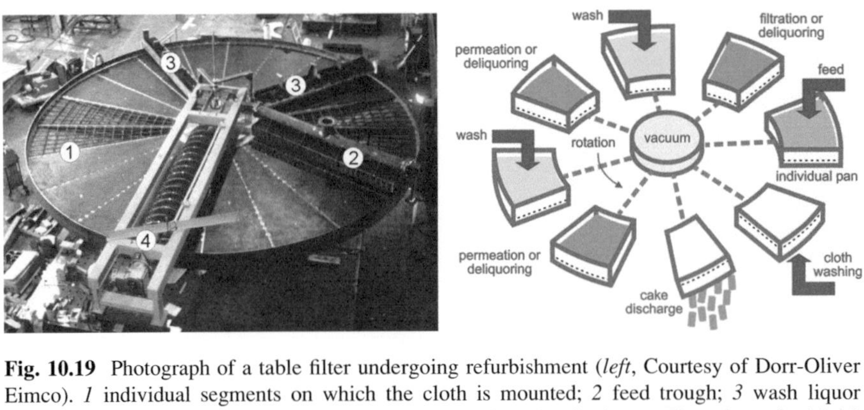

Fig. 10.19 Photograph of a table filter undergoing refurbishment (*left*, Courtesy of Dorr-Oliver Eimco). *1* individual segments on which the cloth is mounted; *2* feed trough; *3* wash liquor delivery or additional feed points; *4* screw conveyor for cake discharge. The schematic (*right*) shows a tilting pan filter (Courtesy of Filtration Solutions, UK)

unwanted mixing of the wash and mother liquors. Although cloth area can be up to 320 m^2, alignment difficulties usually restrict machine size and the available area to below 100 m^2.

Tilting pan: A filter that is similar in general form to the table filter, except the cloth is replaced by a series of annular sectors or pans, each one of which is lined along its perforated bottom by an individual filter cloth. After suspension is introduced to a pan, filtration, deliquoring and washing can take place under the applied vacuum before cake discharge is achieved by a relatively complex tilting mechanism that inverts the pan. The discharge procedure, which may be assisted by air blow-back, leaves no heel of cake and thus in-situ cloth medium cleaning is readily performed using sprays. As all liquors are kept separate, there is little unwanted mixing of mother and wash liquors and counter-current washing can be very good. Tilting pan filters are available with total filter cloth areas up to 200 m^2, however, many of their inherent advantages are offset by their mechanical complexity and higher capital cost.

10.3.3 Rotary Drum Filters

Typical uses: Separation of relatively easy to filter suspensions, the efficiency of cake post treatment depends on the type of drum.

The versatile rotary drum filter (or rotary vacuum filter, RVF) is perhaps the most widely used of the continuous vacuum filters (see Figs. 10.1 and 10.20). The generic type is characterised by a rotating, multi-compartment drum covered externally by a fixed filter cloth of total surface area between 0.1 and 180 m^2; most commercial units are in the range 1–80 m^2. The drum on bottom fed units rotates about a horizontal shaft at speeds up to 5 rpm and is partially submerged in a mechanically agitated tank of constantly replenished suspension. A vacuum is applied inside the drum via a control valve to initiate upward filtration over the submerged region. As the drum rotates so the filter cake formed on the cloth is exposed and a limited number of deliquoring and washing procedures can then be performed at the appropriate level of vacuum. Although washing efficiency is reasonable, the restricted horizontal filter area near the top of the drum prevents further efficiency gains. Cake discharge normally occurs at a point where the final cake is almost vertically oriented and variants differ primarily in the manner in which cake is discharged:

Knife/scraper discharge: The most widely used method when avoidance of cloth blinding can be more or less guaranteed, but requires a minimum 6 mm cake formation. The knife is arranged to leave a heel of cake on the cloth and thus avoid potential damaging contact between the knife and drum. If the cake is thinner then air blow-back can be employed to break the vacuum and assist discharge, though with some designs this tends to cause filtrate to re-enter the cake.

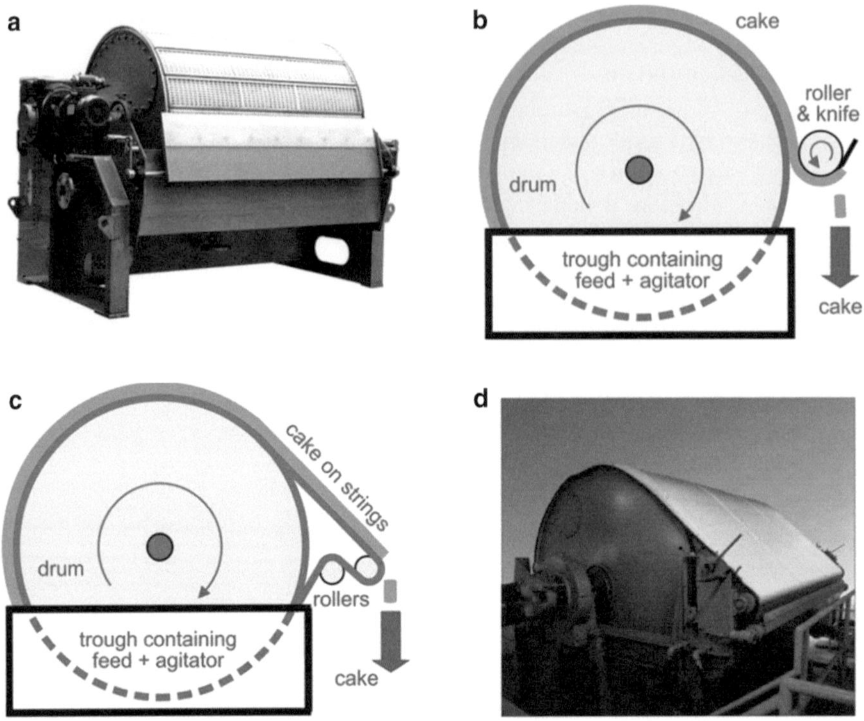

Fig. 10.20 Representations of bottom fed rotary vacuum drum filters. (**a**) Knife/scraper discharge (Courtesy of Filtration Services); (**b**) roller discharge (Courtesy of Filtration Solutions, UK); (**c**) string discharge (Courtesy of Filtration Solutions, UK); (**d**) belt discharge (Courtesy of Dorr-Oliver Eimco)

Roller discharge: Generally used for the complete removal of finer particle, sticky cakes that don't crumble. The 0.5–3 mm cake must preferentially stick to the roller placed adjacent to the drum, a process that is aided by the shearing action of the faster rotating roller. A knife scraper continuously removes the cake from the roller.

String discharge: Suited to the discharge of fairly thick cakes of a fibrous nature that don't crumble. A number of endless strings pass over a series of external rollers and the surface of the filter cloth. In the discharge zone the strings lift away from the cloth to remove the cake completely. The strings may be replaced by endless wires, coils or chains as appropriate.

Belt discharge: Mostly used for the discharge of sticky, thin cakes (<3 mm) whose constituent particles may tend to blind a filter cloth. In this case the endless cloth is not fixed to the drum, instead it passes around its outer periphery and a series of external rollers. At the discharge point the cloth lifts away from the drum and movement over the rollers causes all the cake to be released. The exposed cloth is then cleaned by sprays before returning to the drum. Although relatively expensive to install, belt discharge systems are claimed to raise throughput by up to 30 %.

Drum filters can be enclosed to prevent the escape of heat and/or vapours and typically operate with a submergence equivalent to 30–40 % of the available filter cloth area. Where cake formations are more difficult this may be raised to 60–75 % to produce a 'submerged axis' filter. However, the option is rarely preferred as the trunnion mounts for the drum must be wholly or partly flooded with slurry which necessitates the use of stuffing boxes (sealing units) with their attendant capital and maintenance costs. For faster settling solids the mechanical agitation of suspension in the tank is more problematical and top fed or internally fed drums are options, but rarely favoured as, for instance, horizontal rotary filters are generally better suited.

10.3.4 Rotary Disc Filters

Typical uses: Continuous larger scale separation of relatively free filtering suspensions where washing is not required.

Units comprise up to 12 flat, circular discs mounted vertically on a central horizontal shaft (see Fig. 10.21). The discs, which are themselves permeable, are usually covered externally by sectored filter cloths. Rotation causes them to pass through individual agitated tanks containing the feed suspension(s) and the vacuum is applied inside the discs to promote cake filtration. After this is complete, deliquoring by air-suction can be performed, however, cake washing is essentially impossible due to the vertical cake formation. The final cakes are discharged by blade or wire scrapers on either side of the discs. As an air blow-back system is often employed to aid cake removal, wetter cakes tend to be discharged (in comparison to drum filters) and the discharge of thin cakes can be particularly troublesome. The need to place the discharge scrapers close to the cloth surface frequently leads to cloth damage, though cloth sectoring means only portions need be replaced.

Rotary disc filters are available at a relatively low capital cost with total cloth areas between 1 and 300 m^2. They have an inherently large filter area to floor space

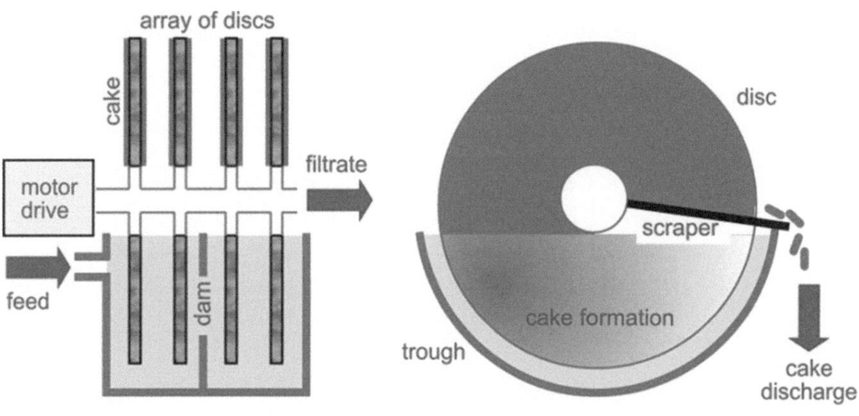

Fig. 10.21 Side and end views of a rotary disc filter (Courtesy of Filtration Solutions, UK)

ratio and their flexibility is enhanced by an ability to process multiple feedstocks at the same time within a single unit.

In a variant, the filter cloths and discs are replaced by sintered alumina membrane discs with near uniform micropores. Although a relatively small vacuum pump is needed to promote cake formation, cake deliquoring proceeds via capillary action with little or no air consumption which significantly reduces the cost of vacuum generation (by *ca.* 90 + %). Whilst the use of a ceramic membrane material allows a clear filtrate to be readily obtained, disc replacement can be expensive. Moreover, backflushing with filtrate and periodic in-situ ultrasonic cleaning must be performed in order to maintain the original permeability of the filtering discs. Ceramic disc filters are available with filter areas of 15–240 m^2 and have been used in both metal and mineral concentrate processing at claimed solids throughputs up to 400 te h^{-1}.

10.3.5 Modelling of Filtration

Cake formation on a vacuum filter takes place at a constant level of vacuum. Thus, process modelling is usually based on a version of the quadratic Eq. (10.14), which is derived from Darcy's Law Eq. (10.1), and the fixed values of cake resistance and solids concentration given by Eqs. (10.10), (10.11), and (10.12). Noting that the configuration of vacuum filters limits the cake formation to a fixed, and generally limited, time period, the volume of filtrate is given by:

$$V_f = \frac{A_f}{c}\left(-\frac{R}{\alpha_{av}} + \sqrt{\left(\frac{R}{\alpha_{av}}\right)^2 + \frac{2ct_f\Delta p_f}{\mu_l\alpha_{av}}}\right) \tag{10.32}$$

and the cake thickness by:

$$L_f = \frac{1}{\rho_s C_{av}}\left(-\frac{R}{\alpha_{av}} + \sqrt{\left(\frac{R}{\alpha_{av}}\right)^2 + \frac{2ct_f\Delta p_f}{\mu_l\alpha_{av}}}\right) \tag{10.33}$$

In these equations the area of filter medium (A_f) devoted to filtration is given by:

$$A_f = \begin{cases} z_f h_B & \text{horizontal belt} \\ \varphi_f \pi D h_D & \text{drum} \\ 0.25\varphi_f \pi\left(d_o^2 - d_i^2\right) & \text{table} \\ \varphi_f n_p A_p & \text{tilting pan} \\ 0.5\varphi_f n_d \pi\left(d_o^2 - d_i^2\right) & \text{disc} \end{cases} \tag{10.34}$$

where z_f is the belt length devoted to filtration, h_B the belt width, φ_f the fraction of filter medium area devoted to filtration, D the drum diameter, h_D the drum width, n_p

the number of pans, A_p the filter medium area in a single pan, n_d the number of discs, and d_i and d_o are respectively the inner and outer diameters of the filter medium on a table or disc. With rotary type filters, such as the rotary drum and tilting pan, the duration of the filtration phase is given by $t_f = 2\pi\varphi_f/\omega$ (ω is the angular velocity), whereas for the horizontal belt $t_f = z_f/v_B$ (v_B is the linear belt velocity).

10.3.6 Modelling of Cake Washing and Gas Deliquoring

Cake washing and gas deliquoring on vacuum filters are modelled in a similar manner to that described for pressure filters in Sects. 10.2.8 and 10.2.9, respectively. However, vacuum filters such as the rotary drum do not always adequately constrain the wash liquor flow in terms of directing it through the filter cake and washing is generally not as efficient as it could be. In such cases it is necessary to correct the calculated dispersion number according to an empirical correlation such as:

$$(D_n)_{corr} = 3.22\log(D_n) + 0.395 \tag{10.35}$$

The value of $(D_n)_{corr}$ is used in place of D_n in design charts.

10.4 Filtering Centrifuges

Filtering centrifuges use centrifugal forces to perform batch and continuous cake filtration on either cylindrical, or conical, semi-permeable surfaces. Displacement washing operations can be accommodated by most centrifuges in addition to efficient cake deliquoring. Several machines are capable of operating in both vertical and horizontal orientations whilst some rely on the favourable sliding and conveying properties of the formed cake for successful operation. The typical characteristics of filtering centrifuges are shown in Table 10.3 and exemplar capital equipment costs can be estimated using equations adapted from Leung [15].

$$\begin{array}{ll} \text{Inverting bag}\,(D = 0.3 - 1\,\text{m}): & 2.3 \times 10^5\exp(1.1D)\ \$ \\ \text{Peeler}\,(D = 0.6 - 1.2\,\text{m}): & 1.8 \times 10^5\exp(0.86D)\,\$ \\ \text{Pusher}\,(D = 0.3 - 1\,\text{m}): & 2.6 \times 10^5 D + 52200\,\$ \end{array} \tag{10.36}$$

10.4.1 Basket

Typical uses: Deliquoring of suspensions with reasonable drainage characteristics.

Table 10.3 Typical characteristics of some industrial filtering centrifuges

	Basket	Pusher	Inverting bag	Slip discharge	Vibratory, oscillatory	Worm screen
Centrifugal force or g-factor (g)	700–2200[c]	800–1700	700–1500	Up to 2900[a]	30–150	500–2600
	200–1200[d]					
Operation	Batch	Continuous	Batch	Continuous	Continuous	Continuous
Max. throughput (te h^{-1})	12–15[e]	5–80[e]	10–450 kg[e]	<150	350	<150
Cake condition	Dry → pasty, granular	Dry, granular	Dry, granular	Dry, granular	Dry, granular	Dry, granular
Bowl diameter (m)	0.4–1.8	0.3–1.3	0.3–1.3	0.5–1.6	0.7–1.5	0.2–0.9
Solids dryness index	9 C	9 C	9 C	7 C	8 C	9 C
Washing index	6	7–8	6	5	5	5
Liquid clarity index	5	4	5	4	4	4
Particle breakage index	5–6	4	6	4	3	4
Particle size in feed (μm)	2–1000	40–7000	2–1000	80–10000	100–10000	60–5000
Feed conc. (%w/w)	4–30	10–40	5–30	10–40	10–40	10–40
Motor drive power (kW)	5–90	3–200[b]	3–90	6–70	8–75	2–90

[a]At the lip (i.e. maximum diameter) of the cone
[b]Including pusher drive
[c]Peeler
[d]Pendulum
[e]Of solids

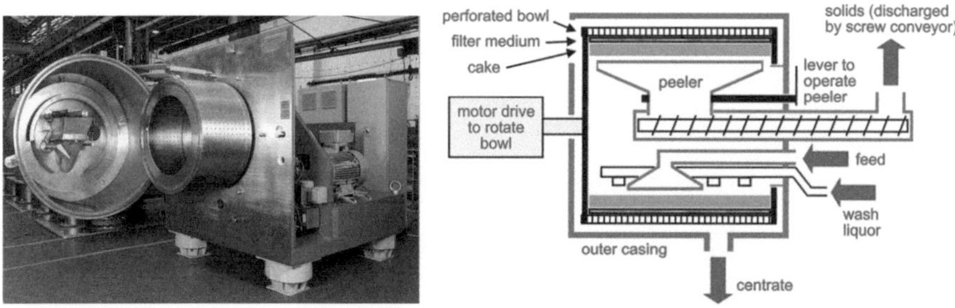

Fig. 10.22 Photograph of a horizontal peeler centrifuge with the front end casing open to facilitate safe operator inspection of the interior (*left*, Courtesy of Thomas Broadbent & Sons) and schematic showing the principal features (Courtesy of Filtration Solutions, UK). In a centrifuge the filtrate is known as centrate

These centrifuges are essentially batch operated and comprise a vertically or horizontally mounted basket with one closed end and one partially open end. The basket, which is perforated and covered by a combination of metal screen(s) and filter cloth, is rotated to give solids throughputs up to 15 te h^{-1}. The induced centrifugal forces allow centrate (filtrate) to pass through the cloth/screen whilst particles accumulate in the form of a filter cake. The cake may subsequently be washed by sprays and/or allowed to deliquor prior to discharge. Variants of the basket filtering centrifuge differ primarily in the process limitations imposed by the axis of rotation.

Horizontal axis: The fully automated horizontal axis basket, or peeler, centrifuge operates for most of its cycle at constant rotational speed to give a g-factor up to 2200 g (see Fig. 10.22). The operating cycle is generally shorter than for vertical axis machines and, with less time lost for acceleration and deceleration, higher throughputs can be achieved. Solids are discharged at moderate speed at the end of the cycle by a rigidly constructed peeler, sometimes with the aid of a compressed gas jet or reciprocating knife. The relatively high speed discharge can induce glazing of the cake heel and hence reduced centrate flow rates in subsequent cycles. Horizontal axis centrifuges tend to be more expensive than equivalent capacity vertical axis machines.

Vertical axis: The vertical axis basket centrifuge, which is also known as the three-column or pendulum centrifuge, allows the feed suspension to be introduced when the basket is either stationary or rotating at a moderate speed. The rotational speed is often varied through a cycle with cake washing and deliquoring being performed at high speed (~1500 rpm) and cake discharge at a lower speed (~60 rpm). On bottom driven machines the basket is usually lifted out manually to allow for cake discharge. Generally more expensive top driven machines are employed for heavier duties and faster filtering feeds. These units are discharged automatically by plough or with the assistance of a gas jet and/or compressed gas blow-back when a residual heel of cake is unacceptable. For cakes that are inherently hard a partial length plough that moves up and down the axis of the basket can be used. Due to uneven cake formation, washing performance can be variable.

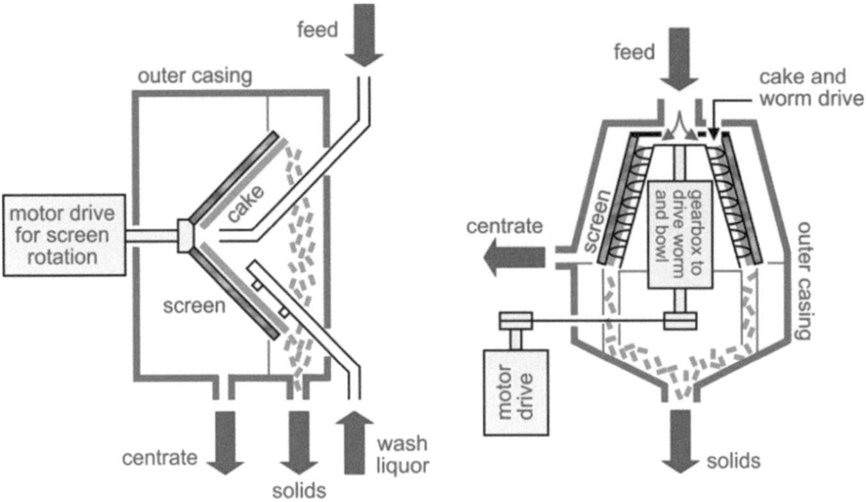

Fig. 10.23 Schematics of two variants of cone screen centrifuge, slip discharge/wide angle cone (*left*) and single stage worm screen (*right*) (Courtesy of Filtration Solutions, UK)

10.4.2 Cone Screen

Typical uses: Continuous deliquoring of suspensions containing relatively free filtering solids.

A cone screen centrifuge comprises of a conical perforated metal screen across which wet solids slide after filtration from relatively high concentration suspension (Fig. 10.23). During their passage, the solids, in the form of a cake, can be washed by sprays and/or deliquored prior to discharge at the wider end of the cone. The four variants of cone screen centrifuge differ primarily in the manner in which the solids are caused to translate along the screen:

Slip discharge/wide angle cone: In these vertical or horizontal axis machines the cake is caused to move by providing a cone with a half-vertex angle in excess of the angle of friction between the cake and the screen. The cone angle is critical for good operation and is typically in the range 25–35°, though the lubrication provided by the liquid in the cake can greatly assist the sliding operation and particularly toward the start of the translation process. Whilst good deliquoring is generally achieved with centrifugal forces up to 2900 g, the rapid transit of solids through the centrifuge means there is a limited time available for washing on the angled surfaces of the cone. Slip discharge centrifuges are best suited to the processing of fairly coarse, fast filtering, granular solids.

Vibratory/oscillatory: These centrifuges work on a similar principle to the slip discharge centrifuge, however, the addition of an eccentric vibratory drive facilitates use of a cone angle lower than the angle of friction between the cake and the screen. Cone angles of 13–18° are common and vibrations in the region of 1700 min^{-1} induce partial fluidisation of the cake which enhance its translation across the screen to give throughputs as high as 350 te h^{-1}. As relatively low centrifugal forces are generated (typically below 120 g), deliquoring and centrate clarity can sometimes be poor and this may in turn lead to lower quality particle and liquid products. Both horizontal and vertical axis machines are available and these are best suited to the processing of relatively coarse, fast filtering solids.

Tumbling: A gyratory motion of the screen bowl about a vertical axis causes the inclination of the cone walls to alter about the angle of friction between the cake and the screen. The tumbling of the cone induces intermittent cake movements and throughputs up to 120 te h^{-1}. Despite relatively modest centrifugal forces, deliquoring can be very good. However, tumbling centrifuges are usually only employed when washing is not required and coarser, faster filtering solids are present in the feed. Variations of cone angle and the speeds of rotation and gyration dictate the compromise between throughput and final moisture content of the cake.

Worm screen: Also known as the conveyor discharge or screen scroll centrifuge, the worm screen centrifuge causes solids to move along the cone via an internal screw conveyor. The conveyor rotates at a differential speed to the cone screen and centrifugal forces below 2600 g facilitate reasonable throughputs. The presence of the conveyor can sometimes lead to particle breakage and abrasion problems, as well as relatively poor washing. There is a compromise between throughput and final cake moisture and this is dictated by the conveyor speed. Typical cake residence times on the screen lie in the range 4–15 s. Worm screen centrifuges are available in either vertical or horizontal orientation and are most frequently used for the processing of fibrous solids/particles. More sophisticated versions employ cones with up to four stages that allow cake formation, two periods of displacement washing (with the potential for segregation of the wash liquors) and final deliquoring to take place.

10.4.3 Pusher

Typical uses: Deliquoring of relatively coarse particulate suspensions where good cake dryness at discharge is required.

The horizontal axis pusher centrifuge is probably the most commonly used design employing a continuous feed of suspension. A single-stage machine comprises of a rotating cylindrical screen bowl into which suspension is introduced and filtered to form a cake. A plate positioned at the closed end of the bowl reciprocates with a 20–80 mm stroke at up to 100 strokes min^{-1} to push the forming cake toward the open end of the bowl and discharge. During transition across the screen, the cake may be washed by sprays and efficiently deliquored as a result of the 500–1700 g centrifugal forces generated and 6–20 s cake residence times. In order for

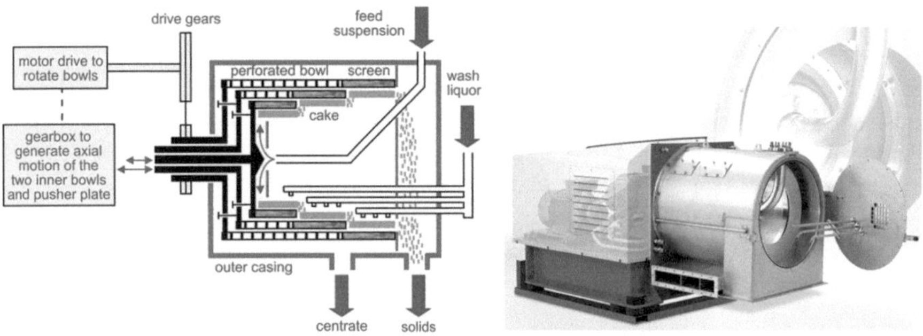

Fig. 10.24 Schematic of a three-stage pusher centrifuge (*left*, Courtesy of Filtration Solutions, UK) and photograph of a pusher centrifuge (*right*, Courtesy of Andritz)

the pusher centrifuge to work correctly the formed cake must have sufficient strength to withstand buckling and slide efficiently across the screen.

For weaker, more friable cakes, or cakes containing finer particles of greater frictional resistance, a multi-stage unit may be needed (see Fig. 10.24). These more expensive machines include a sequence of up to four, relatively short, concentric bowls with progressively increasing diameter. The solids are pushed more readily along the shorter bowls by the reciprocating piston and an ability to separate wash liquors allows for improved washing. The transfer between the bowls tends to lead to cake break-up which can enhance the deliquoring process.

As the open end of a bowl in a pusher centrifuge does not have a retaining lip there is an 'overflow limit' which generally limits solids throughputs to 80 te h^{-1} for single-stage machines and ~45 te h^{-1} for multi-stage machines; some manufacturers claim maximum throughputs of 100 te h^{-1}. The operation of both single and multi-stage centrifuges with solids below 100 μm can be problematical due to blockage of the filtering screen.

10.4.4 Baffle Centrifuge

Typical uses: Deliquoring coarse particulate (e.g. polymer pellet) suspensions where good cake dryness is required.

The family of baffle centrifuges are representative of specialist continuous filtering centrifuges. Both the baffle ring and screen baffle centrifuge can achieve very low residual moistures in granular type materials by causing particles to bounce against (baffle like) obstructions inside the rotating bowl to release additional surface and occluded liquids. Although baffle centrifuges are relatively expensive and restricted to operations with certain types of materials, their use can prove advantageous when other alternatives are unsuitable.

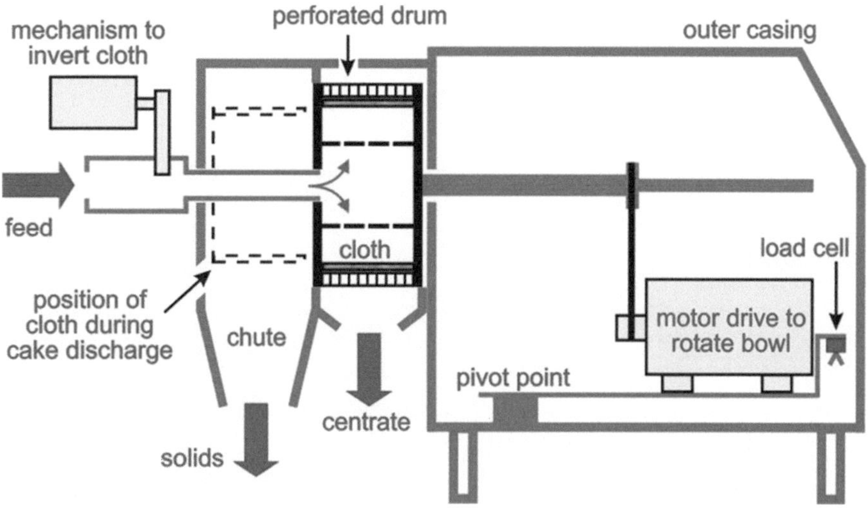

Fig. 10.25 Schematic of the inverting bag centrifuge which can meet clean room requirements, be totally enclosed and gas tight to maintain an inert atmosphere (Courtesy of Filtration Solutions, UK)

10.4.5 Inverting Bag

Typical uses: Semi-continuous deliquoring of suspensions where complete cake discharge and high purity need to be maintained.

The inverting centrifuge (Fig. 10.25) operates semi-continuously via automatic control and features a horizontally mounted, cylindrical drum between 0.3 and 1.3 m diameter which restricts filtration area to ~2 m^2. Suspension is introduced to the drum through gravity by means of a rigid filling pipe that projects through the solids discharge chute. The amount of material delivered is continuously monitored with a non-contact, load cell system resembling a beam type balance.

A typical cycle involves initial cake formation, intermediate deliquoring, rinsing/washing followed by final deliquoring. For discharge, a unique mechanism allows the end of the drum to open through a translational movement and the cake solids are removed completely under rotation as the filter bag inverts through the discharge chute. In this way cloth blinding is avoided and the entire cycle can be performed under high purity conditions. To improve deliquoring the pressure in the filling pipe and internal chamber of the drum can be raised, although any process advantages are offset by the increased mechanical complexity.

10.4.6 Modelling of Filtration

Fundamental theory for fluid flow through porous media has not been used generally for design calculations in filtering centrifuges, instead reliance is often placed

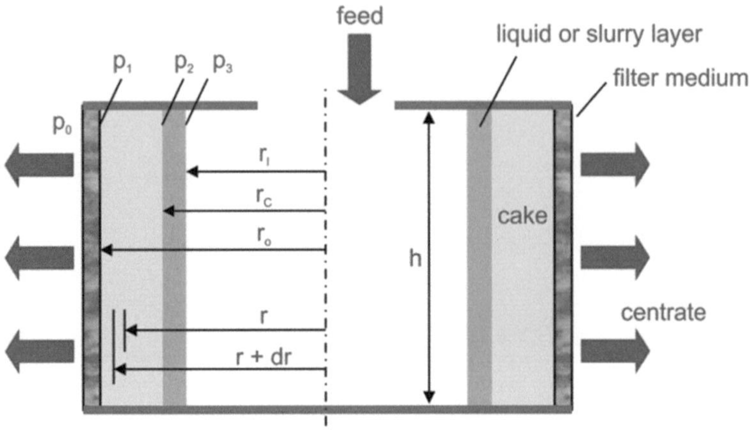

Fig. 10.26 Schematic cross-section through a centrifuge basket showing the filter medium, cake and liquid layers, and defining the notation used for the analysis of centrifugal filtration (Courtesy of Filtration Solutions, UK)

on practical experience and empirical approaches. Modelling difficulties arise from, for instance, the sometimes short processing times, uneven cake formation, variable feed properties, movement of the filter cake through the centrifuge and premature settling of particles. However, the general principles involved can be illustrated with a simple geometry like that associated with a basket centrifuge (see Fig. 10.26); the axis of rotation may be either horizontal or vertical.

By considering a differential cylindrical element of filter cake between radii r and $r + dr$, and assuming that Darcy's Law is valid for a basket radius r_0, height h and angular velocity ω, the volumetric flow rate of centrate can be expressed as:

$$q = \frac{\frac{\rho_l \omega^2}{2}\left(r_0^2 - r_l^2\right)}{\frac{\mu \rho_s \alpha_{av} C_{av}}{2\pi h} \ln\left(\frac{r_0}{r_c}\right) + \frac{\mu R}{2\pi r_0 h}} = \frac{\frac{\rho_l \omega^2}{2}\left(r_0^2 - r_l^2\right)}{\alpha_{av} \mu M_s \frac{\ln(r_0/r_c)}{2\pi h\left(\pi h\left(r_0^2 - r_c^2\right)\right)} + \frac{\mu R}{2\pi r_0 h}} \tag{10.37}$$

where r_0 is the inner radius of the basket, r_l the radius of the liquid layer, r_c the time variant radius of the cake and M_s the mass of solids in the cake. Values for r_c can be obtained from a volumetric balance that equates the solids in the centrifuge feed to the increase of cake volume in the basket

$$r_c = \left(r_0^2 - \frac{V_s}{\pi h C_{av}\left(1 - V_s\right)} qt\right)^{0.5} \tag{10.38}$$

where V_s is the solids volume fraction in the feed suspension. Use of Eqs. (10.37) and (10.38) allows the amount of liquid removed and solids accumulation in the bowl to be determined as a function of time. Similar theoretical analyses are available for pusher centrifuges [10, 35].

10.4.7 Modelling of Cake Washing and Deliquoring

Cake washing and deliquoring on filtering centrifuges is modelled in a similar manner to that described in Sects. 10.2.8 and 10.2.9. For deliquoring the principal differences relate to a design chart where the plot of S_R vs. θp^* involves a family of curves with each curve corresponding to a specified value of r_c/r_o, and θ and p^* are defined slightly differently to account for the centrifugal driving force. There is also no corresponding air/gas flow design chart, again because of the different manner in which the process is driven.

10.5 Centrifugal Sedimenters

Sedimenting centrifuges, which typify this category of equipment, employ centrifugal forces to accelerate the settling of particles (or liquid droplets) within rotating, solid walled equipment. With no filtration occurring, a density difference must exist such that the denser material in the feed preferentially settles to the wall where it is removed in a concentrated form. The clarified phase is also discharged, often at the opposite end to the feed position. Both batch and continuous types are available and typical duties range from the clarification of dilute suspensions to the thickening of fast settling slurries. When fast acting flocculants are used to aid separation, the shear resistance of flocs is an important factor as centrifugal forces can be high depending on the type of machine. Some typical characteristics of sedimenting centrifuges are shown in Table 10.4 and exemplar capital equipment costs for variants made from stainless steel can be estimated using equations adapted from Leung [15]:

$$\begin{array}{ll} \text{Tubular bowl}(D = 4 - 15\,\text{cm}): & < \$140,000 \\ \text{Disk}(D = 0.15 - 0.9\,\text{m}): & \$60,000 \; - \; \$350,000 \\ \text{Decanter}\,(D = 0.3 - 1.3\,\text{m}): & -1.2 \times 10^5 D^2 + 5.6 \times 10^5 D - 36000\,\$ \end{array} \tag{10.39}$$

10.5.1 Tubular Bowl Centrifuge

Typical uses: Batch clarification and (occasionally) particle classification.

The tubular bowl centrifuge shown in Fig. 10.27 is usually regarded as the most efficient of the industrial sedimenting centrifuges with separating forces in the range 14,000–65,000 g. The circular bowl, which typically has a diameter between 4 and 15 cm and an aspect ratio between 4:1 and 8:1, is vertically mounted and rotates at up to 50,000 rpm. The relatively dilute feed suspension is introduced at the bottom of the bowl via a distributor and centrate overflows from the top. The normally denser solids accumulate at the wall of the bowl from where they are

Table 10.4 Typical characteristics of some industrial sedimenting centrifuges

	Tubular bowl	Basket	Disk stack	Scroll decanter
Centrifugal force or g-factor (g)	14,000–65,000	Up to 1600	Up to 14,000	2000–6000
Rotational speed (rpm)	50,000 (max)	450–3500	3000–10,000	1600–6000
Mode of operation	Batch	Batch	Batch or continuous	Continuous
Throughput ($m^3\ h^{-1}$)	<5	6–10	1–100	<150
Sediment condition	Pasty, firm	Firm	Pasty, flowable → firm	Pasty, granular
Bowl diameter (m)	0.04–0.15	0.3–1.8	0.15–0.9	0.15–1.4
Bowl height or length (m)	0.2–0.75	0.2–0.8	Depends on no. disks	0.35–4.5
Solids dryness index	3 S	2 S	2 S	4 C
Washing index	–	–	–	3
Liquid clarity index	6	5	–	4
Particle breakage index	5	5	6	3
Particle size in feed (μm)	0.1–100	0.1–100	0.1–100	1–5000
Feed conc. (%w/w)	<5	<5	0.05–10[a]	4–40
Power requirement (kW)	<4	10–40	1–90	4–450

[a]Depends on variant

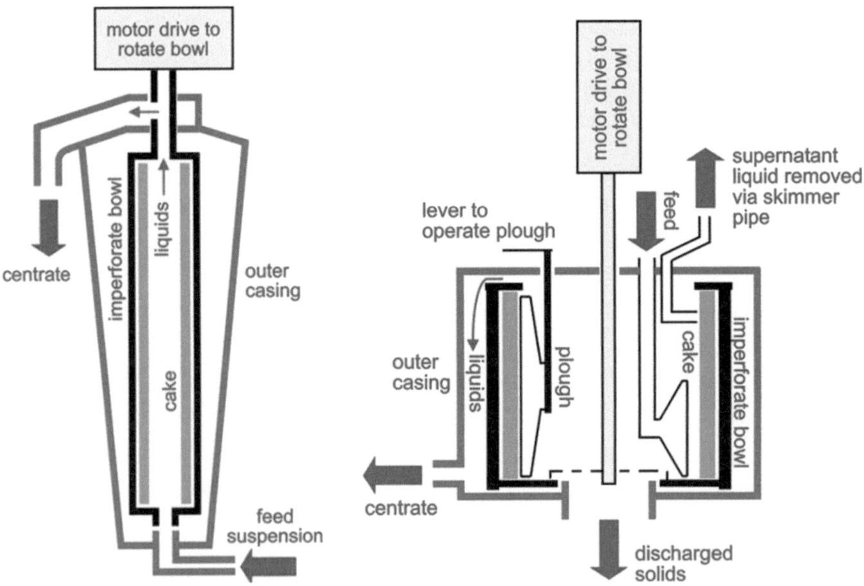

Fig. 10.27 Operating principle and general form of the tubular bowl centrifuge (*left*) and basket sedimenting centrifuge (*right*) (Courtesy of Filtration Solutions, UK)

manually discharged at the end of a batch cycle; the discharge process is sometimes aided by the inclusion of paper liners. Due to the narrowness of the bowl the efficiency of separation is significantly influenced by solids accumulation at the wall and throughputs are restricted to <5 m^3 h^{-1} with dry solid yields of up to 4 kg per batch.

10.5.2 Basket Centrifuge

Typical uses: Recovery and concentration of solid sludges.

The basket bowl centrifuge (Fig. 10.27) works on similar principles to the tubular bowl centrifuge. Semi-continuous operation is achieved, however, by using a bowl diameter between 25 and 180 cm and a much lower aspect ratio of ~0.6:1. The solids accumulate on the wall of the imperforate bowl and liquid overflows via a weir at the top. Cake discharge is performed manually on small machines with the bowl stationary. With larger machines the supernatant liquid remaining in the bowl is siphoned-off using a skimmer pipe and the solids on the wall are removed automatically with a plough, sometimes at a reduced bowl speed.

Basket bowl centrifuges operate at rotational speeds of 450–3500 rpm to allow throughputs in the region 6–10 m^3 h^{-1}. The use of relatively low g-forces in the single bowl centrifuge (<1600 g) has led to the development of multi-bowl basket centrifuges which comprise a series of concentric bowls mounted on a common vertical shaft to facilitate higher g-factors of *ca.* 5000–9000 g. During rotation the feed moves progressively from the inner to the outer bowl with ever finer particles being removed each time. Multi-bowl basket centrifuges offer greater efficiency for a given speed of rotation with the disadvantages of increased capital and operating costs.

10.5.3 Disc Stack Centrifuge

Typical uses: Clarification and thickening to produce a solids sludge, useful for bioprocessing.

The disc stack centrifuge is a versatile device which may be used for separating mixtures in continuous, semi-continuous and batch configurations (see Fig. 10.28). All except some batch operated machines are able to handle toxic, flammable and volatile feeds at typical throughputs of up to 100 m^3 h^{-1} (up to 200 m^3 h^{-1} is claimed in some cases). Particle-liquid and liquid-liquid mixtures can be separated, and with more sophisticated units a three phase separation (two liquid and one solid) is achievable. In all cases a sufficient density difference must exist between the phases present in the feed.

Although variants exist, the generic type is characterised by an imperforate bowl surrounding an inverted stack of 30–200 thin conical discs separated by 0.3–3 mm

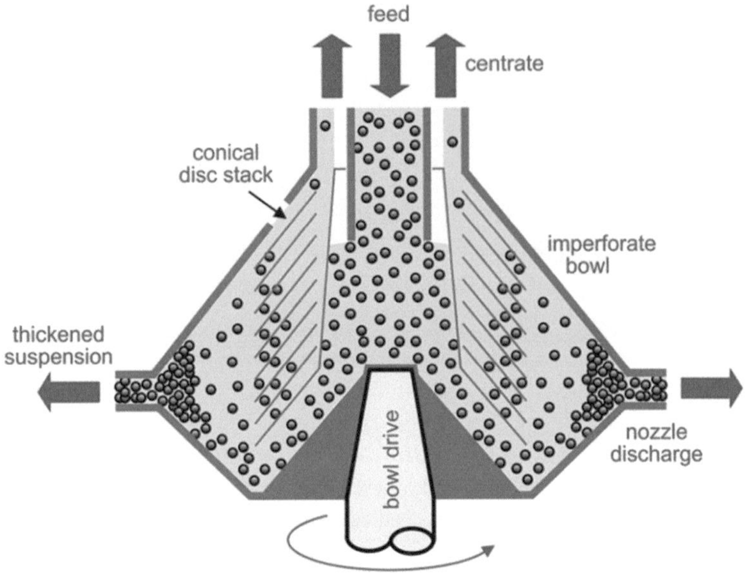

Fig. 10.28 Schematic of a nozzle discharge disc stack sedimenting centrifuge (Courtesy of Filtration Solutions, UK)

spacers. The disc spacing is dependent on the viscosity and solids content in the feed and needs to be fixed accordingly, lower viscosities and solids concentrations favour spacings below 1 mm. As the discs are spun on a common vertical axis the process suspension, which is fed centrally from the top and assumed here to comprise of particles dispersed in a single liquid, travels through the annular spaces between the discs. Centrifugal forces up to 14,000 g cause the particles to accumulate on the underside of the discs from where they slide down toward the outer periphery of the centrifuge bowl to form the sediment.

In batch units the sediment remains in the bowl until the so-termed 'solids handling capacity' of the centrifuge is reached. At this point rotation stops and the basket containing the sediment is manually replaced or a discharge valve on the periphery of the bowl is manually operated to facilitate sediment removal. In continuous units the sediment, which must be flowable, is automatically discharged, sometimes intermittently, through nozzles positioned on the outer periphery of the bowl; a typical centrifuge has between 12 and 24 nozzles of 0.5–3 mm diameter. For sediments that exhibit poor flow characteristics, the 'self-ejecting' design variant allows the bottom portion of the centrifuge to automatically separate at periodic intervals and discharge the accumulated material.

Whilst disc centrifuges are able to accept a wide range of feeds, they are mechanically complex and often expensive. Moreover, the close stacking of conical discs means that mechanical cleaning is a challenge, and periodic chemical cleaning (with its subsequent disposal issues) may be a necessity.

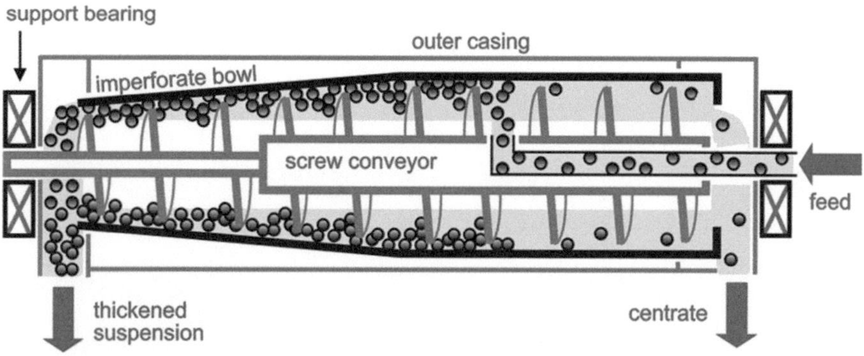

Fig. 10.29 Schematic of a horizontal axis scroll decanter centrifuge. The motor drive for the bowl and the gearbox required to produce the differential rotation speed between the bowl and screw conveyor are omitted for clarity (Courtesy of Filtration Solutions, UK)

10.5.4 Scroll Decanter Centrifuge

Typical uses: Relatively coarse deliquoring and clarification of suspensions.

The scroll decanter is a horizontally or vertically mounted centrifuge which is best suited to the processing of free draining solids from higher concentration feeds (see Figs. 10.3 and 10.29). In extreme cases throughputs of solids can be as large as 100 te h^{-1} whilst liquid throughputs are normally less than 60 m^3 h^{-1}. In a typical unit a cylindrical bowl with a tapered, conical end (the beach) is caused to rotate at speeds between 1600 and 6000 rpm. Inside the bowl a helical screw rotates at a differential speed of up to $\pm$100 rpm. The feed enters through the central axis of the centrifuge where inertial forces of less than 6000 g cause the denser solids to sediment towards the imperforate wall of the bowl. The sediment is conveyed co-currently along the walls of the bowl by the helical screw and moves through the narrower conical end of the centrifuge to discharge. The liquid phase, which may not always be clear due to the presence of fines, leaves the centrifuge via a weir or ports at the broader end of the bowl.

In the alternative screen bowl design the conical section is shortened and a supplementary cylindrical section which is perforated is attached in an effort to promote enhanced deliquoring. Thus, both sedimentation and filtration can be combined within a single unit. Other design variants rely on the addition of baffles, helical discs, vanes, conical disc stacks or fins, all of which alter the flow and/or residence time distributions within the centrifuge (e.g. Fig. 10.3).

When finer particles are being processed the flow properties of the thickened solids can be poor and this leads to high helical screw torques and associated mechanical difficulties. Wear problems on the screw are also caused by more abrasive particles. Scroll decanters can be adapted for use with toxic, flammable and volatile substances and there is some scope to perform (relatively poor) washing.

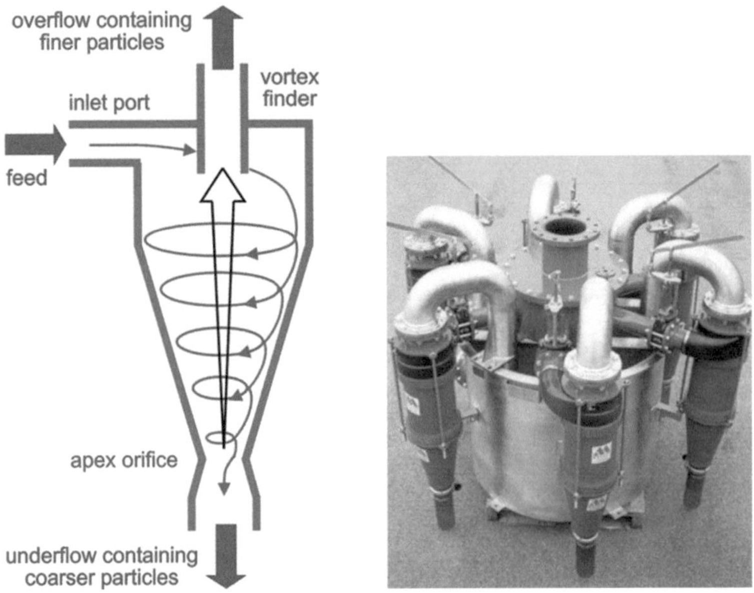

Fig. 10.30 Cross-section through a reverse flow hydrocyclone showing the typical flow patterns (*left*, Courtesy of Filtration Solutions, UK). The *inset* photograph (*right*, Courtesy of Axsia-Mozley) shows a bank of six cyclones connected to a common feed manifold system

10.5.5 Hydrocyclone

Typical uses: Suspension thickening, clarification and particle classification.

Another widely used form of centrifugal sedimenter is the reverse flow hydrocyclone (Fig. 10.30). Either concentration or classification of solids can be performed and the device is particularly attractive because it is relatively cheap, compact, versatile and has no moving parts.

The basic unit comprises an inverted conical bottom section attached to a cylinder containing a tangential inlet port. Feed is pumped through this port at a mean velocity between 10 and 30 m s^{-1} whence geometry induced motion causes the (usually denser) suspended particles to experience centrifugal forces of 70–18,000 g. The combination of these forces and a swirling motion causes the larger particles to exit as a suspension in the underflow stream at the bottom of the hydrocyclone and the finer fractions to leave through the cylindrical vortex finder at the top. With short residence times the particles and liquid move at relatively high speeds and abrasion/particle breakage can sometimes be a problem which necessitates the use of hard, and replaceable, internal linings.

Many standard sizes of hydrocyclone are available with cylinder diameters of 1–30 cm and cone angles of 25–50°. The particle cut size, which is the size equally likely to find its way into the underflow or overflow, is limited to about 5 μm and dependent on several factors including the size and geometry of the hydrocyclone, the inlet flow rate and the pressure drop across the unit. Separation is often more effective (in terms of a lower cut size) with smaller diameter hydrocyclones as higher tangential velocities can be achieved; in such cases overall throughput can be maintained by using several units in parallel (see also Chap. 12).

10.5.6 Modelling

Noting that the modelling of sedimenting centrifuges is inherently difficult and that all published models have their limitations, the most widely used and accessible model for characterising separations is Sigma (Σ) theory which is originally attributable to Ambler [1]. Sigma theory calculates the surface area of a static settling tank that gives the same theoretical performance as the centrifuge. Sigma is defined as

$$\frac{Q}{nu_t} = \sum \tag{10.40}$$

where Q is the volumetric feed flow rate, u_t the Stokes settling velocity given by Eq. (10.6) and n is a constant which is frequently taken as 2. The terms on the left hand side of Eq. (10.40) are solely functions of the process material whilst Σ is related solely to the characteristics of the centrifuge in accordance with the sample formulae given in Table 10.5; alternative formulae and more specific ranges for Σ are given in, for example, Leung [15] and Records and Sutherland [19]. Sigma theory can be used for individual calculations, scaling from one geometrically similar centrifuge to another and assessing relative performance.

Specific aspects for the design of hydrocyclones are provided by Svarovsky [25], but see also Hoffmann and Stein [12].

Table 10.5 Sample formulae and guideline values for Σ

Centrifuge type	Equation for Σ	Σ values (m^2)
Disk stack	$\frac{2\pi\omega^2(N-1)\left(R_2^3 - R_1^3\right)}{3g\tan\theta}$	<150,000
Tubular bowl	$\frac{\pi\omega^2 L}{g}\frac{r_o^2 - r_l^2}{\ln\left(\frac{2r_o^2}{r_o^2 - r_l^2}\right)} \approx \frac{\pi\omega^2 L}{2g}\left(3r_o^2 + r_l^2\right)$	<5000
Decanter	$\frac{\pi\omega^2}{2g}\left(L_1\left(3r_o^2 + r_l^2\right) + \frac{L_2}{2}\left(r_0^2 + 3r_o r_l + 4r_l^2\right)\right)$	<25,000

In the table N is the number of disks, R_1 the inner disk radius, R_2 the outer disk radius, θ the conical half angle, L_1 the length of the cylindrical section and L_2 the length of the conical section

10.6 Gravity Thickeners

A class of solid-walled separator where gravitational forces are used to raise the concentration of a suspension through sedimentation to produce a thickened sludge underflow and a clear liquid. The rate of sedimentation should be as high as reasonably possible to both increase throughput and reduce floor plan area. Sedimentation rates are often artificially increased by the addition of (relatively expensive) coagulants or flocculants. The cross-sectional area of a thickener controls the time available for sedimentation and is important in determining clarification capacity. The physical depth of a separator controls sludge thickening time and is an important parameter in determining thickening capacity. Thickeners and clarifiers can be designed to operate in either batch or continuous modes, although most commercial operations utilise the latter (see Table 10.6). An estimate for the capital cost of gravity thickeners is based on the underflow volumetric flow rate (Q_s) and adapted from ASCE and AWWA [2]:

$$\text{Gravity thickener}\left(Q_s = 0.002 - 0.5\,\text{m}^3\text{h}^{-1}\right) : -3.1 \times 10^5 Q_s^2 + 3.6 \times 10^5 Q_s + 39000\,\$ \tag{10.41}$$

Table 10.6 Typical characteristics of continuous thickeners and clarifiers

	Circular basin	Circular high capacity	Settling tank or lagoon	Deep cone	Lamella separator	Clarifier[a]
Diameter (m)	3–200	4–18	As required	3–25	As noted[b]	4–12[d]
Diameter/depth ratio	2:1–>10:1	3:1	1:1–>10:1	1:1–1:2	N/a	Varies
Solids dryness index	1 S	1 S	1 S	1 S	1 S	1 S
Washing index	2	–	–	–	–	–
Liquid clarity index	5	5	5	5	5	6
Particle breakage index	9	9	9	9	8	9
Particle size in feed (μm)	0.1–500	0.1–300	0.1–500	0.1–500	1–150	1–50
Feed conc. (% w/w)	<20	<15	<20	<20	<15	<15
Rake motor torque (Nm)	3000–8×10^6	n/a	–	$<14 \times 10^6$	–	n/a
Power requirement (kW)	1–35	n/a	–	As noted[c]	–	n/a

[a]Primarily used to recover clear liquor from dilute suspension, but operate on a similar principle
[b]Typically 1/5–1/10th the footprint of a circular basin thickener with an equivalent settling area
[c]Dependent on the rheological properties of the underflow, but tends to be higher for a given diameter compared to a conventional thickener
[d]Rectangular versions are also common

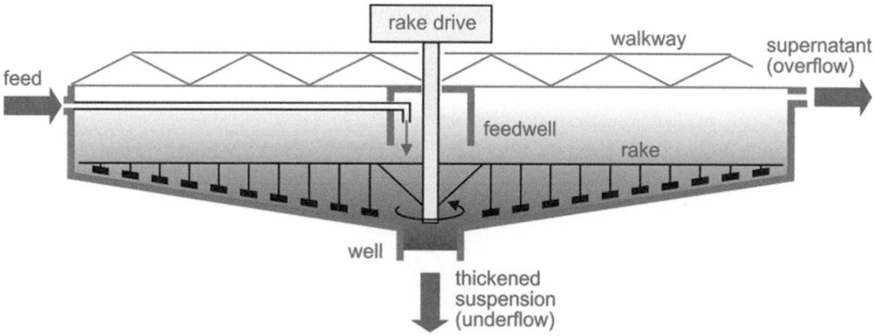

Fig. 10.31 Schematic of a circular basin thickener showing rakes, drive head and walkway (Courtesy of Filtration Solutions, UK)

10.6.1 Circular Basin Thickener

Typical uses: Larger scale thickening and deliquoring of solids from relatively dilute suspension.

The circular thickener comprises a relatively shallow, open top cylindrical tank with either a flat bottom or a bottom shaped in the form of an inverted cone (see Figs. 10.3 and 10.31). The feed mixture is gently and continuously introduced to the feedwell in which exists a pool of settling suspension along with any additional coagulant or flocculant. With settling and thickening proceeding, clear liquid is removed via an annular weir at the top of the unit and solids sludge (sediment) is removed from a 'well' at the bottom. Slowly rotating arms (or rakes) mounted on a central drive head aid the thickening process by directing the sediment towards the well for subsequent discharge, and by creating channels for the release of further liquid from the sediment. The construction and form of the rake are important design parameters as is the rating of the motor in the central drive head which must be capable of moving the rake through the sediment, both during normal operation and during start-up of the rake after a stoppage. Lifting devices are often employed to position the rakes at the optimum height within the basin and thus prevent damage due to excessive torque requirements.

Tanks with a diameter smaller than 25 m are usually formed from steel and have flat bottoms with rake arms at an angle less than 10°. Larger tanks between 25 and 200 m diameter are made from a combination of concrete and steel and employ rakes designed to match the angle of the conical bottom. Circular thickeners are frequently constructed to large scales and can be used to raise suspension concentration prior to another solid/liquid separation process.

Continuous discharge of solids from a gravity settling tank can be achieved without mechanical aid if the tank is shaped so that the sludge flows naturally towards the discharge port. This requires relatively steep sided conical vessels; the angle of the cone is generally 40–60° and thus the diameter of a settling tank is invariably rather less than a thickener. A diaphragm baffle is located near the base to prevent arching of solids across the outlet port.

Particularly large volumes of slowly settling slurries (which are also of low value) may be thickened in lagoons if land is not at a premium. Lagoons usually need to be lined to prevent seepage and are rarely an environmentally friendly option which limits their use.

10.6.2 High Capacity Thickeners

Typical uses: Separation of rapidly settling solids where available space is at a premium.

High capacity thickeners work on a broadly similar principle to conventional thickeners. However, by the correct use of high molecular weight, fast acting flocculants, large flocs that sediment very quickly can be generated to provide thickeners with high solids handling capacities and relatively small floor plan areas. A raised underflow concentration tends to be promoted by an increased thickener depth. Although high capacity thickeners have found many uses, particularly when a significant amount of fines are present, they do not represent a replacement for conventional thickeners as flocculant usage is notably higher and thus more costly.

10.6.2.1 Circular

The circular high capacity thickener is similar in general form to a conventional circular basin thickener (see Fig. 10.31), however, cylinder diameters are generally limited to between 4 and 18 m. Units are constructed from steel and include a cylindrical top portion, an inverted cone bottom section and an angled rake system mounted on a central drive head. Suspension throughputs are limited to about 4000 $m^3\ h^{-1}$.

10.6.2.2 Deep Cone

The deep cone thickener is again broadly similar in form to a conventional thickener (see Fig. 10.31) but the sides of the cylindrical section are longer in relation to the diameter and the inverted cone has much steeper angle in the region of 37°. Although originally conceived on a relatively small scale, units with diameters up to 24 m are now commonly available to process suspensions at throughputs of 850+$m^3\ h^{-1}$; a 40–45 m diameter version is also available. A paddle/rake system rotating at speeds between 0.25 and 2 rpm is usually added to aid the thickening process and facilitate final sludge/paste concentrations of 70+ % w/w. Although deep cone thickeners are relatively cheap to install and occupy a relatively small floor plan area compared to their throughput the operating costs can be higher. The flocculants required to promote efficient settling and operation are generally expensive and raised power inputs may be needed to maintain the stirring action of the paddle through (the often shear thinning) pastes that are characteristic of the high viscosity underflows.

10.6.2.3 Lamella

The lamella separator is characterised by an essentially rectangular tank containing a series of closely spaced rectangular plates inclined at an angle of ~50° to the horizontal (see Fig. 10.32). The plates, which effectively increase the available settling area, allow the sedimenting solids from the feed to slide down their upper surfaces towards a sludge hopper. The clarified liquid overflow is removed from a suitable opening near to the top of the tank. Commercial designs exist for three basic flow arrangements, namely cross-current, co-current and the most popular counter-current where the feed and clarified liquid flows can be most simply arranged. With plate spacings in the region of 50 mm, lamella separators offer a compact design which may be up to 90 % smaller than an equivalent conventional gravity settler. However, maldistribution and solids re-entrainment problems can sometimes limit their effectiveness, a problem which also occurs in an alternative design incorporating inclined tubes rather than plates.

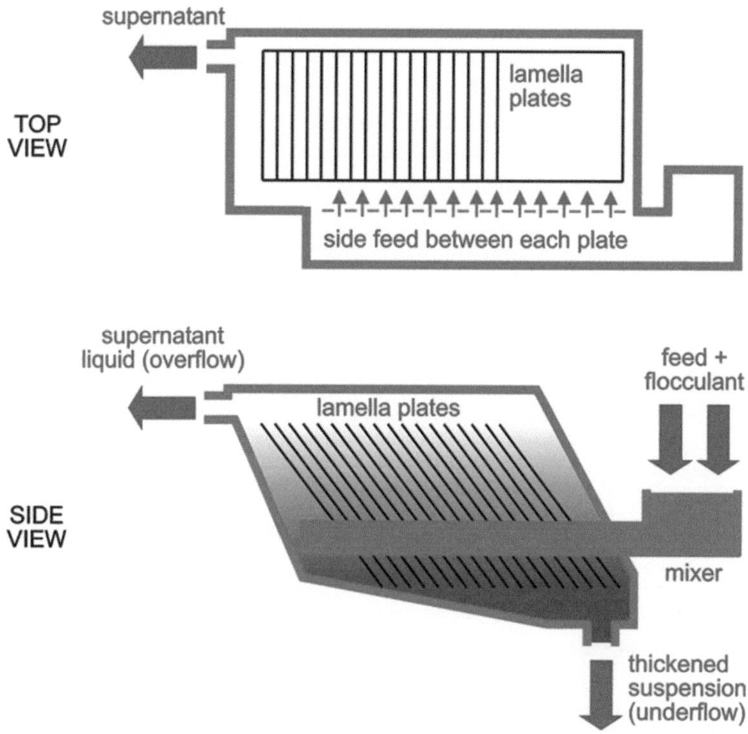

Fig. 10.32 Operating principle of a Lamella separator (Courtesy of Filtration Solutions, UK)

10.6.3 Modelling

Thickener design is traditionally based on either zone sedimentation or compression subsidence in order to establish critical thickener dimensions [9,11].

Zone sedimentation models are characterised by the approaches of Coe and Clevenger [4] and Talmage and Fitch [26], and rely upon graphical analysis of batch settling curves (Fig. 10.4); see also Kynch [13]. Both require jar settling experiments to be performed, although the method of Talmage and Fitch is generally preferred as (i) the amount of laboratory work required is much reduced and (ii) thickener capacity is underestimated such that there is an over-design in terms of floor plan area. Although there are exceptions, the model of Talmage and Fitch is generally best suited to unflocculated suspensions.

In flocculated systems, the flocs are essentially a networked structure of particles, and may have a very high porosity and be of considerable size. Therefore, even at low concentrations flocs may not settle as separate entities. As settling proceeds a compressive stress is developed in the forming sediment, so the flocs are not supported solely by their hydrodynamic drag. A suspension in compression should exhibit a compressive yield value which is a function of the solids concentration. These arguments underpin compression models such as those proposed by Michaels and Bolger [17], and Landman and White [14].

The most appropriate approach is often based on the material being separated. For instance, in the gravity thickening of biological sludges there is a strong emphasis on the compression mode of thickening whereas the design of thickeners for metallurgical and chemical suspensions relies heavily on behaviour in the settling zone. It is common to employ the theory of Talmage and Fitch to predict thickener area requirements from laboratory tests. However, when the feed suspensions are flocculated it is probably more appropriate to model the process on the basis of compression subsidence. The interested reader is directed toward the recently published work of Concha [5] which provides more detail on the modelling of gravity sedimentation.

10.7 Membrane Filters

Whilst it is evident that many of the separators in Sects. 10.2, 10.3, 10.4, 10.5, and 10.6 can effectively separate particles of less than 10 μm, when there is a significant fraction of fine material in the feed suspension, then its separation can become significantly more challenging. To be able to separate sub-micron and nano-sized materials it is common practice to use membranes. The principal characteristics of the most pertinent variants are shown in Table 10.7.

In the context of this chapter, membrane filters can be classified according to whether they are operated in a deadend mode (e.g. clarification and sterile filtration, where the feed moves normal/directly toward the membrane) or a crossflow mode

Table 10.7 Typical characteristics of some membrane separations

	MF	UF	NF	RO
Separation principle	Size	Size, charge	Size, charge, affinity	Size, charge, affinity
Typical rejected species	Silts, bacteria, cysts	Proteins, viruses, endotoxins	Sugars, pesticides	Salts, sugars
Separated size (μm)	0.1–20	0.001–0.1	~0.001	<0.0001
Pressure (bar)	0.2–1	1–5	3–15	10–60
Flux (m^3 m^{-2} day^{-1} bar^{-1})	>2	0.2–3	0.05–0.5	0.02–0.2
Recovery	90–99.99	80–98	50–95	30–90

MF is microfiltration, *UF* is ultrafiltration, *NF* is nanofiltration and *RO* is reverse osmosis

(e.g. thickening, where the feed moves tangential to the membrane to limit material deposition at the surface). Crossflow filters, with membrane areas at the bench scale of a few cm^2 to whole plants containing 1000s of m^2, are then differentiated according to the pore size in the membrane or according to the size of contaminant they will remove from the process stream. In crossflow (or low shear) filters the filter surface is stationary, and these are distinguished from dynamic (or high shear) devices which usually contain a moving surface. MF, UF, NF and RO units are similar in general form and by way of example some features of MF and UF are described in this section. The interested reader is directed to texts such as Cheryan [3] and Schaefer et al. [21] for more specific details.

10.7.1 Low Shear Crossflow

Low shear crossflow filters units usually comprise either a single membrane module or several modules arranged in a series configuration. The feed suspension is pumped at a constant rate and pressure into the module(s) and caused to flow tangential to the stationary semi-permeable membrane surface(s) at a typical linear velocity of 1–2 m s^{-1}. The shearing action at the membrane surface(s) limits material deposition to produce a relatively rapid permeate (liquid) flux decline toward the start of filtration followed by a near constant separation rate. In normal operation the permeate is collected and the retentate of thickened suspension is recirculated until the desired solids concentration is achieved, or pumping can no longer be performed satisfactorily.

Particulate deposition, fouling and adsorption of molecular species at the membrane surfaces often lead to lower than expected permeate fluxes. Whilst chemical cleaning and periodic backflushing/backpulsing with permeate or compressed gas can temporarily increase fluxes, large installed membrane areas may be required to achieve the desired separation rates. This, in conjunction with the pumping duty, means that both capital and operating costs of membrane units can be higher.

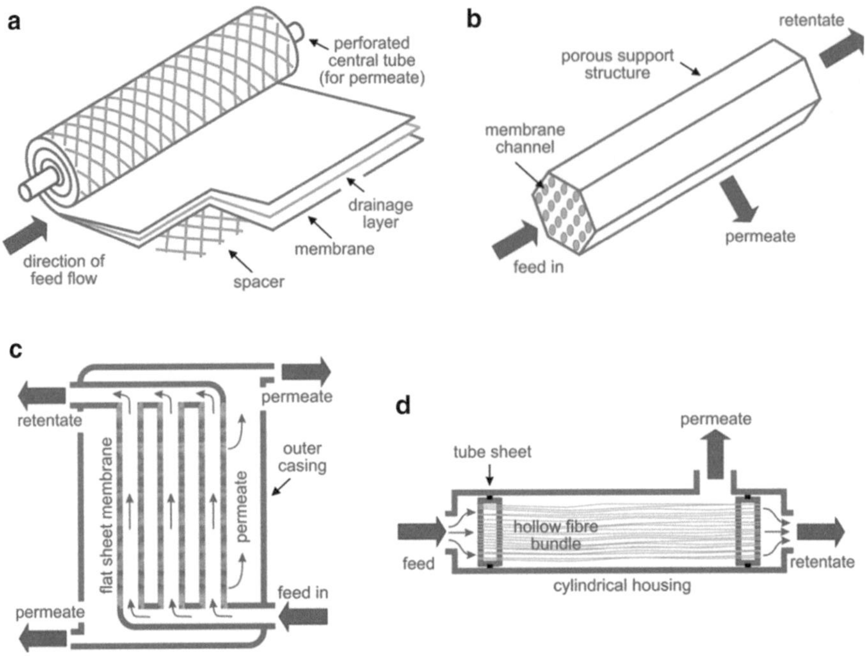

Fig. 10.33 Examples of common crossflow membrane arrangements. (**a**) Spiral wound; (**b**) tubular monolith; (**c**) plate and frame; (**d**) hollow fibre (Courtesy of Filtration Solutions, UK)

However, such costs are frequently offset by the ability to perform separations that are difficult, if not impossible, to achieve economically by other means and both ultra- and micro- filters are becoming the technology of choice in several industrial sectors.

In UF, which is typically used to separate macromolecules, viruses, bacteria, colloids and very fine suspended particles, the membranes are almost exclusively of an asymmetric, microporous construction and available with pore ratings in the range 0.001–0.02 μm. These are manufactured as microporous structures from a range of polymers and ceramics and formed as either flat sheets or tubes for use in one of four basic arrangements (see Fig. 10.33 and Table 10.8). Ultrafilters are usually operated as multiple-pass thickeners in either batch or continuous 'feed and bleed' modes (see Fig. 10.34) where the latter can be cascaded *ca.* three to seven times to produce a multi-stage recycle configuration.

Plate and frame: Flat porous plates covered with polymeric membrane material are assembled with alternate hollow spacers to produce a crossflow system where feed moves through the annular spaces between adjacent membrane surfaces. Although now largely superseded by other designs, this variant is still available with membrane areas up to 80 m^2.

Table 10.8 Relative comparisons between membrane arrangements

Parameter	Spiral wound	Tubular	Plate and frame	Hollow fibre
Availability	UF	UF and MF	UF and MF	UF and MF
Membrane surface per module volume ($m^2\ m^{-3}$)	600	25–50	350–600	600–1200
Investment cost	Medium	High	High	Low
Operating cost	Low	High	Low	Low
Flow control	Fair	Good	Fair	Good
Ease of in-situ cleaning	Poor-fair	Good	Fair	Fair

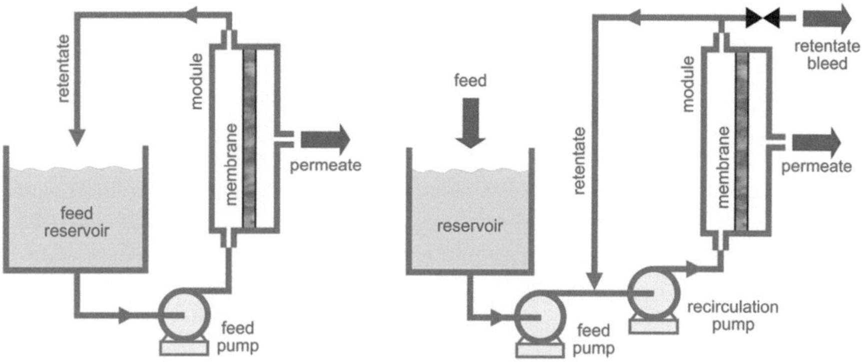

Fig. 10.34 Batch (*left*) and 'feed and bleed' (*right*) membrane plant configurations (Courtesy of Filtration Solutions, UK). The retentate from one 'feed and bleed' loop can be used as the feed to a second loop (and so on) in order to produce a cascaded or multi-stage configuration

Tubular monolith: A thin membrane layer which facilitates the separation is formed on the inside of a more robust, and open, monolith support. In earlier examples the separating layer was formed to constant depth along the length of the monolith, however, newer designs employ a progressively reducing depth that promotes better overall flux performance. A typical monolith is made from alumina or zirconia ceramic and may contain more than 30 individual channels of 4–7 mm diameter and length up to 1 m; some silicon carbide versions can contain over 200 flow channels. The feed passes along the inside of each channel to enable a separation to proceed. As many as 300 individual monoliths can be assembled into a module that is similar in form to a single-pass shell-and-tube heat exchanger.

Hollow fibre: Up to several thousand small diameter (*ca.* 40 μm to 2 mm), hollow tubular membranes are externally sealed at both ends into a larger diameter, solid cylindrical housing. The pressurised feed stream usually flows into this polymeric lumen with the permeate moving radially outward through the fibre walls ('inside-out' filtration). Hollow fibre systems can also be designed to have the feed flow on the outside of the fibres with the permeate collected from the inside of the fibres ('outside-in' filtration). Although hollow fibre ultrafilters offer the advantage of a large membrane area in a small volume, they can be prone to blockage, greater membrane fouling and also cleaning problems.

Spiral wound: Spiral modules are constructed using flat sheet polymeric membranes in the form of a pocket, consisting of two membrane sheets separated by a highly permeable mesh spacer which defines the region for permeate flow. The assembly is sealed using an appropriate epoxy or polyurethane adhesive along three edges. The open side of the pocket is glued to a central perforated tube that is used to collect the permeate flow. Several of these pockets are spirally wound around a single collecting tube using a feed-side mesh as a spacer between the pockets to establish the required feed channel thickness. Like the hollow fibre arrangement, the spiral wound module offers a large membrane area within a small volume but again suffers from potential blocking and cleaning problems.

Microfilters, which are used for the separation of viruses, bacteria, colloids and fine suspended solids, differ from ultrafilters primarily in the pore size range and construction of the membranes used to achieve a separation but are otherwise broadly similar. The polymeric, flat sheet types are usually of a symmetric construction and exhibit either microporous or track-etched forms to facilitate either depth or surface filtration. These membranes are manufactured with pore ratings of 0.02–10 μm. Ceramic and metal microfilters are also available for more extreme duties, e.g. pH range 0–14, as either flat sheet or tubular forms with pore ratings between 0.05 and 8 μm or 0.2 and 20 μm, respectively.

10.7.2 High Shear Crossflow

High shear crossflow filters offer many of the advantages of the low shear filters described in Sect. 10.7.1, but with the potential benefit of higher fluxes. Several variants exist.

A typical unit comprises a cylindrical pressure vessel enclosing 12–15 filter leaves of ~0.5 m diameter. The preferred type uses static circular filter elements with solid discs mounted between. The discs are attached to a central shaft rotating at constant speeds up to 2000 rpm. The rotation ensures the generation of relatively high shear forces and local suspension velocities in excess of 10 m s^{-1}. The feed is pumped into the pressure vessel at a rate dependent on its filtration characteristics and separation proceeds to produce a thickened suspension. As the feed thickens it invariably becomes more viscous with the result that significant rotational energy can be transferred to the feed in the form of heat and, perhaps more importantly, higher motor currents are required to turn the central shaft and discs. These disadvantages are offset by the inherent ability of the filter to decouple the shear generated at the separating surface from the overall suspension throughput. A typical filter utilises flat sheet microfiltration membranes or tightly woven, multifilament filter cloths.

Technical alternatives include versions where the filter media themselves rotate and a variant that employs vibration of the elements to enhance filtration. For the latter up to 100 double-sided, flat disc filter elements separated by thin spacers are clamped together, mounted within a vertically mounted cylindrical vessel and caused to vibrate by a motor drive assembly close to their resonant/natural frequency. Oscillatory motion in the plane of the horizontal elements produces a shear rate up to 150,000 s^{-1} which is many times greater than that observed in typical low shear crossflow filters. In this manner rheologically sensitive feeds can be processed and filter areas up to 200 m^2 can be accommodated.

10.8 Outlook for the Future

Whilst incremental improvements will continue to be made in the mechanical design and operation of separation equipment, many of the recent, and likely future, developments in filter technology centre around improvements to filter media. The introduction of, for instance, composite media which incorporate polymer coatings [16], multi-layer media with complex weaves and combinations of different media types, activated media which facilitate both filtration and adsorptive capacity [24], functionalised media for selective separations (e.g. membranes) and the inclusion of nanofibre webs in otherwise conventional filter media has allowed for improved separation efficiencies at lower pressure drops and reduced operating cost. Although still largely in their infancy, technologies such as 3D printing to create new filter forms by successive layer formation [29] and the use of carbon nanotubes in new membranes for water purification [7] offer great potential going forward.

Developments in filter media have not only been driven by better science and engineering but also by the increased use of computational fluid dynamics (CFD). CFD incorporates a powerful suite of tools for modelling fluid flow that has facilitated not only simulations of filter media [34] but also closer examination of the flow patterns in, for instance, gravity sedimenters and centrifuges. It is likely that CFD will continue to grow in prominence in future years, however, it is important to realise its limits and the ongoing need for experimentation. We are still a considerable way from being able to predict equipment performance from a basic knowledge of equipment characteristics and, more pertinently, fundamental fluid and particle properties and their interactions. This situation is reflected in Sects. 10.2, 10.3, 10.4, 10.5, and 10.6 where there is often a need for empiricism or heuristics in the models presented. Computer programs such as Filter Design Software [8] go some way to providing automated calculations of equipment performance (and selection/scale-up), but until tractable and accurate models of fundamental suspension behaviour are developed it is unlikely that we can progress beyond the widespread use of heuristics and personal experience when designing and specifying filtration and sedimentation equipment.

10.9 Definitions, Abbreviations and Symbols

Reynolds number	dimensionless number describing the ratio of inertial and viscous forces; its value indicates the type of fluid flow (e.g. laminar, turbulent)
Schmidt number	dimensionless number describing the ratio of momentum diffusivity to mass diffusivity

ASCE	American Society of Civil Engineers
AWWA	American Water Works Association
MF	microfiltration
NF	nanofiltration
RO	reverse osmosis
UF	ultrafiltration

A	filtration area, m^2
A_p	area of cloth in a single tilting pan, m^2
c	effective concentration of solids in feed or suspension, kg m^{-3}
C	cake solids volume fraction
C_e	modified consolidation coefficient, $m^2\ s^{-1}$
C_{e0}	modified consolidation coefficient at unit applied pressure, $m^2\ s^{-1}\ kPa^{-\gamma}$
C_0	cake solids volume fraction at unit applied pressure, $kPa^{-\beta}$
d	diameter of a filter candle, m
d_i	inner diameter of filtering surface on a disc or table, m
d_o	outer diameter of filtering surface on a disc or table, m
D	diameter of drum, basket or vessel, m, or molecular diffusivity of solute, $m^2\ s^{-1}$
D_L	axial dispersion coefficient, $m^2\ s^{-1}$
D_n	dispersion number
F	fraction of solute removed from a filter cake by washing
F_D	drag force on a sphere falling through a fluid, N
g	acceleration due to gravity, m s^{-2}
h	height of centrifuge basket, m
h_B	filter belt width, m
h_D	drum width, m
i	number of drainage surfaces
j_{II}	consolidation area factor applied to 2-dimensional expression
k	permeability, m^2
L	thickness of filter cake or length of centrifuge bowl, m
m_{av}	ratio of mass of wet cake to mass of dry cake
m	mass of particle, kg
M	moisture content of a filter cake (ratio of mass of liquid in cake to total mass of wet cake)
M_s	mass of solids in a filter cake, kg

N	number of disks in a centrifuge
n	compressibility index or efficiency factor in a sedimenting centrifuge
n_d	number of discs
n_p	number of tilting pans
p	pressure, usually hydraulic pressure delivered by the pumping system, Pa
Δp	pressure difference, Pa
p_b	threshold pressure or vacuum, Pa
p_0–p_3	pressure at a given position, Pa
q	$= dV/dt$, volume flow rate of liquid (e.g. filtrate), $m^3\ s^{-1}$
Q	volumetric feed flow rate, $m^3\ s^{-1}$
Q_s	volumetric sludge (underflow) flow rate, $m^3\ s^{-1}$
r	radius of rotation, m
r_c	radius of cake layer in centrifuge, m
r_l	radius of liquid layer (or pool/pond) inside a centrifuge, m
r_0	inner radius of centrifuge bowl/basket, m
R	resistance to fluid flow through a filter medium, m^{-1}
R_1	inner radius of a conical disk, m
R_2	outer radius of a conical disk, m
Re	Reynolds number $= \rho_l u_c x_c / \mu_l$, where u_c and x_c are characteristic velocity and linear dimension, respectively
s	mass fraction of solids in a feed suspension
S	saturation, volume of liquid in a cake per unit volume of the voids
Sc	Schmidt number $= \mu_l/(\rho_l D)$
S_R	reduced saturation
S_0	specific surface of particles, $m^2\ m^{-3}$
S_∞	irreducible saturation
t	time, s
T_c	dimensionless consolidation time
u	superficial velocity, $m^3\ m^{-2}\ s^{-1}$, or particle-fluid relative velocity, $m\ s^{-1}$
u_t	terminal velocity, $m\ s^{-1}$
v_B	linear velocity of a filter belt, $m\ s^{-1}$
V, V_f	filtrate volume, m^3
V_s	solids volume fraction in a suspension or solid/liquid mixture
W	wash ratio (amount of wash liquid passed through a cake per unit amount of liquid in the cake at start of washing)
x	particle diameter/size, m
z	co-ordinate direction, or length of filter devoted to a particular process

α	specific resistance of a filter cake, $m\ kg^{-1}$
α_0	specific resistance at unit applied pressure or at zero applied pressure, $m\ kg^{-1}\ kPa^{-n}$
β	compressibility index
γ	compressibility index
ε	porosity, volume of voids per unit volume of filter cake (or porous medium)

λ	compressibility index
μ	viscosity of liquid in a feed or filtrate, Pa s
ν	consolidation index
θ	dimensionless time, or conical half angle of a disk, °
ρ, ρ_l	density of liquid in a feed or filtrate, kg m^{-3}
ρ_c	bulk density of a filter cake, kg m^{-3}
ρ_s	density of solids or particles, kg m^{-3}
σ	surface tension, N m^{-1}
φ	fraction of filter area
ω	angular velocity, s^{-1}
ω_0	volume of solids per unit filter area, m^3 m^{-2}

Subscripts

av	average value
c	referring to consolidation phase
f	referring to filtration phase, or to filtrate
l	referring to liquid phase
s	referring to solid phase
w	referring to washing phase
0	initial value, unless otherwise stated
∞	equilibrium value

Superscripts

*	dimensionless value (unless otherwise stated)

References

1. Ambler, C.M.: Centrifugation. In: Schweitzer, P.A. (ed.) Handbook of Separation Techniques, 2nd edn, pp. 4.59–4.88. McGraw-Hill, New York (1988)
2. ASCE and AWWA:Management of Water Treatment Plant Residuals. USA (1996)
3. Cheryan, M.: Ultrafiltration and Microfiltration Handbook, 2nd edn. CRC Press, Boca Raton (1998)
4. Coe, H.S., Clevenger, G.H.: Methods for determining the capacities of slime thickening tanks. Trans. AIME **55**, 356–384 (1916)
5. Concha, F.: Solid-Liquid Separation in the Mining Industry (Fluid Mechanics and Its Applications). Springer, Heidelberg (2014)
6. Couper, J.R., Penney, W.R., Fair, J.R., Walas, S.M.: Chemical Process Equipment Selection and Design. Butterworth-Heinemann, Burlington (2010)
7. Das, R., Ali, M.E., Hamid, S.B.A., Ramakrishna, S., Chowdhury, Z.Z.: Carbon nanotube membranes for water purification: a bright future in water desalination. Desalination **336**, 97–109 (2014)

8. Filter Design Software:Personal Computer Software for Solid/Liquid Separation Equipment Selection, Scale-up and Simulation, ca. 30,000 lines of code. Filtration Solutions, Exeter, Devon (2005)
9. Fitch, B.: Current theory and thickener design. Ind. Eng. Chem. **58**(10), 18–28 (1966)
10. Hallit, J.: Sugar and sugar centrifuges. Filt. Separ. **12**, 675–680 (1975)
11. Hasset, N.J.: Mechanisms of thickening and thickener design. Trans. IMM **74**, 627–656 (1965)
12. Hoffmann, A.C., Stein, L.E.: Gas Cyclones and Swirl Tubes, 2nd edn. Springer, Berlin (2008)
13. Kynch, G.J.: A theory of sedimentation. Trans. Faraday. Soc. **48**, 166–176 (1952)
14. Landman, K.A., White, L.R.: Determination of the hindered settling factor for flocculated suspensions. AIChE J. **38**, 184–192 (1992)
15. Leung, W.W.F.: Industrial Centrifugation Technology. McGraw-Hill, New York (1998)
16. Lydon R.L.: Developments in filter media, Presentation at the Filtration Society 50th Anniversary Conference, Chester, November 13–14 (2014)
17. Michaels, A.S., Bolger, J.C.: Settling rates and sediment volumes of flocculated kaolin suspensions. Ind. Eng. Chem. Fundam. **1**, 24–33 (1962)
18. Purchas, D.B., Sutherland, K.S.: Handbook of Filter Media, 2nd edn. Elsevier, Oxford (2002)
19. Records, A., Sutherland, K.: Decanter Centrifuge Handbook. Elsevier, Oxford (2001)
20. Richardson, J.F., Zaki, W.N.: Sedimentation and fluidisation, part 1. Trans. IChemE **32**, 35–53 (1954)
21. Schaefer, A., Fane, A.G., Waite, T.D.: Nanofiltration: Principles and Applications. Elsevier, Oxford (2004)
22. Shirato, M., Murase, T., Negawa, M., Moridera, H.: Analysis of expression operations. J. Chem. Eng. Jpn. **4**, 263–268 (1971)
23. Shirato, M., Murase, T., Iritani, E., Tiller, F.M., Alciatore, A.F.: Filtration in the chemical process industry. In: Matteson, M.J., Orr, C. (eds.) Filtration. Marcel Dekker, New York (1987)
24. Smith, A., McGee, D., Lang, B.: Filtration and catalytic VOC destruction. Filtration **10**(3), 190–195 (2010)
25. Svarovsky, L. (ed.): Solid-Liquid Separation, 4th edn. Butterworths, London (2000)
26. Talmage, W.L., Fitch, E.B.: Determining thickener unit areas. Ind. Eng. Chem. **47**, 38–41 (1955)
27. Tarleton, E.S., Wakeman, R.J.: Solid/Liquid Separation: Equipment Selection and Process Design. Elsevier, Oxford (2006)
28. Tarleton, E.S., Wakeman, R.J.: Dictionary of Filtration and Separation. Filtration Solutions, Exeter (2008)
29. Vijayakumar, B., Rennie, A., Burns, N., Travis, D., Battersby, P.: Introducing functionality to filter media. Filtration **14**(4), 217–222 (2014)
30. Wakeman, R.J., Attwood, G.J.: Simulations of dispersion phenomena in filter cake washing. Trans. IChemE **68**, 161–171 (1990)
31. Wakeman, R.J., Vince, A.: Kinetics of gravity drainage from porous media. Trans. IChemE **64**, 94–103 (1986)
32. Wakeman, R.J., Tarleton, E.S.: Solid/Liquid Separation: Principles of Industrial Filtration. Elsevier, Oxford (2005)
33. Wakeman, R.J., Tarleton, E.S. (eds.): Solid/Liquid Separation: Scale-Up of Industrial Equipment. Elsevier, Oxford (2005)
34. Wiegmann, A., Knefel M.:10 years of filtration simulation: what works and challenges, Presentation at the Filtration Society 50th Anniversary Conference, Chester, November 13–14 (2014)
35. Zeitsch K.: Proceedings of the International Symposium, Société Belge de Filtration, Antwerp (1977)

Chapter 11
Flotation

Henk G. Merkus

Abstract The recovery of valuable minerals has developed in the twentieth century into a major economic activity. Flotation processes are now the preferred technique for recovery because of a better understanding and treatment of ore surfaces for selective and improved mineral recovery as well as improvements to the equipment used in the process. Flotation is also applied for treatment of waste water and de-inking of recycled paper. This chapter gives an overview of the flotation process, its background, the equipment used and the important role it plays in the recovery of precious materials from waste.

11.1 Introduction

Modern flotation technology has a long history of more than a century. It typically uses air to remove particles from a suspension and is based on the differences in surface properties. The first idea that differences in surface properties can be used to separate sulfide and non-sulfide minerals appeared in the British patent of 1860 by W. Haynes [7]. The first commercial flotation plant was built in 1877 by the Bessel brothers in Dresden, Germany, to improve the graphite quality from corresponding ores, after adding some oil to the particulate mass and then water, first by boiling the water to use the steam bubbles and later by producing carbon dioxide from alkali carbonate plus acid [2, 3, 6, 10]. Its more wide-spread application in the mining industry began around 1900 when several patents were granted that were still primarily based on empirical findings. The first commercial process in mineral processing was introduced by Potter and Delprat at a mine in Broken Hill, Australia, in 1905. During the twentieth century more knowledge was gained about the mechanisms and chemicals that had been developed, as well as the effect of different particle sizes, and this enabled selective separation of different

H.G. Merkus (✉)
Em. Prof., Chemical Engineering Department, Delft University of Technology, Pijnacker, The Netherlands
e-mail: henkmerkus@hetnet.nl

© Springer International Publishing Switzerland 2016

H.G. Merkus, G.M.H. Meesters (eds.), *Production, Handling and Characterization of Particulate Materials*, Particle Technology Series 25,
DOI 10.1007/978-3-319-20949-4_11

minerals. This and the development of new equipment significantly increased the effective utilization of mineral resources, even of low-grade ores. Nowadays, about two billion tons of ore per annum are processed world-wide by flotation. At the same time, flotation has also successfully been applied in treating waste water (removal of fats, oil, grease and suspended solids) and in paper recycling (de-inking).

11.2 The Flotation Process

Flotation is a process that separates various particles based on differences in their surface properties through using air/gas bubbles.

In the minerals industry, flotation consists of the following steps:

(a) comminution of the ore to small particles
(b) addition of chemicals to enable the selective attachment of target particles to air bubbles and sometimes to cause flocculation
(c) addition of air (gas) bubbles to attach the particles and transport them to an upper froth layer
(d) separation of froth from pulp and gangue, including froth washing
(e) thickening and drying of the flotation product to obtain the valuable mineral.

Ad a. The comminution goal depends on the distribution of the valuable mineral in the ore as well as on the potential of the particles to be collected by the gas bubbles. Comminution is usually executed sequentially by blasting, crushing and milling (see also Chap. 6). Many minerals have an oxide surface, other minerals have, e.g., a sulfide surface, which can oxidize easily in contact with air. When oxide surfaces of such sulfide minerals are to be avoided, then wet milling in a nitrogen environment is often applied. The typical particle size of the minerals in flotation is about 50–120 μm, but smaller sizes (down to about 1 μm) and larger sizes (up to 250 μm) may also occur. In conventional machines, ultrafine particles require very long residence times, while very coarse particles are easily detached from the bubbles. Since comminution requires a lot of energy, optimization of the subsequent steps can save much money. Note that sedimentation of valuable mineral particles may be applied when their size is larger than about 70 μm. The comminution step is usually unnecessary in treatment of wastewater and in recycling of paper.

Ad b. In the flotation process for separation of minerals, surface chemistry is the principal determinant for success. Adhesion of particles to air/gas bubbles selectively occurs if their surface is hydrophobic. The application of specific chemicals and pH can make this process specific for a given mineral. While the chemicals modify the solid surfaces in a way that results in specificity, the surface active agents (surfactants) enable a hydrophobic particle surface to displace water by air and to make the adhesion of the particles to air bubbles attractive. See further Sect. 11.3.

Ad c. Small air or gas bubbles (about 0.3–3 mm size) are created to levitate the particles to an upper froth layer that can be easily separated from the tailings. Particles attach to bubbles only if their surface is sufficiently hydrophobic so that air can replace the water surrounding the particles. Moreover, the bubble size needs to be sufficient to levitate the particles. This gives an upper limit to the size and number of attached particles. Note that flocs are assemblies of small particles connected by weak attractive forces and occupying a large volume, so that they have only a small relative density relative to the real particle density.

Ad d. The separation of froth and pulp is based on the density difference between the two, due to the presence of air in the froth. A stable layer of froth of controlled depth forms in the flotation vessel. A countercurrent flow of clean water can be used in some instances, to wash entrained impurities or gangue back into the cell. The froth flows over the top of the flotation vessel, carrying the valuable particles.

Ad e. The recovery of the valuable material from the froth requires thickening, filtration, drying and smelting. If it is waste material, no further cleaning is done; only its volume is reduced to a minimum – e.g. by thickening and drying – for economic reasons. The rejected gangue material is usually discharged to a dam or sedimentation basin.

11.3 Flotation Background

As indicated in the previous section, adhesion of the particles to air or gas bubbles – for oxide surfaces aided by the addition of various specific chemical substances – is essential to an optimum and selective flotation process. Flocculation may be helpful as it may contribute to improved dewatering of the froth. The chemicals all have their own functions in enabling and optimizing the selective separation of one or more specific minerals from an ore. They can be grouped into [1, 15]:

- frothers: i.e. neutral substances that control the bubble size and the stability of the froth through concentration in the bubble surface. Examples are: polyethylene glycols, polypropylene glycols, (branched) aromatic and aliphatic alcohols and triethoxy butane;
- collectors: i.e. surface-active agents (surfactants) that adsorb to or react with the surface of the particles and cause them to become hydrophobic. Examples are: water-soluble xanthates, dithiophosphates, dialkyldithiocarbamate, various oils, alkyl sulfates and alkyl sulfonates;
- modifiers: i.e. inorganic or organic substances that control the pH of the pulp or modify the electric surface charge. Examples are: inorganic acids, alkalis and sodium carbonate;
- activators: i.e. chemicals that react with a specific mineral surface, thus activating collector adsorption onto the surface, which then enables flotation. Examples are: ferric chloride for quartz and copper sulfate for sphalerite (ZnS);

- depressants: i.e. chemicals that adsorb or react with the solid surface and prevent collector adsorption or bubble attachment to unwanted mineral surfaces through making them hydrophilic. Examples are: starch, sodium silicate, sulfur dioxide, sodium cyanide and alkali sulfides;
- flocculants: i.e. natural and synthetic polymers that assist in dewatering of the flotation concentrates through adsorption to adjacent particles. Flocculation is promoted by low or zero electric surface charge, i.e. near the iso-electric pH.

Thus, adequate treatment of the surface of the target mineral that results in selective hydrophobicity, as well as good contact between particles and air bubbles, are necessary requirements for optimum and selective flotation.

Broadly speaking, it can be said that a particle can have either a polar or a non-polar surface. Oxide surfaces are typically polar; carbonaceous or sulfide surfaces are non-polar when treated with a suitable collector. Polar surfaces usually are hydrophilic: they readily adsorb water and other polar substances, so that particles having such surfaces do not easily attach to air bubbles in an aqueous suspension. Non-polar surfaces are hydrophobic: they repel water, so readily make contact with an air-water interface. Particles that are naturally hydrophobic often do not require collectors for flotation. On the other hand, the surface of sulfides may be oxidized and so easily become polar. The same holds for graphite and coal.

The surfaces of suspended inorganic particles in water typically bear an electric charge. The sign and amount of this charge depends mainly upon the pH, through reaction of surface groups with protons (H^+) or hydroxyl ions (OH^-) and/or through adsorption of oppositely charged ions. Usually potential energy curves are used to represent the interaction between particles as a function of their separation distance. By convention, repulsive forces have a positive sign and attractive forces a negative sign. The net potential energy curve, i.e. the balance between repulsive and attractive forces with varying distances, determines the stability of a dispersion (Fig. 11.1). If there is a maximum, its height indicates the potential barrier for clustering; if there is a minimum, its depth reflects the stability of a cluster. When the attractive forces are relatively strong, the dispersion will be unstable. At increasing repulsive forces, the dispersion first becomes metastable and then stable. A reasonably stable suspension shows a significant primary maximum.

At very short particle distances, strong repulsion occurs by overlap of the electronic clouds of atoms or molecules, which prevents particles from coming within overlapping distance. At a distance between particles of about 1–100 nm, ionization of molecules at the particle surface or adsorbed ions leads to repulsion of particles (electric and/or steric repulsion). Also, the Van der Waals attractive forces are significant. This may result in an agglomerated situation for the particles at the so-called primary minimum. Such agglomeration is undesirable if dispersion is the goal; it may be wanted, though, e.g. for easier collection of particles. The balance of repulsive and attractive forces often also leads to a primary maximum in the net potential energy curve. The height of this maximum should be at least an order of magnitude larger than the kinetic energy (kT) of the particles for adequate stability of a dispersion. Dispersions are unstable if such a primary maximum is too low or

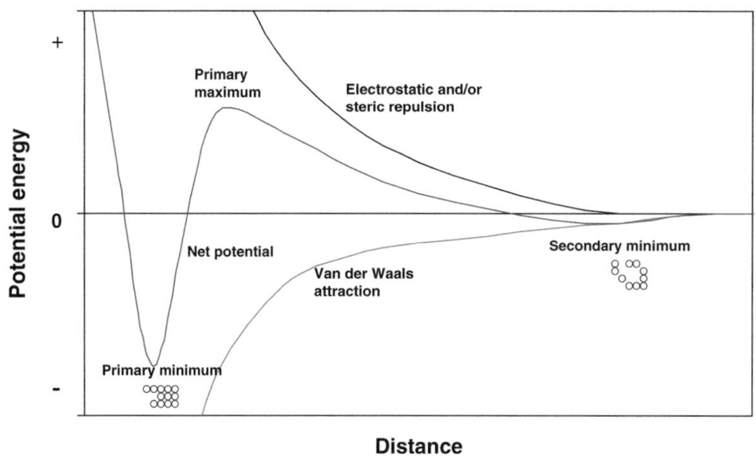

Fig. 11.1 Schematic diagram of potential energy vs. distance of two colloid particles in a reasonably stable suspension [11]; (Copyright Springer; reproduced with permission)

absent. In this case, Brownian motion can bring the particles within attraction distance. A secondary minimum may also occur. Then, the particles will show flocculation, where the attraction is only weak.

Repulsion may have an electrostatic and/or a steric cause. The electric charge of the surface of the particle is usually the origin of repulsion. Sometimes strong adsorption of multivalent ions causes the presence of a double layer near this surface with a net electric charge. Its magnitude strongly depends on concentration and valence of electrolyte ions present in the suspension (Figs. 11.2 and 11.3).

The electric charge at oxidic surfaces of particles determines their properties, when they are dispersed in water. If the water contains dissociated electrolyte species, the ions with an opposite charge will be strongly attracted to the surface of a particle and form the so-called **Stern layer**. The potential change in the Stern layer increases with concentration and valence type of the electrolyte. If the counter ions at the surface of the particle are polyvalent, this may lead to reversal of the sign of the potential in the Stern layer (see lower curve in Fig. 11.3). The thickness of the Stern layer corresponds approximately to a monolayer of ions.

Outside the Stern layer there is a **diffuse ionic atmosphere** in which the potential falls to zero in an exponential manner relative to the inverse of the distance from the Stern layer, with a characteristic distance $1/\kappa$, the statistical thickness of the double layer or Debye length (at half the potential of the Stern layer, or $0.5\psi_{\delta}$). This Debye length is related to the dielectric constant of the medium, the ion concentration in the bulk liquid, the valence of the counter ion and the temperature. In this diffuse layer, thermal agitation permits free movement of the ions. The distribution of positive and negative ions is not uniform, but is governed by the sign and magnitude of the potential at the boundary of the Stern layer.

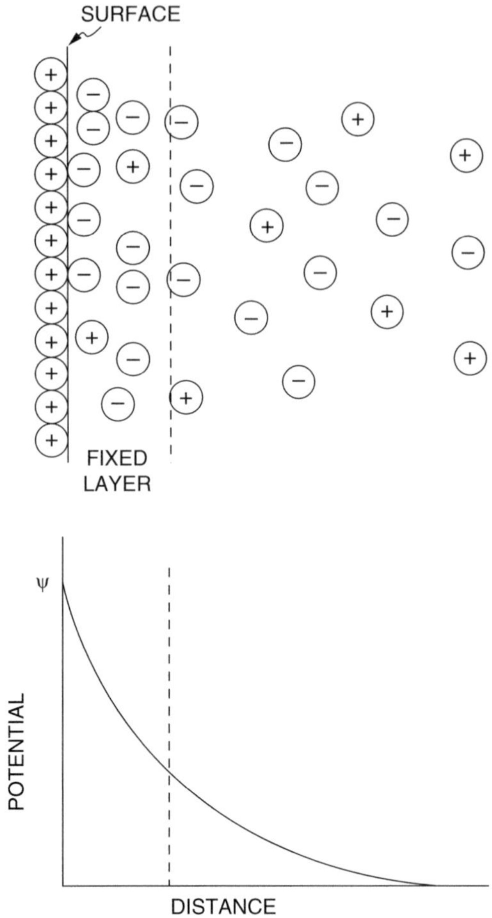

Fig. 11.2 Schematic representation of a diffuse electrical double layer [11]; (Copyright Springer; reproduced with permission)

Additionally, solvation may occur at the surface, probably by charge-dipole interaction. The thickness of this **water layer** is one to two molecules. This layer of adsorbed ions and water is so strongly bonded that it moves with the particle. Thus, it is included in the hydrodynamic diameter of a particle. It can also be concluded that the **surface of shear** is outside the Stern layer. The potential at the surface of shear is called the **zeta potential (ζ)**. It is measured as the potential of moving particles in an electrical field. **The zeta potential is considered to be the relevant parameter for description of the interactions between charged particles.** An increase in the electrolyte concentration produces a decrease of the ζ-potential, since more of the potential drop occurs within the immobile part of the double layer, i.e. the solvation layer. The working distance for these electrostatic repulsive forces may exceed 100 nm at low electrolyte concentrations (e.g. 10^{-5} M). High electrolyte concentrations decrease this working distance significantly (e.g. to about 1 nm at 0.1 M).

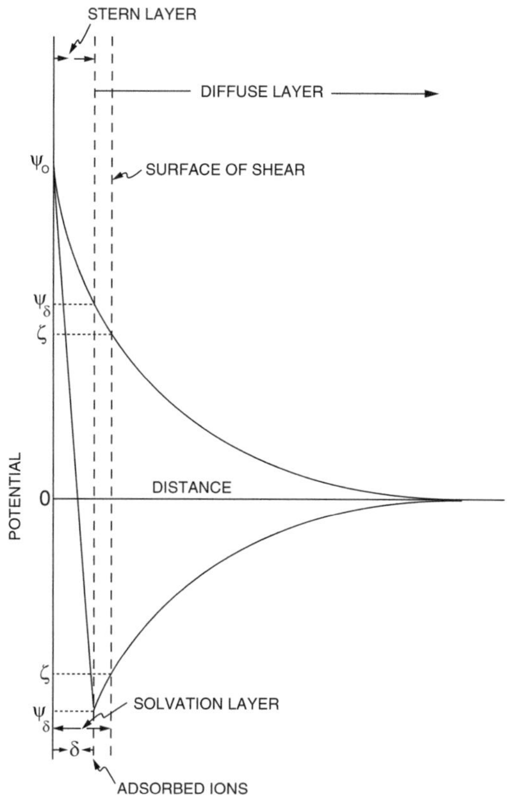

Fig. 11.3 Potential vs. distance from the boundary of a colloid particle in suspension; in the lower curve the potential is reversed due to strong ion adsorption [11]; (Copyright Springer; reproduced with permission)

With regard to flotation, the most important parameter for describing the behavior of minerals in water is their value of surface charge in relation to the pH. This holds especially for the pH for the iso-electric point (IEP), also called the point of zero charge (PZC), where the zeta-potential is considered to be zero. This IEP is quite different for different minerals (see Table 11.1).

It is generally assumed that flotation *collectors* are active as counter ions in the electrical double layer. Thus, their hydrophilic head is adsorbed to the surface and the hydrophobic tail is directed outwards. For oxidic minerals, flotation recovery with adsorbed anionic collectors is appreciable only at pH values below the IEP, with cationic collectors only at pH values above the IEP. For example, in the case of corundum, the anionic sodium dodecylsulfonate gives best flotation recovery below pH 7, the cationic dodecylammoniumchloride at pH 10–12. At very high pH, there are insufficient ammonium ions available to bind the collector to the surface since the pKa of dodecylamine is 10.4 [4]. Note that the length and type of the alkyl group is also of importance for the optimum pH range.

It is interesting to see that the bi-functional surfactants, containing a polar, hydrophilic head and a non-polar, hydrophobic tail, are being used for making hydrophilic oxidic surfaces hydrophobic in order to enable flotation, as well as for

Table 11.1 Iso-electric points for different minerals[a]

Material		Iso-electric pH
Silica (quartz)	SiO_2	2.2
Talc	Mg-silicate	3.5
Cassiterite	SnO_2	4.5
Hematite, natural	Fe_2O_3	4.8–6.7
Anatase	TiO_2	5.9
Rutile	TiO_2	6.0
Geothite	FeOOH	6.7
Boehmite	AlOOH	7.8
Diaspore	AlOOH	8.8
Zinc oxide	ZnO	9.0
Corundum	α-Al_2O_3	9.1
Magnesia	MgO	12

[a]The exact IEP values vary to some extent as they depend upon impurities and surface structure

making hydrophobic surfaces hydrophilic for better dispersion in water, both at very low concentrations.

The *activation* for flotation of oxidic minerals results from reaction or strong adsorption of multivalent ions to the mineral surface that influence or even reverse the sign of the potential (see Fig. 11.3, lower line). It will be clear that the surface needs sufficient charge to enable this adsorption. The collector molecules then adsorb to this intermediate layer of counter ions within the Stern layer.

Depression is the opposite of activation. It prevents adsorption of the collector applied at surfaces of minerals for which flotation is not desired.

Frothers are usually non-ionic surfactants or alcohol type molecules that at very low concentrations (often at 5–50 ppm) stabilize bubbles, control their size and are stable over a wide pH range. Fairly low molecular weight (MW) is regarded as important because it increases flotation kinetics. The main advantage of the higher-MW polyglycols is that they exhibit selectivity towards specific chemical surfaces of minerals. Sometimes mixtures of low- and high-MW frothers are used e.g. when coarser particles must be floated [14].

The wettability of a surface depends upon the balance of interfacial energies (γ) between the solid (S), liquid (L) and gas (V) phases – γ_{SL}, γ_{SV} and γ_{LV} – as expressed in the Young-Dupré equation [13]:

$$\gamma_{SV} - \gamma_{SL} = \gamma_{LV} \cos\theta \tag{11.1}$$

where θ is the contact angle between a liquid droplet and the solid surface.

Wetting occurs when the surface energy of the solid exposed to the atmosphere is greater than that of the liquid, i.e., when the interfacial surface energies are in the order $\gamma_{SV} > \gamma_{SL} > \gamma_{LV}$. The driving force for wetting is $\gamma_{SV} - \gamma_{SL}$. The resisting force is the energy required for increasing the surface area of the liquid drop. The same equation is valid for the ease of removal of water by air/gas. In qualitative terms, ease of wetting is opposite in effect to ease of liquid displacement by air.

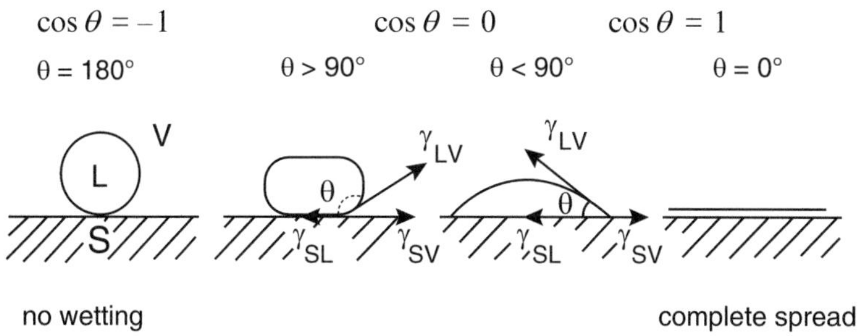

Fig. 11.4 Spreading of liquid on a solid surface [11]; (Copyright Springer; reproduced with permission)

As indicated in Fig. 11.4, there is no wetting when the contact angle θ is 180° ($\cos\theta = -1$). Neither is the liquid able to penetrate into inter- or intra-particle pores. Thus, displacement of water by air is easy. Wetting starts to occur and water removal by air becomes more difficult when the driving force approaches γ_{LG}, i.e., with decreasing contact angle θ.[1] For θ greater than 90° ($\cos 90° = 0$), water displacement by air is (fairly) easy and for smaller θ becomes increasingly difficult, although for flotation purposes, it is known that best performance takes place when the contact angle is in an intermediate range between 45 and 70°. The free energy change (ΔG) accompanying the replacement of a unit area of the solid-liquid interface by solid-gas interface is given by Dupré's equation [5]:

$$\Delta G = \gamma_{\mathrm{SV}} - \gamma_{\mathrm{SL}} - \gamma_{\mathrm{LV}} \tag{11.2}$$

or

$$\Delta G = \gamma_{LV}\ (cos\theta - 1) \tag{11.3}$$

Note that the stability of air bubbles is defined by their size and the interface tension between liquid and air. In general, the presence of particles on the surfaces of colliding bubbles leads to coalescence, although in some circumstances the particles may also prevent coalescence when an individual particle of a particular contact angle acts as a stable bridge between them.

In addition to the delicate control of the mineral interfaces, the flotation process requires efficient capture of specific mineral particles by the air bubbles. The size of

[1] Note that in this Fig. 11.4 a liquid droplet is drawn on the solid surface and the contact angle is taken here at the droplet side; in mineral flotation technology an air bubble is typically drawn at the solid surface so that the contact angle seems to be positioned at the other side (which is still at the side of the liquid).

these bubbles in conventional equipment is about 1–3 mm, in more modern equipment about 300–600 μm. The contact between particles and bubbles is governed by the available surface area of the air bubbles and the hydrodynamics of the mixing equipment. It has been found that particle collection efficiency by bubbles increases significantly by the above given decrease of the bubble size through the increased number and total surface area of these bubbles per unit volume of air [9] (see Table 11.2).

Conventionally, gentle stirring was applied to promote contact (aggregation) between particles and air bubbles. Generally two types of aggregation are discriminated, orthokinetic aggregation, originating from external forces, and perikinetic aggregation, originating from Brownian motion. In the flotation process, usually the orthokinetic type dominates; and only for sub-micrometre particles the perikinetic aggregation may play a role. Contacts between particles and larger bubbles are established somewhat more easily than with smaller bubbles because the liquid stream lines around the rising larger bubbles are typically weaker. Thus, the mixing of small bubbles with particles should be done while using larger forces, e.g. by using mechanical agitators or jets.

As soon as contact is established, adsorption of the particle to the bubble surface can start by displacement of any liquid film remaining in between them. Given good conditions for a hydrophobic surface of the particles, this occurs within 100 msec [12]. Adsorbed particles can be removed again from the bubble surface if the shear forces are stronger than the adhesion forces due to turbulence in the equipment.

The number of particles that can be collected and levitated by air bubbles depends on the size of the bubble as well as on the particle size and density. The mass of the levitated particles must be smaller than the mass of the liquid displaced by the bubble and particles together. If we neglect the mass of the air bubble itself and the extra displacement of liquid by the particle attached to the air bubble, this results in:

$$n \cdot {D_p}^3 \cdot \rho_p < {D_b}^3 \cdot \rho_L \tag{11.4}$$

If we now assume that some 50 % of the density difference is to be used for levitation, then the maximum size of a particle to be levitated by an air bubble (n = 1) as well as the maximum number of particles of a given size can be easily calculated from Eq. (11.4) through:

$$D_{p,\ max} = D_b \cdot \left(0.5 \cdot \rho_L / \, n \cdot \rho_p\right)^{1/3} \tag{11.5}$$

where:

D_p = particle size
$D_{p,\ max}$ = maximum particle size for entrainment by an air bubble
D_b = bubble size
ρ_p = particle density
ρ_L = liquid density
n = number of particles attached to an air bubble.

Table 11.2 Relation of bubble size, volume and surface area with maximum particle size and maximum particle number for $D_p = 50$ and 10 μm ($\rho_p = 4{,}000$ kg/m^3)

Bubble size, μm	30	100	300	1000	3000
Max. particle size, μm	15	50	150	500	1500
Max. particle number/bubble for $D_p = 50$ μm; $\rho_p = 4000$ kg/m^3	0	1	27	1000	10,000[a]
Max. particle number/bubble for $D_p = 10$ μm; $\rho_p = 4000$ kg/m^3	3	130	2800[a]	$3.1 \cdot 10^4$ [a]	$2.8 \cdot 10^5$[a]
Bubble volume, ml	$1.4 \cdot 10^{-8}$	$5.2 \cdot 10^{-7}$	$1.4 \cdot 10^{-5}$	$5.2 \cdot 10^{-4}$	$1.4 \cdot 10^{-2}$
Bubble area, cm^2	$2.8 \cdot 10^{-5}$	$3.1 \cdot 10^{-4}$	$2.8 \cdot 10^{-3}$	$3.1 \cdot 10^{-2}$	$2.8 \cdot 10^{-1}$
Number of bubbles/ml	$7 \cdot 10^7$	$2 \cdot 10^6$	$7 \cdot 10^4$	$2 \cdot 10^3$	71
Spec. surface area of bubbles, cm^2/ml	2000	600	200	60	20

[a]The available surface area of the bubble limits the number of attached particles since their cross sectional area exceeds the available bubble area for a monolayer of particles (both particles and bubbles are assumed to be spheres)

Table 11.2 gives some examples for the relation between bubble size and maximum size of a single particle and maximum number of 50 and 10 μm particles that can be levitated by a single bubble according to Eq. (11.5). The table also presents the volume and area of a single air bubble of the given size, the number of bubbles per ml air and the specific surface area of these bubbles.

The data in the table clearly show that the maximum number of particles that can be levitated by a single air bubble for small bubbles is limited by particle mass, in accordance with Eq. (11.5). However, for larger bubbles the limitation is set by the surface area of the bubble (πD_b^2) in relation to the cross sectional area of the particles ($1/4\pi D_p^2$), assuming a maximum coverage of 78 % of the bubble surface by a monolayer of particles. Note that this coverage may change significantly in two directions: to lower percentages if the residence time is insufficient to reach this maximum coverage, and to higher percentages if the adsorption of particles to the bubble surface is not limited to a monolayer but occurs in more layers.

The froth may not only contain the mineral particles that are adsorbed to the air bubbles but also some small gangue particles, which are entrained by the liquid that surrounds the froth bubbles. Coalescence of bubbles may also occur. Altogether, some washing of the froth has to be executed to remove entrained gangue, to reach a higher product grade.

The process water employed is typically polluted with chemicals and particles. Both for economic and environmental reasons, it is almost completely recycled after removal of the contaminants.

11.4 Flotation Equipment

Commercial flotation is a continuous process. Optimum conditions for recovery of a given mineral from a given ore are usually determined in a laboratory. A simple design is sketched in Fig. 11.5.

Following comminution of the ore to a pre-determined size distribution and mixing with water, the required chemicals are added to the feed in the correct concentration and order in a conditioner and sufficient time is given for conditioning. The mixture is then fed into a series of ‘rougher’ cells, where air is introduced for flotation. Here the first part of the froth concentrate is collected with the objective of removing the maximum amount of valuable mineral from as coarse a particle size as practical. The remaining pulp part is then transferred to so-called ‘scavenger’ cells, where more of the target mineral is collected by flotation. The objective of this scavenger section is to maximize recovery of the target mineral. The froth coming from the scavenger cells is usually of low grade and is recycled either to the conditioner or the first rougher cell. When treating a ‘complex’ ore, i.e. one that contains a number of metals as sulfides, the tailings may enter into another circuit for recovery of a second mineral, after adjustment of the pH and addition of suitable activators and collectors.

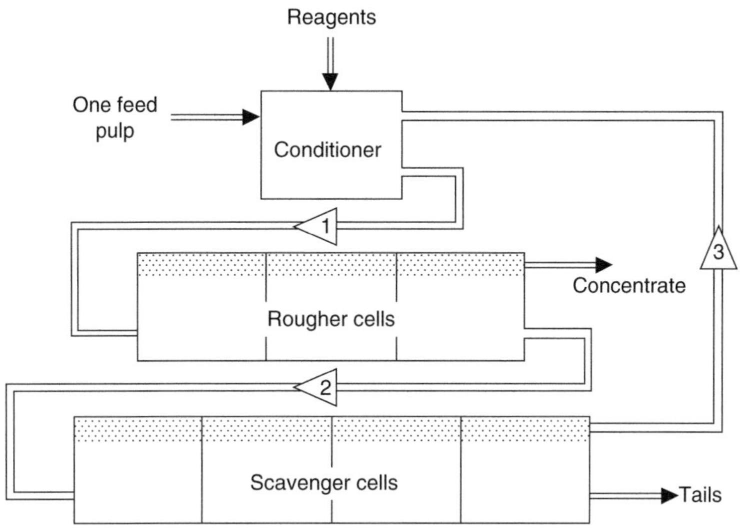

Fig. 11.5 Simple flotation circuit

To improve the quality of the product by removing entrained tailings, the froth is usually diluted with water and introduced into a series of 'cleaner' cells, where it is refloated. The objective of these cleaner cells is to produce as high a concentrate grade as possible. The cleaner tailings are combined with the froth from the scavenger cells and recycled to the head of the flotation system.

The energy used in the comminution of ore is a major cost factor in mineral processing. Therefore, often a fairly coarse fraction of ore is used first and some intermediate comminution of concentrates and/or tailings is carried out prior to their introduction into the next section in order to improve the recovery of the wanted mineral at minimum cost.

Initially, mainly mechanical cells were used in the flotation process. A basic mechanical cell is sketched in Fig. 11.6.

Pulp and air are introduced separately through narrow openings at the bottom of the cell, which is equipped with an impeller. Air bubbles are generated by the shear forces near the impeller. Their size is typically 1–3 mm. The target mineral particles attach to the air bubbles, which levitate them to the froth layer, which is then collected. The remaining pulp is fed into the next flotation cell for further recovery of the mineral. The pulp exit is separated from the rest of the system by a baffle to avoid the froth from exiting here as well.

Over the years, the complexity of the commercial equipment has increased to improve its functionality, and other types of cell, such as flotation columns, have been introduced. The first columns were designed so that the pulp was introduced at the top, while air bubbles were formed using spargers at the bottom. The counter current flow of pulp and air improved the contact between the two, but the columns were very tall, typically 10–15 m in height. Later, the Jameson cell was developed

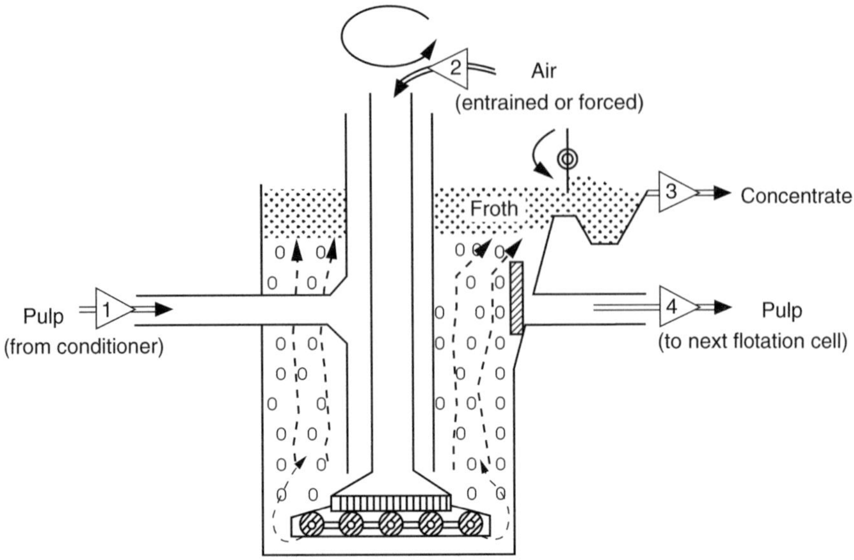

Fig. 11.6 Basic mechanical flotation cell

[8], which uses a jet of combined pulp and air feed to improve mixing, as well as requiring less space than the earlier designs. In a still later design, conical diffusers were placed below the downcomers in order to reduce the turbulence and the resulting detachment of particles from the bubbles. The Jameson cell produces much smaller bubble sizes, 300–600 μm on a number-basis and 360–950 μm when area-weighted (Sauter diameter). Moreover, an external recycle system is included, which compensates for fluctuations in the feed stream as well as gives about 40 % of the tailings a second pass through the system. Wash water can also be added to improve the quality of the froth. Improved mixing, the production of smaller bubbles, recycling and inclusion of washing have together improved the efficiency of the flotation process as well as the grade of the recovered mineral. Figure 11.7 gives an example of the Jameson cell.

It can be concluded that selective, economic flotation requires a complex and highly interesting combination of surface chemistry, with chemical and mechanical engineering:

- comminution of solids to an optimum particle size
- chemical modification of a target solid surface to enable selective attachment of bubbles
- adequate understanding of interface behavior between solids, liquid and gas
- generation of air bubbles of optimum size in the pulp
- provision of optimum mixing and contact time between solid particles and air bubbles
- minimization of detachment of particles from bubbles due to excessive turbulence

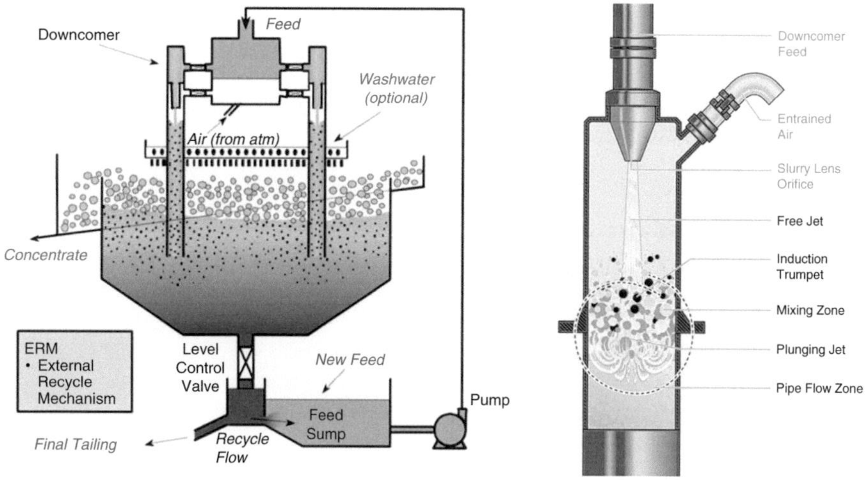

Fig. 11.7 Sketch of an improved flotation cell (Jameson cell)

– sufficient stabilization of the froth
– efficient separation and purification of the froth to reach a high-grade product
– good consideration of the economics of the various stages of the process.

It is to be expected that flotation technology will play an important role in the recovery of precious, rare metals from new waste streams.

11.5 Definitions, Abbreviations and Symbols

Activator	reagent that enhances the selectivity of the collector in respect of a particular mineral
Collector	reagent that attaches selectively to a mineral surface, making it hydrophobic
Depressant	reagent added to a mixture of minerals to prevent a particular mineral from becoming hydrophobic
Frother	surfactant added to the pulp to prevent bubble coalescence in the froth and the pulp
Iso-electric point	pH of a suspension when the zeta-potential is zero
Pulp	mixture of (ore, etc.) particles and water
Sauter mean diameter	area-weighted mean size; size of a 'particle' that has the same area-to-volume ratio as a mixture of particles of different sizes
Surface active agent	chemical substance containing a polar, hydrophilic head and a non-polar, hydrophobic tail
Surfactant	surface active agent
Zeta-potential	potential at the surface of shear of a particle in suspension

IEP	iso-electric point
L	liquid
MW	molecular weight
PZC	point of zero charge (identical to IEP)
S	solid
V	gas
D_b	bubble size
D_p	particle size
$D_{p,max}$	maximum particle size for collection by an air bubble
ΔG	free energy change
k	Boltzman constant
n	number of particles attached to an air bubble
T	absolute temperature
γ_{LV}	interface tension between liquid and gas
γ_{SL}	interface tension between solid and liquid
γ_{SV}	interface tension between solid and gas
θ	contact angle between a liquid droplet and a solid surface
ρ_L	liquid density
ρ_p	particle density
ζ-potential	zeta-potential

Acknowledgement The author gratefully acknowledges the contribution of Prof. Graeme J. Jameson to the text of this chapter.

References

Suggested Reading

1. Fuerstenau, M.C., Jameson, G., Yoon, R.-H (eds.): Froth Flotation – A Century of Innovation. Society for Mining, Metallurgy and Exploration, Inc., Englewood (2007) USA

Further References

2. Bessel, G.: German Patent Berlin. **42** 2 July (1877)
3. Bessel, G.: German Patent Berlin 39, 369 12 May (1886)
4. Fuerstenau, D.W.: A century of developments in the chemistry of flotation processing. In: [1]
5. Fuerstenau, D.W. Raghavan, S.: Some aspects of flotation thermodynamics. In: [1]
6. Graichen, K. Hanish, J. Schubert, H. Steiner, K.D. Tanneberger, C. Wachtler, E.: Die Gebrüder Bessel und die Anfange der flotativen Aufbereitung; Neue Bergbautechnik **7**, 10 October (1977)

7. Haynes, W.: British Patent 488, 20 February (1860)
8. Jameson, G.J.: A new concept in flotation column design. Miner. Metall. Process. **5**(1), 44–47 (1988)
9. Jameson, G.J., Nguyen, A.V., Ata, S.: The Flotation of Fine and Coarse Particles. In: [1]
10. Lynch, A.J., Watt, J.S., Finch, J.A., Harbort, G.E.: History of Flotation Technology. In: [1]
11. Merkus, H.G.: Particle Size Measurements – Fundamentals, Practice, Quality. Springer, Dordrecht (2009)
12. Nguyen, A.V., Evans, G.M., Jameson, G.J.: Bubble-particle attachment in froth flotation, 6th World Congress Chemical Engineering, Melbourne, Australia (2001)
13. Parfitt, G.D. (ed.): Dispersion of Powders in Liquids. Elsevier, Amsterdam (1969)
14. Pugh, R.J.: The Physics and Chemistry of Frothers. In: [1]
15. Wills, B.A.: Mineral Processing Technology, 7th edn, Pergamon Press, Oxford (2006) UK

Chapter 12
Classification

Henk G. Merkus

Abstract Particulate materials are produced intentionally or unintentionally in many different processes. Classification is often required to produce a product that has a given particle size range. In other cases particles have to be removed as far as possible from a dispersion in a liquid or air/gas either to recover valuable materials, or to eliminate health or explosion hazards, or to produce a clear fluid. This chapter gives an overview of the available techniques, especially the application and characteristics of (hydro)cyclones, since they are used in many applications. The application of computational fluid dynamics in combination with (new) flow measurement techniques begins to provide new insight into the influence of the details of cyclone design on the flow patterns, the collection efficiency, the cut size, the pressure drop and potential erosion problems, and, thus, may relate to optimum designs in relation to required tasks.

12.1 Introduction

Particles are produced intentionally or unintentionally in many different processes. The processes occur in the liquid phase by e.g. crystallization and precipitation, or in the gas phase by e.g. partial oxidation, or through comminution. The production processes of many particulate materials produce a particle size distribution (PSD) that is wider than the PSD required for application. Therefore, usually some kind of classification process equipment is necessary to obtain the required PSD and, thus, to optimize product quality. Often similar equipment is applied for removal of suspended particles in air or liquid that are produced unintentionally. The goal is then to provide for a clean fluid, to recover valuable materials, to ascertain optimum human health, to eliminate explosion hazards and to avoid pollution of the environment. This is, for example, necessary in the case of flue gases polluted with

H.G. Merkus (✉)
Em. Prof., Chemical Engineering Department, Delft University of Technology, Pijnacker, The Netherlands
e-mail: henkmerkus@hetnet.nl

© Springer International Publishing Switzerland 2016

H.G. Merkus, G.M.H. Meesters (eds.), *Production, Handling and Characterization of Particulate Materials*, Particle Technology Series 25,
DOI 10.1007/978-3-319-20949-4_12

Table 12.1 Classification processes

Carrier fluid	Classifier type	Cut size range, μm
Gas	Filter screens	0.01–100
	Electrostatic precipitators	0.01–10
	Magnetic separators	0.1–10
	Wet scrubbers	0.1–100
	Cyclones	1–200
	Gravitational sedimentation	20–>1000
Liquid	Filter screens	0.01–100
	Particulate beds	0.01–100
	Hydrocyclones	5–200
	Gravitational sedimentation	50–>1000
	Sieves	50–>1000

small particles. Note that nanometer sized particles in air or liquid (if insufficiently stabilized) easily form agglomerates, which can be removed more easily in relation to their much larger size.

Various processes and equipments are used for this classification in relation to the type of fluid and the particle size (see Table 12.1) [6, 7, 14, 15, 20, 21]. Typically, each equipment for particle removal has a defined operational size range and collection efficiency. Their mechanisms depend upon inertia, diffusion, interception, gravity and electrostatic attraction of particles. The efficiency strongly depends upon the particle characteristics, but also upon equipment design and operating conditions. In general, the required efficiency plays an important role in the decision on process and equipment. In case very wide size distributions have to be handled, often different types of equipment are applied in series. Another important point for the decision in selecting the process equipment is the pressure drop and the related energy consumption. Finally, the maximum operating temperature may be of importance, if high temperatures are employed in the preceding gas phase processes.

Filter screens consist of woven cloth of given thickness. They collect particles at the surface and/or in the narrow channels of the voids (depth) (Fig. 12.1). Usually the coarse particles are collected at the surface, depth filters usually also collect finer particles in the void channels. Fibrous depth filters often have a fiber percentage of only 10 % by volume, the rest being voids in tortuous channels. Then, the combination of diffusion, interception, impaction and adsorption at the fibers take care of fine particle removal. The ratio of these different mechanisms depends upon particle size and filter type. As soon as the particles have formed an accumulated layer on the filter, this layer becomes more important in the removal. Most often, filter screens are used to clean gas and liquid streams, not for classification. After some time of operation, they must be cleaned and, if wanted and possible re-used. The maximum temperature depends on the type of cloth. Often only ambient temperature is possible, some materials allow elevated temperatures. Filter screens can have very high collection efficiency, e.g. HEPA filters more than 99.999 %.

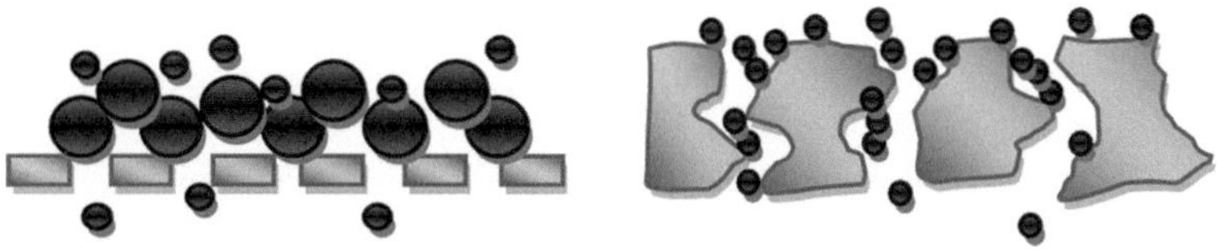

Fig. 12.1 Sketch of surface filtration (*left*) and depth filtration (*right*)

Note that the pressure drop is strongly dependent upon filter type and dust loading/cake formation (see further Chap. 10).

Electrostatic precipitators collect particles, after they have been given an electric charge (typically induced by a high-voltage corona discharge), at electrodes between which a strong electric field is set (high voltage). These precipitators are mainly used for cleaning gases from entrained particles. Their main advantages are the low pressure drop and the high collection efficiency even for very small particles. Both solid particles and liquid mists can be removed. Elevated temperature of operation is possible.

Magnetic separators operate in a similar way as the electrostatic equipment, except that no charging occurs and the separation by strong magnets depends upon the magnetic properties of the particles. The pressure drop is low. Elevated temperature of operation is possible.

Wet scrubbers use the collision of the particles with water droplets for particle removal. It results in a high efficiency even for very small particles. Often, droplets of about 0.5 mm are brought in contact with the gas to be cleaned, which can be removed easily and with high efficiency from the gas. An advantage for flue gases is that some gaseous pollutants are removed as well. A disadvantage for obtaining dry particles is that the particles exit in the form of a (sometimes contaminated) slurry that has to be dried (and cleaned).

Gravitational classifiers are in the most simple form large tanks, in which particles can settle. Their operation fully depends upon (1) the sedimentation rate, governed by the particle size, the density difference between the particles and the fluid and the fluid viscosity, and (2) the residence time given in the tank for sedimentation. Sometimes, different locations of the tank collect particles having different sizes or densities. Stokes' law can be used for calculation of the terminal sedimentation rate for viscous-flow conditions at low particulate concentrations:

$$v_p = \frac{(\rho_p - \rho_f) g D_{St}^2}{18\eta_f} \tag{12.1}$$

where:

v_p = terminal settling velocity of the particle
ρ_p = effective particle density
ρ_f = fluid density
g = gravitational acceleration constant
D_{St} = Stokes' diameter of the particle
η_f = fluid viscosity

At increased particulate concentrations of suspensions in liquids, above a few % v/v, particle-particle interactions start causing hindered settling and the liquid viscosity increases in relation to the concentration [11] (see also Chap. 10).

Most gravitational classifiers operate at laminar flow conditions where turbulence is avoided by using very simple equipment. The coarse particles settle to the bottom of a settling tank, where they can be removed, the fines fraction moves with the overflow. A multi-compartment design containing subsequent sections with different residence times also offers the possibility of collecting different size fractions.

Counter-current classification is also employed, in elutriators. Then, there is a (narrow) section with counter-flow of suspension and clear, make-up water. Again, the coarse particles settle if their sedimentation rate is greater than the counter flow, finer particles move with the make-up water. Different designs exist with increasing complexity.

Drawbacks of gravitational sedimentation are that only fairly coarse particles can be removed easily and that the equipment to remove finer particles is generally large in order to provide for sufficient settling time. For example, the settling rate in water for a 10 μm particle having a density of 2500 kg/m^3 is only 0.15 m/h, for a 50 μm particle 3.7 m/h.

Sieves collect particles that are larger than the apertures of the sieve. Similar to the screens, collected particles can form a layer that contributes to the particle collection, if the sieves have insufficient inclination or are insufficiently vibrated. A parallel fluid flow may also be applied to remove and collect the particles continuously. Sieves are used for removal of unwanted particles as well as for collection of desired particles above a given size. The pressure drop is fairly low. In industrial separations, both perforated-plate sieves (≥500 μm) and woven-wire sieves (≥50 μm) are applied. Note that the shear by sieve screens also may be applied for de-agglomeration of particle clusters in stirred suspensions.

Particulate beds operate similar to filters. They are e.g. applied in the form of sand beds to remove particulate matter from drinking water in a similar way to screens. An advantage is that they can be easily cleaned by application of a back flush. The pressure drop is medium high.

Cyclones and hydrocyclones will be described in the next sections. Their advantage over gravitational classifiers is that the larger centrifugal forces applied enable higher collection efficiencies and much lower cut sizes. Also, they are compact and take much less floor space. Thus, they are present at many different locations in quite different processes.

Often, different techniques are applied in series. Reasons may be the reduction of erosion of equipment by coarse particles and decreasing the particulate load of equipment that requires low particulate concentrations, in relation to the particle size and concentration range of the material produced. Examples are gravitational settling chambers in front of cyclones, and cyclones in front of electrostatic precipitators.

12.2 (Hydro)cyclones

Cyclones have the advantages of high versatility, low investment costs and large capacity range, in relation to the size of the cyclone (diameters from centimeters to meters). The equipment is simple, robust and can be cleaned easily. The energy required to overcome the pressure drop is low in the most simple equipment at ambient conditions, where a simple fan or pump suffices. Gas cyclones can be operated up till temperatures of 1000 °C and pressures up till 500 bar.

A cyclone consists of a cylindrical part at the top and a conical part at the bottom. The feed enters, preferably without meeting any restriction, tangentially into the cyclone at the top of the cylindrical part, causing a spiraling movement. There are two exits: one at the bottom for the coarse particles (underflow) and one in the center at the top for clear fluid and (very) fine particles (overflow).

Gas cyclones and hydrocyclones operate according to the same principles. The only differences are the fluid density and viscosity. The most common type of gas cyclone, the reversed flow cyclone, is depicted in Fig. 12.2, in combination with the symbols for its relevant dimensions.

The optimum cyclone dimensions strongly depend upon feed characteristics and requirements for efficiency and throughput. In addition to collection efficiency, they determine the pressure drop and the energy consumption. The dimensions are typically stated as a ratio of size to the internal diameter D. Some examples of the geometrical dimensions are given in Table 12.2 for two conventional, but still popular Stairmand gas cyclones (high efficiency HE and high flow rate HR) [13, 15, 18–20].

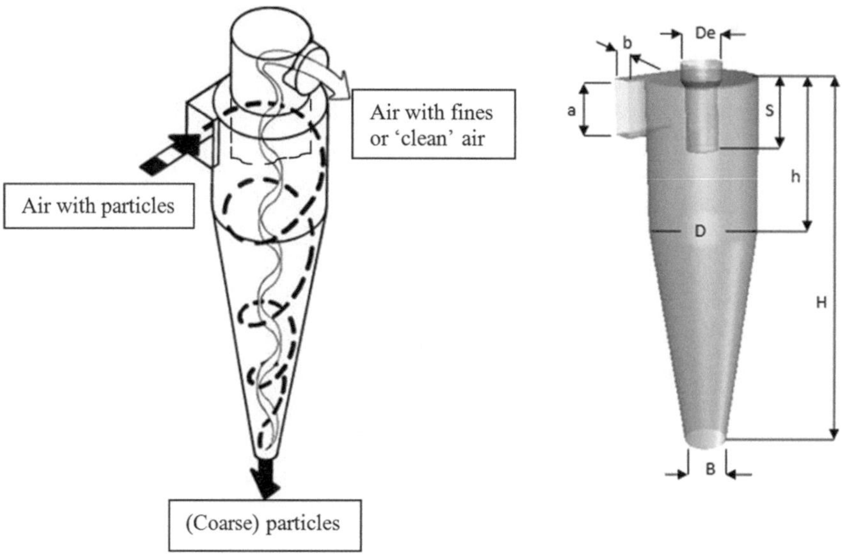

Fig. 12.2 Schematic of a gas cyclone separator (Figure at *right* reproduced with permission by Prof. R. Utikar [24])

Table 12.2 Examples of cyclone dimensions

Type	*H/D*	*h/D*	*a/D*	*b/D*	*B/D*	*De/D*	*S/D*	Θ, °	*Eu*	Stk_{50}
HE	4.0	1.5	0.5	0.2	0.375	0.5	0.5	15	320	$1.2 \cdot 10^{-4}$
HR	4.0	1.5	0.75	0.375	0.575	0.75	0.875	10	46	$6 \cdot 10^{-3}$

Note that these dimensions are, though popular, not the absolute optimum; other authors have published slightly different preferences for the dimensions of general purpose and high-throughput cyclones. The exit duct length *S* is usually slightly greater than the inlet height *a* in order to prevent shortcuts of the inlet gas plus particles directly to the exit. Further fine tuning of performance can be achieved for specific cases by dedicated careful design through fairly small changes of *a/D*, *b/D*, *De/D* and *h/D*, through using computational fluid dynamic modeling in combination with modern measurement techniques [1–4, 8, 9].

The gas laden with particles is brought tangentially into the cylindrical section of the cyclone, usually through a rectangular entrance. It causes a high tangential velocity, which in turn creates a strong vortex inside the cyclone body. Thus, the particles are subjected to centrifugal forces which move them outwards to the cyclone walls, in relation to their inertia and the opposed drag force, and downwards to the bottom, provided that they have a greater density than the fluid. Typically, the centrifugal forces are an order of magnitude larger than the gravitational forces. Coarser particles move closer to the wall than finer particles. The gas flow in the outer cyclone section transports all particles to the bottom of the cyclone (the apex of the conical part). Here, the coarse particles exit the cyclone, where they are collected in a hopper, which is connected to the bottom outlet in an open but air tight manner. The fluid with uncollected particles is now diverted into an upward flow direction in the central part of the cyclone, spiraling at increased tangential velocity. Then, after some interaction between the upward and the downward flow due to the turbulent nature of the two flows, the air laden with uncollected particles or clean air leaves the upper (air) exit through the so-called vortex finder, which has its entrance slightly below the feed entrance. Note that high tangential velocities of particles may result in significant erosion of the equipment, especially in case of fairly coarse or abrasive particles.

Most commonly, gas cyclones are mounted vertically for easier discharge of the particles from the cone; sometimes, they are used in a different orientation.

Hydrocyclones operate in a similar way. Here, only a liquid phase, having a much greater viscosity and density, takes the position of the gas phase. Another difference is that a greater proportion of the liquid phase exits with the particles underflow than in case of gas. An amount of fine particles will always go automatically with this liquid, thus decreasing the apparent separation efficiency [26]. The measured grade efficiency curve is always corrected for this misplacement of fines to yield the characteristic grade efficiency curve of the cyclone.

Some pressure drop exists over the cyclone, in relation to the cyclone dimensions and the feed flow [5, 8, 12, 16–18, 27]. It is one of the major points in the design of a gas cyclone as it involves its energy requirement. It consists of frictional pressure losses, viz. pressure loss at inlet and outlet and losses due to changes in

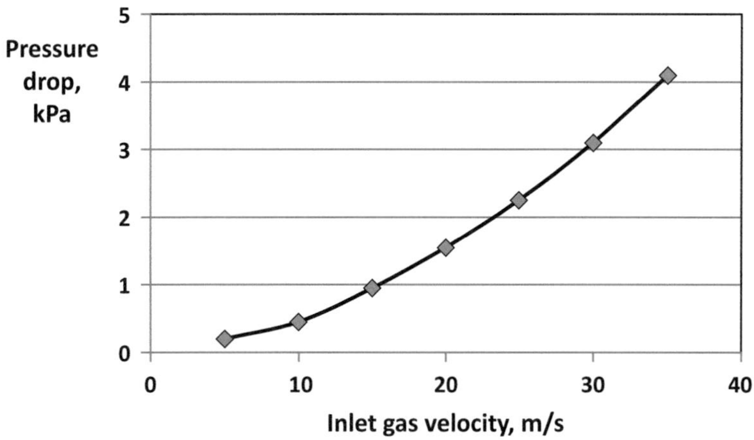

Fig. 12.3 Typical pressure drop against gas flow rate in a cyclone (Adapted from Ref. [25])

flow direction and a radial pressure gradient due to the rotational flow in the forced vortex within the cyclone. The static pressure drop between inlet and outlet is usually proportional to the gas flow rate squared. Their relationship is approximated in a resistance coefficient, the Euler number:

$$Eu = 2 \cdot \Delta P / \rho_f \cdot u^2 \tag{12.2}$$

where

Eu = Euler number
ΔP = pressure drop
ρ_f = fluid density
u = characteristic linear flow rate

$$= 4Q/\pi \cdot D^2 = Q/A \tag{12.3}$$

Q = volumetric flow rate of the feed
D = inside diameter of cylindrical part of cyclone
A = cross sectional area of cylindrical part of cyclone

Figure 12.3 exemplifies the typical pressure drops observed at different flow rates in cyclones.

A first indication for the pressure drop is often obtained from measurements while using only gas without particles.

If such data is not available, the Euler number is often approximated from the dimensions of the inlet and the outlet [13, 20]:

$$Eu = \pi^2 \cdot (D/a) \cdot (D/b) \cdot (D/De)^2 \tag{12.4}$$

The drawback of this equation is that it does not explicitly consider the vertical dimensions of the cyclone. Shepherd et al. relate the pressure drop to the number of inlet heads [16, 17]. The above application of the Euler number seems more appropriate, since it leads to a better comparison of different designs [22].

The Euler number represents the ratio of pressure forces and inertial forces acting on a fluid element. Its value is practically constant for a given cyclone geometry, regardless of its diameter and operating conditions. Above a few g/m^3 dust concentration, its value decreases by the increasing particle-particle interactions with concentration that reduce the degree of turbulence:

$$Eu^* = Eu \cdot \left(1 - 0.02 \cdot c^{0.6}\right) \tag{12.5}$$

where

Eu^* = Eu at increased dust load
c = dust concentration

This empirical correlation holds for coal dust up till a certain level of dust concentration, where higher levels would lead to negative values of Eu^* [20]. Note that the constants given in Eq. (12.5) depend upon the PSD and the density of the solids. Note also that the same equation can be used for the pressure drop if the gas velocity is kept constant, due to the direct relationship between Eu and ΔP.

In case operation is required at increased pressure and/or temperature, the gas density has to be adapted to the actual conditions. This can be approximated by using ideal gas theory:

$$\rho_g = P \cdot M / R \cdot T \tag{12.6}$$

where:

ρ_g = gas density
P = pressure
M = molecular weight of gas
R = gas constant
T = absolute temperature

Thus, the gas density increases in direct relationship with the ratio of pressures, and inversely with the ratio of the absolute temperatures.

The gas viscosity can be assumed to be independent of the pressure and proportional to the square root of the absolute temperature [15].

12.3 Collection and Separation Efficiency

Ideally, all particles above a given size – the cut point – would be collected, whereas all smaller particles would pass. However, this is far from reality, where the collection efficiency depends upon particle size and ranges around the cut point

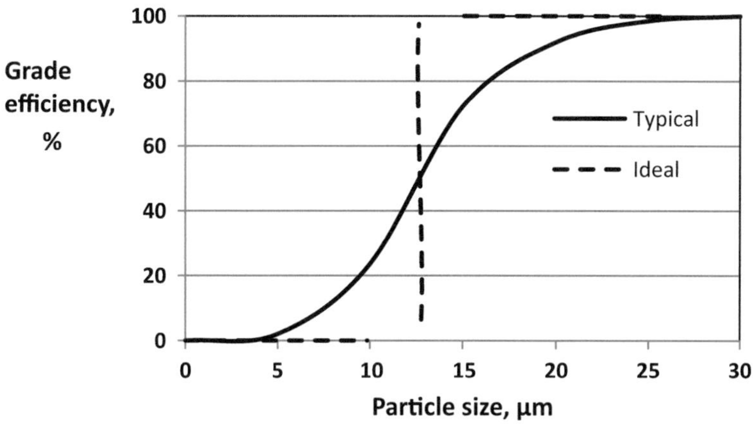

Fig. 12.4 Example of a grade efficiency curve of a gas cyclone

from 0 to 100 %. This is due to particle-particle interactions, particle shortcuts, turbulence and fluctuations of the gas velocity. Thus, the collection efficiency is usually expressed in so-called grade efficiency curves or Tromp curves (see Fig. 12.4), which can be calculated from the measured mass flows and size distributions of the feed and the coarse and fine fractions [7, 13, 20, 23]. The measured grade efficiency curves used for characterization of hydrocyclone operation are most often corrected for the amount of fines that leaves the cyclone through entrainment with the fluid in the coarse particles exit, which may be substantial (about 10–25 % of liquid).

Figure 12.4 clearly shows that the collection efficiency increases with particle size. As this is related to mass, it also increases with particle density.

In the collection efficiency, a distinction should be made between the total and the fractional collection efficiency. The total collection efficiency is the ratio of the mass of the coarse product collected to that in the feed, while assuming no breakage, growth or losses of particles:

$$E_t = M_c/M \tag{12.7}$$

where:

E_t = total collection efficiency
M_c = mass of the coarse product
M = mass of the feed

Similarly, the fractional grade efficiency G expresses this ratio for the size fractions. The cut point is the particle size where the particles have equal probability (50 %) for going into the coarse and the fines fraction (or passage to the air outlet). In other words, 50 % of the particles having this size appear in the fines fraction and 50 % in the coarse product. It is usually expressed as the D_{50} value

(13 μm in Fig. 12.4). Note that this cut size differs from the so-called analytical cut size, which corresponds to the point where the amount of coarse material in the fines fraction equals the amount of fines in the coarse fraction.

The ideal separation point can be derived by an analysis of the forces acting on the particles, viz. centrifugal force, drag and buoyancy. The equilibrium between these forces determines the orbit taken by a particle in the cyclone. It can be derived from Stokes' law that the critical particle size for ideal separation is [13]:

$$D^2_{p,crit} = \frac{18\eta_f}{\left(\rho_p - \rho_f\right)} \frac{U_R}{U^2_{\theta,R}} R_i \tag{12.8}$$

where

$D_{p,crit}$ = critical particle size for separation
η_f = fluid viscosity
U_R = radial velocity component at cyclone wall
R_i = inner radius of cylindrical part of cyclone
ρ_p = effective particle density
ρ_f = fluid density
$U_{\theta,R}$ = tangential velocity component at cyclone wall

Note: In some literature, the aerodynamic size of the particles (particle density 1000 kg/m^3) is used instead of the size with real particle density. This is highly confusing if it is not mentioned.

As said above, ideal separations do not occur in practice. Instead, the grade efficiency depends upon particle size, as exemplified in Fig. 12.4, and the cut point is described in the D_{50} value where particles have equal probability of resulting in the fine or the coarse fraction. The steepness of the grade efficiency curve, i.e. the sharpness of separation, determines the amount of misplaced material, i.e. coarse particles in the fines product and fines in the coarse product. It can be expressed e.g. as the D_{25}/D_{75} ratio (about 0.6 in Fig. 12.4; maximum typically 0.8 for high-efficiency cyclones), or in case log-normal size distributions are used as the geometric standard deviation $\left(\text{GSD} = \sqrt{(D_{84}/D_{15})}\right)$.

Reasons for an apparent non-ideal separation may be:

- shortcut between the inlet and gas outlet streams near the vortex finder
- shortcuts between the downward and upward fluid streams through turbulent eddies
- deviation from the optimum flow and pressure range of the cyclone
- entrainment of small particles into the coarse particle exit; this is especially important in case of hydrocyclones where about 10–25 % liquid (containing an equivalent of the small particles) leaves the cyclone with the coarse particles
- influence of particle-particle interactions on fluid flow at increased particulate concentrations

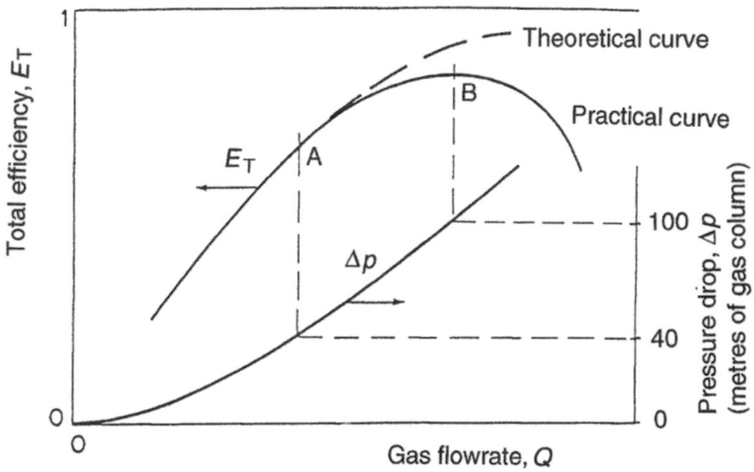

Fig. 12.5 Correlation of total separation efficiency and pressure drop with gas flow rate in a reverse flow cyclone [13] (Copyright Wiley; reproduced with permission)

- adsorption of fine particles to the surface of coarse ones
- agglomeration of particles before or within the cyclone, especially at high particulate concentrations and of very small particles in humid gases
- attrition of coarse particles
- re-entrainment of particles settled at the bottom section of the cyclone (due to inadequate removal of these particles to the solids container and the strong upward vortex in the solids exit)
- incorrect sampling, i.e. sampling the three flows (feed, underflow and overflow) at different periods of time and fluctuating conditions
- better dispersion of particles during PSD measurement than in cyclone.

For a given cyclone geometry, both the total collection efficiency and the pressure drop increase with increasing flow rate up to a certain point (B). Then, the increased turbulence starts causing re-entrainment of separated particles and the efficiency decreases upon further increase of the flow rate. However, the pressure drop keeps increasing (see Fig. 12.5). Moreover, the tendency for erosion increases with increasing flow rate. Thus, it is counterproductive and a waste of energy to operate cyclones above this point and the most desirable operation is just below this maximum point (range A – B). This is also the reason that a difference is made in the designs of high-efficiency and high-flow rate/high-throughput cyclones (*cf.* Table 12.2). The nominal flow rates in these cyclones are $1.5D^2$ m^3s^{-1} for HE cyclones and $4.5D^2$ m^3s^{-1} for HR cyclones. In both types this leads to a linear entrance velocity of approximately 15 $m \cdot s^{-1}$ [15].

12.4 Cyclone Design and Scale-Up

The dimensionless Stokes' number, which represents the ratio of the centrifugal force (less buoyancy) to the drag force, can be used for the scale-up of cyclones. The relationship between the characteristic Stokes' number and the cut size is given by:

$$Stk_{50} = D_{50}^2 \rho_p u / 18\eta_f D \tag{12.9}$$

where

Stk_{50} = characteristic Stokes' number
D_{50} = cut size of cyclone
ρ_p = effective density of particles
u = characteristic linear velocity of fluid
η_f = fluid viscosity
D = diameter of the cylindrical part of the cyclone

The cyclone diameter is regarded as the most important variable for the design in relation to capacity and cut size. The rest of the geometry is usually chosen in relation to this diameter, and to specific refinement of cut size or capacity, or to specific properties (e.g. rheology) of the feed [7, 20].

As can be seen in Eq. (12.9), the cut size decreases at decreasing the diameter of the cyclone of a given geometry while keeping the volumetric fluid flow constant. The collection efficiency increases accordingly. This is due to the inverse relationship of the linear fluid flow rate with the cyclone diameter squared (Eq. 12.3), and to the fact that higher flow rates result in stronger centrifugal forces. Increasing the volumetric flow rate of a cyclone has the same effect, up till the point of maximum efficiency (see above). Decreasing the diameter and increasing the length of the inlet and the gas outlet also slightly decrease the cut size and increase the collection efficiency, through their influence on the linear fluid velocity.

If high volumetric flow rates and high throughputs are required, then typically the diameters are relatively large in order to limit the linear velocity and to keep the pressure drop low. This, however, results in lower grade efficiency. Generally, an inverse relationship exists between the Euler resistance coefficient and the characteristic Stokes' number that can be approximated by:

$$Eu = \sqrt{12/Stk_{50}} \tag{12.10}$$

Thus, high values of the Euler number correspond to low values of the Stokes' number, i.e. low cut sizes and high efficiencies, and vice versa.

The particulate concentration in gases influences the cut size and the total collection efficiency above about 2 g/m^3. At increasing concentration, the particle-particle interactions are increased, which may lead to agglomeration of particles, where the agglomerates are collected more easily due to their increased size than the primary particles that are taken as reference. Thus, the cut size may decrease and the efficiency increase with increasing concentrations. It also means that a design that does not account for increased particulate concentrations is generally conservative with respect to its practical operation.

An empirical correlation for the decreasing Stokes' value with the dust concentration (c) is given by [10, 20]:

$$Stk_{50}{}^{*} = (c_0/c)^{0.4} \cdot Stk_{50} \tag{12.11}$$

where

$Stk_{50}{}^{*} = Stk_{50}$ value at increased dust concentration
c_0 = low dust concentration
c = actual dust concentration.

Note that the constant power (0.4) given in Eq. (12.11) may vary somewhat from material to material, depending upon its tendency for agglomeration.

In hydrocyclones, the cut size usually increases and the collection efficiency decreases at inlet solids concentrations higher than about 1 % v/v due to the increased liquid viscosity [7]. Note that the viscosity of suspensions may strongly increase above particulate concentrations of about 5–10 % v/v. Near maximum packing of the solids in a suspension, the rheological behavior often becomes non-Newtonian [11]. This, of course, makes predictions for the cut size more complex.

For large industrial gas cyclones, both the characteristic Stokes' number and the Euler number are practically independent of the Reynolds number. Also, both numbers can be regarded constant for a given set of geometric proportions of cyclones relative to their diameter. For small diameter cyclones, changes in Reynolds number may affect both the Stokes' number and the Euler number, depending upon cyclone characteristics.

The erosion of gas cyclones may become important when the particulate materials are abrasive, their particles larger than about 200 μm or their density high, and gas velocities in the cyclones are high. Then, gravitational settlers are often installed in front of the cyclones to remove the coarsest particles, for protection against cyclone erosion.

In the last decades, the application of computational fluid dynamics (CFD), in combination with flow measurements by Laser Doppler Anemometry (LDA), have greatly increased the insight in the influence of the cyclone geometry on the flow patterns and the efficiency (see example in Fig. 12.6) [1, 2, 24]. Two weak points for using such simulations in the design are that there is not yet a universal theory for all geometries and that high grid resolution and, thus, much computer time are required for obtaining good results. This is caused by the fact that the turbulent flow and the presence of particles make the simulations very complex, which requires critical optimization of both models and parameters. Due to the fact that the simulation results can be compared with measured data, good progress is made in

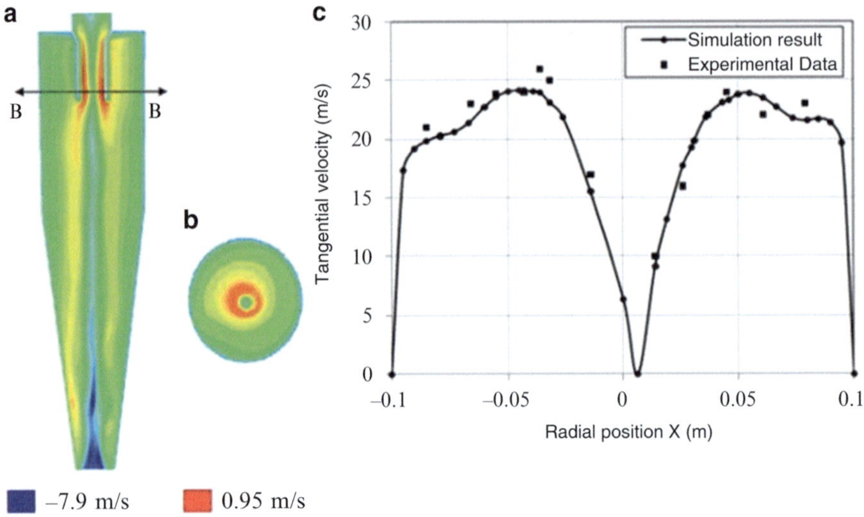

Fig. 12.6 Example of simulated flow pattern in cyclone [24] (Reproduced with permission of Prof. R. Utikar)

this field, however. For example, Elsayed concludes from his simulation results that the relation of the height and width of the cyclone inlet and the diameter of the gas outlet to the diameter of the cylindrical part are the main geometrical dimensions governing the collection efficiency and the pressure drop, in agreement with experimental evidence. Moreover, he states that the results of his mathematical simulations suggest that the cyclone pressure drop of the Stairmand HE model can be significantly decreased and its efficiency increased by slight modifications of the dimensions of the inlet, the length of the conical part and the diameter of the vortex finder [2–4]. The simulation results can indeed be very informative. See, for example in Fig. 12.6 at left, the very high, central upward flow rate near the solids exit, which may cause re-entrainment of any particles that remain there, and the indication for a shortcut near the vortex finder in the upper section [24].

Sometimes, only a single cyclone is used for particle removal. This is, for example, the case if fairly innocent dusts are to be removed, e.g. saw dust. In many other cases, however, it is advantageous to use cyclones in series or in parallel. Reasons for application in series may be to collect the particles in different size fractions and to increase the particles' recovery. A reason for application in parallel is if a high recovery is required at a high overall flow rate. In case of very wide size distributions, other equipment can be applied as well in combination with cyclones. As an example, gravity settling tanks placed before a cyclone have been mentioned earlier for protection against erosion.

The economics and capabilities of the various types of classifiers can be compared through plotting their electrical energy cost equivalent (EECE), which often are a major part of the running costs, against their cut size for a plant of given throughput of dust (see Fig. 12.7) [18, 20]. Although the data are dated (1965), the

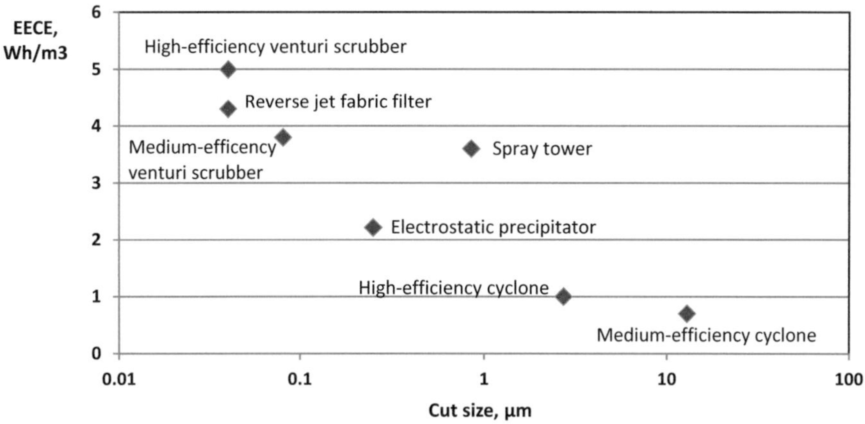

Fig. 12.7 Energy requirement vs. cut size relationship for 100,000 m^3/h classifiers (Adapted from Ref. [20]; data from Ref. [19])

figure still offers a nice indication of the electrical running costs (through multiplication with the actual electrical energy costs per kWh) and cut sizes of the different types of equipment.

12.5 Definitions, Abbreviations and Symbols

Aerodynamic size	size of a particle that has the same aerodynamic properties as the particle of interest, while using a density of 1000 kg/m^3 instead of the real particle density
Classifier	device that separates solids according to size (or density)
Collection efficiency	efficiency for collecting coarse particles
Cut size	particle size that has equal probability of reporting to overflow and underflow
Euler number	dimensionless number characterizing the pressure drop in a cyclone (resistance coefficient)
Grade efficiency	probability of any size fraction of reporting to the classifier underflow
Reynolds number	dimensionless number describing the ratio of inertial forces to viscous forces; its value relates to the degree of turbulence in particle movement (low values: laminar flow; high values: turbulent flow)
Stokes' number	dimensionless number relating inertial and hydrodynamic forces of a moving particle in a fluid
Tromp curve	curve showing the probability of any size fraction of reporting to the classifier underflow (grade efficiency curve)

CFD	Computational Fluid Dynamics
EECE	Electrical Energy Cost Equivalent for running costs of equipment
GSD	Geometric Standard Deviation
HEPA	High Efficiency Particulate Arrestance
LDA	Laser Doppler Anemometry
LES	Large Eddy Simulation
% v/v	percentage by volume
RANS	Reynolds Averaged Navier-Stokes equation

a	height cyclone inlet
b	width cyclone inlet
D	inside diameter of cylindrical part of cyclone
De	diameter of gas exit
$D_{p,crit}$	critical particle size for separation
D_{St}	Stokes' diameter of the particle
Eu	Euler number (Eq. 12.2)
g	gravitational acceleration constant
h	height cylindrical part cyclone
H	total height cyclone
M	molecular weight of gas
P	pressure
ΔP	pressure drop
Q	volumetric flow rate
R	gas constant
R_i	inner radius of cylindrical part of cyclone
Re	Reynolds number
S	length vortex finder
Stk	Stokes' number
Stk_{50}	characteristic Stokes' number used for cyclone design (Eq. 12.9)
T	absolute temperature
$U_{\theta,R}$	tangential velocity component at cyclone wall
U_R	radial velocity component at cyclone wall
u	characteristic linear flow rate
v_p	terminal settling velocity of the particle

η_f	fluid viscosity
η_L	liquid viscosity
θ	angle between cylindrcal and conical part of cyclone
ρ_f	fluid density
ρ_g	gas density
ρ_L	liquid density
ρ_p	effective particle density

References

1. Boysan, F., Ayers, W.H., Swithenbank, J.: Fundamental mathematical modeling approach to cyclone design. Trans. Inst. Chem. Eng. **60**, 222–230 (1982)
2. Elsayed, K.: Analysis and Optimization of Cyclone Separators Geometry Using RANS and LES Methodologies. PhD thesis Free University, Brussels (2011)
3. Elsayed, K., Lacor, C.: Modeling, analysis and optimization of air cyclones using artificial neural network response surface methodology and CFD simulation approaches. Powder Technol. **212**(1), 115–133 (2011)
4. Elsayed, K., Lacor, C.: Optimization of the cyclone separator geometry for minimum pressure drop using mathematical models and CFD simulations. Chem. Eng. Sci. **65**(22), 6048–6058 (2010)
5. Faulkner, W.B., Shaw, B.W.: Efficiency and pressure drop of cyclones across a range of inlet velocities. Appl. Eng. Agric. **22**(1), 155–161 (2005)
6. Gotoh, K. Masuda, H. Yoshida, H. Hidaka, J.: Ch. 5.2 Classification. In: Hagashitani, K., et al. (eds.) Powder Technology Handbook. CRC Press, Boca Raton, USA (2007)
7. Heiskanen, K.: Particle Classification, Powder Technology Series. Chapman & Hall, London (1993)
8. Hsu, C.-W., Huang, S.-H., Lin, C.-W., Hsiao, T.-C., Lin, W.-Y., Chen, C.-C.: An experimental study on performance improvement of the stairmand cyclone design. Aerosol Air Qual. Res. **14**, 1003–1016 (2014)
9. Leith, D., Mehta, D.: Cyclone performance and design. Atmos. Environ. **7**, 527–549 (1973)
10. Matsen, J.M.: Fluidized beds (Ch. 10). In: Bisio, A., Kabel, R.L. (eds.) Scale-Up of Chemical Processes, p. 383. Wiley, New York (1985)
11. Merkus, H.G.: Basic information for the design of particulate products. In: Merkus, H.G., Meesters, G.M.H. (eds.) Particulate Products – Tailoring Properties for Optimal Performance. Springer, Dordrecht, NL (2014).
12. Pell, M. Dunson, J.B.: Gas-solids separations, Ch. 17. In: Green, D.W., Maloney, J.O. (eds.) Perry's Chemical Engineers Handbook. McGraw-Hill, New York (1999)
13. Rhodes, M.: Introduction to Particle Technology. Wiley, Chichester, UK (1998)
14. Richardson, J.F. Harker, J.H. Backhurst, J.R. In: Coulson & Richardson's Chemical Engineering, vol. 2. Butterworth & Heinemann, Oxford, UK (2002)
15. Seville, J. Tüzün, U. Clift, R.: Processing of Particulate Solids, Ch. 7, Gas/Solid Separation. Chapman & Hall, London, UK (1997)
16. Shepherd, C.B., Lapple, C.F.: Flow pattern and pressure drop in cyclone dust collectors. Ind. Eng. Chem. **31**, 972–984 (1939)
17. Shepherd, C.B., Lapple, C.F.: Flow pattern and pressure drop in cyclone dust collectors. Ind. Eng. Chem. **32**, 1246–1248 (1940)
18. Stairmand, C.J.: The design and performance of cyclone separators. Trans. Inst. Chem. Eng. **29a**, 356–383 (1951)
19. Stairmand, C.J.: Chem. Eng. Lond. **194**, 310–326 (1965)
20. Svarovsky, L.: Solid-gas separations, vol. 3. In: Williams, J.C., Allen, T. (eds.) Handbook of Powder Technology. Elsevier, Amsterdam, NL (1981)
21. Svarovsky, L.: Solid-Liquid Separations. Butterworths-Heinemann, Oxford (2001)
22. Svarovsky, L.: Gas Cyclones. GASCYC.pdf internet
23. Tromp, F.: Neue Wege für die Bestimmung der Aufbereitung von Steinkohle, Glückauf Berg- und Hüttenmännische Zeitschrift. **73**(6), 125–156 (1937)
24. Utikar, R. Darmawan, N. Tade, M. Li, Q. Evans, G.: Hydrodynamic simulation of cyclone separators, Ch. 11. In: Hyoung Woo Ho (ed.) Computational Fluid Dynamics. InTech (2010)
25. Wang, B., Xu, D.L., Chu, K.W., Yu, A.B.: Numerical study of gas-solid flow in a cyclone separator. Appl. Math. Model. **30**, 1326–1342 (2006)
26. Wills, B.A.: Mineral Processing Technology, 5th edn. Pergamon Press, Oxford (1992)
27. Yoshida, H., Fukui, K., Yoshida, K., Shinoda, E.: Particle separation by Iinoya's type gas cyclone. Powder Technol. **118**, 16–23 (2001)

Chapter 13
Storage and Discharge of Bulk Solids

Dietmar Schulze

Abstract In most industrial situations, powders and bulk solids have to be stored in and discharged from bins and silos. This can result in – often unexpected – problems, e.g., flow obstructions or segregation. Most of these problems can be avoided by proper design of equipment with respect to the flow properties of the bulk material to be stored. Especially important is the design of the hopper and all downstream equipment (e.g., feeder) having an influence of the flow in the hopper.

In the present chapter the typical problems occurring in bins and silos are addressed, and the design of hoppers and related equipment with the goal to avoid these problems is outlined. Most important point is the determination of the hopper slope required to avoid stagnant zones, and the outlet dimensions to avoid flow obstructions like stable arches or ratholes. These quantities are calculated based on the flow properties of the respective bulk solid. The flow properties are quantities like, e.g., compressive strength and internal friction, and have to be measured with appropriate shear testers.

13.1 Introduction

Bulk solids (in the following the term "bulk solid" is used as the general term covering powders, granulates, etc.) have to be handled or stored in nearly all industries. Especially the discharge of bulk solids from silos, hoppers, transport containers etc. may result in severe problems.

Figure 13.1 shows several problems which can emerge during the operation of a silo. If the outlet opening is too small, a stable arch can form above the outlet and the flow stops (Fig. 13.1a). If particles are not much smaller than the opening, the reason for arch formation is the interlocking of particles. In contrast to this, materials of very fine particle size can form cohesive arches as a result of the compressive strength caused by consolidation and interparticle adhesive forces.

D. Schulze (✉)
Institute for Recycling, Ostfalia University of Applied Sciences,
Robert-Koch-Platz 8a, 38440 Wolfsburg, Germany
e-mail: d.schulze@ostfalia.de

© Springer International Publishing Switzerland 2016

H.G. Merkus, G.M.H. Meesters (eds.), *Production, Handling and Characterization of Particulate Materials*, Particle Technology Series 25,
DOI 10.1007/978-3-319-20949-4_13

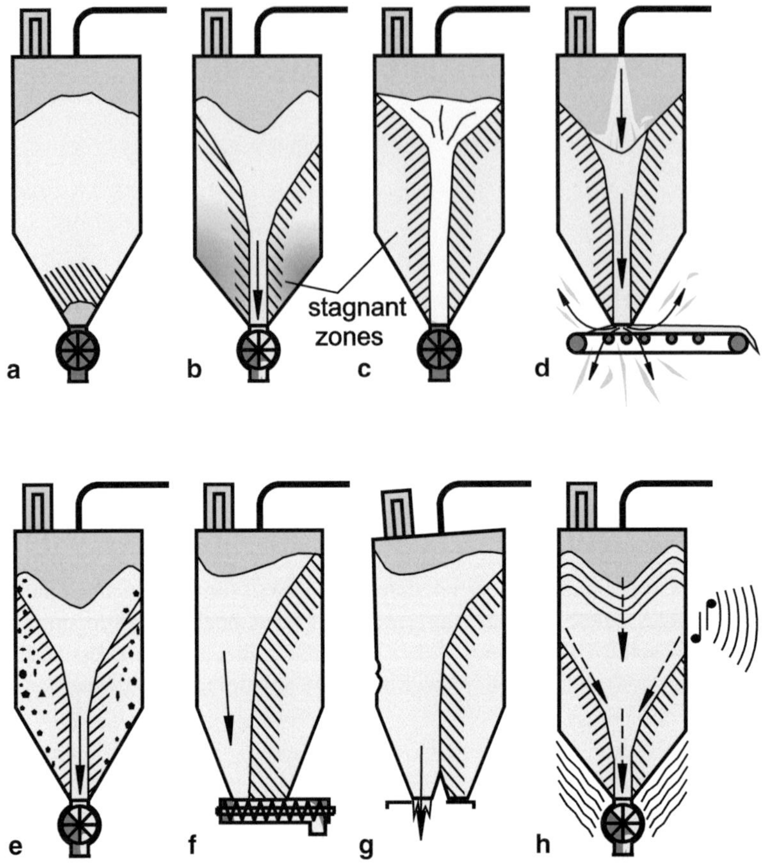

Fig. 13.1 Possible problems during the operation of silos; (**a**) arching; (**b**) funnel flow with wide residence time distribution and deterioration in product quality; (**c**) ratholing; (**d**) flooding; (**e**) segregation; (**f**) non-uniform discharge with screw feeder causing eccentric funnel flow; (**g**) buckling caused by eccentric flow; (**h**) vibrations (silo quaking and silo noise) [37]

Another possible source of problems is funnel flow (Fig. 13.1b). One reason for funnel flow is a hopper wall which is too shallow or too rough. In this case in a filled silo the bulk solid cannot slide downwards along the hopper walls. Thus, so-called stagnant zones build up and the material flow is limited to a flow zone above the outlet opening. In a silo used as a buffer and never discharged completely, bulk material can remain in the stagnant zones over long periods of time and change its properties (e.g., decomposition of food products). Furthermore, the bulk solid in the stagnant zones can consolidate with time to such an extent that it will not be able to flow out after the flow zone has emptied out. The latter results in a "pipe" or "rathole" reaching from the outlet opening to the top of the filling (Fig. 13.1c).

The residence time of the bulk solid in the flow zone of a funnel flow silo can be extremely short (Fig. 13.1d), so that the material, which has just been fed into the silo, is immediately discharged. Within this short time an easily fluidized bulk solid (e.g., flour, fine chalk) cannot sufficiently deaerate. Hence, it will flood out of the outlet opening like a fluid, resulting in increased dust generation and flooding of the feeder.

Funnel flow can also result in reduced product quality due to segregation (Fig. 13.1e). When filling a silo, one has to take into account that the product can segregate across the cross-section of the silo. When the silo is filled centrically, one often will observe an increased amount of fines close to the silo axis, and more coarse particles close to the silo wall. If funnel flow takes place at discharge, at first the material from the center (the fines) flows out, followed by the coarser material from the silo periphery. The time-dependent composition of the discharged bulk solid can be, for example, a quality problem during the filling of small packages, or when steady-state downstream processes have to be charged.

Less well-designed feeders can result in a non-uniform withdrawal of the bulk solid and, thus, in funnel flow (note that funnel flow comprises all types of flow where a part of the contents remains stationary). The screw feeder shown in Fig. 13.1f discharges the bulk solid at its rear end and then conveys it horizontally to its outlet, i.e., the screw is filled already at the rear end of the feeder so that along the rest of the outlet length no further bulk solid can enter the screw. Another reason for eccentric flow is, for example, the discharge from only one of multiple outlets (Fig. 13.1g). Eccentric flow results in unfavorable non-symmetrical loads on the silo structure thus increasing the danger of buckling of thin-walled metal silos and cracking of reinforced concrete silos that have only a single layer of reinforcing steel.

For different reasons the flow of a bulk solid in a silo can result in vibrations and shocks (Fig. 13.1h). Depending on the frequency, the effect is called silo noise (audible) or silo quaking (low frequency, individual shocks).

To avoid such complications, solutions have to be found considering the flow properties of the bulk solid. These flow properties, defining the mechanical behavior of the materials, can be rather different: Particle sizes can range from nanometers to meters, products can be dry or moist, and different chemical compositions may result in rather different behavior. Thus, before a hopper or silo is designed to match the properties of the bulk solid, a detailed characterization of its flow properties is inevitable.

A detailed description of the characterization of bulk solids and the design of silos is given by the author in [37] from which, with respect to the aim of the present book, an extract of the most important topics is taken for the present chapter. Since flow properties are the base for silo design for flow, the most important flow properties and appropriate test methods will be explained in the next sections. Afterwards, flow regimes and stress distributions in silos will be outlined followed by the description of the design of hoppers according to Jenike's theory [18], i.e., determination of hopper slope to achieve mass flow and outlet size to prevent arching (or ratholing in the case of funnel flow). Finally, information on feeder design, discharge aids, and general design precautions are presented. Thus, the necessary steps to achieve an appropriate design of a bin or silo with the scope of avoiding the problems outlined above are as follows:

- Measure flow properties of the material under consideration
- Determine hopper inclination and minimum outlet size
- Specify silo shape based on findings listed above
- Find appropriate feeder
- Add discharge aids if outlet dimensions are smaller than minimum outlet size
- Provide all details listed above for structural design

13.2 Properties of Bulk Solids Related to Silo Design

13.2.1 Stresses

Figure 13.2 shows a bulk solid element in a container (assumptions: infinite filling height, frictionless internal walls). In the vertical direction, positive normal stress ($\sigma_v > 0$, compressive stress) is exerted on the bulk solid. Within the bulk solid (Fig. 13.2) the horizontal stress, σ_h, is a result of the vertical stress, σ_v, where the resulting horizontal stress is less than the vertical stress exerted on the bulk solid from the top. The ratio of horizontal stress to vertical stress is the stress ratio, K (also known as λ):

$$K = \sigma_h / \sigma_v \tag{13.1}$$

Typical values of K are between 0.3 and 0.6 [21, 37].

It follows that in a bulk solid – in analogy to solid bodies – different stresses can be found in different cutting planes. Using a simple equilibrium of forces at a volume element with triangular cross-section cut from the bulk solid element shown in Fig. 13.2 (Fig. 13.3, on the left), the normal stress, σ_α, and the shear stress, τ_α, acting on a plane inclined by an arbitrary angle α, can be calculated.

The pair of values (σ_α, τ_α) for all possible angles α can be plotted in a σ,τ-diagram (normal stress, shear stress – diagram); see Fig. 13.3 on the right. If one joins all plotted pairs of values, a circle emerges; i.e., all calculated pairs of values form a circle in the σ,τ-diagram. This circle is called "the Mohr stress circle". Its

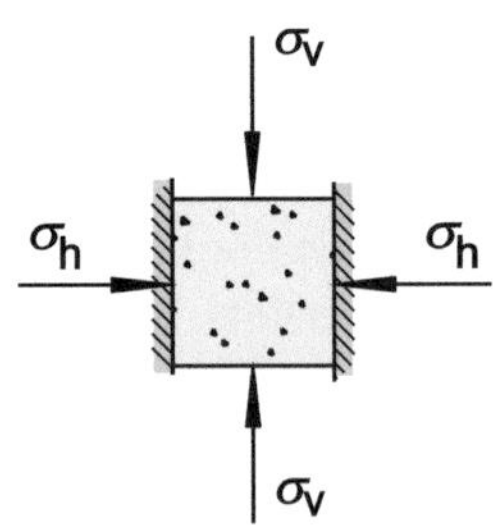

Fig. 13.2 Element of bulk solid [37]

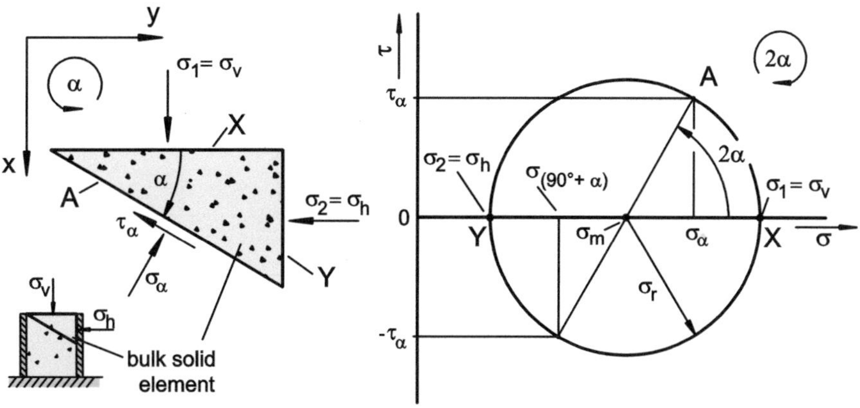

Fig. 13.3 Force equilibrium on an element of bulk solid, the Mohr stress circle [37]

centre is located at $\sigma_m = (\sigma_v + \sigma_h)/2$ and $\tau_m = 0$. The radius of the circle is $\sigma_m = (\sigma_v - \sigma_h)/2$. The Mohr stress circle represents the stresses on all cutting planes at arbitrary inclination angles α, i.e., in all possible cutting planes within a bulk solid element.

Since the centre of the Mohr stress circle is always located on the σ-axis, each Mohr stress circle has two points of intersection with the σ-axis. The normal stresses defined through these points of intersection are called the principal stresses. The larger principal stress – the major principal stress – is designated as σ_1 and the smaller principal stress – the minor principal stress – is designated as σ_2. If both principal stresses are given, the Mohr stress circle is well defined.

In the example of Fig. 13.2 both the horizontal and the vertical plane are free from shear stresses ($\tau = 0$) and are thus principal stress planes. In this case the vertical stress, σ_v, which is greater than the horizontal stress, σ_h, is the major principal stress, σ_1, and the horizontal stress, σ_h, is the minor principal stress, σ_2.

An important qualitative result of the above analysis is that shear stresses can occur in bulk solids at rest. Thus, the state of stress in a bulk solid cannot be completely described by only a single numerical value.

13.2.2 Compressive Strength

The phrase "good flow behavior" usually means that a bulk solid flows easily, i.e., it does not consolidate much and flows out of a silo or a hopper due to the force of gravity alone and no discharge aids (Sect. 13.6.5) are required. Products are "poorly flowing" if they experience flow obstructions or consolidate during storage or transport. An important quantity to describe the ability of a bulk solid to consolidate is the unconfined yield strength (compressive strength) in relation to the consolidation stress. This will be explained in the following paragraphs.

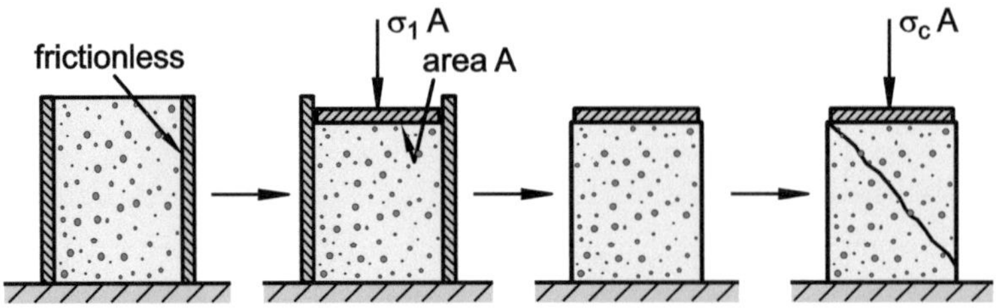

Fig. 13.4 Uniaxial compression test [37]

Figure 13.4 shows a hollow cylinder filled with a fine-grained bulk solid (cross-sectional area A; internal wall of the hollow cylinder assumed as frictionless). The bulk solid is loaded by the stress σ_1 – the consolidation stress – in the vertical direction. This results in a decrease of the specimen's volume and, thus, an increase of the bulk density. The more the volume of the bulk solid specimen is reduced, the more compressible the bulk solid is. In addition to the increase in bulk density, one will observe also an increase in strength of the bulk solid specimen. Hence, the bulk solid is both consolidated and compressed through the effect of the consolidation stress.

After consolidation, the bulk solid specimen is relieved of the consolidation stress, σ_1, and the hollow cylinder is removed. If subsequently the consolidated cylindrical bulk solid specimen is loaded with an increasing vertical compressive stress, the specimen will break (fail) at a certain stress. The stress causing failure is called compressive strength or unconfined yield strength, σ_c.

Uniaxial compression tests (Fig. 13.4) conducted at different consolidation stresses, σ_1, lead to different pairs of values (σ_c, σ_1) and (ρ_b, σ_1). Plotting these pairs of values as points in a σ_c,σ_1-diagram and a ρ_b,σ_1-diagram, respectively, and drawing in each diagram a curve through these points, usually results in curves like those for product A in Fig. 13.5a, where bulk density, ρ_b, and unconfined yield strength, σ_c, typically increase with consolidation stress, σ_1. The curve $\sigma_c(\sigma_1)$ is called the flow function.

Some bulk solids increase in strength if they are stored for a longer time at rest under a compressive stress (e.g. in a silo or an intermediate bulk container). This effect is called time consolidation or caking. Typical causes of time consolidation are, e.g., chemical processes, crystallizations between the particles, enlargement of the contact areas through plastic deformation, capillary condensation, or biological processes such as fungal growth [37].

Time consolidation can be determined with the test shown in Fig. 13.4, in order, e.g., to simulate long-term storage in a silo. For this one loads the specimen with consolidation stress, σ_1, not only for a short moment, but for a defined period of time, t_1. Then the unconfined yield strength is determined following the principle explained above (Fig. 13.4). If a bulk solid exhibits time consolidation, one obtains curves $\sigma_c(\sigma_1)$ for storage periods $t > 0$ (curves A_1, A_2 in Fig. 13.5b) which are called time flow functions. Here each curve emerges from the connection of several pairs of values (σ_c, σ_1), which were measured at identical storage periods, t, but at different consolidation stresses, σ_1.

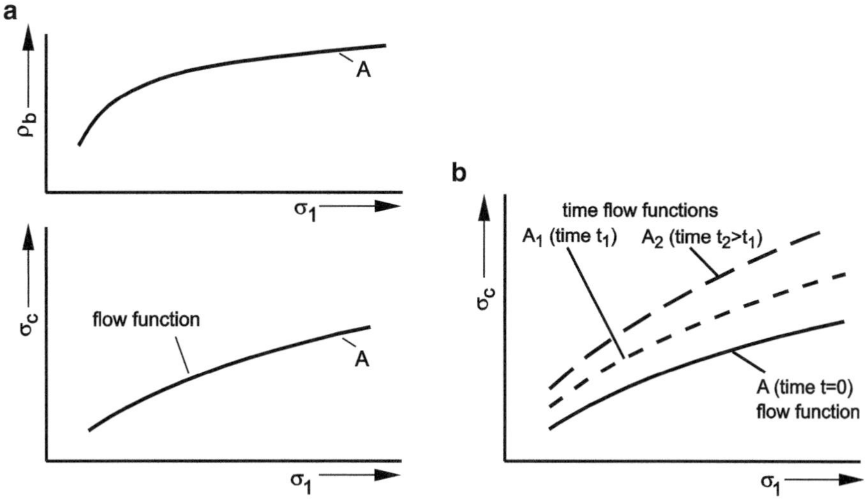

Fig. 13.5 (**a**) Bulk density, ρ_b, and unconfined yield strength, σ_c, vs. consolidation stress, σ_1; (**b**) Flow function and time flow functions for two different storage times t_1 and $t_2 > t_1$ [37]

The uniaxial compression test presented in Fig. 13.4 is shown in a σ,τ-diagram (Fig. 13.6). If one neglects the force of gravity of the bulk solid specimen and assumes that no friction is acting between the wall of the hollow cylinder and the bulk solid, both vertical stress, σ_v, as well as horizontal stress, σ_h, are constant within the entire bulk solid specimen. Therefore at each position in the specimen the state of stress is identical.

During consolidation the vertical normal stress, σ_1, acts on the top of the bulk solid specimen. Perpendicular to the vertical stress the smaller horizontal stress prevails according to stress ratio K (see Sect. 13.2.1). Neither at top nor at bottom of the specimen, nor at the internal wall of the hollow cylinder, which is assumed as frictionless, will shear stresses be found; i.e., $\tau = 0$. The pairs of values (σ, τ) for vertical and horizontal cutting planes within the bulk solid specimen are plotted in the σ,τ-diagram (Fig. 13.6). Both points are located on the σ-axis because $\tau = 0$. The vertical stress is identical to the major principal stress, σ_1 (see Fig. 13.3), and the horizontal stress is equal to the minor principal stress, σ_2. Thus, the Mohr stress circle (A) representing the conditions during at the end of consolidation is defined (note that usually only the upper semicircle needs to be drawn).

The consolidation step causes a certain strength of the bulk solid specimen. This "strength" is characterized by a yield limit, called a "yield locus" in bulk solids technology. A yield locus drawn in a σ,τ-diagram represents pairs of values (σ, τ) causing failure, or flow, of the bulk solid. In simple words, it can be stated that a yield locus gives the shear stress required to move the particles in a particular cutting plane across each other as a function of the normal stress. The yield locus is dependent on previous consolidation: The greater the consolidation stress, σ_1, the more the yield locus is shifted towards higher shear stresses.

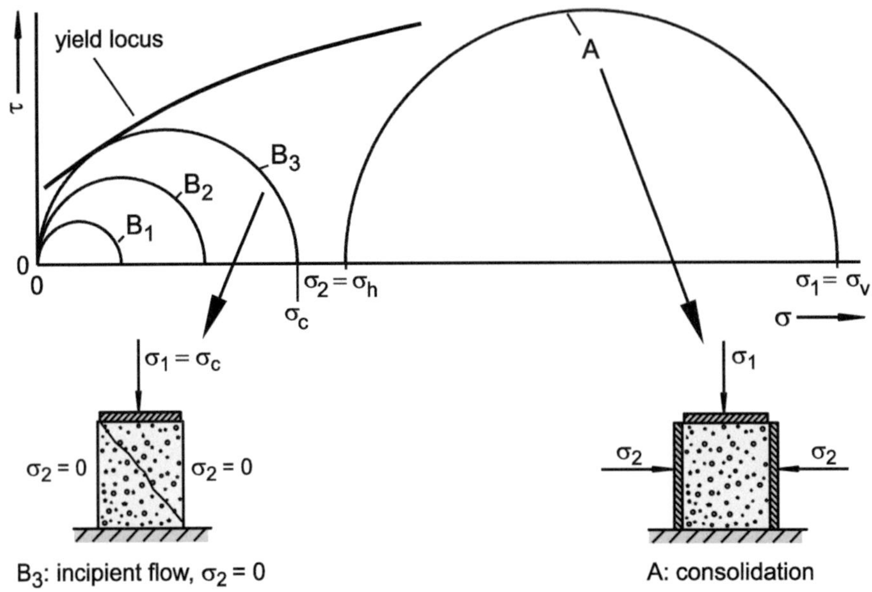

Fig. 13.6 Measurement of unconfined yield strength in a σ,τ-diagram [37]

In the second part of the test shown in Fig. 13.4, the specimen is loaded with increasing vertical stress after it has been relieved of the consolidation stress and the hollow cylinder has been removed. Again, vertical and horizontal stresses are principal stresses. The horizontal stress is equal to zero since the lateral surface of the specimen is uncovered. During the increasing vertical load, the stress states at different load steps are represented by stress circles with increasing diameter (stress circles B_1, B_2, B_3 in Fig. 13.6). Mohr stress circles B_1 and B_2, which are completely below the yield locus, cause only an elastic deformation of the bulk solid specimen, but no failure and/or flow. At failure of the specimen the Mohr stress circle B_3 represents the stresses in the bulk solid specimen. Since the load corresponding to this Mohr stress circle causes incipient flow of the specimen, the yield limit of the consolidated bulk solid must have been attained in one cutting plane of the specimen. Thus, Mohr stress circle B_3 must touch the yield locus in the σ,τ-diagram. The unconfined yield strength, σ_c, is represented by the right point of intersection of Mohr stress circle B_3 with the σ-axis.

At the beginning of the present chapter words like "poor" and "easy" were used to describe the flow behavior of a bulk solid. The link between the strength of a bulk solid and its flow behavior is given by the ratio of unconfined yield strength, σ_c, to consolidation stress, σ_1, which is called the flowability, ff_c:

$$ff_c = \sigma_1 / \sigma_c \quad (13.2)$$

The larger ff_c is, i.e., the smaller the ratio of the unconfined yield strength to the consolidation stress, the better a bulk solid flows. Similar to the classification used by Jenike [18, 19], one can define flow behavior based on ff_c as follows:

- $ff_c < 1$: not flowing
- $1 < ff_c < 2$: very cohesive
- $2 < ff_c < 4$: cohesive
- $4 < ff_c < 10$: easy-flowing
- $10 < ff_c$: free-flowing

Flowability, ff_c, allows for a comparison of different bulk solid samples, e.g., a bulk solid mixed with different amounts of a flow agent, or produced in different ways. The advantage of this method is that it is based on a well-defined physical quantity, namely the unconfined yield strength which is a result of adhesive forces.

However, since unconfined yield strength, σ_c, is not proportional to consolidation stress, σ_1, flowability varies with consolidation stress (usually it increases with increasing consolidation stress). This is not a disadvantage, but just the reflection of the real behavior of bulk solids: In a silo's hopper the stresses are proportional to local hopper diameter (this will be discussed in detail in Sect. 13.5.2.1). Thus, if the outlet opening of a hopper is very small, both stresses and flowability close to the outlet are small and vice versa if the outlet opening is large. Experience has shown that flow is more difficult to achieve if an outlet is smaller, so obviously flowability is worse at a smaller outlet (small stresses) compared to a larger one (greater stresses). In consequence of the stress dependence of flowability, comparative tests have to be conducted at identical consolidation stresses [37].

13.2.3 Practical Determination of Flow Properties

13.2.3.1 Yield Locus

The use of the uniaxial compression test with fine-grained, cohesive bulk solids is problematic, because one obtains unconfined yield strength values that are too low [45], and preparation of the hollow cylinder to obtain frictionless walls is very time-consuming. In addition, further important parameters (e.g., internal friction and wall friction) cannot be determined with this test.

In advanced bulk solids technology shear testers are used to measure yield loci of bulk solids. For a shear test, a bulk solid specimen is loaded vertically by a normal stress, σ (Fig. 13.7a). Then a shear deformation is applied on the specimen by moving the top platen with a constant velocity, v. This results in a horizontal shear stress, τ, originated by the friction between the particles (Fig. 13.7b).

When a point of a yield locus is measured, in analogy to the uniaxial compression test, two steps are necessary: First the bulk solid specimen is consolidated. This is called "preshear". Subsequently a point of the yield locus is measured. This step

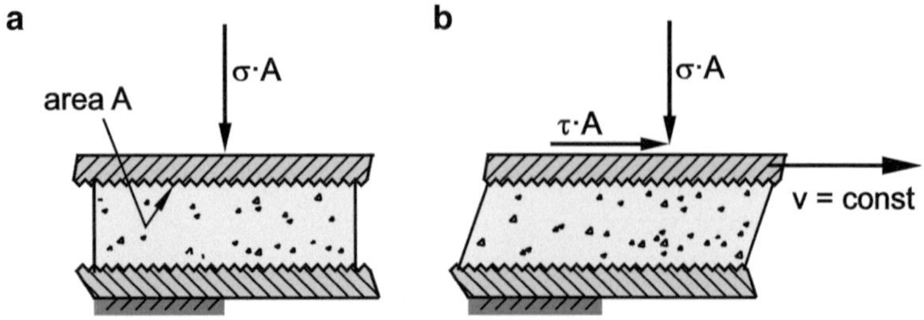

Fig. 13.7 Bulk solid specimen: (**a**) initial loading with normal stress σ; (**b**) shear deformation (velocity $v = \text{const}$) [37]

is called "shear" or "shear to failure". A detailed description of the procedure is given in [18, 37, 48].

After several points of a yield locus have been measured, the yield locus is drawn as a line or curve through these points as shown in Fig. 13.8. This allows for the determination of the parameters representing the flow properties. The consolidation stress, σ_1, is equal to the major principal stress of the Mohr stress circle which is tangential to the yield locus and intersects at the preshear point defined by the normal and shear stress at the end of consolidation. Thus, this stress circle represents the stresses in the specimen at the end of the consolidation procedure (similar as the stress circle at the end of consolidation at the uniaxial compression test, Fig. 13.6). The unconfined yield strength, σ_c, results from the stress circle which is tangential to the yield locus and which runs through the origin (minor principal stress $\sigma_2 = 0$). This stress circle represents a similar stress state as the one which prevails in the second step of the uniaxial compression test (stress circle B_3, Fig. 13.6). With consolidation stress, σ_1, and unconfined yield strength, σ_c, one can calculate flowability, ff_c, as defined in Eq. (13.2).

A straight line through the origin of the σ,τ-diagram, tangent to the greater Mohr circle (representing the stresses at the end of consolidation), is the effective yield locus as defined by Jenike [18] (broken line in Fig. 13.8). It encloses the σ-axis with the angle φ_e (effective angle of internal friction). Because the largest Mohr stress circle indicates a state of steady-state flow (i.e., flow at constant stresses and constant bulk density), the angle φ_e can be regarded as a measure of the internal friction at steady-state flow. This angle is required for silo design according to Jenike's theory. Another important parameter is the slope of the yield locus. Since the yield locus is often slightly curved, usually the slope angle, φ_{lin}, of the linearized yield locus is taken (Fig. 13.8).

If several yield loci are measured at different stress levels, i.e., with different normal stresses at preshear, σ_{pre}, each yield locus represents another state of consolidation and another bulk density. The above mentioned flow properties (unconfined yield strength, effective angle of internal friction) can be indicated as a function of the consolidation stress, σ_1, similar to Fig. 13.5a where bulk density and unconfined yield strength are plotted vs. consolidation stress.

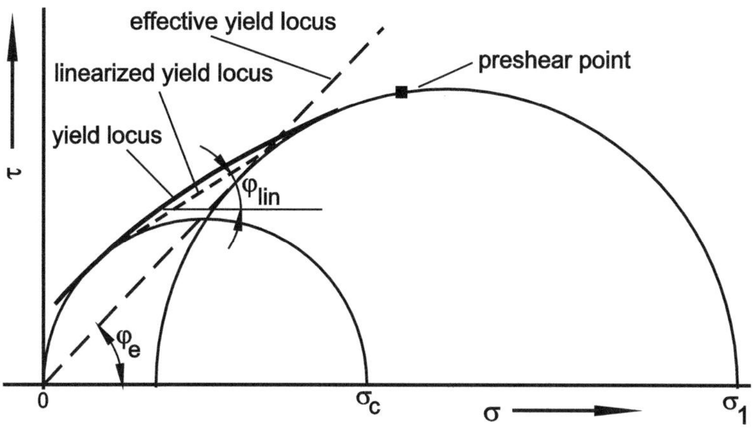

Fig. 13.8 Yield locus and parameters [37]

13.2.3.2 Time Yield Locus

The time consolidation, which describes the increase of the unconfined yield strength with time during storage at rest, is measured with a shear tester similar to the measurement of a yield locus. The difference to a yield locus test is that after preshear the specimen is stored for consolidation time period t under static vertical load equal to consolidation stress, σ_1, of the corresponding yield locus. After the time consolidation period t, the specimen is sheared to failure. If consolidation time affects the bulk solid under consideration, after the consolidation period the shear stress maximum will be larger than it would have been without a consolidation period between preshear and shear.

With the measured shear points a time yield locus can be approximated similarly to the approximation of a yield locus (Fig. 13.9). Compared to the yield locus, the time yield locus is shifted towards greater shear stresses, τ (if the bulk solid shows an increase of strength with time). The unconfined yield strength, σ_c, is determined in the same way as for a yield locus by drawing a Mohr stress circle through the origin and tangent to the time yield locus. In Fig. 13.9 the values of the unconfined yield strength for the consolidation periods, t_1 and t_2, are designated as $\sigma_c(t_1)$ and $\sigma_c(t_2)$.

13.2.3.3 Wall Yield Locus

Wall friction is the friction between a bulk solid and the surface of a solid, e.g. the wall of a silo or a hopper. The wall friction angle is important both for silo design for flow and silo design for strength, but also for the design of chutes and other equipment, where the bulk solid will flow across a solid surface.

The principle of a wall friction test, where the kinematic angle of wall friction is determined, is shown in Fig. 13.10. The bulk solid specimen is subjected to a

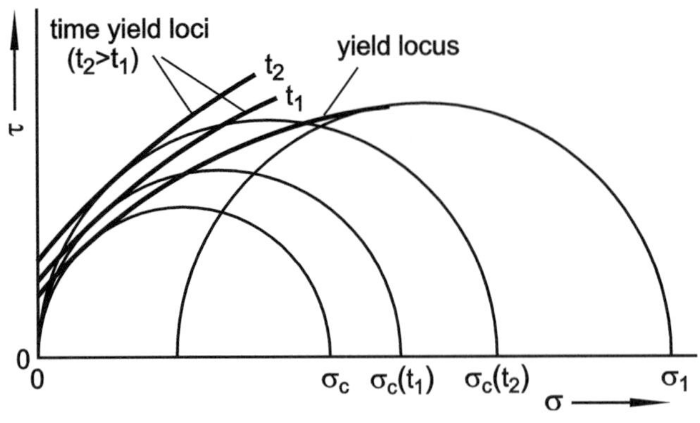

Fig. 13.9 Yield locus and time yield loci [37]

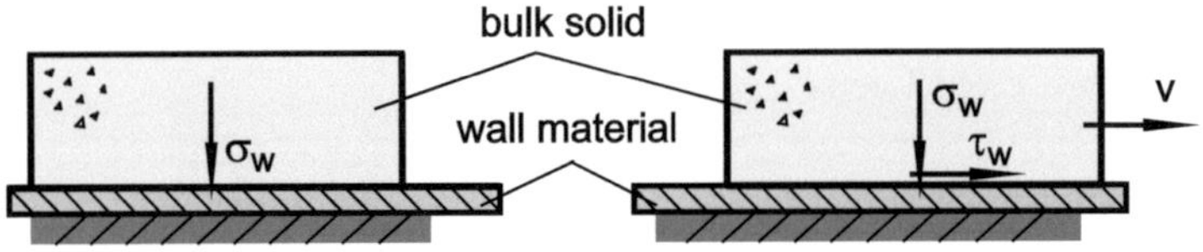

Fig. 13.10 Principle of a wall friction test [37]

vertical normal stress. The normal stress acting between bulk solid specimen and wall material is called the wall normal stress, σ_w.

The bulk solid specimen is then shifted relative to the wall material surface with a constant velocity, v. The shear stress acting between bulk solid specimen and wall material is measured. This is done at different levels of normal stress. All pairs of values of wall normal stress and wall shear stress are plotted in a σ_w,τ_w-diagram (Fig. 13.11). The curve (or line) running through the measured points is called the wall yield locus.

The wall yield locus is a yield limit like the yield locus. The wall yield locus describes the wall shear stress, τ_w, necessary to shift a bulk solid continuously across a wall surface under a certain wall normal stress, σ_w. To quantify wall friction, the wall friction angle, φ_x, is used. It is the slope of a line running through the origin of the σ_w,τ_w-diagram and a point of the wall yield locus. If the wall yield locus is a straight line running through the origin (Fig. 13.11), one obtains the identical wall friction angle, φ_x, for each point of the wall yield locus. Otherwise, φ_x had to be determined from the ratio of wall shear stress, τ_w, to wall normal stress, σ_w, for the desired wall normal stress [37].

13.2.3.4 Shear Testers

Around 1960 Jenike [18] published his fundamental work on silo and bulk solids technology and introduced the Jenike Shear Tester (Fig. 13.12a). As outlined in the

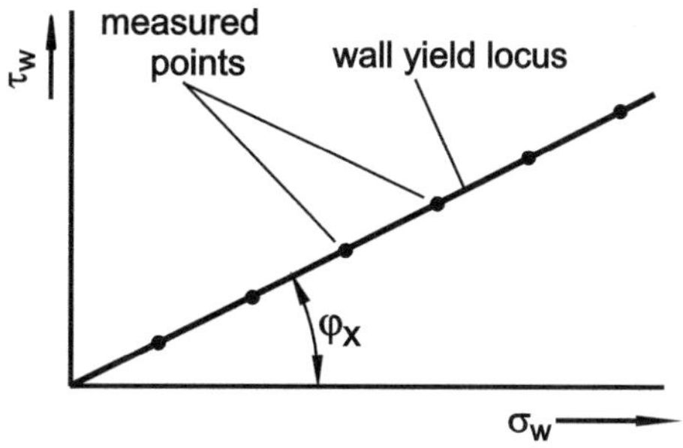

Fig. 13.11 Course of wall shear stress in a wall friction test; wall yield locus [37]

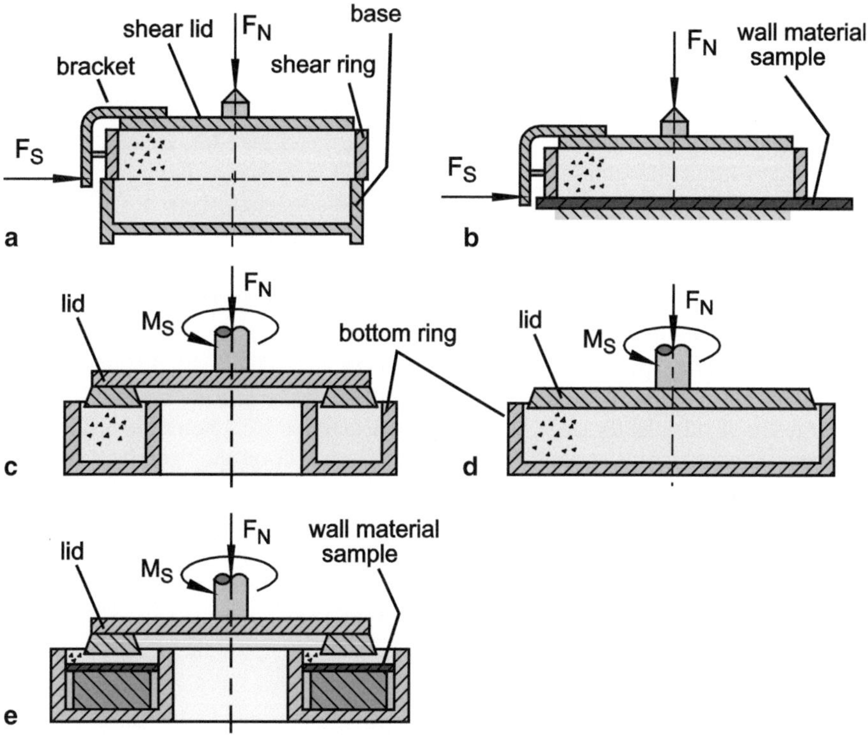

Fig. 13.12 Shear cells with powder specimens (schematic); (**a**) Jenike shear tester, (**b**) Set-up of Jenike shear cell for wall friction tests, (**c**) Ring shear tester, (**d**) Torsional shear tester; (**e**) Ring shear cell for wall friction tests

previous chapters, for a shear test a powder specimen is subjected to a vertically acting normal stress, σ, and then sheared by shifting its upper and lower side relative to each other in the horizontal direction. This principle is realized with the Jenike tester (Fig. 13.12a) as follows: The specimen is contained in a shear cell consisting of a base of circular cross-section, a shear ring of the same diameter lying above the base, and a shear lid. The shear lid is loaded centrally with a normal force,

F_N. To shear the specimen, the upper part of the shear cell is moved horizontally by a stem that presses against a bracket attached to the shear lid. Force F_S required to do this is measured. The normal stress, σ, and shear stress, τ, acting in the horizontal plane between shear ring and base are determined by dividing normal force, F_N, and shear force, F_S, by the cross-sectional area of the shear cell, A. For the measurement of wall friction, the base of the shear cell is replaced by a sample of wall material (Fig. 13.12b). The test procedure for the Jenike shear tester is described in detail in the "Standard Shear Testing Technique" [48], which is an internationally agreed standard on the operation of the Jenike shear tester, and in ASTM standard D6128 [2]. A shorter description is given in [37].

The Jenike shear tester was the first one designed for the purposes of powder technology, and still today shear testers are compared to the Jenike Shear Tester. However, the operation of the Jenike tester is difficult and time-consuming (e.g., due to the limited shear displacement for each preshear/shear to failure cycle a new specimen has to be filled into the shear cell and subjected to a manual preconsolidation). Therefore, it was rarely applied on other tasks than silo design. Progress was made with the development of rotational shear testers, e.g., the first ring shear tester for powders and bulk solids [10], or the torsional shear tester [3, 29] (Fig. 13.12c, d), where the relative displacement is achieved by rotation of the top of the powder specimen relative to the bottom. The major difference between these two types of testers is that the cross-section of the specimen in a torsional shear cell is circular (Fig. 13.12d), while it is annular in a ring shear cell (Fig. 13.12c). This difference is important because due to the rotational movement, the shear deformation of the bulk solid specimen varies with radius. With a ring shear tester of sufficiently large ratio of inner to outer ring diameter the differences in shear deformation are reduced so far that their effect on the test results can be neglected [12, 25, 49].

The specimen in a rotational shear tester (Fig. 13.12c, d) is loaded vertically with a normal force, F_N, by the (circular or annular) lid. The lid is provided with a rough underside, realized, e.g., by bars protruding into the bulk solid specimen. By rotating the bottom ring relative to the lid around the vertical axis of the shear cell, the specimen is subjected to shear deformation. The moment, M_S, resulting from the shear stress, τ, developing in the specimen is measured. Normal stress, σ, and shear stress, τ, are determined from normal force, F_N, and moment, M_S. For the measurement of wall friction with a ring shear tester, a bottom ring provided with a sample of the wall material under consideration is used (Fig. 13.12e).

The fundamental advantage of rotational shear testers (Fig. 13.12c, d) is that the shear displacement is not limited by the apparatus as demonstrated with the Jenike shear tester. In the last decades automatic rotational shear testers were developed resulting in a significant reduction of operation time (e.g., [28, 38]). Round robin tests on two types of ring shear testers developed by the author (ASTM standard D6773 [4]) confirm their ability to obtain consistent test results independent of location and operator as well, and demonstrate that results comparable to the Jenike shear tester, but with less scatter, are obtained [39]. As a result of these developments, ring shear testing has become more and more popular, not only for silo design, but also for product development or quality control.

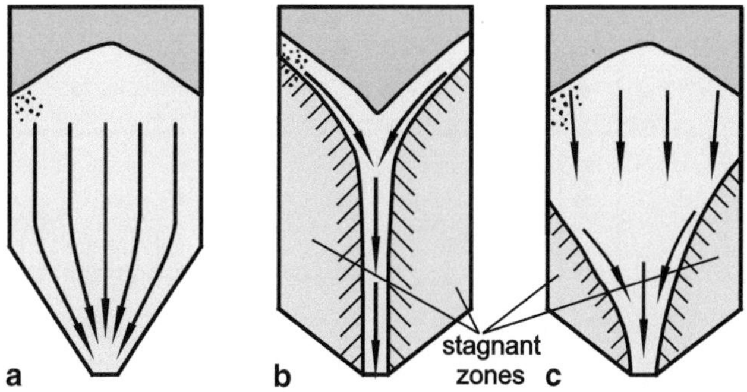

Fig. 13.13 Flow profiles: (**a**) Mass flow; (**b**) Funnel flow with stagnant zones up to the level of filling; (**c**) Funnel flow with stagnant zones in the lower silo part; here the stagnant zones are asymmetric [37]

13.3 Flow Profiles: Mass Flow and Funnel Flow

When a bulk solid discharges from a silo, one must distinguish between mass flow and funnel flow, see Fig. 13.13. In a mass flow silo every particle of the bulk solid in the silo is moving whenever the outlet is opened provided that arching does not occur. Mass flow is only possible if the hopper walls are steep and/or low enough in friction. If the hopper wall is too flat or too frictional, funnel flow will occur. In a funnel flow silo (Fig. 13.13b, c) at first only the bulk solid in a channel above the opening flows downwards. The bulk solid located in the stagnant zones, which develop at the silo periphery starting at the hopper walls directly above the opening, can be discharged only if the silo is emptied completely. The stagnant zones can reach the top level of the filling with the result that a flow funnel is formed at the surface. It is also possible that the stagnant zones exist only in the lower part of the silo, because the boundary between the flowing bulk solid and the stagnant zones intersects the silo wall beneath the top level of the filling. In this case a clear recognition of the flow profile is not possible by only looking from above on the filling (Fig. 13.13c). Additionally the stagnant zones can be asymmetrical (even if the silo is symmetric with a centric outlet) thus causing disadvantageous loads on the silo walls.

13.4 Stresses in Silos

As outlined in Sect. 13.2, bulk solids at rest can transmit shear stresses (contrary to Newtonian fluids). While pressure increases linearly with depth in a liquid container (Fig. 13.14), the shear stress exerted from the bulk solid onto the container wall – i.e. the friction at the container wall – carries part of the bulk solids weight.

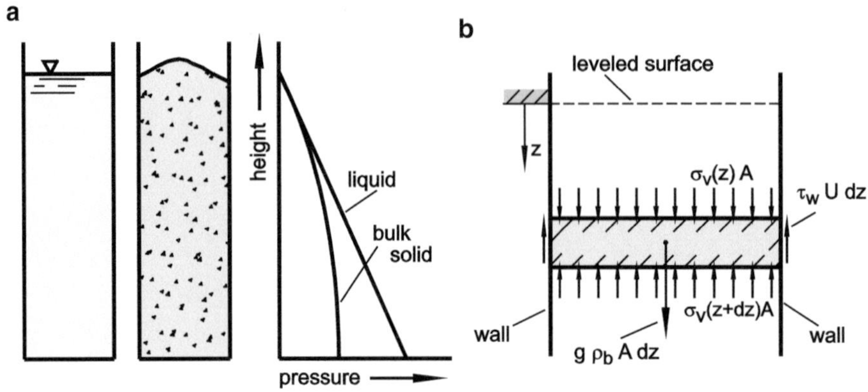

Fig. 13.14 (**a**) Pressure (stress) in liquids and bulk solids; (**b**) Vertical forces acting on one slice element of bulk solid within the vertical section of a silo [37]

As a consequence the stress, or pressure, respectively, in a container filled with a bulk solid is increasing less and less in downwards direction.

The stresses in a silo's vertical section (vertical walls) can be calculated according to a derivation by Janssen [17]. He considered a slice element, i.e., an elemental section of the vertical section, of infinitesimal height, dz (Fig. 13.14b). With the assumptions of constant vertical stress, σ_v, acting across the cross-section, and constant bulk density, ρ_b, the equilibrium of forces in z-direction results in a differential equation describing the vertical stress as a function of coordinate z. For the case that the stress on the top surface is zero ($\sigma_{v0} = 0$ at $z = 0$), the solution of the differential equation is the well-known "Janssen equation":

$$\sigma_v = \frac{g \rho_b A}{K \tan\ (\varphi_x)\ U} \cdot \left[1 - e^{\frac{-K\ \tan\ (\varphi_x)\ U\ z}{A}} \right] \qquad (13.3)$$

where

A: cross-sectional area of vertical section
g: acceleration due to gravity
K: stress ratio (Sect. 13.2.1)
U: perimeter of vertical section
φ_x: wall friction angle (Sect. 13.2.3.3)
ρ_b: bulk density

The quantities describing the bulk solid's properties have to be determined with appropriate shear testers. It is recommended to measure these properties for the largest vertical stress expected in the vertical section. If the influence of the stress is very strong, it is more accurate to solve the above mentioned differential equation numerically while considering the stress dependence of bulk density and wall friction angle (A program for the assessment of stresses in silos can be downloaded from the author's website [40]).

For large values of z the exponential function in Eq. (13.3) approaches zero. Thus, the expression in front of the brackets is the final (asymptotic) stress, which is attained for $z \rightarrow \infty$. The final value of the vertical stress is:

$$\sigma_{v\infty} = \sigma_v(z \rightarrow \infty) = \frac{g\,\rho_b A}{K \tan\,(\varphi_x)\,U} \tag{13.4}$$

$\sigma_{v\infty}$ is independent of silo height and surcharge stress, σ_{v0}. It depends on the bulk solid's properties and the ratio A/U, where $A/U = D/4$ in a cylinder of diameter D. Thus, according to Eq. (13.4), the maximum possible stress in a cylindrical section, where usually no significant surcharge load is applied, is proportional to the silo diameter. Therefore slim silos can be high without getting too large stresses in the silo. A "rough" wall of larger angle of wall friction, φ_x, results in smaller stresses because it carries a larger part of the bulk solids weight than a "smooth" wall characterized by a small angle of wall friction.

A typical silo consists of a vertical section (vertical section, cylinder section) and a hopper. The stress conditions in the hopper are more complex than in the vertical section. If a previously empty silo is filled with a bulk solid, a stress distribution results as shown in Fig. 13.15a [23, 37]. In the vertical section both the wall normal stress, σ_w, and the mean vertical stress, σ_v, are increasing in the downward direction and tend to approach an asymptotic value. The ratio of the wall normal stress to mean vertical stress is given by the lateral stress ratio, K (Sect. 13.2.1). The major principal stress, σ_1, is oriented vertically along the silo axis and deviates more and more from vertical towards the silo walls. This state of stress is called "active state of stress" or "active stress field". The direction of the major principal stress is represented in Fig. 13.15 by major principal stress lines.

At the transition to the hopper the wall normal stress has a discontinuity caused by the sudden change of wall inclination. Further downwards in the hopper both the vertical stress and the wall normal stress are decreasing and approach zero at the hopper apex (the outlet is assumed as infinitely small), but depending on the vertical stress at the transition, the silo shape, and the bulk solid's properties the stresses in the hopper either increase in the first instance and then decrease, or decrease continuously from the transition to the apex as in Fig. 13.15a (the stress distributions plotted in Fig. 13.15 shall be regarded as qualitative examples). In general the stresses in vertical direction are larger than those in horizontal direction. Along the hopper axis the major principal stress is oriented vertically, i.e., an active state of stress (active stress field) prevails as in the vertical section. The stress field in the hopper prevailing after filling is also referred to as "filling state of stress" or just "filling conditions".

When material is discharged from a mass flow silo the first time after it has been filled, after a short transition period the entire contents of the silo moves downward. Due to the convergent flow zone in the hopper the bulk solid is compressed horizontally, while it dilates in the vertical direction due the downwards flow.

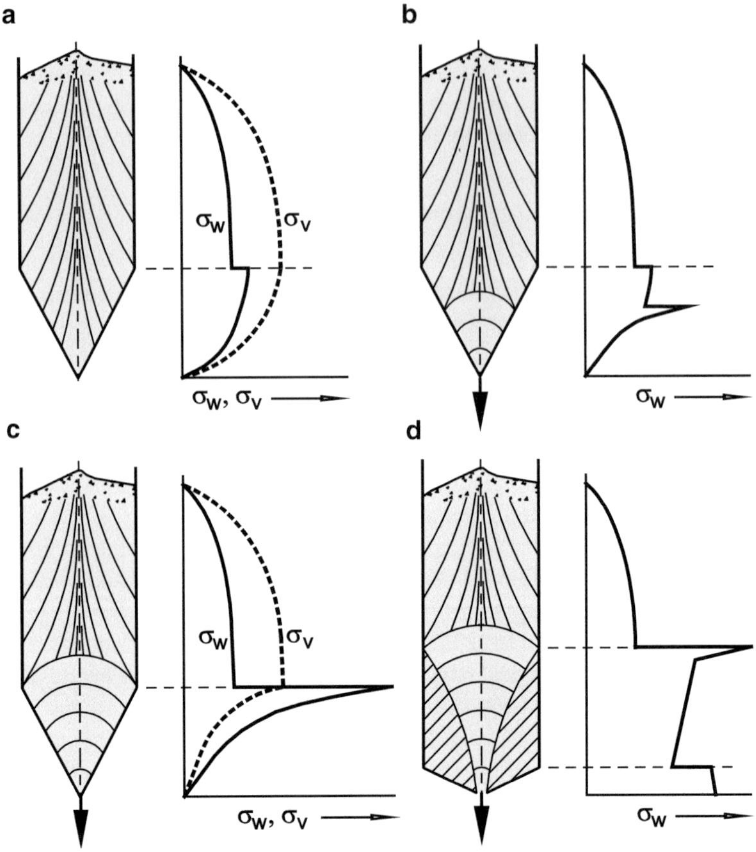

Fig. 13.15 Qualitative distributions of wall normal stress, σ_w, and mean vertical stress, σ_v (*dashed*, only in **a** and **c**), vs. the vertical coordinate (within the silos *lines* of major principal stress, σ_1, are plotted; stagnant zones are hatched [1, 23]; the hopper outlet in (**a**–**c**) is assumed to be infinitely small). (**a**) Filling conditions prevailing during initial filling of an empty silo; (**b**) emptying conditions in the lower part of the hopper; (**c**) emptying conditions in the entire hopper; (**d**) emptying conditions in a funnel flow silo [37]

As a result the larger stresses act in the horizontal direction, and the major principal stress along the hopper axis is oriented horizontally. This stress field is called "arched" (according to the shape of the lines of major principal stress in the hopper, which are increasingly inclined towards the wall; see Fig. 13.15c) or "passive" stress field. Other designations are "passive state of stress", "emptying state of stress" or just "emptying conditions". Figure 13.15b shows the situation a very short time after the onset of discharge, where the passive stress field has developed only in the lower part of the hopper. A bit later compared to the situation in Fig. 13.15b, the passive stress field is fully developed (Fig. 13.15c). Here the stresses in the hopper decrease remarkably towards the apex. In the lower part of the hopper the so-called "radial stress field" develops where the local stress is nearly proportional to the distance from the hopper apex. In the emptying state the stresses close to the outlet are independent of the stresses in the upper part of the hopper and, therefore, also independent of the silo's dimensions or level of filling.

In the silo's vertical section the active state of stress remains during emptying, as long as no local convergences exist (reduction of the cross-section due to inserts, dents, etc.). The transition from the active to the passive state of stress is called "switch". Considering the arched principal stress lines in the hopper, the switch takes place in an arched region originating from the cylinder/hopper transition (see Fig. 13.15c). The switch is accompanied by a local peak of the wall normal stress. Since the passive stress field and the stress peak are not a result of dynamic forces, they are preserved even when discharge is stopped.

The switch stress peak develops while starting discharge from a silo which has been freshly filled from completely empty (Fig. 13.15a). From the beginning of discharge the passive stress field develops in the hopper starting at the outlet and, thus, the switch region and the stress peak travel up (Fig. 13.15b) until they become caught at the cylinder/hopper transition (Fig. 13.15c).

In funnel flow stagnant zones are formed which remain at rest while the material in the flow zone is flowing downwards (Fig. 13.15d). If the boundary between a stagnant zone and the flow zone meets the silo wall within the vertical section, a stress peak occurs due to the "switch" from active to passive stress field caused by the convergent flow zone beneath the top of the stagnant zone. For the sake of completeness it has to be mentioned that the large wall normal stress along the hopper walls of the funnel flow silo in Fig. 13.15d results from the shallow slope of the hopper walls.

13.5 Silo Design for Flow

13.5.1 Problems

Problems often occurring during the storage of bulk solids in silos were described briefly in Sect. 13.1 (Fig. 13.1). Most of the mentioned problems are connected with funnel flow (Fig. 13.13). Therefore, many of the problems shown in Fig. 13.1 can be avoided if the silo is designed for mass flow:

- Ratholes (also called pipes) are formed from consolidated stagnant zones. Therefore, they are only possible in funnel flow (Fig. 13.1c).
- The residence time distribution in a mass flow silo is narrow ("first in – first out"). Thus, the unfavorable long and unknown residence times occurring in a funnel flow silo (Fig. 13.1b) can be avoided.
- Flooding often is the result of an insufficient deaeration of the bulk solid, and, therefore, it occurs when the residence time is too short. Especially in a funnel flow silo the residence time can be very short when the bulk solid is discharged during filling. As a result of the stagnant zone, the bulk solid just fed in appears

after a very short time at the silo outlet (Fig. 13.1d). Compared to a funnel flow silo, at identical mass of filling and identical discharge rates the residence time in a mass flow silo is much longer.

- Segregation during filling, where different fractions separate across the surface of the silo filling (e.g., while filling a silo with a not too fine bulk solid from the center of the top, often the coarser particles will collect near the silo walls whereas the fines collect in the silo axis; more segregation mechanisms are described in [37]), has a strong influence on the transient composition of the bulk solid discharged from funnel flow silos (Fig. 13.1e). In contrast to this, in a mass flow silo the bulk solid is often sufficiently remixed in the hopper section.

Thus, in mass flow silos only the potential problem of arching remains where a stable arch forms above the silo outlet so that discharge is stopped. Coarse grained bulk solids can build up arches due to interlocking and wedging of particles (Fig. 13.16a). This kind of arching can be avoided if the diameter of the circular outlet of a conical hopper is at least six to ten times the maximum particle size, x_{max}. The width of the rectangular outlet of a wedge-shaped hopper should be at least three to seven times x_{max} (outlet dimension depends on particle size distribution and particle shape [44]). With fine-grained and cohesive bulk solids the reason for arching is the cohesive strength (compressive strength, unconfined yield strength) of the bulk solid due to adhesive forces between individual particles. Even in this case arching can be avoided by a sufficiently large outlet opening. The calculation procedure will be explained in this section.

In case of a funnel flow silo, all problems mentioned above, which are characteristic of funnel flow, can occur. To ensure at least unobstructed flow, arching as well as ratholing (piping) (Fig. 13.16c) must be avoided. A stable rathole develops in a funnel flow silo if only the bulk solid vertically above the outlet discharges, whereby the rest of the bulk solid – the material in the stagnant zones – remains stationary in the silo due to its cohesive strength, thus building the walls of a stable rathole. If the bulk solid tends to time consolidation, the strength of the stationary material in the stagnant zones increases with time, and as a result, the tendency for ratholing is also increased. In the extreme case the bulk solid can be discharged only

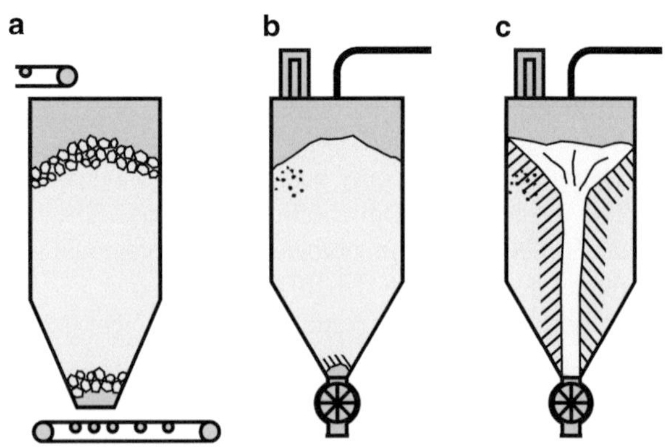

Fig. 13.16 (**a**) Arching due to interlocking and wedging of coarse-grained bulk solids; (**b**) arching due to the cohesive strength of a bulk solid; (**c**) ratholing (piping) [37]

with major effort (e.g., manually loosened and discharged from the silo top with truck mounted vacuum equipment). Like arching, ratholing can be avoided with a sufficiently large outlet opening. The calculation procedure will also be explained in this section.

13.5.2 Jenike's Design Procedure

The design procedure developed by Jenike [18] has been applied with great success since the 1960s. It allows determination of the hopper slope required for mass flow and outlet dimensions for unobstructed gravity flow, i.e., no arching in mass flow silos and neither arching nor ratholing in funnel flow silos.

For the design procedure the quantities characterizing the flow properties of the bulk solid have to be known (Sect. 13.2). Essentially, these are the bulk density, ρ_b, the effective angle of internal friction, φ_e, characterizing the internal friction of the bulk solid, the slope of the linearized yield locus, φ_{lin}, the unconfined yield strength, σ_c, and the angle of wall friction, φ_x. The angle of wall friction is the major property for mass flow hopper slope design, whereas the unconfined yield strength is the decisive property considering arching. All flow properties mentioned above depend on the stress level being represented by the major consolidation stress, σ_1. They can be measured with shear testers [18, 37] (see Sect. 13.2).

13.5.2.1 Design of Mass Flow Silos

The basic geometrical shapes investigated by Jenike are the conical hopper and the wedge-shaped hopper (Fig. 13.17). Also asymmetric wedge-shaped hoppers are treated and instructions are given for asymmetric conical hoppers. However these asymmetric shapes have several significant disadvantages and offer no advantages considering discharge behavior or best use of space compared to the equivalent symmetric hoppers. To neglect the influence of the front walls, it is supposed for the wedge-shaped hopper that the length, L, of the rectangular outlet is at least three times its width, b ($L > 3b$). The goal of the design process is the determination of the hopper slope necessary for mass flow and the minimum outlet size to prevent flow problems due to arching or ratholing.

Jenike's approach is based on the calculation of stresses in hoppers. With the conditions of equilibrium applied to an infinitesimal volume element of bulk solid in the hopper, he derived two partial differential equations. The properties of the bulk solid are taken into account by parameters obtained from yield loci. For the flowing bulk solid steady-state flow conditions are assumed where the ratio of major to minor principal stress is defined by the effective angle of internal friction, φ_e, or the effective yield locus, respectively. Incipient flow (being important for arching and ratholing not to occur) is characterized by an equation for the linearized yield locus. With further simplifying assumptions Jenike postulated that the major

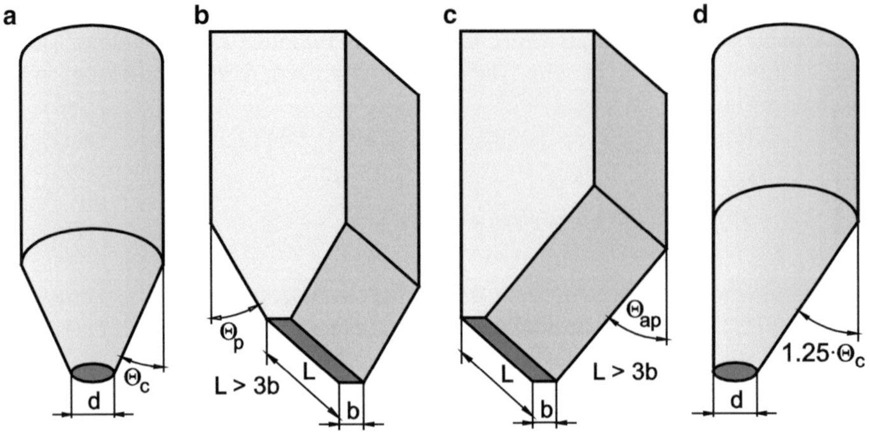

Fig. 13.17 Basic silo shapes [37]: (**a**) conical; (**b**) wedge-shaped; (**c**) asymmetric wedge-shaped; (**d**) asymmetric conical

principal stress σ_1 in the lower hopper part is proportional to the distance, r, from the virtual hopper apex ("radial stress field") [18]:

$$\sigma_1 = r \cdot g \cdot \rho_b \cdot s(\Theta', \Theta, \varphi_x, \varphi_e) \cdot (1 + \sin \varphi_e) \tag{13.5}$$

As an approximation the bulk density, ρ_b, which in reality is a function of the local stress in the hopper, is assumed to be constant. Angle Θ' and radius r determine the position of the bulk solid element in the hopper (polar coordinates; see Fig. 13.18), whereas Θ is the inclination of the hopper wall to vertical. Thus, function s depends on the bulk solid's properties, φ_e and φ_x, the hopper slope, Θ, and coordinate, Θ'.

A solution of the system of differential equations exists only for specific combinations of parameters, Θ, φ_e, and φ_x. Thus, only for these conditions mass flow will occur. If the hopper wall is not steep enough, no solution exists fulfilling the condition of bulk solid moving along the hopper wall (mobilization of wall friction), and, thus, a stagnant zone will form (funnel flow). With the solution of the system of differential equations function s in Eq. (13.5) is also known. Thus, besides others, the major principal stress in the hopper can be calculated.

In Fig. 13.19 the combinations of parameters which lead to solutions of the system of differential equations, and, thus, to mass flow for conical hoppers, are indicated as "mass flow" region. The mass flow region is separated from the funnel flow region (where no solutions of the differential equations are found) by the mass flow boundaries, which depend somewhat on the effective angle of internal friction, φ_e, characterizing the internal friction at steady-state flow. In Fig. 13.20 the boundaries for wedge-shaped hoppers are shown which are not theoretically determined boundaries, but modified based on Jenike's practical experience (the theoretically obtained boundaries are located more to the right, but are not sufficient to

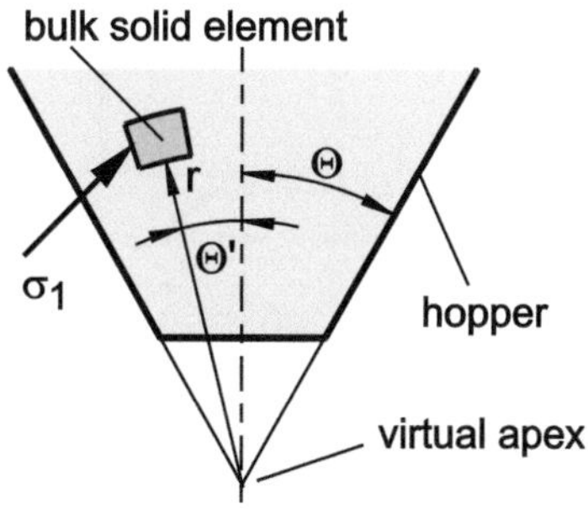

Fig. 13.18 Polar coordinates in a hopper [37]

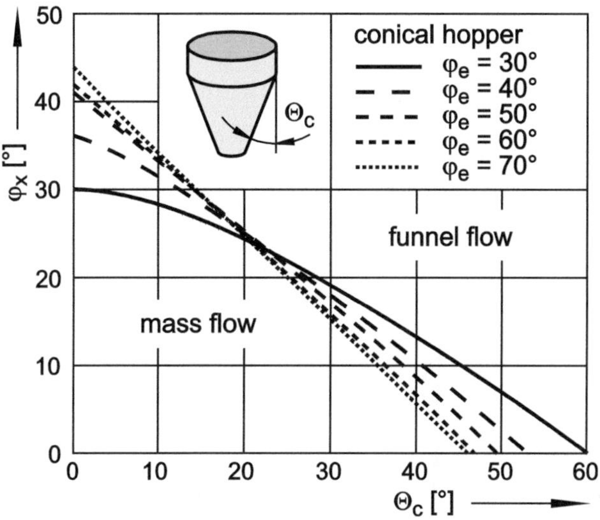

Fig. 13.19 Mass flow diagram (conical hopper) [37]

obtain mass flow at all conditions). In both figures the angle of wall friction, φ_x, is plotted versus the slope of the hopper wall, Θ, from vertical (Θ_c for conical hoppers, Θ_p for wedge-shaped hoppers).

Conditions within the mass flow boundaries (Figs. 13.19 and 13.20) lead to mass flow whereas conditions outside represent funnel flow. Knowing the angle of wall friction, φ_x, and the effective angle of internal friction, φ_e, the maximum inclination, Θ, of the hopper wall for mass flow to occur can be determined from the mass flow boundaries. For the case of a conical hopper, a safety margin of 3–5° [18] should be subtracted from the maximum inclination because the mass flow boundaries have been calculated for ideal conditions (a safety margin of 2–3° may be sufficient if the properties of the bulk solid and the wall surface are well-defined and do not change with time, and the maximum wall friction angles have been determined accurately). All boundary lines have a similar shape indicating that the maximum hopper slope for mass flow is steeper (Θ smaller) the larger the angle of wall friction, φ_x.

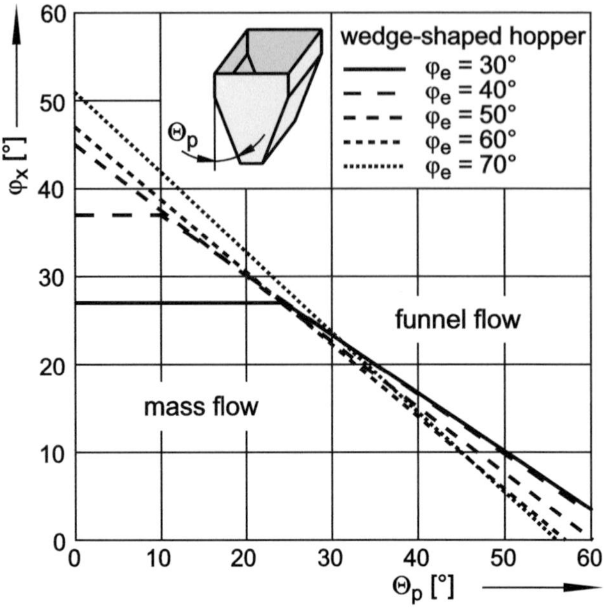

Fig. 13.20 Mass flow diagram (wedge-shaped hopper, $L > 3b$) [37]

For identical material properties (φ_x, φ_e), the wedge-shaped mass flow hopper can be flatter (larger angle Θ) than a conical mass flow hopper; the difference is normally in the range of 8–12°. This can be explained by the fact that at identical hopper inclination the cross-section of a conical hopper is reduced more in the direction of flow than that of a wedge-shaped hopper. The inclined wall of an asymmetric wedge-shaped hopper can even be flatter (Fig. 13.17c; diagram with mass flow boundary: see [18]). The maximum wall inclination of an asymmetric conical hopper (Fig. 13.17d) follows from a multiplication of the maximum angle of inclination of a symmetric conical hopper, Θ_c, by 1.25 [18].

For the next step, i.e., determination of the minimum size of the outlet opening to avoid arching, the stresses in the hopper are considered. When bulk solid is discharged from a mass flow silo, the radial stress field develops in the hopper (Sect. 13.4). The major principal stress, σ_1, of the radial stress field is (at least at a sufficient distance from the silo's vertical section) proportional to the local hopper diameter which in turn is proportional to the distance, r, to the virtual hopper apex. Thus, σ_1 tends towards zero at the hopper apex (Fig. 13.21). The major principal stress, σ_1, is acting as the consolidation stress which determines the local properties of the bulk solid.

For each major principal stress (= consolidation stress), σ_1, the unconfined yield strength, σ_c, and other relevant properties (bulk density, effective angle of internal friction) can be measured (Sect. 13.2). Thus, these quantities are known as a function of consolidation stress, σ_1 (Fig. 13.22). The relation $\sigma_c(\sigma_1)$ is called the

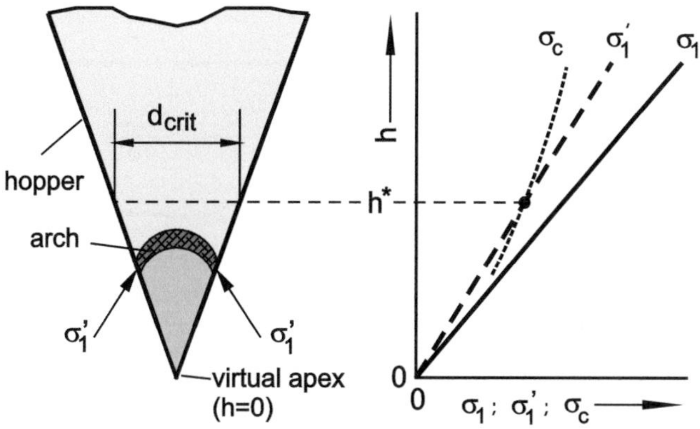

Fig. 13.21 Model for determination of minimum outlet size, d_{crit}, to avoid arching [37]

(instantaneous) flow function. According to the flow function, in Fig. 13.21 for every consolidation stress, σ_1, the corresponding value of unconfined yield strength, σ_c, is plotted.

If a cohesive arch has formed in a hopper (Fig. 13.21), a force resulting from the weight of the bulk solid is transferred to the hopper walls. This effect is represented by the major stress required to support a stable bulk solid arch, σ_1', which is comparable to the bearing stress of a road bridge. Jenike calculated σ_1' by assuming that the bulk solid arch has a smooth shape with a constant thickness in vertical direction, and that the arch must carry only its own weight, i.e., any load from the material above the arch is neglected. With further assumptions it follows for the major stress in a stable arch [44]:

$$\sigma_1' = \frac{2r \cdot \sin\Theta \cdot g \cdot \rho_b}{1+m} \tag{13.6}$$

m is a parameter describing the hopper shape: $m = 0$ for wedge-shaped hoppers and $m = 1$ for conical hoppers. The term ($2r \cdot \sin\Theta$) represents the local hopper diameter, d, for the conical hopper and the local width, b, for the wedge-shaped hopper (silo shapes: see Fig. 13.17). Coordinate r measures the distance from the hopper apex to the support of the arch (similar to Fig. 13.18, but along the hopper wall, i.e., $\Theta' = \Theta$).

A stable arch is only possible in that part of the hopper where the unconfined yield strength is greater than the stress that would exist in a stable arch ($\sigma_c > \sigma_1'$), i.e., beneath the point of intersection of the σ_c curve with the σ_1' line (Fig. 13.21). Above the point of intersection the unconfined yield strength is smaller than the major stress in the arch, i.e., the material will flow. The intersection point defines that position in the hopper (height h^*, Fig. 13.21) where the hopper diameter is equal to the so-called critical outlet diameter, d_{crit}, which must be exceeded if

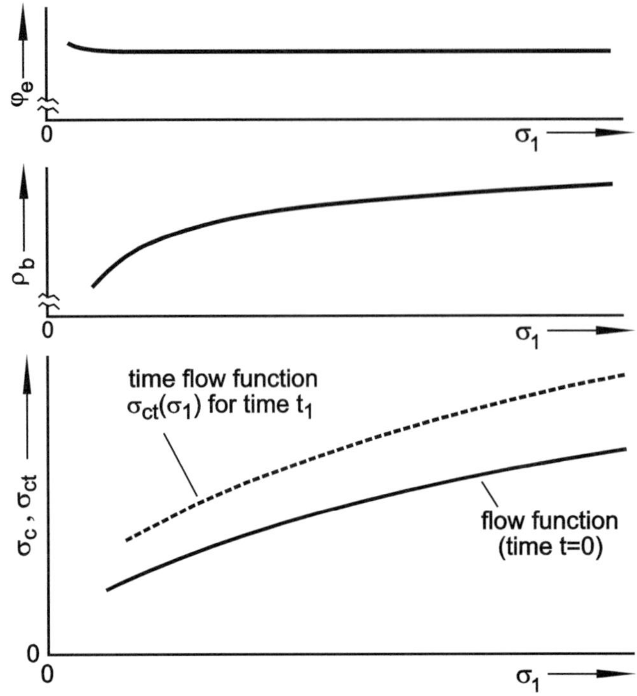

Fig. 13.22 Unconfined yield strength, σ_c (flow function and time flow function for time t_1); bulk density, ρ_b, and effective angle of internal friction, φ_e, as functions of consolidation stress, σ_1 [37] (Note that for the sake of simplification the effective angle of internal friction, φ_e, is assumed to be constant in the relevant stress range)

arching is to be avoided. If the outlet opening is smaller than d_{crit} (i.e., lower in the hopper), discharge aids (Sect. 13.6.6) have to be installed between the actual outlet opening and the critical diameter, d_{crit} (height h^*). The equivalent procedure for a wedge-shaped hopper with a rectangular outlet leads to a critical outlet width, b_{crit}. Since here the influence of the front walls (friction!) is neglected, the length, L, of the outlet must be at least three times its width, b.

Some bulk solids have the tendency to consolidate with time when stored at rest, for example, in a silo (time consolidation, Sect. 13.2.3.2). In analogy to the (instantaneous) flow function which characterizes the instantaneous behavior (unconfined yield strength directly after consolidation, i.e., after the storage period $t=0$), time flow functions $\sigma_{ct}=\sigma_c(\sigma_1,t)$ exist which represent the unconfined yield strength after certain consolidation periods. Two time flow functions are plotted in Fig. 13.22 being valid for two storage periods at rest, t_1 and $t_2>t_1$. If the time flow function line would be transferred into Fig. 13.21, a point of intersection of σ_1' and σ_{ct} would result located further upwards, i.e., at $h>h^*$. Thus, a larger critical diameter, d_{crit}, would result. This leads to the well

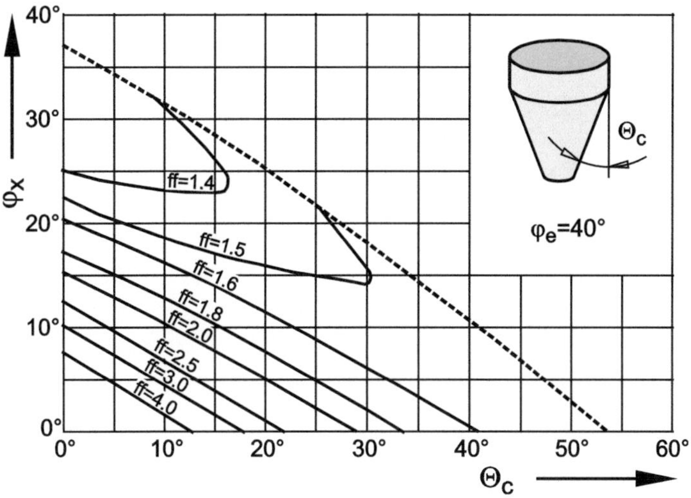

Fig. 13.23 Flow factor, *ff*, for conical hoppers and $\varphi_e = 40°$ [18, 37]

known effect that with increasing time of storage at rest larger outlet diameters are necessary to avoid arching.

For the practical determination of critical outlet dimensions Jenike proposed a procedure outlined in the following. It is based on the fact that both stresses σ_1' and σ_1 are proportional to the local hopper diameter which in turn is a function of the distance, *r*, to the hopper apex (see also Eqs. 13.5 and 13.6). Thus, at the hopper wall ($\Theta' = \Theta$) the ratio $\sigma_1/\sigma_1^{'}$, called the flow factor, *ff*, is constant.

$$ff = \frac{\sigma_1}{\sigma_1^{'}} = const = (1+m) \cdot s(\Theta, \varphi_x, \varphi_e) \cdot \frac{1+\sin\varphi_e}{2\sin\,\Theta} \tag{13.7}$$

The flow factor *ff* is dependent on the flow properties (φ_x, φ_e) and the hopper shape (m, Θ). Knowing function *s* (Eq. 13.5), the flow factor can be calculated. Jenike provided diagrams for an easy determination of the flow factor (All diagrams required for silo design are published in Bulletin 123 "Storage and Flow of Solids" [18]. It is strongly recommended to use this Bulletin if silos have to be designed). Each of the diagrams is valid for a specific hopper geometry (e.g., conical) and a fixed value of the effective angle of internal friction, φ_e, being varied from diagram to diagram in steps of 10°. Two of these diagrams (valid for conical hoppers) are shown in Figs. 13.23 and 13.24. The solid curves represent constant values of *ff*, as given by the labels. For values in between and φ_e values which are not multiples of 10°, the flow factor values must be obtained by interpolation. The mass flow boundary is also plotted in each diagram (dashed curve). The complete set of diagrams is given in Jenike's Bulletin 123 [18]. Equations approximating *ff* and the maximum hopper wall angles for mass flow are published in [24].

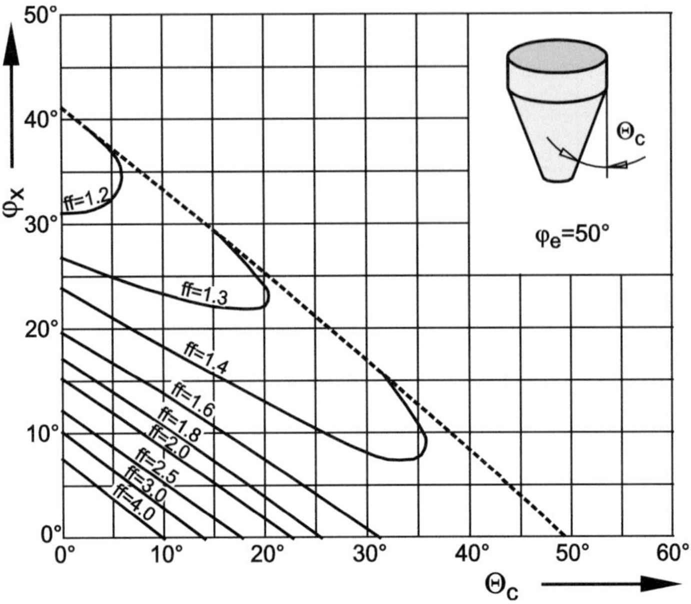

Fig. 13.24 Flow factor, *ff*, for conical hoppers and $\varphi_e = 50°$ [18, 37]

For determination of critical outlet dimension to avoid arching a multiple diagram as shown in Fig. 13.25 is used where unconfined yield strength, bulk density, and effective angle of internal friction are plotted according to Fig. 13.22. The major part is the σ_c,σ_1 diagram where in addition to Fig. 13.22 the major stress in the arch, σ_1', is plotted, which, according to Eq. (13.7), is directly proportional to σ_1:

$$\sigma_1^{'} = \frac{\sigma_1}{ff} \tag{13.8}$$

With the measured flow properties, the hopper inclination, Θ, determined at the mass flow design, and Jenike's diagrams [18] (e.g. Fig. 13.23), the flow factor, *ff*, can easily be determined (If the effective angle of internal friction varies with consolidation stress, it has to be estimated first, and to be corrected later by an iterative process. This is explained at the end of the present section).

The point of intersection of the major stress in the arch, σ_1', and the flow function in the σ_c,σ_1 diagram (Fig. 13.25) is equivalent to the point of intersection in Fig. 13.21. Thus, the consolidation stress, σ_1, at this point represents the consolidation stress in the hopper where the diameter is equal to the minimum outlet diameter, d_{crit} (Fig. 13.21). Therefore, coordinates of the intersection point are provided with the index "crit" (unconfined yield strength, $\sigma_{c,crit}$, and consolidation stress, $\sigma_{1,crit}$). Knowing critical consolidation stress, $\sigma_{1,crit}$, bulk density, ρ_b,

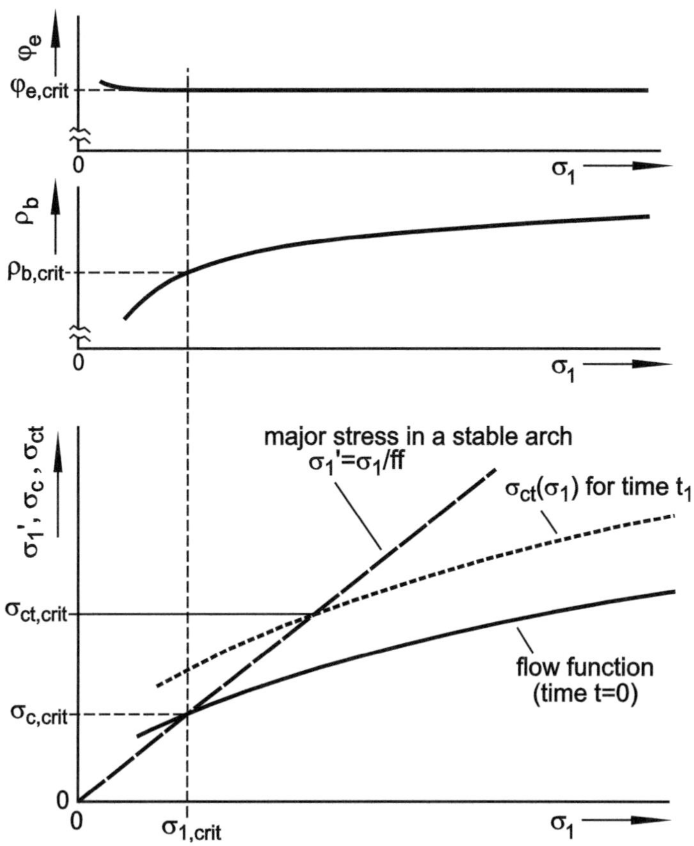

Fig. 13.25 Diagram with flow function, time flow function and major stress in a stable arch, σ_1' (*bottom*) [37], and diagrams of bulk density, ρ_b, and effective angle of internal friction, φ_e, as a function of consolidation stress, σ_1

$_{crit}$, and effective angle of internal friction, $\varphi_{e,crit}$, can be determined from the functions plotted in the diagrams on top in Fig. 13.25.

Finally the equivalent local hopper diameter, the critical diameter, d_{crit}, has to be calculated. To do so Eq. (13.6) is rearranged, and the major stress in the arch, σ_1', is replaced by the unconfined yield strength, $\sigma_{c,crit}$, (at the point of intersection is $\sigma_1' = \sigma_{c,crit}$). With geometrical relationships between local hopper diameter and width, respectively, and polar coordinates according to Fig. 13.18

$$d = 2r \cdot \sin \Theta \qquad (13.9a)$$

and

$$b = 2r \cdot \sin\Theta \tag{13.9b}$$

it follows for the critical outlet width, b_{crit}, of a wedge-shaped hopper:

$$b_{crit} = \frac{\sigma_{c,crit}}{g \cdot \rho_b} \tag{13.10a}$$

The critical outlet diameter, d_{crit}, of a conical hopper is:

$$d_{crit} = 2\frac{\sigma_{c,crit}}{g \cdot \rho_b} \tag{13.10b}$$

A more precise calculation yields the following equations:

$$b_{crit} = H(\Theta_p)\frac{\sigma_{c,crit}}{g\rho_{b,crit}} \tag{13.11a}$$

and

$$d_{crit} = H(\Theta_c)\frac{\sigma_{c,crit}}{g\rho_{b,crit}} \tag{13.11b}$$

For bulk density, ρ_b, the value $\rho_{b,crit}$ corresponding to $\sigma_{1,crit}$ is used (Fig. 13.25). The function $H(\Theta)$ takes into account the hopper geometry (conical or wedge-shaped; wall inclination Θ_c and Θ_p, respectively). $H(\Theta)$ is plotted in Fig. 13.26 (for approximate equations for $H(\Theta)$ refer to [24]).

In a similar way the outlet dimensions can be determined which are required to avoid arching after longer periods of storage at rest (= no bulk solid is discharged). For this case the point of intersection of the major stress in the arch and the time flow function for the storage time in question, $\sigma_{ct} = \sigma_c(\sigma_1,t)$, has to be determined (Fig. 13.25). For this point the critical consolidation stress and critical flow

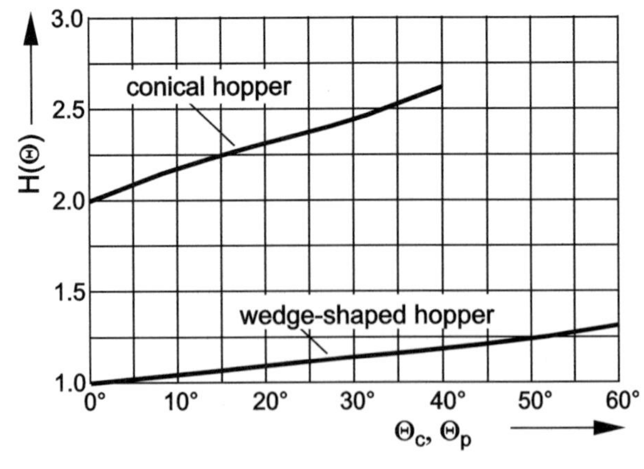

Fig. 13.26 Function $H(\Theta)$ [18, 37]

properties have to be determined in the same way as explained above for the instantaneous flow function.

If a silo has been designed for a specific storage time at rest, some bulk solid must be discharged after this time to avoid arching. During discharge, the particles in a mass flow hopper are moved relative to each other so existing particle contacts are replaced by new ones. Thus, any time consolidation that prevailed during the storage at rest diminishes, i.e., the bulk solid behaves again as prior to the storage period. To obtain this effect, only a portion of the silo contents has to be discharged, because in a mass flow silo the whole contents are in motion during discharge (this is discussed further in Sect. 13.6.2).

Since the design procedure outlined above is based on the radial stress field, i.e., stresses prevailing in the hopper after the bulk solid has moved downwards (emptying state, Fig. 13.15c), the calculations are not valid for the filling state which develops when bulk solid is fed into an empty silo without any discharge. In the filling state the stresses in the hopper can be larger than in the emptying state (see Fig. 13.15) thus leading to a stronger consolidation of the bulk solid.

In practice often an iterative procedure is necessary for silo design. Both the hopper angle for mass flow and the flow factor, *ff*, depend on the effective angle of internal friction, φ_e, and the angle of wall friction, φ_x, which often are stress-dependent. At the start of the design process the outlet dimensions and, thus, the major principal stress at the outlet, are not yet known. Therefore, values for φ_e and φ_x have to be estimated. After the critical outlet dimension has been determined and the major principal stress at the outlet is known, φ_e and φ_x can be determined for the conditions at the outlet. If these values are not equal to the estimated values, the design process has to be repeated with the new values, and so on until finally the estimated values are equal to the calculated values. The same procedure must be followed if a critical outlet dimension is determined with a time flow function. An example for this case is given in Chap. 15 of [37].

The design procedure using diagrams as outlined above may appear old-fashioned. However, a translation of the design process to a computer program may be possible, but can be dangerous since during the design process one often has to extrapolate data, or to use test results of unfriendly materials which do not deliver nice curves as shown here or in other textbooks. Thus, experience and sure instinct as well are required what may not be covered by software which someday will be operated by a person without sufficient experience in silo design.

13.5.2.2 Design of Funnel Flow Silos

To avoid the problems typically occurring in funnel flow silos (e.g., segregation, degradation, ratholing; see Sect. 13.5.1), mass flow should be preferred. Funnel flow may only be acceptable if "friendly" bulk solids are stored which do not exhibit such problems. In the following it will be shown how funnel flow silos have to be designed for trouble-free operation.

The walls of a funnel flow silo are flatter than those of a mass flow silo, but should be steep enough to ensure that the silo can be cleared completely alone by gravity. For cohesive bulk solids, the maximum inclination of the hopper walls to vertical, Θ_{cd}, can be assessed according to Jenike [18] as a function of the angle of wall friction, φ_x:

$$\Theta_{cd} = 65^\circ - \varphi_x \tag{13.12}$$

This value is only a rough estimate for poorly flowing bulk solids. Having easy-flowing bulk solids (i.e. granules), even flatter hoppers can be sufficient (e.g., with an inclination from horizontal of about $\varphi_x + 10^\circ$). Prerequisite for complete clearance by gravity is that the strength of the bulk solid is too small to support a stable rathole.

For the design of funnel flow silos to avoid ratholing two procedures are outlined in the following. The approach according to Jenike's first edition of Bulletin 123 (1964 edition [19]) assumes that the stresses at the wall of a stable rathole are independent of the filling level. This procedure is referred to as giving a lower bound of the maximum diameter of a stable rathole which may only be approached when the bulk solid is withdrawn during filling [32]. If the latter is not the case, the procedure may significantly underestimate the maximum rathole diameter. Thus, Jenike presented another approach in the 1980 edition of Bulletin 123 [18]. This approach leads to an upper bound of the rathole dimension. It is based on the assumption that the bulk solid is consolidated by filling stresses which depend on the filling height.

The "lower bound" approach of Jenike from 1964 [19] is based on the following considerations: If a stable rathole of diameter D is formed within the bulk solid above the outlet (Fig. 13.27), a circumferential compressive stress, σ_1'', is acting close to the surface of the rathole wall. The reason for this stress is that the bulk solid tends to flow into the interior of the rathole. If the unconfined yield strength, σ_c, of the bulk solid exceeds the circumferential stress, σ_1'', the bulk solid cannot start flowing, and, thus, the rathole remains stable.

The circumferential stress, σ_1'', depends on the rathole diameter, D, and can be calculated as follows [18, 44]:

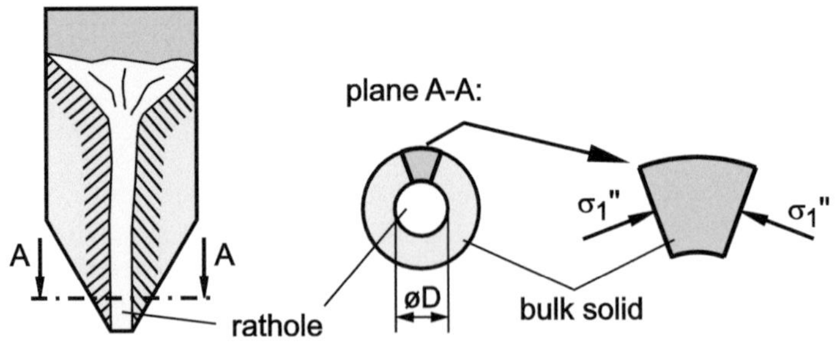

Fig. 13.27 Stable rathole and circumferential stress, σ_1'' [37]

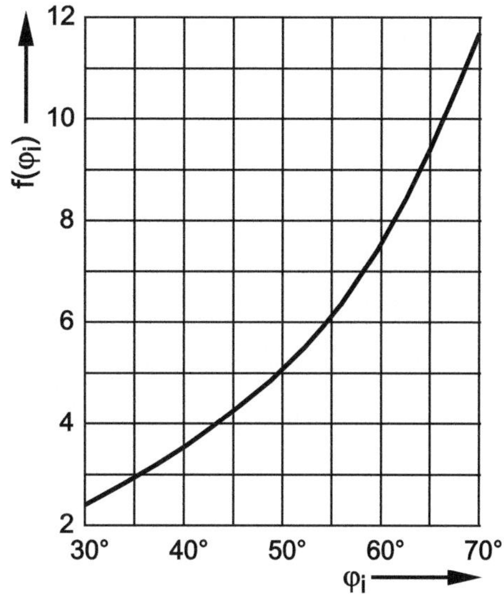

Fig. 13.28 Function $f(\varphi_i)$ [19, 37]

$$\sigma_1^{''} = \frac{D \cdot g \cdot \rho_b}{f(\varphi_i)} \tag{13.13}$$

The function $f(\varphi_i)$ is depicted in Fig. 13.28. The angle φ_i is the local slope of the yield locus, which is approximated by the slope of the linearized yield locus, φ_{lin}. As can be seen in Eq. (13.13), the circumferential stress, σ_1'', is proportional to the rathole diameter, D.

To be able to compare the circumferential stress to the unconfined yield strength, σ_c, the consolidation stress σ_1 (major principal stress) acting at the surface of the rathole has to be known. Then with the flow function the corresponding unconfined yield strength can be determined. For this procedure the lower part of the rathole, i.e., the part directly above the outlet opening, is considered. If the rathole is not stable in this region, the bulk solid will also flow in the upper silo parts. The consolidation stress above the outlet has been derived [44] to be:

$$\sigma_1 = \frac{1 + \sin \varphi_e}{4 \cdot \sin \varphi_e} \cdot D \cdot g \cdot \rho_b \tag{3.14}$$

The major principal stress, σ_1, and the circumferential stress, σ_1'', near the surface of a rathole are proportional to the rathole diameter, D. The ratio of σ_1, to σ_1'' is called the "flow factor for ratholing", ff_p:

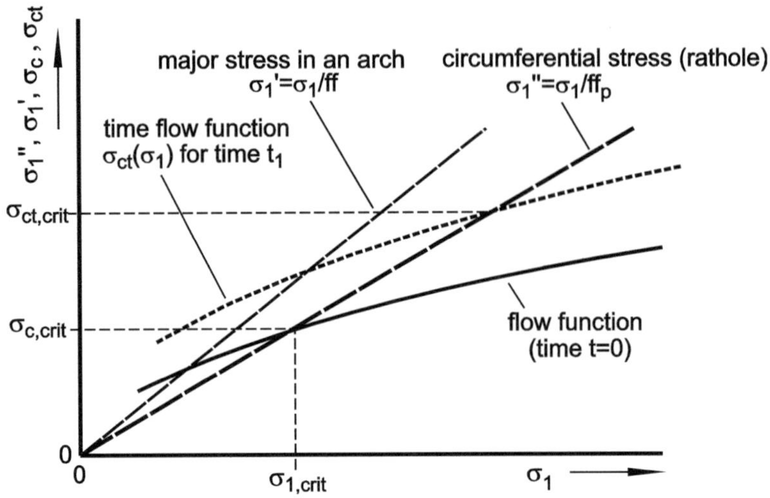

Fig. 13.29 Flow function and time flow functions; circumferential stress, σ_1'' [37]

$$ff_p = \frac{\sigma_1}{\sigma_1^{''}} = \frac{1 + \sin\varphi_e}{4 \cdot \sin\varphi_e} \cdot f(\varphi_i) \tag{13.15}$$

If the calculation leads to a value of $ff_p < 1.7$, a fixed value of $ff_p = 1.7$ has to be used.

For the design of a funnel flow silo to avoid ratholing the circumferential stress, σ_1'', has to be drawn in a σ_c,σ_1 diagram (Fig. 13.29). The flow factor, ff_p, needed for this is calculated with Eq. (13.15):

$$\sigma_1^{''} = \frac{\sigma_1}{ff_p} \tag{13.16}$$

The coordinates of the point of intersection of the circumferential stress with the unconfined yield strength are designated by the index "crit" (unconfined yield strength, $\sigma_{c,crit}$, and consolidation stress, $\sigma_{1,crit}$). Finally the appropriate outlet diameter, D_{crit}, is calculated. In doing so the circumferential stress, σ_1'', in Eq. (13.13 is replaced by the unconfined yield strength, $\sigma_{c,crit}$, at the point of intersection. Considering the bulk density, ρ_b, the "critical" value $\rho_{b,crit}$ corresponding to $\sigma_{1,crit}$ is used ($\rho_{b,crit}$ is determined in the same way as explained in Fig. 13.25):

$$D_{crit} = f(\varphi_i) \frac{\sigma_{c,crit}}{g \cdot \rho_{b,crit}} \tag{13.17}$$

The points of intersection of the circumferential stress with time flow functions lead to critical outlet dimensions required to avoid ratholing after the equivalent storage times at rest.

The flow factor, ff_p, depends on the effective angle of internal friction, φ_e, and the slope of the linearized yield locus, φ_{lin}, which is entered in Eq. (13.17) as the slope of the yield locus, φ_i. These angles may depend on consolidation stress, σ_1. At the start of the design process values for φ_e and φ_{lin} have to be estimated because the major principal stress at the outlet, $\sigma_{1,crit}$, is not yet known. After the critical consolidation stress, $\sigma_{1,crit}$, has been is known, φ_e and φ_{lin} can be determined for the conditions at the outlet. If these values are not equal to the estimated values, the design process has to be repeated with the new values, and so on until finally the estimated values are equal to the calculated values.

Since the design procedure outlined above is based on stresses prevailing in the emptying state, i.e. after some bulk solid has been discharged, the calculations are not valid for the filling state. The filling state develops when bulk solid is fed into an empty silo while no material is discharged from the silo. In this case larger stresses prevail close to the outlet leading to a higher degree of consolidation. As outlined above, a second design procedure [18] allows for estimation of the maximum rathole diameter under consideration of the stresses prevailing in the filling condition (upper bound of rathole diameter) [32]. This procedure is described in the following.

First the maximum consolidation stress (major principal stress) for filling conditions is estimated with Janssen's equation (Sect. 13.4, Eq. (13.3)) [17]. Although this equation is valid for vertical sections only, it is also applied here for reasons of simplification to estimate stresses in the hopper. This results in a distribution of vertical stress as depicted in principle in Fig. 13.30. Since for filling conditions the major principal stress, σ_1, is oriented more or less vertically in the entire silo (see Sect. 13.4), it is approximately equal to the mean vertical stress, σ_v. Thus, the maximum consolidation stress at the bottom of the silo, which is the critical stress, $\sigma_{1,crit}$, is obtained with Janssen's equation with height of filling, h_f, as shown in Fig. 13.30.

For the major consolidation stress, $\sigma_{1,crit}$, according to σ_v in Eq. (13.3), the corresponding values of the unconfined yield strength, $\sigma_{c,crit}$, may be determined from the measured instantaneous flow function or time flow function (Fig. 13.29).

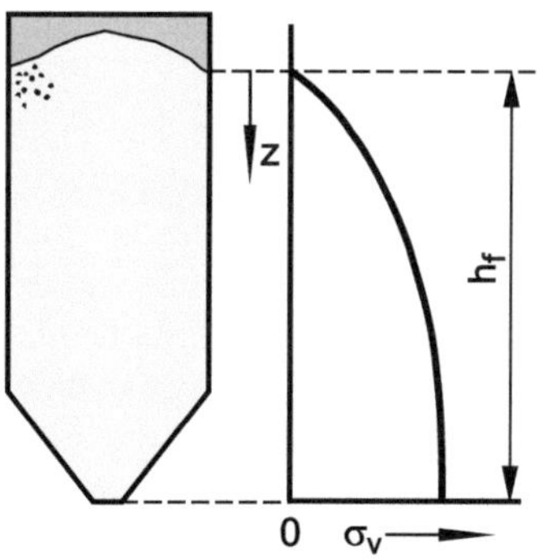

Fig. 13.30 Distribution of vertical stress, σ_v, in a funnel flow silo for filling conditions (stresses in the hopper approximated with Janssen's equation, see Sect. 13.4) [37]

Also the values of ρ_b and φ_{lin} corresponding to $\sigma_{1,crit}$ have to be determined. Then the upper bound value of the critical outlet diameter, D_{crit}, to avoid ratholing can be calculated with Eq. (13.17).

Since for a conical funnel flow hopper the critical outlet diameter to avoid ratholing is always larger than the one to avoid arching ($D_{crit} > d_{crit}$), only a design to avoid ratholing is necessary. For a wedge-shaped funnel flow hopper the calculated critical value D_{crit} has to be applied on the diagonal of the rectangular outlet. Therefore, the width, b, of the outlet is not necessarily large enough to avoid arching. Thus, the critical width, b_{crit}, of the rectangular outlet must be determined with the design procedure to prevent arching (see Sect. 13.5.2.1) where the values $ff = 1.7$ and $H(\Theta) = 1.15$ have to be used [18].

The description of the determination of critical outlet dimensions for ratholing not to occur might lead to the impression that a funnel flow silo can be operated without any problems as a mass flow silo. This, however, is not the case because of the potential problems related to funnel flow (see Sects. 13.1 and 13.5.1; e.g. segregation). Additionally, the time consolidation in a funnel flow silo cannot be limited by discharge of a part of the silo contents in regular time intervals, because in contrast to a mass flow silo the particles in the stagnant zones are not moved during discharge. Thus, time consolidation is not reduced. Therefore it can be stated: If a bulk solid that is affected by time consolidation is stored in a funnel flow silo (which generally should be avoided), the silo has to be cleared completely after the storage period for which the outlet opening has been designed.

13.5.3 Maximum Discharge Rate

The unrestricted rate of discharge of a bulk solid under gravity (i.e., when an outlet is opened) is dependent on several parameters, e.g., the bulk solid's properties, the size and shape of the outlet, the hopper inclination, and, in the case of funnel flow, the shape of the stagnant zones [9, 44]. The conditions of unrestricted discharge can also be fulfilled in the presence of a feeder if the desired discharge rate exceeds the unrestricted discharge rate. Thus, if a feeder is installed, it should be ensured that the unrestricted discharge rate exceeds the range of adjustable feeding rates.

With fine-grained bulk solids the influence of air flow has to be considered [14, 20, 26, 31, 33, 35, 46]. Since in the hopper the stresses decrease in the downward direction (see Fig. 13.15), the bulk solid dilates when flowing downwards, i.e., the bulk density, ρ_b, decreases and the porosity, ε, increases. Thus, the gas in the voids expands which results in the development of a negative pressure. The negative pressure causes gas to flow following the pressure gradient. At sufficiently large filling heights most of the gas will enter the hopper through the outlet opening since the flow resistance of the bulk solid above the hopper is definitely larger. The pressure gradient in the hopper acts opposite to the gravity force resulting in a decrease of the mass flow rate. Often a pulsating outflow can also be observed.

Due to the many influencing parameters it is impossible to exactly predict mass flow rates. Especially for fine-grained materials where the gas pressure plays a role predictions are hardly possible. For coarse-grained materials (roughly: particle size >200 μm and solid density >1000 kg/m^3; better: groups B and D according to Geldart's diagram [13]), where the gas pressure can be neglected, empirical equations have been found [9, 27, 44]. For circular outlet openings of diameter d and coarse-grained bulk solids of particle diameter x (very narrow particle size distribution) Beverloo [5] stated:

$$\dot{m} = C\rho_b\sqrt{g}(d - kx)^{2.5} \tag{13.18}$$

where C and k are fitting parameters. With different bulk solids Beverloo found a range of about $C = (0.55 \ldots 0.65)$ and $k = (1.5 \ldots 3.0)$.

A similar estimate is given by the British Materials Handling Board [7, 50] for coarse-grained (less than 3% < 250 μm), free-flowing bulk solids. It is distinguished between circular and rectangular outlet openings:

Circular outlet openings (diameter d):

$$\dot{m} = 0.58\rho_b\sqrt{g}(d - kx)^{2.5}k_\Theta \tag{13.19}$$

Rectangular outlet openings (width b, length $L > 3b$):

$$\dot{m} = 1.03\rho_b\sqrt{g}(L - kx)(b - kx)^{1.5}k_\Theta \tag{13.20}$$

In Eqs. 13.19 and 13.20 k is dependent on the particle shape ($k = 1.6$ for spherical and $k = 2.5$ for non-spherical particles). The hopper inclination, Θ, is taken into account by the factor k_Θ: $k_\Theta = (\tan\Theta)^{-0.35}$ for hopper inclinations $\Theta < 45°$; $k_\Theta = 1$ for $\Theta \geq 45°$. x represents the mean particle diameter. The kind of "mean diameter" is not specified in [7], but since usually $d \gg x$, the influence of x is small.

13.6 Silo Configurations

13.6.1 Influence of Flow Properties

Figure 13.31 shows a diagram with different combinations of the quantities determined with the silo design procedure due to Jenike [18], i.e., the maximum hopper wall slope for mass flow, Θ_c, and the minimum outlet diameter, d_{crit}, for arching not to occur in a conical hopper. According to the values of Θ_c and d_{crit}, different silo configurations are appropriate.

The easy-flowing bulk solid A (e.g., plastic granules) having a small angle of wall friction needs only a relatively shallow hopper (large value of Θ_c) for mass flow, and a small outlet opening to avoid arching. Here for feeding and dosing a

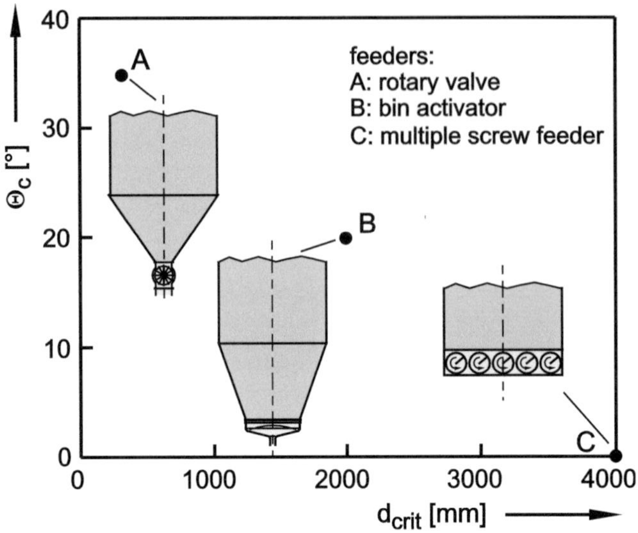

Fig. 13.31 Silo configurations as a function of the results of silo design for flow [37]

rotary valve feeder may be useful. Bulk solid B requires a larger effort: For mass flow the hopper has to be steeper, and a larger outlet diameter (2000 mm) is necessary to avoid arching. With respect to the large outlet diameter, a bin activator might be an appropriate discharger. Bulk solid C is characterized by an extremely large angle of wall friction with the result that even in the steepest hopper mass flow cannot be achieved. Thus, a silo without hopper is required ($\Theta_c = 0°$), and the feeder must be directly attached to the cylindrical section. The feeder must be able to discharge bulk solid from the entire cross-section of the silo which can be accomplished, for example, with a properly designed large multiple screw feeder.

Often it makes sense to compare different wall materials regarding their influence on hopper slope, i.e., one can determine hopper slope angles and outlet dimensions for different hopper shapes and different wall materials. Thus a comparison of costs is possible considering different shapes and materials [34, 42]. For example, it can be stated whether lining of the hopper walls (e.g., with cold-rolled stainless steel) is advantageous considering costs in a specific application.

13.6.2 Critical Outlet Dimensions

In Fig. 13.32a exemplary critical outlet dimensions are plotted versus storage time at rest, t, for a bulk solid showing time consolidation. Typically the values of d_{crit} (outlet diameter of a conical hopper) are about twice as large as the values of b_{crit} (outlet width of a wedge-shaped hopper). The critical rathole diameter, D_{crit}, is often clearly larger than d_{crit}, especially if the upper bound values are considered.

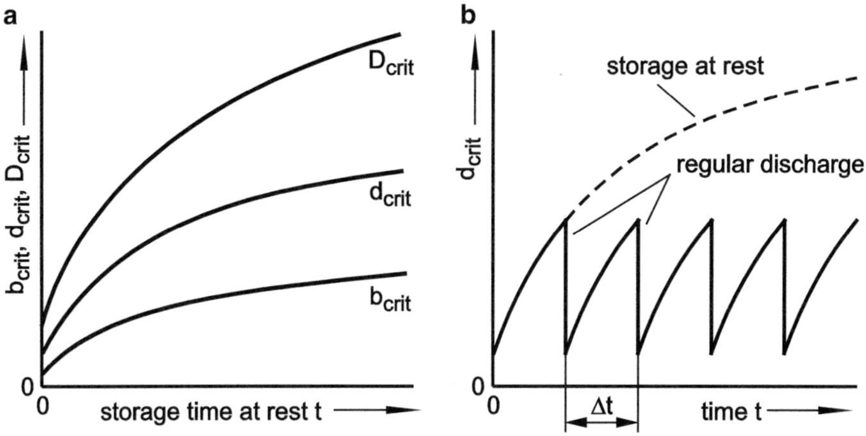

Fig. 13.32 (**a**) Exemplary critical outlet dimensions versus storage time at rest, t; (**b**) critical outlet diameter, d_{crit}, vs. time when bulk solid is discharged in regular time intervals, Δt; each discharge cycle reduces the strength of the bulk solid to its instantaneous value, and, thus, the critical diameter to $d_{crit}(t=0)$. For comparison the critical diameter for storage at rest, i.e. without intermediate discharge, is plotted (*dashed curve*) [37]

With many bulk solids exhibiting time consolidation the increase of critical outlet dimensions with time becomes smaller with increasing storage time (as shown in Fig. 13.32a).

If the critical outlet dimensions as shown in Fig. 13.32a are known, it is possible to quantitatively predict the necessary outlet size as a function of storage time at rest. It is also possible to determine the maximum storage time at rest for given outlet dimensions so that arching and ratholing do not occur. After this maximum storage time the material has to be set in motion, for example, in the case of a mass flow silo, by discharge of a portion of the silo filling, which can be fed again into the silo. Thereby the effect of the time consolidation is completely reduced, and, thus, the behavior of the bulk solid is characterized by its instantaneous flow properties (i.e., the flow properties according to the storage time at rest $t=0$) so that the critical dimensions for $t=0$ are valid again (Fig. 13.32b). It should be noted that this procedure does not work with funnel flow silos, because the material in the stagnant zones remains stationary during discharge of only a portion of the silo filling. Thus, time consolidation in the stagnant zones continues and the critical rathole diameter, D_{crit}, increases further. Therefore a funnel flow silo must be completely discharged in order to limit the effect of time consolidation and, thus, the critical rathole diameter, D_{crit}.

Another way to deal with time consolidation without providing a sufficiently large outlet opening is the application of discharge aids (Sect. 13.6.6) in the hopper region below the critical outlet dimension. If after a longer period of storage at rest arches are formed, flow can be initiated by the action of the discharge aids (e.g., by air injection). In a mass flow silo, all contents is set in motion this way so that the effect of time consolidation is completely reduced as explained above (Fig. 13.32b). Once flow is achieved, further operation of the discharge aids is not required as long as discharge is not paused for a longer period so that the bulk solid can consolidate again.

13.6.3 Hopper Shape

The design procedure due to Jenike [18] yields the basic geometrical data for conical and wedge-shaped hoppers (Fig. 13.17a, b), i.e., the wall slope, Θ_c, and the outlet diameter, *d*, of a conical mass flow hopper, and the wall slope, Θ_p, and outlet width, *b*, of a wedge-shaped hopper. In addition, guidelines for the design of asymmetric wedge-shaped hoppers are given for a reduced parameter set (Fig. 13.17c: slope Θ_{ap}, outlet width *b*). Considering the wedge-shaped hopper it is assumed that the length, *L*, of the rectangular outlet is at least three times its width, *b*. Only then can the influence of the end walls close to the outlet be neglected.

The maximum wall slope of an asymmetric conical hopper (Fig. 13.17d) should not exceed $1.25 \cdot \Theta_c$ (Sect. 13.5.2.1) [18]. The relatively steep inclination is necessary since this hopper shape tends to funnel flow with a flow zone along the steeper side of the hopper.

The design quantities for the basic hopper shapes as shown in Fig. 13.17 can also be applied to other hopper shapes. In Fig. 13.33 further possibilities are shown [18, 50]. If a cylindrical vertical section is combined with a wedge-shaped hopper, designs according to Fig. 13.33a, b are possible. Also in these applications the length, *L*, of the rectangular outlet has to be at least three times its width, *b*. If the end walls are inclined as in Fig. 13.33a, their slope must not exceed Θ_c, i.e., the maximum wall slope of a conical hopper.

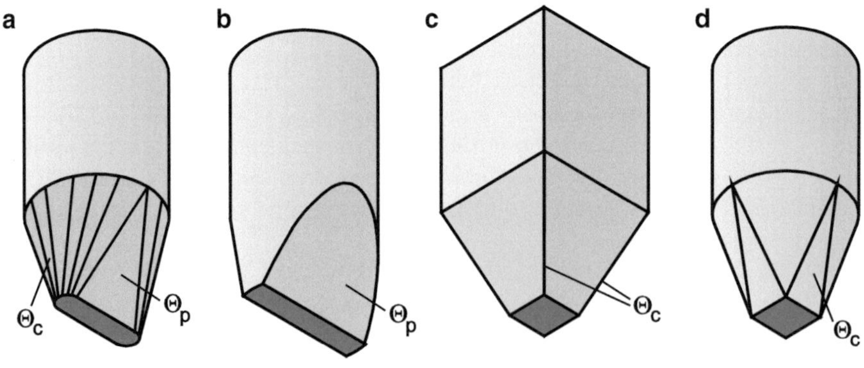

Fig. 13.33 Modified hopper shapes [7, 37, 50]; angles (Θ) indicate the maximum slope to the vertical for mass flow

The pyramidal hopper (Fig. 13.33c) is disadvantageous if mass flow is to be obtained, because the bulk solid has to flow downwards in the valley formed by pairs of adjacent hopper walls, where friction has to be overcome on two sides. To get mass flow in a pyramidal hopper, the valleys should be rounded, and the inclination of the valleys to the vertical (i.e., the valley angle, Θ_v) must conform to the maximum wall slope of a conical mass flow hopper, Θ_c. Since the walls of a pyramidal hopper are steeper than the valleys, a pyramidal hopper designed for mass flow is always steeper and, thus, higher than an equivalent conical hopper. The valley angle, Θ_v, can be calculated from the angles of inclination of the adjacent walls, Θ_1 and Θ_2 (all measured to the vertical):

$$(\tan\Theta_v)^2 = (\tan\Theta_1)^2 + (\tan\Theta_2)^2 \tag{13.21}$$

A mass flow hopper according to Fig. 13.33d also has to be steeper than a conical mass flow hopper. This hopper realizes the transition from a cylindrical section to a square outlet. The slope of the plane hopper walls to the vertical should not be larger than Θ_c.

Comparing the hoppers shown in Figs. 13.17 and 13.33 considering their wall steepness required to achieve mass flow, variants c and d in Fig. 13.33 have to be steepest. The conical hopper (Fig. 13.17a) can be somewhat flatter. The lowest steepness is possible for a symmetric wedge-shaped hopper (Fig. 13.17b) and its modifications (Fig. 13.33a, b).

13.6.4 Transitions and Inclined Walls

Even if the hopper walls are steep enough for mass flow, stagnant zones may build up due to poorly designed transitions to downstream equipment, inappropriate wall surfaces, or eccentric discharge by feeders (regarding feeder design refer to Sect. 13.6.7).

In Fig. 13.34 two transitions from a hopper to a feeder (or any other downstream equipment) are shown. In case of Fig. 13.34a the cross-section is reduced abruptly thus forming a horizontal edge on which stagnant zones can form – even if the hopper slope fulfils the requirement for mass flow. Also less obvious stepwise reductions of the cross-section must be avoided, because a protrusion with a width of only a few particle diameters is sufficient for the formation of a stagnant zone. Therefore, possible reasons for stagnant zones can be gaskets or welds protruding inside, poorly fitted flanges, partially closed slide gate, etc. Fig. 13.34b illustrates a more expensive solution by using an intermediate transition hopper. The transition hopper is designed for mass flow. Thus, stagnant zones cannot form. At flange connections it is always advisable to make the inner diameter of the lower flange (e.g., the transition diameter of the downstream element) larger than the

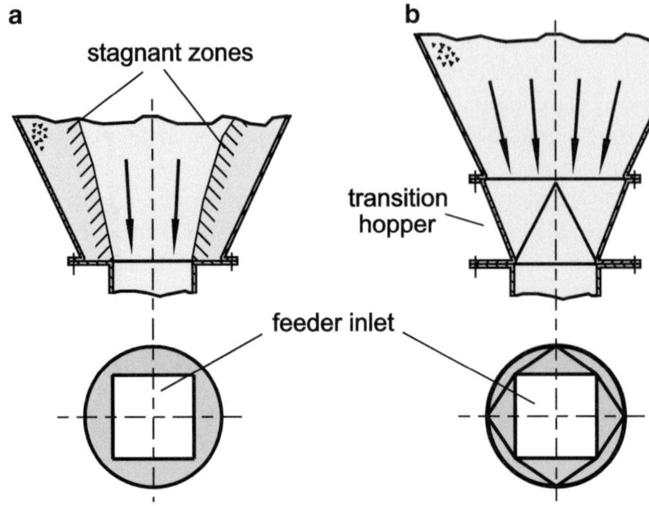

Fig. 13.34 Transition from hopper to feeder; (**a**) abrupt reduction of cross-section leads to stagnant zones; (**b**) transition hopper designed for mass flow [37]

inner diameter of the upstream flange so that a protrusion of the downstream flange can be excluded from consideration of possible manufacturing tolerances.

When a hopper is provided with a liner, care has to be taken that the liner does not cause stagnant zones due to protrusions. If metal sheets are butt welded, it is important that the welds, especially if they are not parallel to the flow direction, do not protrude into the bulk solid, but are smooth and flush with the liner surface. If necessary, the welds should be ground fine, but it is important that the liner's surface not become roughened due to the grinding action, because this might increase the angle of wall friction, and, thus, could be a reason for mass flow not to occur. Similar to protruding screws, devices for the measurement of temperature, humidity, pressure etc. can have a detrimental effect on flow if placed on sloping surfaces. Thus, such devices should, if possible, be placed on vertical surfaces.

13.6.5 Discharge Aids

Discharge aids are required to initiate flow if the outlet dimension is smaller than the critical outlet size (Sect. 13.5.2). If the outlet size is smaller than the critical outlet size determined with the instantaneous properties of the bulk solid, they are needed permanently to maintain flow. If in the case of a bulk solid exhibiting time consolidation the outlet size is larger than that according to the instantaneous properties, but smaller than that required to initiate flow after a certain storage time at rest, discharge aids are needed to initiate flow after the storage period, but

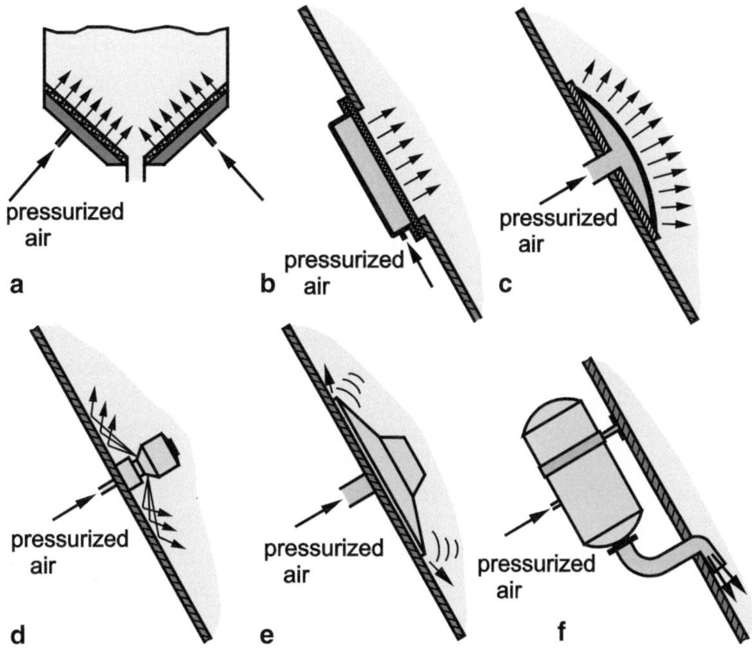

Fig. 13.35 Pneumatic discharge aids (examples); (**a**) porous hopper; (**b**) aeration box; (**c**) aeration pad; (**d**) aeration nozzle; (**e**) vibrating aeration nozzle; (**f**) air cannon [37]

not to maintain flow because time consolidation is no longer a consideration after the material has been caused to flow.

Discharge aids which initiate flow by transfer of energy can be roughly divided into the pneumatic discharge aids (e.g., air injection) and mechanical discharge aids (e.g., knockers, vibrators, agitators).

Pneumatic discharge aids (Fig. 13.35) initiate flow by the injection of air. In the systems according to Fig. 13.35a–e air is injected continuously at low flow velocities. This can be done as shown in Fig. 13.35a through hopper walls made of porous material, but also through aeration elements placed at the hopper walls as in the Fig. 13.35b–e. Having a simple aeration box or aeration pads, the air enters the bulk solid through a porous material, for example, fabric weave or sintered material. Some types of aeration nozzles blow the air parallel to the wall into the bulk solid in order to activate a larger area of the silo wall. The vibrating aeration nozzle is provided with a flexible cap pressed against the silo wall. Air is flowing between wall and cap into the silo while the cap is excited to vibrate, i.e., the bulk solid is stimulated by vibration in addition to air flow.

Air injection causes local overpressures. Thus an air flow prevails in direction to lower pressures, for example, the ambient pressure at the outlet opening or at the surface of the filling, so the major portion of the air takes the path with the smallest air resistance. Thus, if the silo's filling level is sufficiently high, and air is introduced into the lower part of the hopper (which is usually the case), the majority of the injected air flows towards the outlet opening. Thus, the air flow creates additional forces on the bulk solid in the direction to the outlet so that wall friction may be overcome, and arches and ratholes may be destroyed. Furthermore the air flow encourages bulk solids flow towards the outlet due to the action of a gas overpressure in the hopper. This method of aeration is mostly advisable for fine-grained, easy-to-fluidize materials (group A materials according to Geldart's classification [13], e.g., cement, flour). To get a similar effect with distinctly larger particles very high air velocities and thus uneconomically large air volume rates would be necessary [50]. Having very cohesive bulk solids (e.g., titanium dioxide being composed of submicron particles), the injection of air can also be problematic, since those bulk solids tend to channeling, i.e. voids and fissures are created in the bulk solid (e.g., close to the wall) through which the air escapes, while the rest of the bulk solid remains stationary.

The optimal position for air injection (e.g., for the devices according to Fig. 13.35b–e) is just above the possible arch. Thus the air flow resistance is limited which reduces the energy consumption.

Besides continuous injection of air at small flow rates, the possibility exists to impulsively release compressed air stored in a pressure vessel (volumes of 2–500 L, pressure up to 10 bars) through one or more nozzles into the silo. Equivalent discharge aids (Fig. 13.35f) are known under different commercial names; often the term "air cannon is used (survey and applications see [51]). The pressure impulse created by the air cannon generates a stress wave in the bulk solids structure propagating faster than the air pressure wave expanding in the pore structure of the bulk solid. Clearly a stress wave can be sufficient for the collapse of arches and ratholes [30].

Contrary to continuous air injection, air cannons can also be used for coarse-grained bulk solids (e.g., for flow problems resulting from time consolidation), since due to the impulsive injection of air a positive action can be achieved even if the air flow resistance of coarse-grained bulk solids is low. Also for the storage of poorly flowing bulk solids in large silos air cannons will be used since their range of action is large compared to the devices with a slow air injection (note that the required number and size of air cannons depends on bulk solid properties and silo dimensions).

When using air cannons it has to be kept in mind that problems can arise if the air cannons are not correctly applied and designed (dust generation due to too large air volumes; extremely high local stressing of the silo walls due to the pressure impulse [47], especially when the bulk solid fails to move; excessive loads on the hopper walls or the feeder due to large amounts of material suddenly collapsing, the latter may even cause negative pressure above the silo filling and, thus, failure of the silo structure). A typical application of air cannons is the retrofitting of silos which exhibit flow problems and where no other means are possible or practicable. For new silos air cannons are significant if bulk solids with a strong time consolidation effect have to be stored.

The simplest mechanical discharge aids are rods and similar means inserted into a hopper through the outlet opening or poke holes to encourage the bulk material to flow, or a hammer used for beating the hopper wall. Further development of these archaic methods led to the application of knockers and agitators. Generally the application of agitators is limited to small diameters because the driving torque increases progressively with agitator diameter.

Vibrators are widely applied as mechanical discharge aids, and different forms of devices are available. Some of them transfer vibrations into the silo wall like knockers, but with higher frequencies. Others excite inserts located within the bulk solid. Vibrations induced at the silo wall may aid to overcome wall friction. Additionally, vibrations momentarily increase the stress in the bulk solid. If thereby the yield limit is attained, the vibration breaks stable arches or ratholes. However the design of vibrating devices mounted directly on a silo wall is difficult, because the properties of the silo structure have to be taken into account (e.g., wall thickness and stiffness), and a (conical) hopper is particular stiff close to its outlet (small radius of curvature) where stable arches are formed predominantly. With vibrating inserts it has to be ensured that they do not act as a flow obstruction or as the basis for a stagnant zone, which is often the case at least when they are not active.

All discharge aids should be used only during discharge. Vibration of a stagnant bulk solid as well as the action of agitators and knockers, or a continuous releasing of air cannons is likely to cause further consolidation and, thus, increasing flow problems. Apart from air cannons, most discharge aids seem to be most suitable for the elimination of smaller flow problems in silos or hoppers that are not too large [37]

13.6.6 Discharge Devices

Discharge devices are well known devices as, for example, belt feeders, apron feeders, en-masse feeders, vibratory feeders, rotary valves, and screw feeders being attached to the outlet of a silo. Shut-off devices (e.g., slide gates, clamshell gates) can also be regarded as discharge devices since they allow a certain regulation of the discharge (often poor and accompanied by problems, see Fig. 13.37).

Even in an appropriately designed mass flow hopper with sufficient steepness (Sect. 13.5.2.1), mass flow can be achieved only if the bulk material is withdrawn from the entire outlet opening. Therefore, mainly two rules, which are described in the following, have to be followed when designing feeders:

- Any planes which are not sufficiently steep to avoid the formation of stagnant zones have to be avoided.
- The feeder must be able to discharge bulk solid from the entire outlet opening.

Feeders have to be designed so that all surfaces being in contact with the bulk solid are sufficiently steep to avoid the formation of stagnant zones. An example of this situation is the double gate shown in Fig. 13.36a. If one of the butterfly valves is closed, a stagnant zone will form on the stationary saddle between the valves, and the plate of the closed valve. An alternative is the design according to Fig. 13.36b

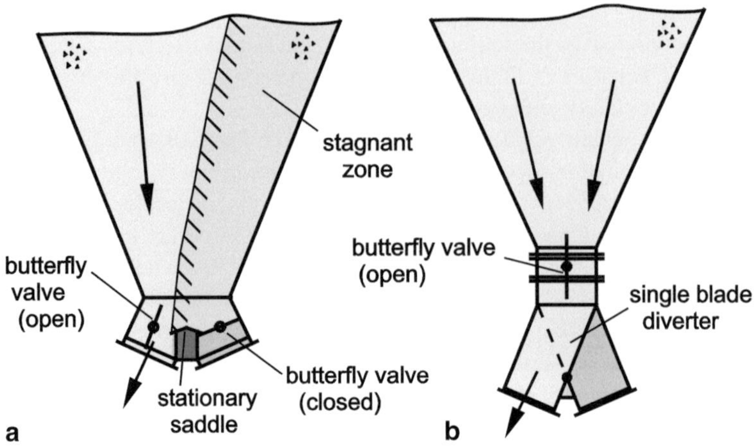

Fig. 13.36 (**a**) Diverter with two butterfly valves; stagnant zone forms on the closed valve; (**b**) butterfly valve with downstream single blade diverter [37]

Fig. 13.37 Formation of stagnant zones due to partly opened gates: (**a**) slide gate; (**b**) clamshell gate [37]

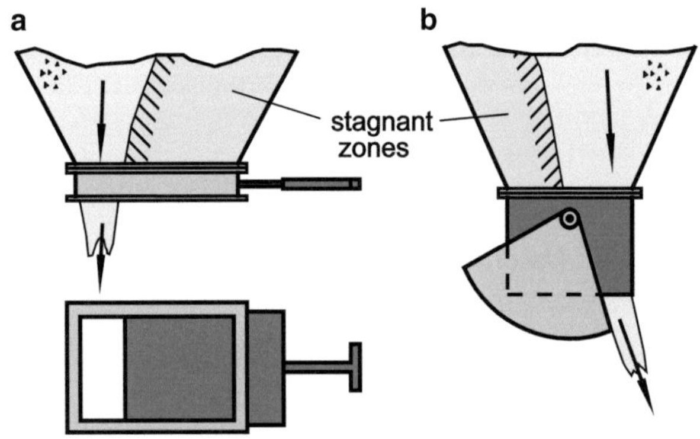

where a single blade diverter is installed below a butterfly valve. All surfaces being in contact with the bulk solid are sufficiently steep to avoid stagnant zones.

Another example of the formation of stagnant zones is the slide gate shown in Fig. 13.37a being only partly opened. Originating from the slide plate a stagnant zone is formed leading to funnel flow with all known disadvantages (flow problems, segregation,. . .). The same problem will occur with any configuration of valves and gates (e.g., clamshell valve, Fig. 13.37b), which are not opened completely during discharge. Therefore, if mass flow is desired, gates have to be opened completely for discharge, and it must be ensured that there are no protrusions into the flowing bulk solid.

Many types of feeders tend to discharge the bulk solid only from a portion of the outlet. An example is the screw feeder (Fig. 13.38). The major element of a screw feeder is the rotating screw, consisting of a central shaft to which a helical flight is fitted, conveying the bulk solid in a trough with circular or U-shaped cross section.

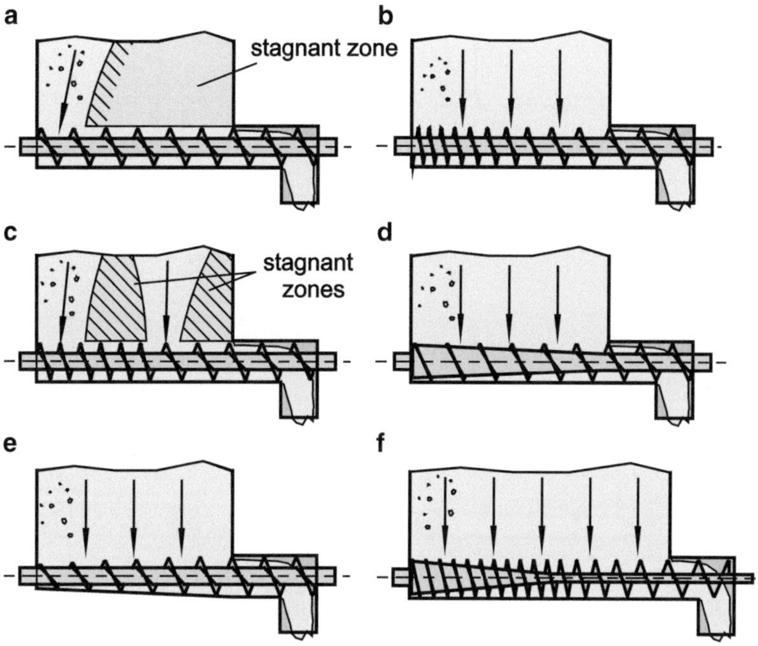

Fig. 13.38 Configurations of screw feeders [18, 22, 37, 43]

The screw feeder shown in Fig. 13.38a withdraws the bulk solid preferentially only from the rear end of the outlet slot so a stagnant zone is formed (funnel flow) even if the silo is designed as a mass flow silo. The reason for this behavior is the fact that the screw fills completely at the rear part of the outlet, and the screw conveys this bulk solid towards the exit beneath the bulk solid resting above the front part. This behavior can be avoided by designing the helical flights in such a way that the conveying capacity is increasing in the conveying direction, for example, by an increasing pitch of the screw helix as shown in Fig. 13.38b (design guidelines in [15]). Using this method the screw can discharge bulk solid across the entire length of the outlet slot. The increase of the pitch of the screw should be continuous. If there are too long sections with constant capacities (Fig. 13.38c), stagnant zones can be formed [43].

Other possibilities to realize increasing capacity are to decrease the shaft diameter in the conveying direction (Fig. 13.38d), or increase the flight diameter (Fig. 13.38e). The last option is less convenient because the width of the trough also has to be enlarged in the conveying direction which itself requires an equivalent fitting of the hopper geometry. For outlet slots with a large length-to-width ratio it is useful to start with a decreasing shaft diameter at constant pitch followed by an increasing pitch at constant shaft diameter (Fig. 13.38f) [22].

Belt feeders (Fig. 13.39) show the tendency to withdraw bulk solid preferentially from one end of the outlet. A smooth belt and a tall gate opening at the front end result in preferential withdraw from the front part of the outlet slot (Fig. 13.39a,

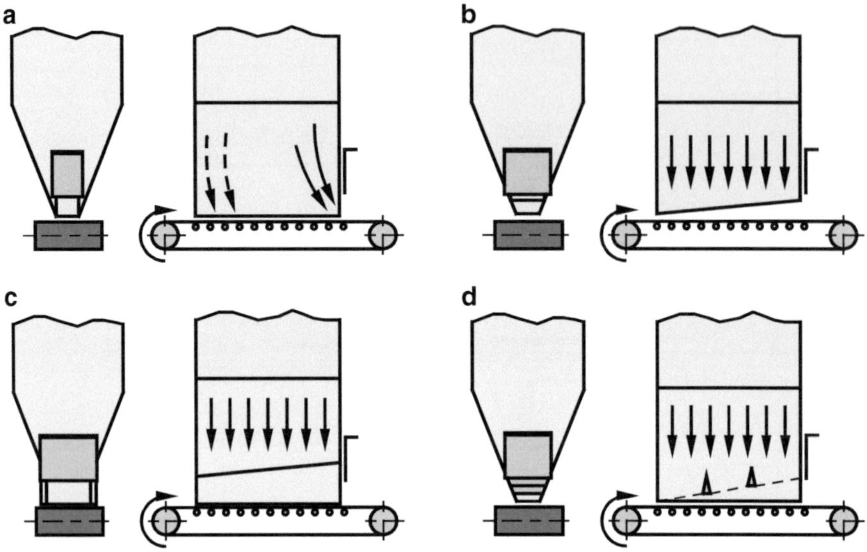

Fig. 13.39 Discharge with a belt feeder [37]: (**a**) non-uniform discharge; a smooth belt and a large layer height preferentially results in discharge from the front end; a rough belt leads to discharge from the rear end (*dashed arrows*); (**b**) uniform discharge due to a tapered outlet; (**c**) uniform discharge due to a hopper with tapered lower edge and vertical walls; (**d**) uniform discharge due to layer-height limiters (rough belt)

solid arrows). The bulk solid is more discharged from the rear part of the silo if the conveyor belt is rough and/or the gate is opened less (dashed arrows) [6, 36, 41]. Other parameters determining the flow pattern are the angle of internal friction (large internal friction promotes discharge of the bulk solid from the front part), and for the case that the outlet is provided with vertical side skirts, the wall friction angle of the side skirts (large friction angle at the side skirts promotes discharge of the bulk solid from the front part of the outlet slot). The large number of influencing parameters makes it impossible to predict exactly the flow pattern, which is, as stated above, usually non-uniform for a standard interface as shown in Fig. 13.39a. Thus, to withdraw material from the entire outlet slot, it is essential to apply the principle of increasing capacity. This can be achieved by sloping the lower edge of the hopper thus realizing a taper in both plan and elevation, Fig. 13.39b. As a result, an increasing amount of bulk solid can be transported by the belt feeder in direction of conveying (increasing capacity).

In the example of Fig. 13.39c vertical walls (non-parallel side skirts) are attached to the hopper of Fig. 13.39b restricting the taper in plan and resulting in uniform withdrawal if the gate is properly adjusted [6]. Since a misadjusted gate can result in non-uniform withdrawal even if the interface is properly designed as in Fig. 13.39c, in general the discharge rate should be adjusted by belt speed rather than by the position of the gate. Attention has to be paid when discharging cohesive bulk solids since their flow pattern is very sensitive to gate adjustment, and often it takes a very long time until a steady flow pattern is attained [36, 41].

Another possibility to get a more uniform withdrawal from a long outlet slot is the placement of inserts acting as layer-height limiters (Fig. 13.39d). Due to the inserts the height of the horizontally transported layer increases in conveying direction thus providing the desired increasing capacity. This method can also be applied to apron feeders or en-masse conveyors [36, 41]. The inserts have to be designed so that neither arches nor stagnant zones can be formed on them.

Prerequisite for success of the means shown in Fig. 13.39b–d is that the angle of friction between the belt and the bulk solid is sufficiently large to withdraw material from the rear part of the outlet slot. If the friction is too small, the material will still be discharged preferentially from the front part (Fig. 13.39a, solid arrows).

Also feeders not discussed here, e.g., vibratory feeders, en-masse feeders, apron feeders, or air slides for air-assisted discharge, may result in uneven withdrawal if not properly designed. More examples are given in [37].

13.7 Structural Silo Design

Silos or bins are structures and, thus, in addition to fulfilling the functional requirements, they have to be designed for strength like, e.g., buildings or bridges. They can be constructed of aluminum, steel, wood, fiber reinforced plastic, or reinforced concrete. Silos can be ground supported, supported on columns, or hung from floors. Even the structural form of a silo can be rather different, e.g., steel silos can consist of rolled steel plate constructions (shells), stiffened plate constructions, or corrugated sheetings [8, 16, 23]. Structural design of bins and silos is commonly done by civil engineers based on appropriate codes.

For structural design the loads on bins or silos have to be evaluated. These loads depend strongly on bulk solid properties, as shown in Sect. 13.4 by the calculation of loads in a cylinder using the Janssen equation. Wall friction, stress ratio, and bulk density are important quantities to be determined for the entire stress range being expected to occur in the respective silo. However, even the knowledge of the flow profile (mass flow or funnel flow) is of great importance since it determines the position of the stress peak acting on the silo wall (Fig. 13.15c, d): In a mass flow silo the stress peak is at the transition from the cylindrical section to the hopper, while in a funnel flow silo the location of the stress peak cannot be predicted. Especially eccentric flow, e.g., caused by an inappropriately designed feeder (Fig. 13.1f) or eccentric discharge (Fig. 13.1g), is dangerous because it results in an unsymmetrical stress distribution along the silo wall which increases the probability of buckling of the walls of metal silos if not considered at structural design [8, 16, 23]. Therefore, it is very important to provide the civil engineer with sufficient information on the bulk solid's properties and on the type of flow. For the evaluation of loads on bins and silos codes are available, e.g., EN 1991-4 [11], which is a part of Eurocode 3.

13.8 Definitions, Abbreviations and Symbols

Funnel flow	Any type of flow pattern where part of the container contents of bulk solids remains stationary during discharge
Mass flow	Flow pattern where every particle of the container contents of bulk solids is moving during discharge

A	[m^2]	area
b	[m]	width
b_{crit}	[m]	critical outlet width to avoid arching, i.e., the minimum width of a rectangular outlet opening to avoid arching
d	[m]	diameter
d_{crit}	[m]	critical outlet diameter to avoid arching, i.e., the minimum diameter of a circular outlet opening to avoid arching
D	[m]	diameter; silo diameter
D_{crit}	[m]	critical outlet diameter to avoid ratholing, i.e., the minimum diameter of a circular outlet opening or the minimum length of the diagonal of a rectangular outlet opening to avoid ratholing
f_c	[Pa]	compressive strength; unconfined yield strength (also: σ_c)
F	[N]	force
F_N	[N]	normal force
F_R	[N]	resultant force; friction force
F_S	[N]	shear force
ff	[–]	flow factor (depending on flow properties and hopper geometry)
ff_c	[–]	flowability, calculated from the ratio of consolidation stress, σ_1, to compressive strength, σ_c
g	[m/s^2]	acceleration due to gravity, $g = 9.81$ m/s^2
h_f	[m]	filling height
K	[–]	(lateral) stress ratio, i.e., ratio of horizontal stress to mean vertical stress in the vertical section of a silo
L	[m]	length; length of a rectangular outlet opening
m	[kg]	mass
m	[–]	parameter describing hopper geometry ($m = 0$: wedge-shaped; $m = 1$: conical)
M	[Nm]	moment
$\dot{m}$	[kg/s]	discharge rate, flow rate
p	[Pa]	gas pressure
r	[m]	radius; polar coordinate
t	[s]	time; storage time
U	[m]	perimeter
v	[m/s]	velocity
V	[m^3]	volume
x	[m]	particle size

z	[m]	coordinate
α	[°]	angle
ε	[–]	porosity, void ratio
Θ	[°]	inclination of a hopper wall to the vertical
Θ_{ap}	[°]	inclination of the inclined wall of a asymmetric wedge-shaped hopper to the vertical
Θ_c	[°]	inclination of the wall of a conical hopper to the vertical (also: hopper half-angle)
Θ_{cd}	[°]	maximum inclination of a hopper wall for complete clearance due to gravity (measured to the vertical)
Θ_p	[°]	inclination of the side walls of a wedge-shaped hopper to the vertical (also: hopper half-angle)
Θ_v	[°]	valley angle, i.e., slope of the corner of a pyramidal hopper to the vertical
ρ_b	[kg/m^3]	bulk density
σ	[Pa]	normal stress
σ_{sh}	[Pa]	normal stress at shear to failure
σ_{pre}	[Pa]	normal stress at preshear
σ_α	[Pa]	normal stress in a cutting plane inclined to the horizontal by angle α
σ_c	[Pa]	compressive strength; unconfined yield strength (also known as f_c)
σ_{ct}	[Pa]	unconfined yield strength attained after time consolidation; $\sigma_{ct} = \sigma_c(\sigma_1,t)$
σ_h	[Pa]	horizontal stress
σ_v	[Pa]	vertical stress
$\sigma_{v\infty}$	[Pa]	final or asymptotic stress, i.e., the vertical stress in the vertical section of a silo in infinite depth
σ_w	[Pa]	wall normal stress
σ_1	[Pa]	major principal stress
σ_1'	[Pa]	major stress in an arch
σ_1''	[Pa]	circumferential stress in the wall of a rathole
σ_2	[Pa]	minor principal stress
τ_{sh}	[Pa]	maximum shear stress at shear to failure
τ_{pre}	[Pa]	shear stress at the end of preshear (steady-state flow)
τ_{sf}	[Pa]	shear stress at steady-state flow

τ_w	[Pa]	wall shear stress
τ_{xy}	[Pa]	shear stress (parallel to y axis, perpendicular to x axis)
φ_e	[°]	effective angle of internal friction
φ_i	[°]	local slope of a yield locus
φ_{lin}	[°]	slope of a linearized yield locus
φ_{sf}	[°]	internal angle of friction at steady-state flow
φ_t	[°]	local slope of a time yield locus
φ_x	[°]	kinematic angle of wall friction

Indices

c	axisymmetric state of flow; conical hopper
p	plane state of flow; wedge-shaped hopper
ap	plane-asymmetric state of flow; asymmetric wedge-shaped hopper
h	horizontal
crit	critical (representing conditions at the minimum outlet size required to avoid arching or ratholing, respectively)
max	maximum
min	minimum
rel	relative
v	vertical

References

1. Arnold, P.C., McLean, A.G.: Improved analytical flow factors for mass-flow hoppers. Powder Technol. **15**, 279–281 (1976)
2. ASTM Standard D6128.: Standard Test Method for Shear Testing of Bulk Solids Using the Jenike Shear Cell. ASTM International. Available from www.astm.org. 1 Jan 2015
3. ASTM Standard D6682.: Standard Test Method for Measuring the Shear Stresses of Powders Using the Peschl Rotational Split Level Shear Tester. ASTM International. Available from www.astm.org. 1 Jan 2015
4. ASTM Standard D6773.: Standard Shear Test Method for Bulk Solids Using the Schulze Ring Shear Tester. ASTM International. Available from www.astm.org. 1 Jan 2015
5. Beverloo, W.A., Leniger, H.A., van der Felde, J.: The flow of granular solids through orifices. Chem. Eng. Sci. **15**, 260–269 (1961)
6. Bridge, D.T., Carson, J.W.: How to design efficient screw and belt feeders for bulk solids. Powder and Bulk Solids 12th Annual Conference, Rosemont, IL, USA (1987)
7. British Standards Institution (Ed).: Draft Design Code for Silos, Bins, Bunkers and Hoppers. Published by BSI in association with the British Materials Handling Board (1987)
8. Brown, C.J., Nielson, J. (eds.): Silos – Fundamentals of Theory, Behaviour and Design. E & FN Spon, London (1998)
9. Brown, R.L., Richards, J.C.: Principles of Powder Mechanics. Pergamon Press, Oxford (1970)
10. Carr, J.F., Walker, D.M.: An annular shear cell for granular materials. Powder Technol. **1**, 369–373 (1967/1968)
11. EN 1991-4.: Eurocode 1: Actions on Structures – Part 4: Silos and Tanks (2006)

12. Gebhard, H.: Scherversuche an leicht verdichteten Schüttgütern unter besonderer Berücksichtigung des Verformungsverhaltens. Dissertation Univ. Karlsruhe, Germany (1982)
13. Geldart, D.: Types of gas fluidization. Powder Technol. **7**, 285–292 (1973)
14. Gu, Z.H., Arnold, P.C., McLean, A.G.: Modeling of air pressure distributions in mass flow bins. Powder Technol. **71**, 121–130 (1992)
15. Haaker, G., van Poppelen, M.P., Jongejan, M.P., Stokkers, G.J.: Improvement of screw feeder geometry for better draw-down performance. Proceedings of the International Symposium on Reliable flow of Particulate Solids II, Oslo, Norway, pp 551–561 (1993)
16. Hampe, E.: Silos, Band 1 (Grundlagen). VEB Verlag für Bauwesen, Berlin (1987)
17. Janssen, H.A.: Getreidedruck in Silozellen. Z. Ver. Dtsch. Ing. **39**, 1045–1049 (1895)
18. Jenike, A.W.: Storage and flow of solids. Bull. No. 123, 20th Printing, revised 1980. Eng. Exp. Station, Univ. of Utah, Salt Lake City. Available at: The University of Utah, Utah Engineering Experiment Station, 1515 Mineral Square, Room 138, Salt Lake City, UT 84112-1109, USA (1964/1980)
19. Jenike, A.W.: Storage and flow of solids. Bull. No. 123, 1st printing, Eng. Exp. Station. University of Utah, Salt Lake City (1964)
20. Johanson, J.R., Jenike, A.W.: The effect of the gaseous phase on pressures in a cylindrical silo. Powder Technol. **5**, 133–145 (1971/1972)
21. Kwade, A., Schulze, D., Schwedes, J.: Determination of the stress ratio in uniaxial compression tests. Powder Handl. Process. **6**(1), 199–203 (1994)
22. Marinelli, J., Carson, J.W.: Use screw feeders effectively. Chem. Eng. Progress. **88**(12), 47–51 (1992)
23. Martens, P. (ed.): Silohandbuch. Wilhelm Ernst & Sohn Verlag, Berlin (1988)
24. McLean, A.G.: Empirical critical flow factor equations. Bulk. Solids. Handl. **6**, 779–782 (1986)
25. Münz, G.: Entwicklung eines Ringschergerätes zur Messung der Fließeigenschaften von Schüttgütern und Bestimmung des Einflusses der Teilchengrößenverteilung auf die Fließeigenschaften kohäsiver Kalksteinpulver, Dissertation, University of Karlsruhe, Germany (1976)
26. Murfitt, P.G., Bransby, P.L.: Deaeration of powders in hoppers. Powder Technol. **27**, 149–163 (1980)
27. Nedderman, R.M., Tüzün, U., Savage, S.B., Houlby, G.: The flow of granular materials. Chem. Eng. Sci. **37**, 1597–1609 (1982)
28. Peschl, I.A.S.Z.: Shear test for process control and engineering. Powder Handl. Process. **11**(1), 43–48 (1999)
29. Peschl, I.A.S.Z., Colijn, H.: New rotational shear testing technique. J. Powder Bulk Solids Technol. **1**, 55–60 (1977)
30. Raabe, T., Schwedes, J.: On the operation of air blasters as discharge-aids – experimental investigations. Proceedings of the "6th International Conference on Bulk Materials Storage, Handling and Transportation", pp. 507–514. Wollongong (1998)
31. Rathbone, T., Nedderman, R.M., Davidson, J.F.: Aeration, deaeration, and flooding of fine particles. Chem. Eng. Sci. **42**, 725–736 (1987)
32. Roberts, A.W.: Modern Concepts in the Design and Engineering of Bulk Solids Handling Systems. TUNRA Ltd., The University of Newcastle, Callaghan (1990)
33. Royal, T.A., Carson, J.W.: Fine Powder Flow Phenomena in Bins, Hoppers and Processing Vessels. Bulk 2000, London (1991)
34. Schäfer, R., Schröer, H., Schwedes, J.: Auslegung einer Großsilo-Anlage für die Einlagerung eines feuchten Schüttgutes und praktische Erfahrungen. Zement Kalk Gips **75**, 587–592 (1986)
35. Schulze, D.: A theoretical model for the prediction of the gas pressure distribution in silos at filling and discharge. Proceedings of the "International Symposium on Reliable Flow of Particulate Solids", pp. 331–340. Porsgrunn (1999)
36. Schulze, D.: Untersuchungen zur gegenseitigen Beeinflussung von Silo und Austraggerät. Ph. D. thesis, Technische Universität Braunschweig, Germany (1991)

37. Schulze, D.: Powders and Bulk Solids – Behavior, Characterization, Storage and Flow. Springer, Berlin (2008)
38. Schulze, D.: Measurement of flow properties of particulate solids in food and pharmaceutical technology using a new automated ring shear tester. Preprints PARTEC "1st European Symposium on Process Technology in Pharmaceutical and Nutritional Sciences", pp. 276–285. Nuremberg, 10–12 Mar 1998 (1998)
39. Schulze, D.: Round robin test on ring shear testers. Adv. Powder Technol. **22**(2), 197–202 (2011)
40. Schulze, D.: Silo Stress Tool, Freeware, available from www.dietmar-schulze.de. 1 Jan 2015. (1999)
41. Schulze, D., Schwedes, J.: Experimental investigation of silo stresses under consideration of the influence of hopper/feeder interface. Kona **8**, 134–144 (1990)
42. Schulze, D., Schwedes, J.: Examples of modern silo design. Bulk Solids Handl. **11**, 47–52 (1991)
43. Schumacher, W.: Zum Förderverhalten von Bunkerabzugsschnecken mit Vollblattwendeln. Ph.D. thesis, TH Aachen, Germany (1987)
44. Schwedes, J.: Fließverhalten von Schüttgütern in Bunkern. Verlag Chemie, Weinheim (1968)
45. Schwedes, J., Schulze, D.: Measurement of flow properties of bulk solids. Powder Technol. **61**, 59–68 (1990)
46. Spink, C.D., Nedderman, R.M.: Gravity discharge rate of fine particles from hoppers. Powder Technol. **21**, 245–261 (1978)
47. Terziovski, A., Arnold, P.C.: On the effective sizing and placement of air blasters. Bulk Solids Handl. **10**, 181–185 (1990)
48. Institution of Chemical Engineers (Eds.).: Standard Shear Testing Technique for Particulate Solids Using the Jenike Shear Cell. Institution of Chemical Engineers, Rugby (1989)
49. Wilms, H., Schwedes, J.: Interpretation of ring shear tests. Bulk Solids Handl. **5**, 1017–1020 (1987)
50. Woodcock, C.R., Mason, J.S.: Bulk Solids Handling. Leonard Hill, Glasgow (1987)
51. Zimmer, W.: Luftkanonen und Luftinjektoren – Was ist der Unterschied? Schüttgut **12**, 262–268 (2006)

Chapter 14
Solids Transport and Handling

George E. Klinzing

Abstract Solid transport and handling is a crucial part of any process that involves solids. This topic sometimes is forgotten until the very end of the process design and retrofits are needed and decisions are made which will not be optimum for the overall process. There are many types of techniques that can be employed for the transport and handling of solids and the basic decision is one that depends intimately on the process and materials being processed. Unfortunately there is not an overall economic strategy that could be employed to guide the designer on which technique for transport and handling is appropriate and best. Sometimes, several procedures can be employed and the process will operate without problems. Often, however, an arbitrary choice of equipment leads to severe problems. This chapter gives an overview of the types of transport and handling operations and consideration that are available. A more in depth analysis is given for the pneumatic transport operation since this mode of transport is probably the most common spanning both long and short distance and small and large capacity operations. The topics of electrostatic generation, dust explosions, segregation, mixing and flow measurements are included in this chapter.

14.1 Introduction

The processing and handling of solids in industrial operations present challenges to design and operate various units and component parts. Studies have shown that if a plant processes or handles only liquids or gases, these operations are more reliable both in coming on-line and keeping operational [59]. One of the big problems of engineers who have not been exposed to the uniqueness of solids is to apply the same principles as employed in liquid and gas operations. Sometimes with a little luck this works but more often it only leads to further problems and inoperability. One simple example is that solids have lateral pressures that can stress the

G.E. Klinzing (✉)
University of Pittsburgh, Pittsburgh, PA, USA
e-mail: klinzing@pitt.edu

© Springer International Publishing Switzerland 2016

H.G. Merkus, G.M.H. Meesters (eds.), *Production, Handling and Characterization of Particulate Materials*, Particle Technology Series 25,
DOI 10.1007/978-3-319-20949-4_14

containers in which they are stored and utilized. We know that in liquids and gases pressure is transmitted in all directions equally causing us to use different physical concepts than in the solids cases. Solids have sizes, distributions, shapes and a myriad of other specific properties unlike liquids and gases that depends on their molecular structure to give bulk properties. We are now faced with a whole new viewpoint when processing and handling solids some of which are unique to a particular material and some phenomena that still remained unexplained today.

As we try to regularize the solids processing and handling field, we have applied more science and new measurements to try to guarantee that our designs will function and stay functioning. Many studies have been implemented to probe the solid behaviors hopefully to get to the point where we will have the same reliability of designs as we have with liquids and gases.

The first thing to consider in solids operations is the particle size and its distribution. When these parameters fail to give us sufficient information, we go naturally to the shape which is a different realm of complexity. Statistics comes into play immediately with the size and distribution consideration and carries over to the shape as well. In event that will still run into problems the surface and bulk properties of the solids must be addressed. Particle strength, elasticity, electrostatic charge and tackiness may be important. All these factors influence the bulk properties of the solids.

In the processing field engineers must store and move solids and of course blend them to have uniformity of the product. Chemical reactions and temperature and pressure effects come into play quickly as we start to handle both solids with liquids and gases. All of a sudden we have two or three phases to consider. It appears that we have opened a Pandora's Box of phenomena and issues that tax our abilities to design and perform operations. The engineers have responded creatively to these issues developing models, correlations and more recently trying to develop reliable computer simulations. There still remains much to do and many hurtles to cross to achieve our overall objective of reliable design and operation. We are always looking to some unifying approach that will give us the confidence one has in operating with liquids and gases. Particles as they become smaller often take on the properties of being the trigger for explosions which can have serious consequences in financial and human loss.

14.2 Various Modes of Bulk Material Transport

Woodcock and Mason [101] in their book on Bulk Solids Handling discuss a wide variety of bulk solids transport modes everything from gravity driven means to vibratory means. Mechanical conveyors and their selection and operation were discussed by Fayeed and Skocir [25]. The least expensive of these modes is using the gravity force to drive the flow. While this technique may be inexpensive it is dependent on the containing vessel, carrying chute and particle properties in relationships that are not obvious and can be problematic. Belt conveyors and their vertical derivative of bucket elevators have a high degree of reliability being used in long distance conveying for the belt conveyor and short vertical lifts for the bucket elevator. Screw conveyors are usually limited to short distances conveying and are coupled with other modes of transport such as pneumatic conveying which is a large component of all solids transports being employed in industry. Using vibratory motion, bulk solids can move particles on a trough or belt. Using a liquid as the motive force for hydraulic conveying or slurry transport has been applied successfully in long distance conveying.

14.3 Transport Economics

One approach in evaluating the various types of solids handling operations is to try to approach the topic from an economic viewpoint. While this seems like a logical approach to the subject, one quickly finds that this is a very difficult task which depends on the particular physical situation dependent on the properties of the particles that are to be transported. There is no easy "figure of merit" that has been developed in such terms as cost/unit length of transport that provides the engineer a guideline to apply.

One analysis worth mentioning was carried out by Soo et al. [88] in both short and long distance conveying of coal. They considered both high and low capacity transfer with slurries, barges, rail, and pneumatics. The least expensive mode of transport was found to be barges which had a range of transport of 100–2000 miles in distance. It is noteworthy that at the time that this study was carried out the use of belt conveyors were not generally considered for long distance conveying. Subsequent developments in belt conveyors have placed this mode of transport second only to slurry transport in costs for long distance conveying. Belt conveying has become very reliable with many long distances, large installations around the world in some very remote locations.

The Soo study was the first to consider long distance conveying using pneumatic transport. While the study showed the feasibility of pneumatic transport being employed up to 100 miles in distance, a test facility was never built. For shorter transport distance up to 2 or 3 miles pneumatic transport was shown to be very cost effective. In fact Soo has shown that up to 14 miles in length pneumatic transport came in at the lowest cost per ton of transport per mile.

14.4 Gravity Driven Processes

Continuing to look at economics at first glance it appears that the processes that are driven by gravity have an advantage that external power sources are not needed to actuate the particle flow.

Bates [4] has prepared an excellent summary of the discharge rates and conditions of bulk materials from storage and chutes which are the less energy dependent mode of moving and transporting particles since it depends mainly on gravity as the driving force. Having such a discharge function continuously and without stoppage is not an easy task and depends on the particle properties as well as the vessel in which it is contained. If there is an excess amount of air entrapped in a powder the condition of "flushing" occurs which is very rapid and often difficult or impossible to control.

Several individuals has suggest volumetric flow rate which can be delivered from an orifice of a vessel [8, 10, 11, 18, 26, 27]. Nedderman et al. [65] complied a thorough review article of these flow rates and equations.

Bates [4] has noted that the Beverloo's equation is usually adequate for free flowing materials from funnel flow hoppers. The Beverloo equation is given as:

$$W = C\rho_B\sqrt{g}\left(D - kd_p\right)^{5/2} \tag{14.1}$$

where:

W = solids flow rate
ρ_B = bulk density of the solid
g = gravitational constant
D = diameter of the outlet
d_p = particle size
C and k are fitting parameters; parameter k depends upon the particle shape and has an average value of 1.4 for sand but it can vary between 1.0 and 2.0.

Ariza-Zafram et al. [1] did a more recent review of the proposed models for discharge rates of bulk particulate from silos and bins.

The book by Benjamin [7] brings together the way transfer chute design has evolved and relates it to underlying design principles that must be applied if the transfer is to operate with minimum of maintenance consequences.

Table 14.1 gives an overview of pros and cons of gravity feeders.

14.5 Belt Conveyors

Belt conveyor design and testing has been extensively implemented by Roberts and his team at the University of Newcastle in Australia. In a conversation with Emeritus Professor Alan Roberts, AM. Roberts has commented that one of the most challenging applications of belt conveyors that he worked on was the passage of ore from a copper mine at 4300 m to a processing plant at 2500 m with 4 m in diameter pipe all taking place in a remote wilderness area [66]. Belt conveyors now that are 20 km in length at speeds of 10 m/s are a proven reality.

Table 14.1 Advantages and disadvantages of chutes and gravity feeders

Advantages	Disadvantages
Inexpensive	Little control
Used to provide accumulation to shipping area	Exposed and can be contaminated
Loss energy by relying on gradient force	Dependent on particle properties for smooth operation
Can be used between floors	Dependent of outlet size of bin or hopper

In analyzing the details of belt conveyer design Wheeler et al. [98] determined the flexure resistance of bulk solids transported on belt conveyors. Plate theory was used to approximate the deflection of the troughed conveyor belt between idler sets with the flexure resistance being calculated from the difference between the work done in deflecting the bulk solids during induced active and passive stress states.

Roberts [76] explored a cost model for belt conveyor systems that include the energy cost and annual equivalent cost of conveyor components. He noted that in comparing the continuous modes of transport as slurry pipeline, belt conveyors, screw conveyors and pneumatic systems along with discontinuous modes such as ship or barge, road or rail, the variation in costs may differ by several orders of magnitude. For any specific mode chosen there are many combinations and choices that can vary in the overall costs seen.

The distance travelled by the material is a crucial parameter in the cost model. Belt conveyors are cited as being the second lowest cost for long distance continuous conveying. Slurry pipelines are cited as being the lowest.

The Conveyor Equipment Manufacturers Association (CEMA) design methods have been analyzed by Roberts [76] showing its utility when using commercial software design packages. He shows that specifying the overall operation conditions are essential. Specific inputs to the commonly accepted loss categories are identified paralleling the range of friction coefficient recommended. The loss category equations are refined [15].

Reicks [73] explored belt conveyor models using quantitative platforms for belt conveyor energy options. He emphasized as Roberts the importance of the CEMA fully quantified design methods when used in belt conveyor design software. A wide range of possible influences can influence optimization. Energy losses depend on the specific application.

For short conveyor belt transport of 50 tons/h and lengths up to 100 m, Nanetoe and Sopongpongpipat [64] considered the optimum design and cost. They considered a 500 mm wide belt at various inclinations angles.

Table 14.2 gives an overview of pros and cons of belt conveyors.

Table 14.2 Advantages and disadvantages of belt conveyors

Advantages	Disadvantages
Long distance conveying possible	Angle of inclination should be less than 22°
Can be open or covered to protect against environment	Difficult to transport hot materials
Can convey almost any type of solid and size	Small scale testing needed to test
Can adapt to plant layout	Many moving parts

14.6 Screw Conveyors

Bates [3] has written a comprehensive book on the use of screw conveying in the particulate field. The technology has ancient roots but more modern applications started about 1750 in flour mills. The handling of powders and fine particulates is the dominant usage of screw conveying. The screw conveyor is reliable and is used often to lift solid particles vertically for short distances. The screw conveyor is used as a feeder in other modes of solids transport such as pneumatic conveying. The intersection of a feed tank or bin and a pneumatic conveying line with screw feeders is a common mode of incorporating this screw. The analysis of screw conveyors uses the same principles employ is powder mechanics. Screw conveyors can be used to meter the amount of particulates fed to a unit because of its controlled mode of operation.

Roberts [77] has addressed screw conveying design looking at the torque and power which are influenced by the vortex motion of the bulk solids being transported. The vortex motion and the degree of fill of the screw determine the volumetric efficiency and throughput. This effect influences the torque, power and efficiency. Using his theory he was able to show agreement to the operation of screw conveyors in elevating wheat.

Table 14.3 gives an overview of pros and cons of screw conveying.

14.7 Vibratory Conveyors

Colijn has written a book [16] doing a detailed review with design procedures for mechanical conveyors of bulk solids. He stresses a system's approach to the application in the procedure to choosing a mechanical conveyor. In the case of vibratory conveyors he notes that this type of conveyance is used usually for short distances up to 200 ft in length. The external vibratory drive imparts energy to the bulk solids causing movement. The interaction of the vibration produced and imparting to the solid particle is complex and of course depends on a number of parameters of the carrier trough and the material from size, shape to bed thickness.

The frequency and amplitude of the vibration as well as the angle of inclination are imperative to the vibratory conveyor design. Colijn notes that the particles can be forced to move at velocities of 0.005–0.5 m/s with vibrations going from 4 to

Table 14.3 Advantages and disadvantages of screw conveying

Advantages	Disadvantages
Economical	Short distances
Low maintenance	Attrition
Totally enclosed	Wear with abrasive materials
Multiple discharge points	Particles must be smaller than the pitch
Easy to install	Small scale testing preferred before usage
Quiet operation	Needs to run full of particles
No return run	As the angle of inclination increases, the allowable capacity decreases
Can operate at inclines	
Volume control	
Can handle bulk materials from sluggish to free-flowing	

Table 14.4 Advantages and disadvantages of vibratory feeders

Advantages	Disadvantages
Very low and very high feed rates	Can create dust
Uniform material flow can be obtained with gentle handling	Can cause segregation of particles
Low installation cost	Can't provide positive material flow
Few moving parts	Adhesive material and fines can build up
Low power consumption	Can't provide a linear feed rate

60 Hz and stroke lengths of 5×10^{-4} to 0.075 m. The lower vibration frequency is applicable to longer stroke length.

Molerus [61] has performed experiments on the movement of particles in a rubber tube under the influence of vibratory actions. He employed an off-center motor to impart the vibrations to the rubber tube. This technique would be an idea for the conveying of fragile materials.

In the tire industry the conveying of carbon black with its very fine size characteristics a rubber pipe was been chosen to move this material with motive air. Vibrations are imparted to the pipe from the overall operation of the unit. These vibrations have the effect of reducing the caking of the carbon black on the pipe walls of the rubber pipe.

Table 14.4 gives an overview of pros and cons of vibratory feeders.

14.8 Slurry or Hydraulic Transport

Slurry or hydraulic transport uses a liquid medium to move the solids. In most cases the liquid is water but there have been applications with other liquids and even liquid carbon dioxide. The movement must employ a pump that has to be roughed enough to handle the two phase flow imposed by varying concentration of solids. Long distance conveying can be achieved in slurry transport with intermediate pumping stations which can have over 1000 kW of power required. This mode of transport competes with trucks, trains and barges to moving vast mineral resources. The Black Mesa slurry pipeline system is used to carry coal to a powder plant 273 miles away. One of the problems of long distance slurry transport is securing the right of way of the land over which it is constructed and operated. The availability of water at the source of the pipeline can be a challenge in arid climates. At the end of every slurry pipeline there needs to be a method to remove the liquid from the slurry. The finer the solids handled the more difficult it is to remove this liquid. Filtration is put to a severe test with these fine solids. Hydrocyclones are often used in the separation process as the first stage of separations. In handling of fine coal 50 % of solids or higher have been achieved. The higher the concentration of solids the less Newtonian is the behavior of the slurry. Many of the concepts used in pneumatic conveying are also applicable in slurry conveying especially with minimum conveying velocity and the concentration gradients across the cross-sectional area of the pipe.

14.9 Mixing and Segregation

Mixing and segregation go hand in hand in solids processing and handling (see also Chap. 9). The very operation of movement of particulate solids enhances the process of segregation or de-mixing. When there are size differences between the particles, one is faced with the proper way to deliver the material that has a homogenous composition at constant concentration. Considerable effort and money can be expended on trying to produce a uniform mixture in any of the various mixers that are available. Once a mixture standard is achieved in the mixer the material has to be transported to another operation and in a few minutes in the transfer process de-mixing occurs. The transport line and distances between these operations should be as short as possible. One curious operation that has been cited and argued as to the mechanics is the Brazil nut phenomenon where on vibrating a mixture of large and fine particles the large particles rise to the surface of the container giving segregation.

Sommer [83] has produced an excellent account of how to sample a mixture of particulates providing detailed statistical analysis to try to guarantee the true representation of the total vessel or container. One of the first researcher to address the mixing and sampling issues was Williams [99], who identified four mechanisms for the segregation of particles:

1. Trajectory – which is driven by Stokes law
2. Percolation – movement of fine powders in a vessel
3. Rise of coarse particles on with vibrations – Brazil nut phenomenon
4. Elutriation – fine particles remaining in suspension for long periods of time without settling.

In the mixing technology area mixing can occur by convection, diffusion and shear. Different mixers take advantage of these different phenomena. Miyanami [60] has noted as the mixing time increases the mechanisms of mixing goes from convective to convective and shear and then finally to diffusive mixing. He also compiled a listing of the various expressions that can be used to assess the degree of mixedness where a value of 0 is completely segregated and a value of 1.0 is complete mixedness.

Statistics of random binary mixtures is a way to measure the quality of a mixture in solids processing [57, 74]. The true standard deviation (σ) and the true variance (σ^2) of the composition of the (quasi-) two-component mixture are indicators of its quality. These true values are usually not known but can be estimated by s^2 defined as:

$$s_p^2 = \frac{\Sigma(y_i - \mu)^2}{N} \tag{14.2}$$

where

s_p^2 = estimated variance of p
p = proportion by number of one of the two components
y_i = measured composition of sample i
μ = true composition
N = number of analyzed samples

The true composition μ is generally unknown. Then, it should be replaced by the calculated mean composition $<y>$:

$$s_p^2 = \frac{\Sigma(y_i - <y>)^2}{N - 1} \tag{14.3}$$

An upper and a lower limit exist for the variance of (quasi-) two-component systems. The upper limit holds for a fully segregated system:

$$\text{Upper Limit} = \sigma_0^2 = p(1 - p) \tag{14.4}$$

The lower limit is also named fundamental error. It means a fully random mixture where the variance only depends upon the number of particles n involved:

$$\text{Lower Limit} = \sigma_R^2 = p(1 - p)/n \tag{14.5}$$

It has been found that the mixing index suggested by Poole et al. gives good discrimination for practical mixtures. It approaches unity for completely random mixtures.

$$\text{Poole mixing index} = s_p/\sigma_R \tag{14.6}$$

Kehlenbeck [40] studied continuous mixers to mix entering components radially and to smooth fluctuations of the entering mass flow in the axial direction. He found as Sommer predicted that the system could be modeled by the Fokker-Planck equations. He also found that flow fluctuation is the main influencing parameter of the mixing quality, independent of the mixer setting for the average residence time, normalized with the period length of the entering mass.

Relatively simple static mixers have been employed in gas-solid systems. These units are dependent on the feed and discharge rates as well are the internal mixer structure. Gyenis et al. [30] explored the gravity flow of particulate solids in a vertical motionless mixer to define various mixing regimes.

14.10 Pneumatic Conveying

Pneumatic conveying is one of the most commonly employed solids transport in industry. This is the reason that it is more thoroughly handled in this chapter.

Pneumatic conveying can handle a wide range of particle sizes from the submicrometre to 8 cm rocks. Obviously the designs will have to be different for these two materials. Because of the flexibility in design and handling a wide spectrum of materials of various densities can be employed. Pneumatic conveying has even been tested under reduced gravity conditions for considering building such systems on the Moon and other planets [63, 90].

The distances conveyed for these systems have generally been employed in transport distance less than 2500 ft although in constructing the Grand Coolie Dam cement was blasted in conveying like for several miles of distance. S.L. Soo [87, 89] did a theoretical study for the pneumatic conveying of hundreds of miles showing such a design is possible.

There is a wide variety of advantages and disadvantages for using pneumatic conveying. Sticky and cohesive materials generally are not recommended for such systems. Inserting more energy into a system by increasing the pressure and velocity of transport can usually get the solids to convey but not economically. Doing such increases is not advisable since energy will be wasted and often time the particles will end up as a fine dust which may not be what the customer wanted to purchase from you. See Table 14.5 for a general overview of pneumatic conveying capabilities.

There are five different components to a common pneumatic conveying system. One has to deliver the material into the conveying line; one must convey the material in a line; a transport gas must be inserted and the transported material must be collected at the end of the system. Finally there should be some sort of controls built into the system even though these may relatively simple in concept. For some specialized materials other components may be added and novel moving devices have been designed [42, 78]. Table 14.6 lists pros and cons of pneumatic conveying.

Pneumatic conveying has been around as a technology for at least over a hundred years although some of my oriental friends contend that the ancient Chinese used

Table 14.5 Overview of pneumatic conveying capabilities

Great range of materials can be conveyed
Powder
All less that 700 μm (24 mesh–0.028 in.)
Minimum of 25 % less than 75 μm (200 mesh–0.0029 in.)
Note: if 100 % less that 10 μm (1250 mesh–0.00039 in.) material may agglomerate
Granular
Uniform particle sizes
Minimum of fines
Free flowing
Almost all materials can be conveyed pneumatically with few technical limitations
Sponge iron balls 9.52 mm (3/8) diameter
Rock up to 76.2 mm (3 in.)
Explosives
Manufactured parts
Live chickens
Capacity
Small capacity
Difficult to control feed
High cost for small return
Small diameter conveying lines are less effective.
Large capacity – almost unlimited
Common up to 100 tons/h
Some approaching 500 tons/h
Distance
Short distance – less that 10 m (32 ft)
High cost for small return
Not economically competitive with mechanical systems
Long distance
Common conveying to 300 m (1000 ft)
Some systems approaching 2500 m (8000 ft)

Table 14.6 Advantages and disadvantages of pneumatic conveying

Advantages	Disadvantages
Lower initial cost	Requires higher horsepower
Clean, totally enclosed	Requires higher technology to operate
Easy to automate	Pipe erosion
Savings in bulk shipments	
Less maintenance	

the technology in their early society. Grain unloading from ships was used in Russia in the late 1800s.

The motive force to move the solids in pneumatic conveying can come from application of a pressure or by a vacuum. The vacuum systems have an overall limitation due to the maximum forces that can be achieved with these systems while the pressure systems can always seem to just up the power input. Sometimes it may be advantageous to use a hybrid system of both vacuum and pressure.

The dilute phase mode of transport is the most common in pneumatic conveying applications. Older designs mostly used excessive air hoping to eliminate operational problems. Such systems consumed too much energy and lead to pipe wear and particle erosion and degradation. Keeping the air flow at a minimum above the saltation velocity is the optimum point of operation to save energy.

Dense phase conveying systems were introduced commercially in the early 1970s and these designs were thought to be the answer to a maiden's prayer in the conveying world. Lower air velocities eliminated the need to be above the saltation velocity as well as cut down on the amount of particle breakage and pipe erosion. These objectives have been achieved but the operation of the dense phase system can be problematic since unstable flow situations may occur. There are many unique designs for dense phase systems devised by various manufacturers with a wide range of pricing [42].

Dense phase is an ill-defined word in itself and can represent a wide variation in the flow condition from dune type flows to moving piston flow [97]. These systems in general operate at higher pressures than the dilute phase flow systems.

14.10.1 Pickup and Saltation Design Parameters

In order to insure that the solids are transported over the whole piping system one must establish the proper and optimum velocity for transport. The parameters of pickup velocity and saltation velocity have been employed in this context. Figures 14.1 and 14.2 show the mechanism of these two phenomena. An interrelationship between the velocity has also been probed and is depicted in Fig. 14.3.

The hysteresis seen between the two velocities on increasing and decreasing is most noteworthy. The pickup velocity has two definitions; one is pickup at the feed point of the introduction of solids and the other is pickup from a deposited layer on the bottom of the pipe. The saltation velocity is the velocity at which particles start to fall out of the transport line and deposit on the bottom of the pipeline. One needs to be sure that the transport velocity is sufficient to pick up the solids and sufficient not to let deposition occur in transport. There is a consideration for the transport velocities for materials that are highly fluidizable and are transported in waves rather than be totally suspended. These materials require a lower transport velocity than expected. Rizk [75], Cabrejos and Klinzing [14], and more recently Rabinovich and Kalman [71] who developed a more general classification for pickup and saltation velocities correlations. The Rizk correlation which was developed for plastic pellets is given as:

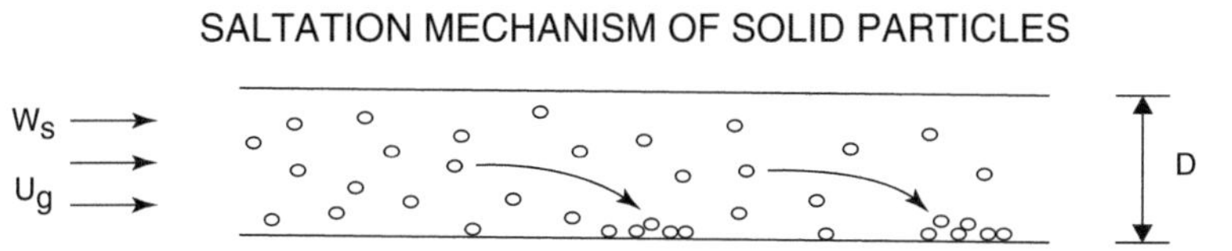

Fig. 14.1 Saltation velocity cartoon [14]

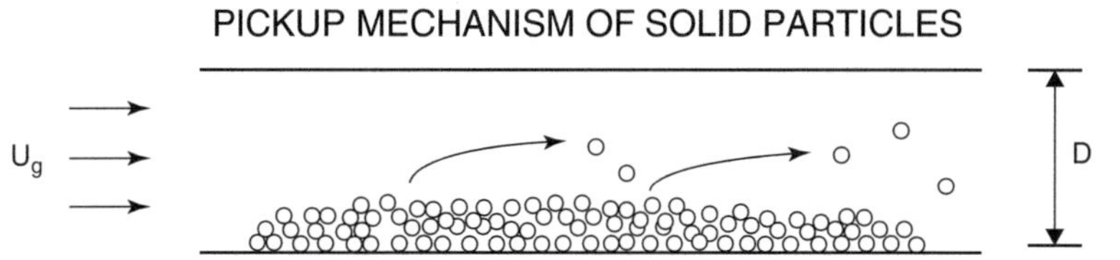

Fig. 14.2 Pickup velocity cartoon [14]

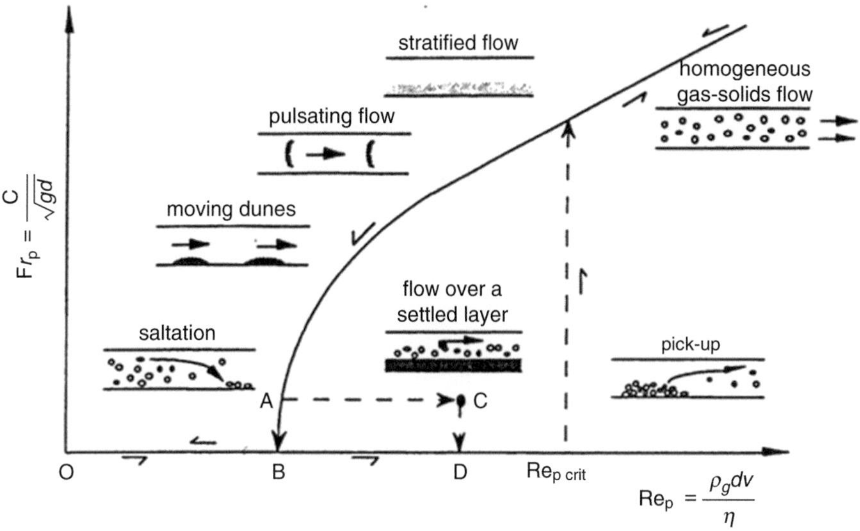

Fig. 14.3 Dimensionless diagram of Froude and Reynolds numbers at pickup and saltation [14]

$$\mu = \left(1/10^{\delta}\right)\left(u_p/\sqrt{gD}\right)^{\chi} \tag{14.7}$$

where:

μ = solids loading
$\delta = 1.44d_p + 1.96$
$\chi = 1.1d_p + 2.5$

The Cabrejos [12] analysis provides two equations; one for pickup and one for saltation:

$$\left(u_{gpu}/\sqrt{gd_p}\right) = 0.0428\left(Re_p\right)^{0.175}\left(D/d_p\right)^{0.25}\left(\rho_p/\rho_g\right)^{0.75} \tag{14.8}$$

and

$$\left(u_{gs}/\sqrt{gd_p}\right) = \left(u_{gs0}/\sqrt{gd_p}\right) + 0.00224\left(\rho_p/\rho_g\right)^{1.25}\mu^{0.5} \tag{14.9}$$

where:

u_{gpu} = pickup velocity of the gas
u_{gs} = saltation velocity of the solids
u_{gs0} = saltation velocity of a single particle
Re_p = Reynolds number based on the particle diameter
μ = loading = W_s/W_g = solids flow rate/gas flow rate
$1200 < \rho_p/\rho_g < 3200$
$\mu < 50$

Over the years industrial experience has developed some guidelines on the appropriate conveying velocities. Table 14.7 is a listing of these recommendations [82].

Figure 14.4 shows the pressure drop in a pneumatic conveying system plotted versus the average air conveying velocity at different solids flow rates. It is often called the Zenz plot since he was the first to use this depiction [104]. The figure is from the work of Rizk [75] at Karlsruhe University using polymer pellets. The line through the minimum pressure drop points is often called the division line between dilute phase to the right and dense phase to the left. Having a system convey at this minimum pressure drop point is considered the economical. As one moves to the left into the more dense phase regime certain instabilities arise such that the region is often called the unstable region, which can be problematic in carrying out steady state operations. It should be noted that this Zenz plot has an unique behavior for each individual material transported and for most reliable design the full plot should be available. In this way various regions of flow can be explored. In combining the motive air information with the Zenz plot one can see readily the ranges of operability of the system.

14.10.2 Dense Phase Conveying

In 1969 Muschelknautz and Krambrock [62] did a fundamental experimental study on different types of dense phase conveying from the stratified and wave flow to the plug flow situation. They developed two correlations for these two types of flow. For the stratified flow they present the pressure loss as:

$$\Delta p = (1 - A_c)L\rho_s g f_{sr} \tag{14.10}$$

Table 14.7 Pick-up velocities from industrial experience [82]

Minimum recommended velocities at the feed point for various materials			
Material	Pressure feed	Vacuum feed	Vacuum feed/rest
	m/s	m/s	m/s
Alumina	18.3	22.4	32
Calcium carbonate	9.8	23.4	33.5
Clay	18.3	22.4	33.5
Coffee beans	15.3	17.3	22.9
Corn grits	21.3	23.9	30.5
Corn shelled	16.8	20.8	32.0
Lime hydrated	12.2	16.3	27.5
Lime pebbles	21.3	27.0	32.0
Malt	16.8	20.3	30.5
Phosphate	22.9	26.0	33.5
PE pellets	21.3	22.4	24.4
Salt	25.4	28.5	36.6
Sand	33.0	34.6	38.1
Soda ash	19.8	23.4	33.6
Starch	16.8	19.8	27.5
Sugar	18.3	22.4	33.6
Wheat	16.8	20.9	32.0

Note: Pressure Feed is a positive pressure conveying system; Vacuum Feed is a vacuum conveying system; Vacuum Feed/Rest refers to a vacuum conveying system where the particle initial velocity is zero or at rest

where:

f_{sr} = friction factor for sliding and rolling of the particles (varies from 0.4 to 0.8 with an average value of 0.6)
A_c = cross sectional area of the wave layer.

For the plug flow case they present an exponential increase in pressure drop through the plug as the length of the plug increases:

$$\Delta p = exp(f_{sr}\mu\Delta L)/(RTu_p/u_g) \tag{14.11}$$

where the friction factor employ is the same as for the strand flow case and

R = gas law constant
T = absolute temperature

In his book Weber [96] developed similar expressions to calculate the pressure drop in dense phase flow. Again using a force balance on a plug he used solid mechanics to consider the stresses in the radial and axial direction and their ratio termed the stress transmission ratio or coefficient. He was able to solve for the conveying pressure of the plug. In addition he approached the plug flow where the

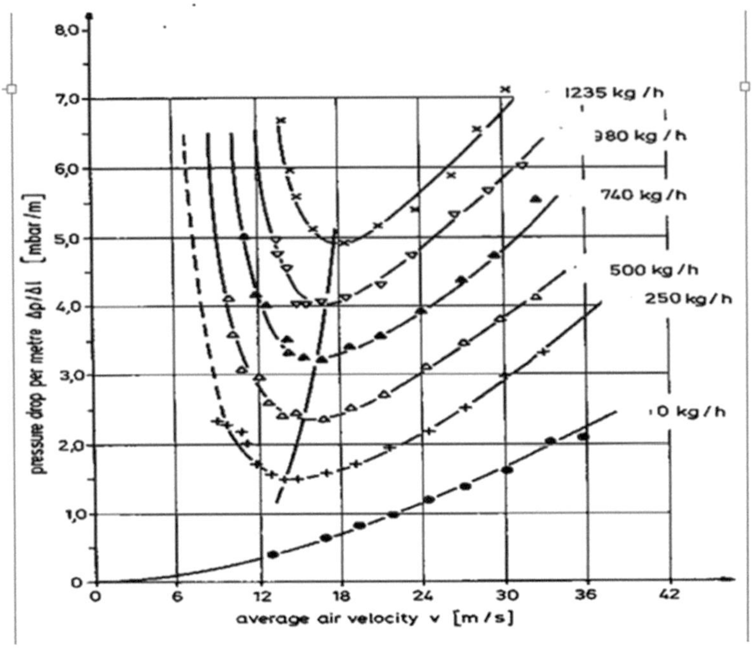

Fig. 14.4 Zenz diagram for polymer pellet conveying at varying solids flow rates [75] (Reproduced with permission of F. Rizk)

secondary air source was the motive force obtaining an exponential pressure increase with the length of the plug as Muschelknautz and Krambrock. Konrad et al. [44] again used the solid mechanics approach and Ergun's equation as a moving packed bed. The Konrad approach discussed the stress behavior in the plug using the concepts of passive and active failures employed in solid mechanics. The stress transmission coefficient was a key factor to the plug behavior. Since that time, several researchers have explored this plug regime [22, 36, 37, 42, 55]. This history is rich with information and approaches. More recently Sommer [49, 67] with his students Lecreps and Nied have probed the plug formation and structure in detail showing novel internal structures and processes.

In many of the studies on plug flow the frictional term between the material and the wall has been a crucial parameter. Several researchers have probed this friction and offered suggestions on its values and importance [47–49, 68, 95].

Dense phase conveying can be categorized in a number of ways. Injection of solids into the conveying air or injection of conveying air into the solids are two general techniques. The blow tank has become one common way for insuring dense phase conveying. Some materials mostly plastic pellets can form plugs naturally if they are handled and injected in the proper manner.

From the force balance analysis one finds that there will be an exponential rise in pressure across a plug which has a significant effect on the transport of long plugs in a pipeline:

$$p_1 = p_2 exp\left(\beta g \mu_R L / R' T u_p / u_g\right) \tag{14.12}$$

where:

$\beta = \sin\delta + \mu_R \cos\delta$
μ_R = coefficient of sliding friction
δ = angle of inclination of pipe
R' = gas law constant/*MW*

Tied up in this expression are assumptions about the particle-wall interactions, particle-particle interactions and the structure of the plug. Most of these parameters are unique to the type of material being conveyed and the pipe in which they are conveyed.

As mentioned before Wen and Simon [97] and Muschelknautz and Krambrock approached the wave like transport of solids as one of the techniques for dense phase conveying. More recently Pan and Wypych [69], Williams and Jones [100], [37], Behera et al. [6], and Setia et al. [80] also delved into this regime, studying mostly fly ash or similar materials that can be transported in a wave like format.

14.10.3 Plug Flow

A plug can break into fragments mostly on the topic of the horizontal flowing plug because the internal structure is weak or the plug is too short to support the applied pressure.

There is a variety of different methods to classify flows in pneumatic conveying. These graphical depictions are meant to help us understand and visual the process of conveying.

One of the common classifications of particles and powders in motion is the Geldart classification [29]. This classification using the density difference between the solid particles and the gas plotted against the diameter of the particle (Fig. 14.5).

The principal characteristics of handling of materials under Geldart classification can be seen in Table 14.8.

In applying the Geldart classification the boundary regions between the A, B, C, and D classes hold some challenges in trying to understand their behaviors in bulk solid handling and pneumatic conveying. Table 14.9 lists materials found to lie in these boundary zones.

Choking is a phenomenon that arises in vertical flow when the energy is not sufficient to sustain a steady flow of materials and thus chokes. Yang [103] has explored this region and developed a correlation:

$$2gD\left(\varepsilon_c^{-4.7} - 1\right) / \left(u_{gc} - u_t\right)^2 = 6.81 \times 10^5 \left(\rho_g / \rho_p\right)^{2.2} \tag{14.13}$$

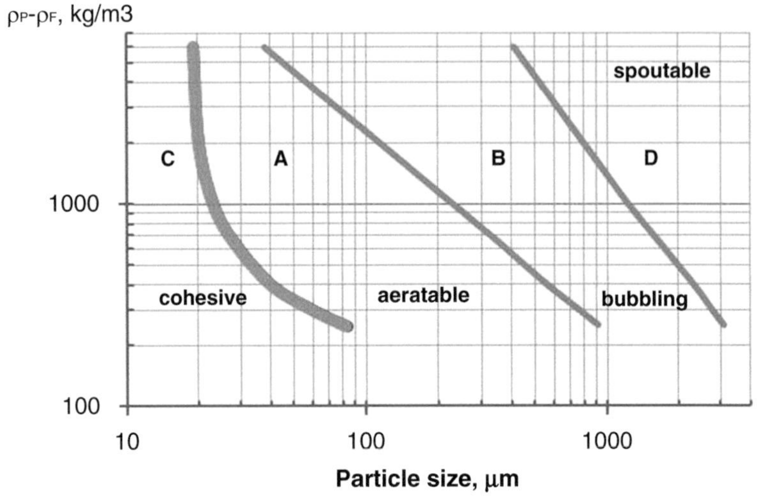

Fig. 14.5 Geldart fluidization diagram [58] (Copyright Springer; reproduced with permission)

and

$$W_s/A\rho_p = \left(u_{gc} - u_t\right)(1 - \varepsilon_c) \tag{14.14}$$

where

ε_c = voidage at choking
u_{gc} = gas velocity as choking
A = cross-sectional area of the pipe

14.10.4 Flow Measurements and Indicators

Measurements in pneumatic conveying or in solids handling both have many challenges. The devices that are common in single phase do not function well when gas and solids phases are present at the same time. Some devices such as transducers can be applied to obtain basic measurements in pneumatic conveying even though the conditions are more rugged than the single phase flow. Some clever observations of gas-solid flows can be explored to measure flow rates and conditions of flow. The finding of Gaesterstadt [28] showing that the pressure loss for dilute phase pneumatic transport could be represented by a linear increase in pressure drop was further expanded and refined by Cabrejos and Klinzing [13] to provide a convenient way to measure the solids flow rates. Use of pressure transducer signals which generate pressure fluctuations related closely to the condition of the flow all the way from dilute to dense phase has been developed to provide an index of the kind of flow

Table 14.8 Examples of materials in the Geldart classification and their handling and conveying properties [78].

Properties	Group A	Group B	Group C	Group D
Type of material	Powder	Coarse	Cohesive fine powder	Granular
Mean diameter, μm	20 to 50–100	50–100 to 500–1000	<20	>600–1000
Density, kg/m^3	1000–4000	1000–5000	>2000	<3000
Fluidization	Considerable bed expansion	Naturally occurring bubbles start at u_{mf}	Normal fluidization is very difficult	Naturally occurring bubbles start at u_{mf}
	Before u_{crit}	Bed expansion is small		Bed expansion is small
				Plugs form at large flow rates
Pressure drop at minimum fluidization, mbar/m	<30	>80	50–130	5–150
Permeability, m^2/bar-s	0.1	0.01–0.1 to 1	0.1–1	>1.0
De-aeration	Collapses slowly	Collapses rapidly	Collapses slowly -good air retention	Collapses very rapidly
Type of flow in a conventional system	Moving bed	Hard to convey in dense phase	Can convey in dense phase with difficulty	Possible candidate for plug flow
Examples	Fly ash, pulverized coal, flour, PVC powder, alumina, sugar	Sand, granulated sugar, alumina, semolina, PVC granules, mineral powder, glass beads	Cement, dolomite, pulverized coal, titanium dioxide, fly ash, alumina powder	Plastic pellets, polyethylene, wheat, glass beads, coarse sand, seeds

Table 14.9 Geldart boundary properties

A/C	A/B	B/D
Cement (14 μm)	Alumina (50–60 μm)	Alumina (435 μm)
Coal (20 μm)	Coal (146 μm)	Granulated sugar (720 μm)
Pulverized fly ash (20 μm)	Copper ore (55 μm)	Polyethylene powder (825 μm)
	Sugar (157 μm)	Pulverized fly ash (700 μm)
		Slate dust (500 μm)

patterns occurring in a pneumatic conveying system [20]. Signal analysis procedures used in electrical engineering, such as powder spectral analysis, fractal analysis, Hurst coefficient and wavelet analysis, have provided information on the finer details of the pressure fluctuations [42]. Using the concepts of electrostatic generated by moving charged particles and the interaction of charge particles and polarized light fields also provides basic information to develop pneumatic conveying measuring devices. The use of electromagnetic tomography has given further information into the structure of complex gas-solid flows by Dyakowski et al. [23] and Nied et al. [67]. New and exciting phenomena have been observed helping to develop more precise theories of transport. The use of sonic vibrations interacting with gas-solid flows is another technique to measure solids flows [17, 91, 92]. Using the principle of the interaction of polarized light with a magnetic field generated by electrostatics in pneumatic conveying Radar et al. [72] showed a flow meter could be developed from this concept.

14.10.5 Electrostatics in Pneumatic Conveying

In solids processing electrostatics can have dramatic effects which can cause the loss of property and lifes. The pneumatic conveying operation generates the most electrostatic effects in all of solids processing. Learning how to control or reduce these effects is most important. Often times when a particular behavior remains unexplained, we find reference to possible electrostatic effects. In pneumatic conveying it is imperative to ground the transport system and insure in time that this ground is maintained and the connections are not broken, rusted or painted over. The use of nonconductive piping, connectors or gaskets are prime points to the electrostatic generation. Conveying under the conditions of greater than 70 % relative humidity will reduce electrostatic effects as well as the addition of material to reduce electrostatic charge such as a tetra-ammonium salt. This material contaminates the product, so its application is limited. Insertion of a conducting wire inside the pipe or wrapped around the outside of the pipe has little effect in reducing electrostatic effects. Jones and King [38] have developed an excellent book on electrostatics discussing the details of generation and reduction.

Dust explosions which are usually triggered by the presence of dust in facilities that handle particles are particularly prone to occur. In order to have the right conditions for an explosion there needs to be a fuel, an oxidizing agent and an ignition source. Fine powders serve as the fuel, the oxygen in air serves as the oxidizing agent and an electrostatic spark serves as the ignition [2, 24, 50].

The minimum ignition energy (MIE) is a term that is often employed in electrostatics. This is termed the minimum energy to trigger a spark and is related to the capacitance of the material and minimum sparking potential. Having a value larger than 5 mJ is to enter a dangerous zone for ignition. The minimum ignition energy can be written as:

$$U_e = 0.5 C_{object} V_{max}^2 \approx 5\text{mJ} = \text{MIE} \tag{14.15}$$

where

C_{object} = capacitance of the object
V_{max} = maximum discharge voltage

When hydrocarbon vapors are present this MIE is lowered to less than 0.25 mJ. When one see sparks occurring in pneumatic transport, the breakdown voltage of the air is 30 kV/cm which is a large charge.

Once a material has been charged by conveying or other means it takes a considerable amount of time for that charge to dissipate. Charges that do not dissipate in less than 100 s are considered dangerous.

Some unusual flow patterns in pneumatic conveying can be observed because of the presence of electrostatic charging. Figure 14.6 show these flow patterns [19, 41]. He et al. [31] have recently looked at charged particulate flow with an unique dual-material probe. Some modeling of the electrification of particles in pneumatic conveying has been presented by Korevaar et al. [46].

Over the years many dust explosions have occurred. Note that all kinds of powders that are combustible may cause them. Examples are: flour, grain dust, powdered milk, starch powder, sugar dust, saw dust, coal dust, resin and metal dust. The first extensive record is a flour explosion in 1785 that destroyed the warehouse of a bakery in Turin, Italy. Another example is the explosion in 1878 of a flour mill and its adjacent buildings in Minnesota, USA. In 1977 a grain elevator in Westwego, USA exploded. In 2008 a dust explosion destroyed a sugar refinery in Port Wentworth, USA. And in 2011 aluminum dust in a Chinese factory in Chengdu was the cause of an explosion. As a result, dust explosions in last 25 years resulted in several hundreds of casualties and over 100 million dollars financial damage, despite all kinds of measures and guidelines for safe operation [50]. Figure 14.7 of ruined sugar refinery facilities shows how devastating dust explosions can be.

14.10.6 Simulations

More and more simulation work has taken place in gas-solid flows. The agreement between the experimental evidence and the numerical simulation is becoming more exact. Realistic phenomena for both the solid and gas phase behaviors have been incorporated into the models. Both continuous and discrete representations have had success in their approaches. Tsuji [94, Sommerfeld and Zivkovic [85], Sommerfeld [86], Bolio et al. [9], Levy and Mason [51], and Yu and his team [52, 53] have carried out realistic modeling of pneumatic conveying.

Tsuji [94] has given an excellent summary of computational activities related to solids processing including pneumatic conveying. He noted that these computations can be separated into four groups:

1. Gas-particle multiphase flow [87]
2. Collision-dominated flow
3. Direct Simulation Monte Carlo method (DSMC) and contact-dominated flow
4. The Discrete Element Method (DEM).

In particle-gas multiphase flow there is an interchange of mass, momentum and energy between the gas and the particles. Dilute flows have only particle gas interactions while dense phase flows also have particle-particle interactions. In

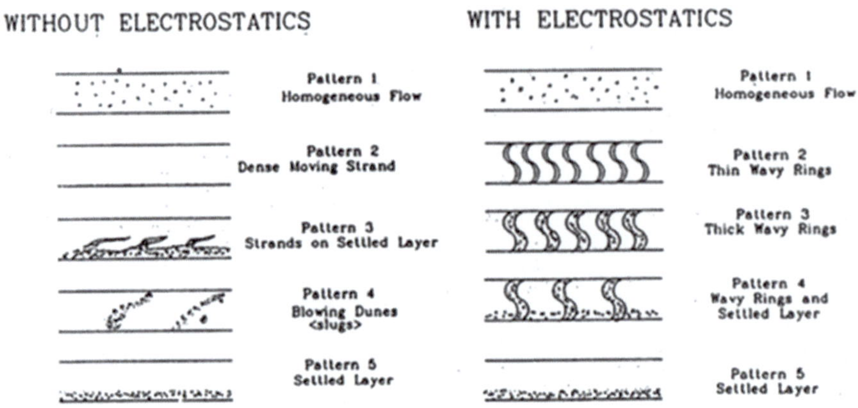

Fig. 14.6 Flow patterns for PVC with electrostatic charging present [19]

Fig. 14.7 Damaged sugar refinery in Port Wentworth by a dust explosion in 2008 (www.foodprocessing.com.au)

the particle-gas multiphase flows there is an interchange of mass, momentum and energy between the gas and the particles.

For the Discrete Element Method (DEM) collisions dominate the flow and can have elastic and inelastic collisions relevant to dense phase flow. Using the micro scale for individual particles, the meso scale for the local region and the macro scale for the system, Tsuji considered particle motion from the trajectory level, the continuum model and the one and multidimensional viewpoints. The fluid motion in the same sequence of scales considers the flow around a single particle, local averages and one and two dimensional flows. The DEM collisions are dominant with either inelastic or elastic collision.

Theuerkauf et al. [93] have given a good summary of modeling granular flows using the discrete element method (DEM). While they concentrate on bins, hoppers, feeders and mixers, the basic techniques have been employed in other fields including pneumatic conveying. They project that DEM will be an integral part of the modeling toolkit for designing solids processing systems in the next 10 years. They break down the approaches into the soft-sphere and hard-sphere models as suggested above by Tsuji. There are several important concepts in using the soft sphere model which begins with Newton's law of motion including translational and rotational forces. The hard-sphere approach is event-drive. In this approach the particle collisions are assumed to be binary and instantaneous. The calculations are based on impulse and momentum transfer having the energy dissipation accounted for by the coefficient of restitution.

There have been a number of attempts to simulate dense phase conveying using modern computational techniques. McGlinchey et al. [56] simulated dense phase flow using the combined computational fluid dynamics and discrete element method in an effort to predict the contact forces. They showed that the particles forming the shell of a plug experience the highest stress intensities in terms of contact pressure and relative velocity. The dominant stress mode was found to be the sliding friction.

For dense phase pneumatic conveying, collision dominated flows are considered for intermediate solid concentrations. Contact dominated flows consider the forces of drag and gravity acting on the particle and interacting with the fluid velocity to produce a respondent particle velocity. Both the hard and soft sphere models can be applied with the damping vibration analysis. These procedures can be applied to dense phase conveying.

Sommerfeld and colleagues [85] have examined particle collisions between the particles themselves and with walls to begin to numerically address realistic representations of gas-solid flows. The simulations studied by Sommerfeld [84] showed good agreement with their experiments on particle-wall collisions. They noted that when the particle size was the same order in size as the roughness of the pipe, the collision process became stochastic in nature. Using pulsed-laser imaging Shaffer and Ramer [81] explored particle-wall collisions. A flat plate with varying angles of impingement were tested. Louge et al. [54] simulated the role of particle-particle collisions of rather larger particles in pneumatic conveying. They calculated and measured (by Tsuji) fully developed velocity profiles. Close agreement is observed between Tsuji's data and their model. Sommerfeld [86] has also simulated particle-wall collisions which showed good agreement with experimental data.

14.10.7 Basic Physics

We look first at the basic physics and then proceed to modify and adapt our analyses to incorporate observations. The remainder of this chapter will focus in on pneumatic conveying since this is one of the most commonly employed solids transport that we find in industry.

A force balance can be written on the pipe or pipe section utilizing the various forces that act on the particles and gas in transport. The balance of forces is obtained by summing up the drag and external forces in the longitudinal direction and noting the gravitational force in the vertical direction.

The force balance with its three forces is written in a differential format. The drag force uses a drag coefficient which will vary depending on the fluid and particle velocities. The electrostatic term, while very important, in some cases is hard to measure and quantify.

The general drag coefficient expressions, C_D, has three regimes.

$$\begin{array}{ll} C_D = 24/\mathrm{Re}_p & \mathrm{Re}_p < 2.0\,\text{Stokes} \\ C_D = 18.5/\mathrm{Re}_p^{0.6} & 0.5 < \mathrm{Re}_p < 500 \ \text{Intermediate} \\ C_D = 0.44 & 500 < \mathrm{Re}_p < 2\times 10^3 \ \text{Newton} \end{array} \tag{14.16}$$

Kaskas [39] has been able to develop one equation for the drag coefficient that would serve for all the three flow regimes experienced by the particle. The first term is essentially the laminar flow or Stokes regime while the second term brings in the intermediate range and as the velocity increases the drag coefficient approaches a constant.

$$C_D = \frac{24}{d_p \rho_f (u_f - u_p)} + \frac{4\mu^{*1/2}}{d_p^{0.5} \rho_f^{0.5} (u_f - u_p)^{0.5}} + 0.4 \tag{14.17}$$

where

d_p = particle size
ρ_f = fluid density
u_f = velocity of the fluid
u_p = velocity of the particle
μ^* = viscosity of the fluid

In the turbulent flow regime Pettyjohn and Christiansen [70] looked closely at the drag coefficient and suggest that the sphericity of the particle is a dominant factor for the drag coefficient. The sphericity is the degree to which a particle deviates from a true spherical shape. For a value of the sphericity of 1.0 the particle is a perfect sphere.

$$C_D = 5.31 - 4.38\psi \tag{14.18}$$

where:

ψ = sphericity of the particle; surface area of a sphere with the same volume as the particle divided by the surface area of the particle.

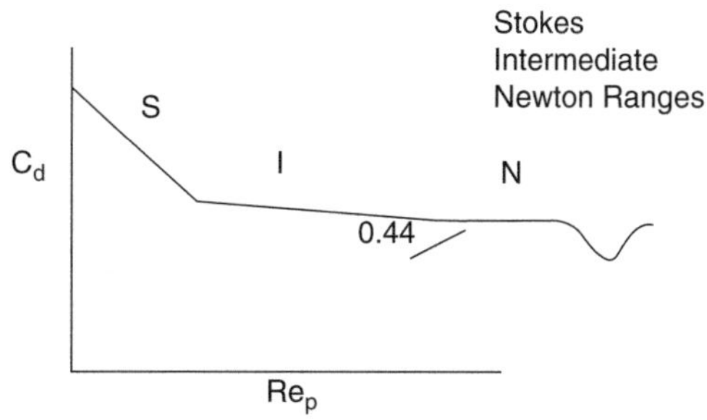

Fig. 14.8 Drag coefficient behavior

The drag coefficient behavior can be graphically depicted in Fig. 14.8.

The force balance can be further amplified and broken down in individual equations for the solid and gas phases including frictional forces, pressure and electrostatics. These equations can be summed to obtain the final working equation for the pressure drop:

$$-\frac{\Delta p}{L} = g\rho_f\varepsilon + g\rho_p(1-\varepsilon) + 2f_s\frac{}{(1-\varepsilon)D}(u_f - u_p)^2 + \frac{2f_g\rho_f\varepsilon u_f^2}{D}$$
$$\varepsilon = voidage = 1 - \frac{4W_s}{\pi D^2 u_p\rho_p} \tag{14.19}$$

where

$\Delta p/L$ = pressure drop per unit length of pipe
g = gravitational constant
ε = voidage
ρ_p = particle density
D = pipe diameter
f_g = gas viscosity
f_s = solids friction factor
W_s = solids flow rate

One should note the term voidage in this equation. The voidage is the degree of filling the space with voids in a two phase gas-solid flow system. The closer the voidage is to 1.0 the closer to a single phase flow one experiences. Packed bed arrangements are low values for the voidage and the more the flow approaches dense phase conveying. The expression given for the voidage depends on the solids flow rate and the particle velocity. It should also be noted that the solids friction factors term is often expressed only in terms of the solid velocity rather than a slip velocity between the gas and the solid as seen in Eq. 14.19.

Determining the voidage of a system requires an estimate of the particle velocity if it is not measured directly. Very few systems have the ability to measure the particle velocity outside the research laboratory. In analyzing the voidage it becomes clear that there can be various kinds of velocities that can be discussed and calculated. The superficial gas velocity (u_g) is the velocity of the gas in an empty pipe with no particles present. The actual gas velocity (u_f) is the velocity in the presence of particles which is higher than the superficial gas velocity since the voidage is less than 1.0. The slip velocity ($u_f - u_p$) is the difference between the actual gas velocity and the particle velocity. One often notes that for particles less than 40 μm the particle velocity can be estimated by the difference between the superficial gas velocity and the terminal velocity of the particle. This expression can be easily calculated.

Other techniques are available in order to estimate the particle velocity in a pneumatic conveying system. Hinkle [32] studied pneumatic conveying in the 1950s and developed an empirical expression for the particle velocity. The expression (14.20) is valid for calculations with the English units. Ideally these expressions should be dimensionless in nature and not depend on units. One has to be careful in the literature to note the author's definitions of velocity, voidage and pressure loss and use the correction units in all expressions. In their extensive work on coal processing IGT [33] developed expression (14.21) for the particle velocity.

Hinkle expression (in fps-units):

$$u_p = u_g\left(1 - 0.179 d_p^{0.3} \rho_p^{0.50}\right) \tag{14.20}$$

where:

u_g = superficial gas velocity

IGT modified Hinkle equation (in SI units):

$$u_p = u_g\left(1 - 0.68\, d_p^{0.92} \rho_p^{0.5} \rho_f^{-0.2} D^{0.54}\right) \tag{14.21}$$

where:

D = pipe diameter

The parameter of particle size and distribution can vary depending on the mode of measurement whether using a mass, length, area or volume techniques. It starts with the simple sieving operation based on mass properties and goes to optical techniques to determine the size and size distribution, based on number or volume. In reviewing the literature care must be taken to understand what size and distribution basis is employed in the analysis. In the presentation given here the mass (or volume) average is employed. Some analysis could employ a size distribution taking contributions from each fraction and summing to obtain the final result.

Yang [102] through his large amount of work in fluidization and pneumatic conveying developed an equation for the particle velocity that is implicit in nature which requires an iterative technique for solution. Note the particle velocity exists on both sides of the equation in different format. Another important and crucial point in dealing with the pneumatic conveying literature is to understand the definition of the solids friction factor employed by the author.

$$u_p = u_g - u_t\sqrt{\left(1 + 2f_s u_p^2/gD\right)\varepsilon^{4.7}} \tag{14.22}$$

where:

f_s = solid friction factor
u_t = terminal velocity of the particle
$f_p = 4f_s$

There is no set standard and thus confusion can arise. The friction factor term (f_p) is four times the friction factor noted as (f_s). The former one is employed in the mechanical engineering literature and the latter comes from the chemical engineering literature. Friction factors for dilute phase flow have been found to vary depending on the investigator going from a constant value of 0.003 to various expressions developed by different researchers [42]. We have investigated the effect of the angle of inclination of the pipeline effect on the particle velocity. Figure 14.9 shows that the pressure drop is largest when the angle of inclination is about 75° because recycling begins to occur especially in the dense phase transport system [43, 82].

The important parameter of friction factor for the single phase gas flow and the friction factor for the solid flow are very important in the design of a pneumatic conveying system. For the single phase flow we are using the Koo et al. [45] Eq. 14.23 which yields the same as the Prandtl correlation developed in the early 1900s. The Koo equation is an explicit format in the friction factor making it easier to use.

Koo equation:

$$f_g = 0.0014 + 0.125 Re_g^{-0.32} \tag{14.23}$$

where

Re_g = Reynolds number of the gas based on the pipe diameter

Solid friction expression:

$$f_s = \frac{\Delta p}{L}\frac{D}{(1-\varepsilon)\rho_p u_p^2} \tag{14.24}$$

As is obvious horizontal and vertical flows are not always possible in pneumatic conveying, the flow must traverse bends. Usually two contributions are assumed with a linear effect, the gas only energy loss across a bend and a solids flow contribution. Ito [34, 35] carried out one of the most comprehensive studies on gas only energy loss. For the contribution to the pressure drop due to the solids presence Schuchart [79] developed the following equation for the pressure loss due to only the solids utilizing plastic pellets:

Fig. 14.9 Pressure drop versus angle of inclination

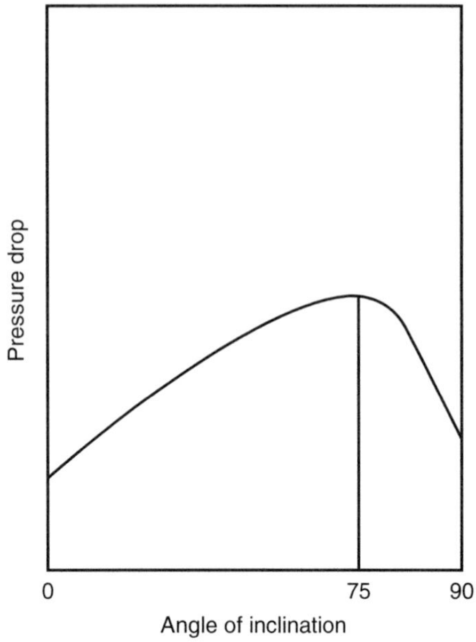

$$\left(\Delta p_{bend}/\Delta p_{straight}\right) = 210(2R_B/D)^{-1.15} \tag{14.25}$$

where:

$\Delta p_{bend} = \Delta p_{bend\text{-}gas} + \Delta p_{bend\text{-}solids}$
R_B = radius of the bend

A comprehensive analysis of pressure losses in bends can be seen in Dhodapkar et al. [21].

Closely related to the bend pressure loss is the pressure loss in the acceleration region at the beginning point of introducing solids into the piping system. A correlation of an extensive data base provide information to present the distance needed to establish steady state in pneumatic transport [41]. This equation is given as:

$$L_{accel}/D = 27.66\left(d_p/D\right)^{0.953}.(\mu)^{-0.0912}.\left(\rho_f/\rho_p\right)^{-0.924} \tag{14.26}$$

where

L_{accel} = acceleration length of the solids
μ = particle loading = solids flow rate/gas flow rate

A quick estimate for the pressure drop in the acceleration region can be obtained from:

$$\Delta p_{acceleration} = \mu u_g \rho_f u_p \tag{14.27}$$

14.10.8 Solid Handling Tips [5]

Unfortunately there is not one single parameter that can characterize bulk materials. As mentioned particle size and distribution as well as bulk density and porosity are obvious parameters. One should also consider the moisture content, quality of the material, health issues in usage as well as hazards posed. The wall friction and shear strength come into play for bins and hoppers and other situations such as dense phase conveying where intimate contacts with the wall exist.

A number of statements has appeared in the literature as well as being transferred in casual conversations, which are good pointers for pneumatic conveying systems:

1. Increasing the air flow will not increase the capacity. Increasing the air flow is in effect decreasing the concentration of the transport particles to an even more dilute state without affecting capacity.
2. Pickup and saltation velocities are not dependent on the bulk properties. The individual particle properties are the controlling parameter for pickup and saltation since all correlation have been found to be dependent on these individual particle properties.
3. Mass flow hoppers do not guarantee plug flow.
4. Larger residence times in a batch mixer do not always produce better mixing. The optimum residence time in a batch mixer is finite, longer times cause no further improvements.
5. Different particle size measuring devices may yield different results. The different principles upon which measuring devices are based cause this. One difference is the size of the equivalent spheres according to the measurement principle, e.g. laser diffraction, microscopy and sedimentation; another is the way for quantification, such as based on number, weight, volume, and charge for example.
6. Draining a charge from an insulating materials does not occur instantaneously. In reality charge can be retained in powders sometimes for days.

Another important point to remember in powder processing technology is that a scale-up of the operation will be required. The operation of these units is dependent on the consistency of the raw material. Besides scale-up, also running at reduced capacity should be addressed as well as process upsets. The overall process must also respond well to control strategies.

To have a reliable flow from a hopper one must insure that the discharge will be consistent over wide changes in conditions and that the units will function well in shut down operations and long production runs. For proper design a representative sample of the material for testing is needed. Care should be taken when handling free flowing materials, low permeable materials. Be vigilant that now flow obstructions occur in the hopper possibly from the installation operation.

To reduce flushing and de-aeration in bins and hoppers one should avoid funnel flow in bins and hoppers, use tangential entry of material to the bin and avoid uncontrolled air injection as a flow aid.

It should be recognized that free flowing particles tend to segregate if they have any physical difference such as size. Transfer operations enhance the opportunity for segregation. If one can remove fines from the operations, segregation issues are reduced. It has been recognized that segregation occurs at the end of any processing operation.

In order to minimize attrition sliding, impacts, dilations and shear flows should be a minimum. Dense phase pneumatic conveying helps to reduce attrition of particles.

Outlet stresses in bins and hoppers influence the feeder choice for the system and can be a factor in the arching phenomenon. To eliminate jamming or structure blockages in hoppers one should be careful the ratio of the particle size to the orifice diameter. Note that jamming increases with particle size increases. Particle jamming is more prevalent when the size distributions of particles are wide. Application of vibrations can sometimes help to eliminate jamming but the application of the vibrations has not thoroughly been studied.

Be careful to realize that cyclones can increase attrition rates of the particles and also be a factor in erosive wear. Cyclones design must consider overall efficiency, pressure drop and sharpness of the cut. Screw feeders can cause discharging problems therefore the screw feeder and the hopper should be considered as one unit in the overall design. Because of their motion screw feeders can cause particle degradation. One should not rely on a screw feeder to constrain a fluidizible powder. Also one should be careful with fibrous materials and screw feeders.

14.11 Definitions, Abbreviations and Symbols

Equivalent sphere	sphere that has the same property as the observed particle in relation to a given measurement principle (this makes the size and size distribution of non-spherical particles dependent on measurement principles)
Particle size	diameter of an equivalent sphere
Pickup	the velocity at which the particles are lifted from the bottom of the pipe.
Saltation	deposition of particles on the bottom of the pipe from the flowing stream of particulate matter.
Segregation	separation of particles in a mixture (e.g. powder) according to differences in size, shape and/or density
Sphericity	surface area of a sphere with the same volume as the particle divided by the surface area of the particle
Voidage	degree of filling of the space with voids in a two phase gas-solid flow system

CEMA	Conveyor Equipment Manufacturers Association

A	area
A_c	cross sectional area of the wave layer in stratified flow
C_D	drag coefficient

C_{object}	capacitance of the object in question for charging
d_p	particle size
D	pipe diameter
f_s	solids friction factor
f_p	solids friction factor ($f_p = 4f_s$)
f_{sp}	friction coefficient for sliding and rolling – wave flow
g	gravitation constant
L	length
L_{accel}	acceleration length
N	number of analyzed samples
p	proportion by number of one of the components of a (quasi-) two-component mixture
R'	gas law constant
Re_p	particle Reynolds number based on particle size
Re_g	superficial gas velocity Reynolds number
R_B	radius of curvature of a bend
s_p	estimated standard deviation of p
U_e	minimum ignition energy
u_{gs}	saltation velocity of the particles
u_{gso}	saltation velocity of a single particle
u_g	superficial gas velocity
u_{gpu}	pickup velocity of the gas for the particles
u_f	fluid velocity
u_p	particle velocity
u_t	terminal velocity of the particle
V_{max}	maximum breakdown voltage
W_s	solids flow rate
W_g	gas flow rate
y_i	measured composition of sample i
$\langle y \rangle$	measured mean composition of the mixture

β	frictional coefficient for a plug
ε	voidage
ε_C	voidage at choking condition
μ^*	viscosity of the gas
μ	solids loading
μ	true composition of a two-component mixture
μ_R	coefficient of sliding friction
ρ_B	bulk density of the solid
ρ_f	fluid density
ρ_g	gas density
ρ_p	particle density
σ	standard deviation of the number-based composition of a (quasi-) two-component mixture
σ_R	standard deviation of the number-based composition of a randomly mixed (quasi-) two-component mixture

References

1. Ariza-Zafra, K., Berry, R.J., Bradley, M.S.A.: Review of models for predicting the discharge rates of bulk particulates from silos and bins. In: Proceedings RELPOWFLOW IV. (2008)
2. Bartknecht, W.: Dust Explosions. Springer, Berlin (1989)
3. Bates, L.: Guide to the Design, Selection and Application of Screw Feeders. British Materials Handling Board. Professional Engineering Publishing, London (2000)
4. Bates, L.: The Discharge Rate and Condition of Bulk Material from Storage. Ajax Equipment Ltd, Bolton (2014) (personal communication)
5. Bates, L., Dhodapkar, S.V., Klinzing, G.E.: Know-Floe's Korner. Part. Technol. Forum, A.I. Ch.E. Newsletters. (2003–2014)
6. Behera, N., Agarwal, V.K., Jones, M.G.: Modeling and analysis for fluidized dense phase conveying including particle size distribution. Powder Technol. **235**, 386–394 (2013)
7. Benjamin, C.W.: Bulk-Online. http://forum.bulk-online.com (2013)
8. Beverloo, W.A., Leninger, H.A., Van de Velde, J.: Chem. Eng. Sci. **15**, 260–269 (1961)
9. Bolio, E.J., Yasuna, J.A., Sinclair, J.L.: Dilute turbulent gas-solid flow in risers with particle-particle interactions. A.I.C.h.E. **41**, 1375–1388 (1995)
10. Brown, R.L., Richards, J.C.: Trans. Inst. Chem. Eng. **37**, 108 (1959)
11. Brown, R.L., Richards, J.C.: Trans. Inst. Chem. Eng. **38**, 243 (1960)
12. Cabrejos, F.J.: Experimental investigation on the fully developed pipe flow of dilute gas-solids suspension. Ph.D. dissertation. University of Pittsburgh (1994)
13. Cabrejos, F.J., Klinzing, G.E.: Solids Mass Flow Rate Measurements in Pneumatic Conveying, ASME 92-WA/FE-3 (1992a)
14. Cabrejos, F.J., Klinzing, G.E.: Incipient motion of solid particles in horizontal pneumatic conveying. Powder Technol. **72**, 51–61 (1992)
15. CEMA Belt Book Chapter 6, Belt Tension, Powder, and Drive Engineering, 6th edn. Conveyor Equipment Manufacturers Association. Naples, Florida (2005)
16. Colijn, H.: Mechanical Conveyors for Bulk Solids. Elsevier, Amsterdam (1985)
17. Davies, C.E., Tallon, S.: A pressure pulse method for monitoring flow rate in dense down flows of Geldart group B particles. Powder Technol. **148**, 102–105 (2004)
18. Deming, W.E., Mehring, A.L.: The gravitational flow of fertilizers and other comminuted solids. Ind. Eng. Chem. **21**(7), 661–665 (1929)
19. Dhodapkar, S.V.: Flow Pattern classification in gas-solid suspensions, Ph.D. dissertation. University of Pittsburgh (1991)
20. Dhodapkar, S.V., Klinzing, G.E.: Pressure fluctuations in pneumatic conveying systems. Powder Technol. **74**, 179–195 (1994)
21. Dhodapkar, S.V., Solt, P., Klinzing, G.E.: Understanding bends in pneumatic conveying systems. Chem. Eng. April, 53–58 (2009)
22. Dixon, G.: Chapter "Pneumatic conveying". In: Butters, G. (ed.) Plastics Pneumatic Conveying and Bulk Storage. Applied Sciences Publisher, Reading (1981)
23. Dyakowski, T., Luke, S.D., Ostrowski, K.L., Williams, R.A.: On-line monitoring of dense phase flow using real time dielectric imaging. Powder Technol. **104**, 285–295 (1991)
24. Eckhoff, R.K.: Dust Explosions in the Process Industries. Gulf Professional Publishing (Elsevier), Oxford (2003)
25. Fayeed, M.E., Skocir, T.S.: Mechanical Conveyors – Selection and Operation. Technomic Publishing Co., Lancaster (1997)
26. Flowler, R.T., Glastonbury, J.R.: The flow of granular solids through orifices. Chem. Eng. Sci. **10**, 150–156 (1959)
27. Franklin, F.C., Johanson, L.N.: Flow of granular materials through a circular orifice. Chem. Eng. Sci. **4**, 119–121 (1955)
28. Gaesterstadt, J.: Die experimentelle Untersuchung des pneumatischen Fordervorganges. VDI Z. **68**(24), 617–624 (1924)
29. Geldart, D.: Types of gas fluidization. Powder Technol. **7**, 285–292 (1973)

30. Gyenis, J., Arva, J., Nemeth, L.: Steady state particle flow in mixer tubes equipped with motionless mixer elements. In: Tatterson, G.B., Calabrese, R.V., Penney, W.R. (eds.) Industrial Mixing Technology, AIChE Symposium Series 90, pp. 144–156 (1994)
31. He, C., Bi, X.T., Grace, J.R.: Contact electrification of novel dual-material probe with charged particulate flow. Powder Technol. **253**, 1–9 (2014)
32. Hinkle, B.L.: Acceleration of particles and pressure drops encountered in horizontal pneumatic conveying, Ph.D. thesis. Georgia Institute of Technology, Atlanta (1953)
33. IGT, Institute of Gas Technology, Department of Energy Contract FE 228632 (1978)
34. Ito, H.: Friction factors for turbulent flow in curved pipes. Trans. ASME J. Basic Eng. **81D**, 123–134 (1959)
35. Ito, H.: Pressure losses in smooth pipe bends. Trans. ASME J Basic Eng. **82D**, 131 (1960)
36. Jones, M.G., Williams, K.C.: Solids friction factors for fluidized dense phase conveying. Part. Sci. Technol. **21**, 45–56 (2003)
37. Jones, M.G., Williams, K.C.: Secondary air in pneumatic conveying, San Francisco Meeting. (2013)
38. Jones, T.J., King, J.L.: Powder Handling and Electrostatics: Understanding and Preventing Hazards. Lewis Publishing, Chelsea (1992)
39. Kaskas, A.A.: Schwarmgeschwindigkeit in Mehrkornsuspensionen am Beispiel der Sedimentation, D. Diss, TU Berlin. (1970)
40. Kehlenbeck, V.: Continuous dynamic mixing of cohesive powers, D. Diss. TU München. (2006)
41. Klinzing, G.E., Dhodapkar, S.V.: Acceleration length correlation of data, unpublished (1991)
42. Klinzing, G.E., Rizk, F., Marcus, R.D., Leung, L.S.: Pneumatic Conveying of Solids: A Theoretical and Practical Approach, 3rd edn. Springer, Dordrecht (2010)
43. Klinzing, G.E., Rohatgi, N.D., Myler, C.A., Dhodapkar, S.V., Zaltash, A.R., Mathur, M.P.: Pneumatic transport of solids in an inclined geometry. Can. J. Chem. Eng. **67**, 237–244 (1989)
44. Konrad, K., Harrison, D., Nedderman, R.M., Davidson, J.F.: Prediction of the pressure drop for horizontal dense phase pneumatic conveying of particles. Pneumotransport 5, Paper E1 (1980)
45. Koo, E.C., Drew, T.B., McAdam, W.H.: The friction factors for clean round pipes. Trans. A.I. C.h.E. **28**, 56 (1932)
46. Korevaar, M.W., Padding, J.T., Van de Hoef, M.A., Kuipers, J.A.M.: Integrated DEM-CFD modeling of contact charging of pneumatically conveyed powders. Powder Technol. **258**, 144–156 (2014)
47. Krull, T., Jones, M.G.: Stress-filed modeling and pressure drop prediction for slug-flow pneumatic conveying in an aerated radial stress chamber. Part. Sci. Technol. **22**, 129–139 (2004)
48. Krull, T., Jones, M.G., Roberts, A.W., Wypych, P.W.: Measurement of the stress transmission coefficient of material slugs in an aerated radial stress chamber. Part. Sci. Technol. **21**, 327–340 (2003)
49. Lecreps, I., Orozovic, O., Erden, T., Jones, M.G., Sommer, K.: Physical mechanisms involved in slug transport and pipe blockage during horizontal pneumatic conveying. Powder Technol. **253**, 710–721 (2014)
50. Lemkowitz, S.M., Pasman, H.J.: Assessment and control of fire and explosion hazards and risks of particulates. In: Merkus, H.G., Meesters, G.M.H. (eds.) Particulate Products – Tailoring Properties for Optimal Performance. Springer, Cham (2014)
51. Levy, A., Mason, D.J.: Two – layer model for non-suspension gas-solids flow in pipes. Powder Technol. **112**, 256–262 (2000)
52. Li, K., Kuang, S.B., Pan, R.H., Yu, A.B.: Numerical study of horizontal pneumatic conveying, effect of material properties. Powder Technol. **251**, 15–24 (2014)
53. Lim, E.W.X., Wang, C.-H., Yu, A.-B.: Discrete element simulation for pneumatic conveying of granular material. Part. Technol. Fluidization **52**, 496–509 (2006)

54. Louge, M.Y., Mastorakos, E., Jenkins, J.T.: The role of particle collisions in pneumatic transport. J. Fluid Mech. **231**, 345–349 (1991)
55. Mainwaring, N.J., Reed, A.R.: An appraisal for Dixon's slugging diagram for assessing the dense phase transport potential of bulk solids materials. In: Pneumatech 3, pp.221–234 (1987)
56. McGlinchey, D., Xiang, J., Cowell, A., Frye, L., Peukert, W.: The prediction of contact forces during dense phase pneumatic conveying. Bulk Solids Powder Sci. Technol. **1**, 44–50 (2004)
57. Merkus, H.G.: Particle Size Measurements – Fundamentals, Practice, Quality. Springer, Dordrecht (2009)
58. Merkus, H.G., Meesters, G.M.H. (eds.): Particulate Products – Tailoring Properties for Optimal Performance. Springer, Cham (2014)
59. Merrow, E.W.: Estimating startup times for solids-processing plants. Chem. Eng. **89**, 89–92 (1988)
60. Miyanami, K.: Chapter 6 Mixing. In: Gotoh, K., Masuda, H., Higashitani, K. (eds.) Powder Technology Handbook, 2nd edn. Marcel Dekker, New York (1997)
61. Molerus, O. (ed.): Principles of Flow in Disperse Systems. Chapman and Hall, London (1993)
62. Muschelknautz, E., Krambrock, W.: Vereinfachte Berechnung horizontaler pneumatischer Forderleitungen bei hoher Butbeladung mit feinkornigen Produkten. Chem. Ing. Techn. **41**, 1164–1172 (1969)
63. Muschelknautz, E.: Untersuchung des Einflusses der Reibungsverluste auf die pneumatische Förderung unter den Bedingungen der Schwerelosigkeit, Contract No. QV543. (1984)
64. Nanetoe, S., Supongonpipat, N.: Optimum design and cost of characteristic of small conveyors. Am. J. Appl. Sci. **10**, 15–23 (2013)
65. Nedderman, R.M., Tuzun, U., Savage, S.B., Houlsby, G.T.: The flow of granular materials -1, discharge rates from hoppers. Chem. Eng. Sci. **37**(11), 1597–1609 (1982)
66. Newcastle Innovation Ltd., Industry Development Centre.: A conversation with Emeritus Professor Alan Roberts (2014)
67. Nied, C., Frank, E., Dauth, H., Sommer, K.: Experimental analysis of dynamic porosity changes during plug flow using electrical capacitance tomography.In: Proceeding of ICBMH 2013 – University of Newcastle, Newcastle, Australia. (2013)
68. Pahk, J.B., Klinzing, G.E.: Frictional force measurement between a single plug and the pipe wall in dense phase pneumatic conveying. Powder Technol. **222**, 58–64 (2012)
69. Pan, R., Wypych, P.W.: Dilute and dense phase pneumatic conveying of fly ash. In: Proceedings of the 6th International Conference on Bulk Materials Storage and Transportation, Wollongong, NSW, Australia, pp. 183–189 (1998)
70. Pettyjohn, E.S., Christiansen, E.B.: Effect of particle shape on free-settling rates of isometric particles. Chem. Eng. Prog. **44**, 157 (1948)
71. Rabinovich, E., Kalman, H.: Pickup, critical and wind threshold velocities of particles. Powder Technol. **176**, 9–17 (2007)
72. Radar, J., Prakash, A., Klinzing, G.E.: A light/charge solids flow meter. AIChE Symp. Ser. Fluidization Fluid Part. Syst. **91**(308), 154–163 (1995)
73. Reicks, A.V.: Conveyor Models as Quantitative Platforms for Belt Conveyor Energy Options. Bulk Solids Europe (2012)
74. Rhodes, M.: Introduction to Particle Technology. Wiley, Chichester (1998)
75. Rizk, F.: Pneumatische Förderung von Kunststoffgranulaten in horizontalen Rohrleitungen unter Berücksichtigung des Gewichteinflusses in Zusammenhang mit Gut- und Rohrwerkstoffeigenschaften in optimalen Förderbereich; Ph.D. dissertation, Universität Karlsruhe (1973)
76. Roberts, A.W.: Economic Analysis in the Optimization of Belt Conveyor Systems. BELTCON 1 (1981)
77. Roberts, A.W.: Design considerations and performance evaluation of screw conveyors. In: Proceedings of the BELTCON 1, pp. 1–20 (2001)
78. Sanchez, L., Vasquez, N., Klinzing, G.E., Dhodapkar, S.V.: Characterization of bulk solids to assess dense phase pneumatic conveying. Powder Technol. **139**, 93–117 (2003)

79. Schuchart, P.: Pressure drop across a bend in pneumatic transport. Chem. Eng. Techn. **41**, 1251 (1969)
80. Setia, G., Mallick, S.S., Wypych, P.W.: On improving solid friction factor modeling for fluidized dense-phase pneumatic conveying systems. Powder Technol. **257**, 88–103 (2014)
81. Shaffer, F.D., Ramer, E.R.: Automated analysis of multiple-pulse particle image velocimetry data. Appl. Optics **31**(6), 779–784 (1992)
82. Solt, P.: Pneumatic Conveying Consultants, Personal Communications (1994)
83. Sommer, K.: Sampling of Powders and Bulk Materials. Springer, New York (1986)
84. Sommerfeld, M.: ASME FED, vol. 180, Book No. G00858, pp. 1–14 (1994)
85. Sommerfeld, M., Zivkovic, G.: In: Hirsch, C.H., Zienkiewicz, O.C., Onate, E. (eds.) Computational Fluid Dynamics Conference and First European Conference on Numerical Methods in Engineering, "Characterization of the cross-sectional particle concentration distribution in pneumatic conveying systems"; Eccomas, Elsevier (1992)
86. Sommerfeld, M.: Modelling of particle-wall collisions in confined gas-particle flows. Int. J. Multiphase Flow **18**, 905–926 (1992)
87. Soo, S.L.: Fluid dynamics of multiphase systems. Blaisdell Pub. Co., Waltham (1967)
88. Soo, S.L., Rieber, M., King, P., King, Y.T., Leung, T., Wu, H., Wu, J.: U.S. Department of Interior Contract Report J0166163 (1977)
89. Soo, S.L.: Long distance pneumatic transportation. J. Pipelines **4**, 79–85 (1984)
90. Sullivan, T.A., Koenig, E., Knudsen, C.W., Gibson, M.A.: Pneumatic conveying of materials at partial gravity. Aerosp. Eng. **7**, 199–208 (1994)
91. Tallon, S.J., Davies, C.E.: Velocity measurements in dense down flow of bulk solids using a non-restrictive acoustic method. Flow Meas. Instrum. **11**, 171–176 (2000)
92. Tallon, S., Davies, C.E.: Advances in fluidization and fluid-particle systems. AIChE Symp. Ser. **317**(93), 136–140 (1997)
93. Theuerkauf, J., Dhodapkar, S., Jacob, K.: Modeling granular flow using discrete element method- from theory to practice: this emerging simulation technique is changing the face of solids processing. Chem. Eng. 69–74 (2007)
94. Tsuji, Y.: Proceedings of the 5th World Congress on Particle Technology, Orlando, FL, A.I. Ch.E., Plenary Lecture, Paper 109 (2006)
95. Vasquez, N., Sanchez, L., Klinzing, G.E., Dhodapkar, S.V.: Friction measurement in dense phase plug flow analysis. Powder Technol. **137**, 167–183 (2003)
96. Weber, M.: Stromungs-Fordertechnik. Krausskopf Verlag, Mainz (1973)
97. Wen, C.Y., Simon, H.P.: Flow characteristics in horizontal fluidized solids transport. AIChE. J. **5**(2), 263–267 (1959)
98. Wheeler, C.A., Roberts, A.W., Jones, M.G.: Calculating the flexure resistance of bulk solids transported on belt conveyors. Part. Part. Syst. Charact. **21**, 340–347 (2004)
99. Williams, J.C.: Mixing and segregation in powders. In: Rhodes, M.J. (ed.) Principles of Powder Technology. Wiley, Chichester (1990)
100. Williams, K.C., Jones, M.G.: Numerical model velocity profile of fluidized dense phase pneumatic conveying, In: Proceeding of the 8th International Conference on Bulk Materials Storage and Transportation, Wollongong, NSW, Australia, pp. 354–358 (2004)
101. Woodcock, C.R., Mason, J.S.: Bulk Solids Handling. Leonard Hill, London (1987)
102. Yang, W.C.: Estimating the solid particle velocity in vertical pneumatic conveying lines. Ind. Eng. Chem. Fundam. **12**, 349–352 (1973)
103. Yang, W.C.: A mathematical definition of choking phenomena and a mathematical model for predicting choking velocity and choking voidage. AICHE. J. **21**, 1013–1015 (1975)
104. Zenz, F.A, Othmer, D.F.: Fluidization and Fluid-Particle Systems. Rheinholt Chemical Engineering Series (1960)

Chapter 15
Sampling and Characterization of Bulk Particulate Materials and Products

Don McGlinchey

Abstract This chapter is aimed primarily at engineers working in industry with little formal training or education in particle technology or bulk solids and aims to provide an overview of simple sampling and characterisation methods and techniques many of which can be undertaken with standard laboratory apparatus. It is written in a fairly informal style, however, reference is made to the academic literature where appropriate. A descriptive distinction between random sampling and systematic sampling is given, followed by advice on sampling methods covering a range of sampling techniques. Once a sample has been obtained, typically, either the properties of individual particles or of the bulk are required. An overview of common characterisation techniques, apparatus and interpretation of results is given. This is by necessity a non-exhaustive review and a list of references and standards is provided for the interested reader.

15.1 The Sampling of Particulate Materials

15.1.1 Introduction

The purpose of taking samples from a batch is to obtain an estimate of certain characteristics of the batch. The characteristics most commonly investigated when dealing with particulate materials are particle size, composition and quality. In order to obtain the 'best' estimate, the sample has to be representative of the whole bulk, that is, the total sample should contain in the correct proportion all the variation in particle size, shape, density, etc. that is present in the bulk. This relates to the 'scale of scrutiny' which is discussed later. This is sometimes not a straightforward task and may require many individual samples which may be analysed separately or combined and subsequently sub-sampled into the (occasionally very

D. McGlinchey (✉)
Department of Engineering, Glasgow Caledonian University, Glasgow, UK
e-mail: D.McGlinchey@gcu.ac.uk

© Springer International Publishing Switzerland 2016
H.G. Merkus, G.M.H. Meesters (eds.), *Production, Handling and Characterization of Particulate Materials*, Particle Technology Series 25,
DOI 10.1007/978-3-319-20949-4_15

small) sample size required by any particular measurement technique. There are techniques for sampling from a batch, from a process stream or even from a 100,000 tonne stockpile, all of which present challenges, but there are some things in common.

Regardless of the particle or bulk property to be measured, e.g. mean particle size or shear strength, the nature and properties of the particulates and the intended processes must be considered as part of a sampling exercise. According to Lyn Bates [2], the first step in this sample selection process is to determine what forms of variation may prevail in the material in the circumstances and at the scale of scrutiny significant to the application. Variations in the material's condition may be considered under three headings:

Uniformity
How heterogeneous is the material? i.e. of varied composition throughout the mass, not subject to local variations of particle size, isotropy, moisture content, temperature or any other feature that may affect its condition or behaviour.

Consistency
Is the material always of a similar nature on a day-to-day, year-to-year or source-to-source basis? Naturally derived products, whether organic or mineral, may be expected to vary to some degree according to their origin and history and some process operations are also prone to have variable product condition outputs.

Stability
Over the relevant plant life of the material, will changes take place in its condition due to age, chemical or thermal action or other internal or external variants to adversely change the nature of the material?

Bates makes the important points that, any variation of a bulk material composition or condition essentially represents a different 'material' and it is essential that the quality of material taken as a sample is representative within a stated confidence interval, and that the testing circumstances reflect the relevant application conditions.

Bates' second step of the sample selection process is to establish the particular condition(s) which give the 'worst' product properties to be accommodated in the equipment or the process under review. The term 'worst' here meaning, for example, those properties giving the most challenging flow problems or having the greatest segregation potential.

And his third step of the sample selection process is to secure or prepare a sample in the condition required.

Great care is needed in the selection of material samples and choosing test conditions to avoid deriving useless or misleading data. Published data must also be used with caution, as too often data are given without mentioning, for example, the sampling/measurement instrument or technique (which equivalent size), or which mean PSD value has been used, or by using the word mean instead of median.

It is very difficult to carry out any selection process without introducing individual preference or bias into the final selection. The effects of biased sampling will be detrimental to any investigation but will be especially pernicious if there are large local variations in the population. Particulate materials are particularly prone to such local variations because of their segregating characteristics, and can pose extremely difficult sampling problems. In general it can be said that, the greater the segregating potential of the material, the greater the care needed to select potential samples.

Two approaches to help mitigate the problems of bias are: random sampling, where, as the term suggests sampling locations are chosen at random; and systematic sampling, where a structure is imposed on sampling location.

15.1.2 Simple Random Selection

Some consider the best way to eliminate personal preferences, convenience and habit from sample selection, is by identifying sample sites by the use of random numbers. In work with powders or other particulate materials the most useful selection method is that of simple random selection, (s.r.s.), in which every unselected sample has an equal chance of selection. The method relies on the assumption that a sample volume can be isolated so that every potential sample site within that volume can be identified by three co-ordinates and that samples are taken without replacement. The technique can be simply illustrated in two dimensions. If the area to be investigated is divided into potential sampling sites, each site can then be identified (Fig. 15.1).

The sites to be used for sampling can then be identified from, for example, a table of random numbers. This method is particularly important as a selection procedure in that it forms the theoretical basis for subsequent evaluations of errors of the estimates of the material characteristics. The s.r.s. technique ensures that personal preferences are removed from the sample selection and that an estimate of precision can usually be placed on the determined values of material properties.

If the samples are of approximately equal size and equal to or larger than the characteristic volume, and the property of interest, be it particle size or mixture composition, from samples withdrawn from the material are assumed to be normally distributed, then the precision of the material variance can be estimated by one of two methods depending on the number of samples taken.

Fig. 15.1 Area to be sampled divided into potential sample sites

	1	**2**	**3**	**4**	**5**
1	1	2	3	4	5
2	6	7	8	9	10
3	11	12	13	14	15
4	16	17	18	19	20
5	21	22	23	24	25

(I) ***If more than about fifty samples are withdrawn from the material.***

We can determine for example, how close the measured mean property, $\hat{u}$, represents the true mean of the batch with a specified degree of confidence by using the two-tailed "*t*" test. Suppose we want to be 95 % confident that the true mean of a given product property lies within a certain range. We must determine this range.

Ninety-five percent confidence is equivalent to the 0.05 level in "*t*" tables.

Then the true mean, $\bar{u}$, lies within the range

$$\bar{u} = \hat{u} \pm t_{0.05,(n-1)} \cdot s/\sqrt{n} \tag{15.1}$$

where

$\hat{u}$ is the measured mean
s is the standard deviation
n is the number of samples
$t_{0.05,(n-1)}$ is the two sided value of "*t*" at the 0.05 level for $n-1$ degrees of freedom

It is assumed that the second stage distribution of estimates of variance is also normal. The second normal curve can be standardised in terms of a single variable "*t*" such that

$$t = \frac{s^2 - \sigma^2}{s.e.(s^2)} \tag{15.2}$$

or

$$\hat{\sigma}^2 = s^2 \pm t \times \text{s.e.}(s^2) \tag{15.3}$$

where $\hat{\sigma}^2$ is the best estimate of the true variance.

For high precision the values of *t* and of the standard error should be minimised. The value of *t* can be obtained from Students *t* distribution tables and its value will largely be determined by the confidence level required of the estimate.

The value of the standard error of the sample variance is estimated by

$$\text{s.e.}(s^2) = s^2\sqrt{\frac{2}{n}} \tag{15.4}$$

where n is the number of samples withdrawn from the material.

It can be seen that the increase in precision is proportional to the square root of the number of samples taken.

(II) ***If less than fifty samples are withdrawn from the material.***

It is no longer safe to assume that the second stage distribution of estimates of variance will be normal and the curve is better described as a χ^2 distribution. As with estimates of precision based on normality, the precision can be increased by requiring a lower confidence level in the precision estimate and by taking a larger number of samples. Unlike the estimates based on normality, the limits of precision will not be symmetrical. The precision of the estimate will decrease as the number of samples decreases. For the χ^2 distribution:

$$\hat{\sigma}^2_{(lower)} = s^2 \frac{(n-1)}{\chi^2_{(lower)}} \tag{15.5}$$

and

$$\hat{\sigma}^2_{(upper)} = s^2 \frac{(n-1)}{\chi^2_{(upper)}} \tag{15.6}$$

The values of $\chi^2_{(lower)}$ and $\chi^2_{(upper)}$ for a given confidence level can be obtained from χ^2 distribution tables (see also Chap. 9).

If it cannot be assumed that the samples withdrawn from the material are normally distributed then no estimate of precision can be made. In practical terms, a non-normal distribution is likely to result from a strongly segregated material and as the segregation decreases so will the distribution curve approach normality. This means that reliable comparisons can be made between "good" (non-segregated) materials but not between "bad" (segregated) materials.

It can be argued that s.r.s. is not necessarily the best technique if the object of sampling a material is stated as being to obtain as much information as possible about the material. The "weakness" of the technique lies in the fact that random numbers are free to cluster. The characteristics of particulate materials frequently change gradually over comparatively large areas or volumes and chance local clusters of sample selections tend to give repetitive information. If samples were taken more systematically over the entire population then they would arguably give more information on the characteristics of the batch as a whole. This is the basis of the technique of systematic sampling. A systematic sample can give a 'better' estimate of the characteristic; however, it is not strictly correct to place precision limits on an estimate of this type.

15.1.3 Systematic Sampling

Considerable care must be taken if systematic sampling is to be used successfully. The technique assumes knowledge of the distribution characteristics of the

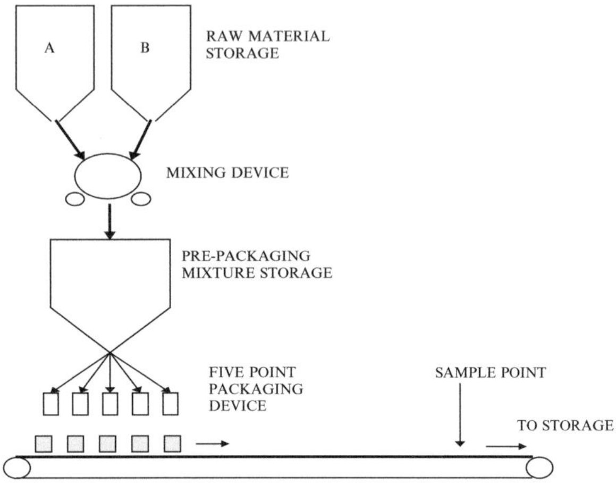

Fig. 15.2 Typical packaging process diagram (Adapted from Ref. [1])

material. The technique and possible pitfalls can be discussed by way of an example. Think about the problems of sampling packages of an instant drink powder material as they are taken from the production line to a storage point, see Fig. 15.2.

A form of systematic sampling is proposed. It may seem reasonable to ask the process operator to remove every hundredth package for quality checking purposes. If a check is made of the process flow sheet, it reveals that the 'reasonable' systematic sampling interval is a simple multiple of the number (five) of packaging devices. Should the operator strictly follow the instructions as to the sampling frequency all the packages he removes for checking will originate from the same section of the packaging device and possibly from the same section of the pre-packaging material storage. It is clear that the systematic sampling used in this example would not give 'better' estimates of the batch characteristics as the systematic sample is heavily biased to a small proportion of the product output. This is a rather obvious example of the sampling interval exactly coinciding with the natural cyclic frequency of the process. Less obvious process cycles exist in many practical continuous and batch processes. One possibility which may be worth considering is the use of sampling based on random time periods, that is sampling from predetermined locations but at random time interval.

15.2 The Retrieval of Samples from a Batch or Process Stream

Most of the methods for describing the state of a material require the measurement of composition in samples of relatively small and uniform size at a large number of locations throughout the material (in order for the result to be representative for the

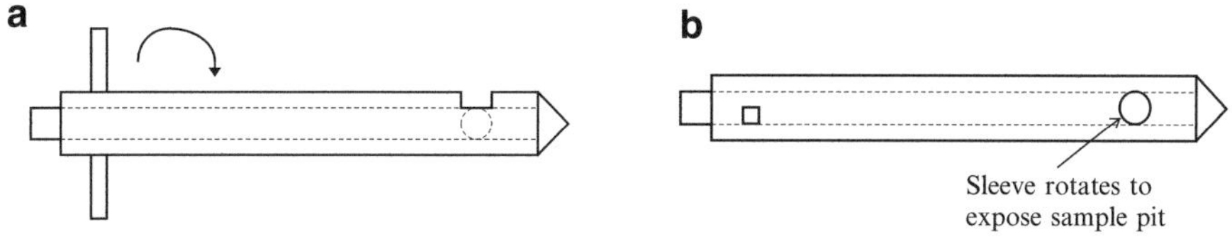

Fig. 15.3 Sample probe. (**a**) Sampler in closed insertion position. (**b**) Sampler in open retrieval position

batch within a stated confidence interval). The methods and techniques differ depending on whether the material is stationary, for example a batch stored in an intermediate hopper, or in motion, for example discharging from the same hopper. It would seem that relatively less effort has been devoted to this practical exercise compared to that which has been applied to the mathematical aspects discussed above. Whatever sampling theory is decided on, the problem remains of how to practically obtain a sample from a bulk of material. It is easy in theory to divide a large heap of powder in to potential sample sites but rather more difficult in practice to retrieve a sample volume without disturbing the remainder.

Several tools have been developed for retrieving samples from within a bulk. Perhaps the most common method of sampling used in research and control is by the side-sampling thief probe, an example of which is illustrated in Fig. 15.3.

Many variations exist but they all operate on similar principles: an aperture is opened in the side of the probe after it has been inserted into the material. Material then flows under its own weight into the aperture (or apertures or slots), which is then closed before removal of the probe from the material. Such probes are limited to relatively free-flowing materials, although fins have been attached to some probes in order to sweep cohesive materials into the aperture.

Doubts must arise when sampling segregating materials because segregation could well occur during the flow into the aperture so that the sample collected does not truly represent the composition of the volume investigated.

By far the greatest drawback of side-sampling thief probes is the disturbance caused in inserting the probe into the material. This sort of disturbance appears to be independent of the profile of the tip, it being caused mainly by friction along the length of the probe. To overcome this problem an end sampling probe has been developed although its use is mainly confined to cohesive powders. The disturbance still occurs but the sample is removed from undisturbed material.

Clearly both these disadvantages can reintroduce a bias into the estimate of the population characteristics and undo all the careful site selection work of the theory.

A variation of the side-sampling thief is the reflectivity probe, which has a transparent window in the place of the usual aperture. A constant light source is directed, by fibre optics, from inside the probe to the material adjacent to the window and the reflected light is measured by a photo cell mounted in the probe. Thus, if used with a material of different-coloured powders, the reflectivity or amount of reflected light can be used to determine the local composition. With the window in the side the problems with material disturbance again arise, but end sampling probes would be a simple modification. Reflectivity probes, with suitable instrumentation, greatly speed up the data gathering, permitting large numbers of samples to be analysed in-situ and at known locations, if auto-correlation is of interest. The use of this type of probe has been confined to research investigations where the powders can be selected to facilitate its use. Probes using variations of this approach of a window, fibre optics, a [laser] light source and a camera to illuminate and capture images of particles are available commercially to obtain particle size information, particularly useful in 'stirred vessels' where a stream of particles pass before the window.

Pneumatic and 'core' probes provide alternative sampling devices but are both subject to bias in many circumstances. Any mechanical sampling device will disturb the arrangement of particles in bulk and potentially will lead to biased sampling.

A technique that has been employed in research programmes for sampling is when the particulate material has been set solid by the addition of an adhesive and then diced into sample volumes for analysis.

Often samples have to be taken in far from ideal conditions, but a little forethought will be helpful. If, for example, a scoop is all that is available (and scoop sampling is far from ideal) to take a sample from powder loaded on a truck it is recommended that the sample be dug out from roughly 30 cm below the surface, because the surface region is usually affected by the segregation. With all probe sampling, even if the sample is taken from undisturbed material at the end of the probe, gross disturbance could be caused if large numbers of samples are taken. This counters the argument for the statistical benefits to be gained from taking many samples.

It has been argued however that measuring the material quality in-situ in a vessel overlooks the possibility of changes in quality during discharge and it is more realistic to take samples from the outlet stream as the vessel is discharged or directly from a process line. This procedure would be recommended for most practical situations and it has the advantage that the sampling is simplified. If possible then, it is practically much simpler and more efficient to sample the stream of powder entering or leaving storage than to attempt to sample from the powder bulk. In general it is thought better to sample at the transport rather than at the bulk holding operation in a process. With continuous processes opportunities of sampling the powder in motion option usually exists. With batch processes such opportunities are less common, though it is frequently possible to sample when bulk storage containers and handling equipment are being filled or emptied.

It should be borne in mind that even when sampling from a moving stream the potential for segregation to occur and produce a biased sample always exists, e.g. due to process fluctuations and segregation during intermediate storage.

Consider the very common situation in which powder is to be sampled as it falls from the end of a conveyor belt. The powder on the belt will probably have undergone two kinds of segregation. Firstly, the deposition of the powder will

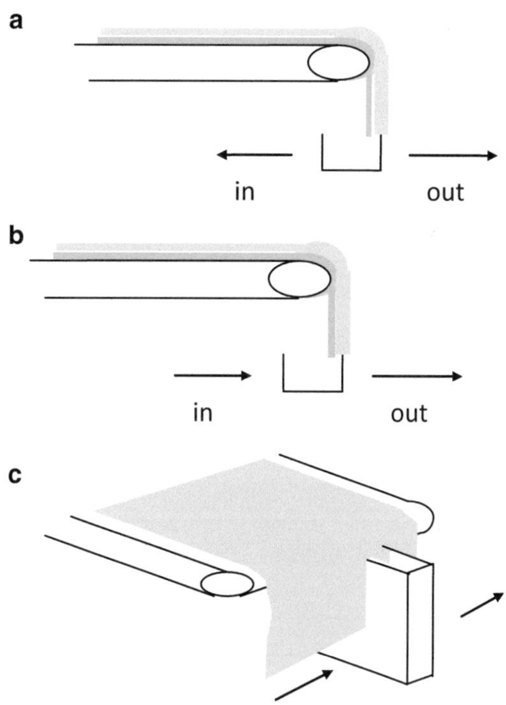

Fig. 15.4 Sampling from moving streams. (**a**) Bad sampling technique. (**b**) Good sampling technique. (**c**) Sampling procedure to be adopted for high mass flow rate (Adapted from Ref. [1])

have been accompanied by a segregation of the fine particles to the centre of the belt with the coarse particles rolling toward the edge. Secondly if the belt is subsequently vibrated on its supporting rollers large particles remaining in the bulk of the belt charge will rise to the free surface and perhaps roll to the edges of the belt. An unbiased sample from such a stream should contain the correct proportion of all particles sizes. With an awareness of the powder segregation mechanisms the design of an effective sampling strategy becomes clear. Any sampling method should uniformly traverse the whole of the steam for a short time rather than part of the stream for a long time.

Consider the schematic sampling devices of Fig. 15.4. The reciprocating action of method (a) would lead to an excess of fine particles within the sample box unless the speed of the traverse was very carefully controlled. Techniques (b) and (c) which avoid bias involve a uniform motion of the sampling box though the entire stream. Overfilling of the sample box could also lead to preferential sampling. If the box did overflow then not only would a section of the flow not appear in the final sample but fine particles would be preferentially retained and coarse particles rejected. Again, as the level of powder rises in the box so the likelihood of particles rebounding clear of the box increases. The ability of particles to escape

Fig. 15.5 Full stream sampler (Adapted from Ref. [1])

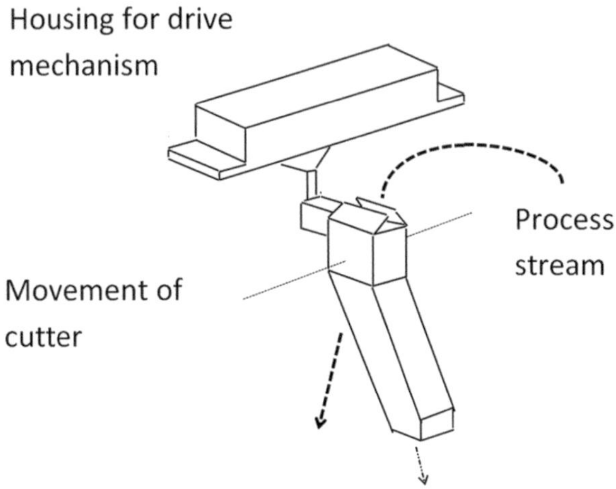

Fig. 15.6 Constant volume sample in a Chute (Adapted from Ref. [1])

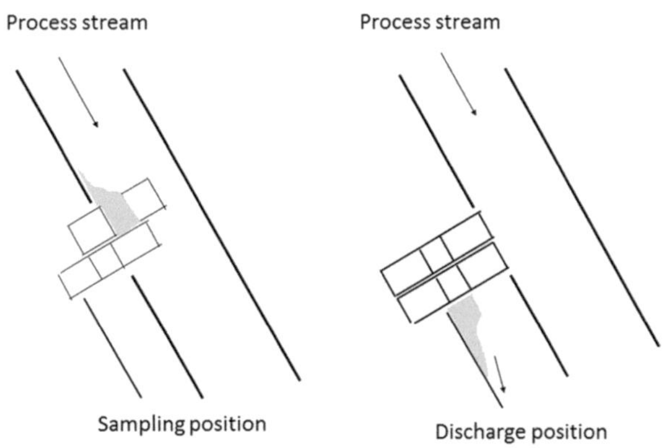

by rebound is a function of their physical characteristics and there could well be a preferential rejection of a component or size range.

The established principles of sampling the entire section of flow are equally valid when sampling from an air slide process stream or a pneumatic conveying system. Any sampling device which samples only a segment of the flow should be viewed with suspicion. Two other practical sampling devices are illustrated in Figs. 15.5 and 15.6.

The full-stream sampler, also named a cutter sampler, comes in various forms, such as the car type, belt, and rotary type. The sampling width (or cutter width) and the sampling speed are adjustable in these samplers. Therefore, it is possible to take out the desired mass of powder. Random sampling is also possible by randomly changing the starting time of the sampler. This type of equipment can meet all the requirements discussed. It traverses the entire stream at controlled rate, has no risk of overflowing and the free surface of particles within the sampler is unlikely to cause rebound.

The constant volume chute sampler, on the other hand, meets none of the requirements. Only the segment of powder flow adjacent to the chute wall has a chance of retention and within that segment of flow there is every possibility of bias occurring due to the overflow effect of the sample pit and the preferential rebound of some of the particles.

15.2.1 Summary of Particulate Material Sampling

Sample identification, retrieval and analysis have been identified as distinct stages in obtaining an estimate of a particulate material characteristic. At each of these stages there is the distinct possibility of producing a biased estimate. Whilst each process presents its own, often unique, problems it can generally be said that the problems of biased estimation will be more important when the particles being processed are of a free-flowing and segregating nature. The extent of segregation within the process will be a function of both the powder and process characteristics. In these common and difficult circumstances it is important that the quality control system be designed with the combined expertise of a statistician and a powder technologist.

Two general guide rules can be quoted in designing such a sample system:

1. A powder should be sampled when in motion.
2. The whole of the process stream should be taken for short increments of time in preference to part of the stream being taken for a prolonged period.

Generally, for the measurement of product properties, it is better to have an analytical apparatus which produces a random scatter of values about a true value than an apparatus that consistently produces a biased result. (This is not necessarily so for process control).

15.2.2 Sub-sampling

The above techniques can result in samples that are too large for analytical apparatus; in this case the gross samples must be divided into sub-samples which themselves may require to be sampled to provide a test sample. The same comments made about obtaining a representative sample are also true for obtaining a sub-sample.

One of the simplest techniques for obtaining a sub-sample is by 'coning and quartering' the gross sample. This technique is commonly used by analytical chemists and can be useful but is subject to operator error. In essence the technique involves pouring the sample to form a small conical heap, and then pressing on the top to flatten it. The sample shape is now a truncated cone and looks like a 'wheel'. The wheel is then divided into quarters. Two of the quarters which sit opposite one another, say east and west, are discarded, while the other two (north and south) are combined to produce a reduced sample. The coning and quartering process is continued until the desired sample size is obtained.

Another technique that is common but is less operator dependent than coning and quartering is the use of a 'riffler' or sometimes known as a sample splitter. There are two basic designs; the riffle box and the spinning riffler. The riffle box is difficult to describe but its operation is obvious when seen in action. The box has a series of divisions which partition the gross sample as it is poured into the box into two sub-samples. One sub-sample is discarded, the other is used for analysis or can again fed through the riffle box, this time to give a quarter of the original gross sample. Riffle boxes can range in size from fairly large and robust for on-site sampling of mined products to small lab scale. The spinning riffler typically consists of a feed hopper where the gross sample is placed, the material is discharged from this hopper onto a vibrating tray which conveys the material to a rotating section containing a number of compartments (sometimes test tubes). The feed rate of material and rotational speed of the collection section are matched to ensure that each compartment is visited several times and the sub-sample is made up of material discharged throughout the discharge time. This is the technique that has the highest precision [1, 11].

Readers who require a more in depth theoretical and practical treatment of sampling of particulate materials are recommended to start their studies with the works of Piere Gy and his 'Theory of Sampling' [5, 6, 13].

15.3 Characterisation of the Particulate Material

Assuming that we have obtained a representative sample we have choices to make on which particle properties or bulk properties will provide the information needed to 'characterise' the material. There are tests that can be undertaken to establish a range of properties, such as, particle size distribution, bulk density, flowability, etc. The following sections provide an overview of the more common properties, how they are measured and how they are used to characterise a material for a particular application or process. See also the list of written standards [14–52].

15.3.1 Particle Size and Size Distribution

> *…, it must be realised that particle size analysis is not an objective in itself but is a means to an end, the end being the correlation of powder properties with some process of manufacture, usage or preparation.* Harold Heywood [7]

It is very common to read the term 'particle size' but what is in fact meant by this term? If the particle is a perfect sphere then the particle size is most sensibly

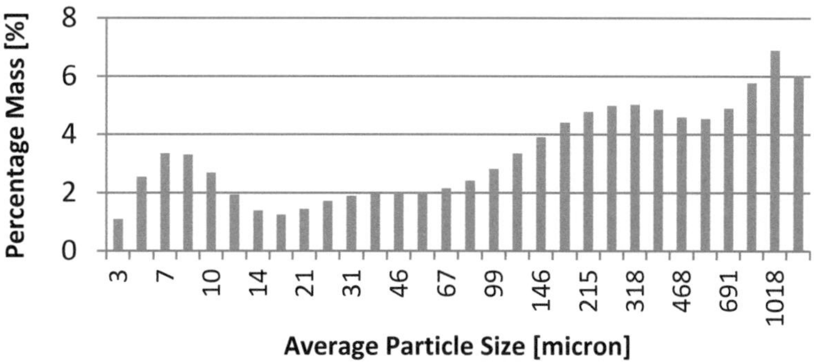

Fig. 15.7 Particle size distribution histogram

described by the diameter (or radius) of the sphere. If the spheres are not 'ideal' but are manufactured, then there is inevitably some uncertainty in the diameter quoted, but if this is a manufactured mono-sized sphere then a mean particle size with some uncertainty due to both the variances in the process and the uncertainty in the measurement made. What, however, if, for whatever reason, there is a distribution of particle sizes about some 'average' value? We need a value of the average particle size, a value which is related to the range of particle sizes and possibly also a value for the uncertainty in the aforementioned average and range. The above has an implicit understanding that there is an even distribution about the average value. This need not be the case and in very many industrial applications the distribution is far from Gaussian and can be bi-modal or indeed multi-modal. In the case of a multi-modal distribution, an 'average' particle size can have no physical equivalent i.e. there is no actual particle of that size. We are still talking about spherical particles here, real particles and the meaning of particle size for non-spherical particles will be addresses later. Graphical representation can be used to describe a size distribution. In order to achieve this representation it is typical to form a histogram of particle sizes within proscribed ranges or bins. The most common histogram is formed on a mass basis, that is, the mass of particles within the bin range (see Fig. 15.7). Less common, but occasionally very useful, is a number based description where the number of particles within the bin range is used.

A smoothed line can be constructed to follow the contour of the histogram. From this line it is possible to get an intuitive understanding of the distribution of particle sizes as well as numeric values for a mean size and the range. An alternative representation is the cumulative size distribution where the distribution can be described as a percentage undersize or oversize a particular size value (see Fig. 15.8). In this context the median particle size is noted as the 'd_{50}' size, is the particle diameter at the 50 % mark on the cumulative distribution. For example, if the $d_{50} = 2$ mm, then 50 % of the particles will have a size larger than 2 mm and 50 % will have a particle size smaller than 2 mm. The d_{50} value is known as the mass median diameter, if the size distribution has a mass basis. Also commonly

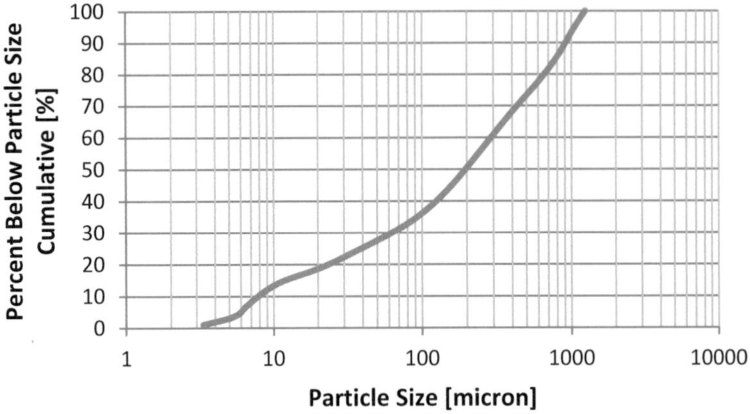

Fig. 15.8 Cumulative particle size distribution

quoted are d_{10} and d_{90} which are the sizes at 10 % and 90 % respectively on the cumulative distribution. It will be recognized that the distributions represented in Figs. 15.7 and 15.8 are of a real material taken by a specific instrument which do not cover the entire size range of the materials i.e. there is some material below 3 μm and some greater than 1018 μm. This highlights a practical difficulty in many industrial cases and gives a level of uncertainty to the values of d_{10}, d_{50} and d_{90} found. There is then a decision to be made on the value of the additional effort required to cover the entire size range or if the measure made is sufficient for the process under consideration. A typical strategy is to sieve off and characterize the larger size fraction and 'combine' this information with the laser sizing of the material passing through the finest sieve. However, this also introduces uncertainties to the total distribution and d_{10}, d_{50} and d_{90} values.

There are many different values e.g. amount of fines (or d_{10}) or amount of coarse (or d_{90}) that can be used in relation to product performance based on "local knowledge" of the process. The "sharpness" of the size distribution can be obtained from the cumulative size distribution and it is usual to quote something like:

$$\frac{d_{50}}{d_{97}} \quad \text{for the coarse fraction,} \tag{15.7}$$

or

$$\frac{d_{10}}{d_{50}} \quad \text{for the fines fraction} \tag{15.8}$$

15.3.1.1 Influence of Particle Shape on Measured Particle Size

Real particles are not spherical, therefore, what does the term particle size actually mean? Where the particles are non-spherical it becomes necessary to define more carefully the parameters used for size and shape. In order to represent the size of an irregularly shaped particle by a single value it is customary to use an 'equivalent diameter', corresponding to the diameter of a sphere that exhibits the same behaviour as the particle under certain conditions or that has the same value of some other descriptive characteristics. For example, the diameter used could be that of a sphere which just passes through the same square sieve aperture, or which falls at the same velocity in a fluid or which has the same projected area. For individual particles:

- Volume diameter = diameter of a sphere that has the same volume as the particle:

$$d_{3,0} = (6V/\pi)^{1/3} = 1.241V^{1/3}$$

- Surface diameter = diameter of a sphere that has the same surface area as the particle:

$$d_{2,0} = (A/\pi)^{1/2} = 0.564A^{1/2}$$

Example Volume of shape $= 1 \times 1 \times 1 = 1$
Volume of sphere $= \pi d^3/6$
Equivalent volume diameter, $d_{3,0} = \left(\frac{6\times 1}{\pi}\right)^{1/3} = 1.24$

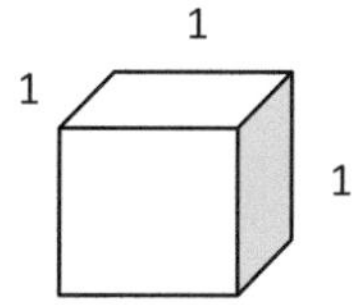

The value given for the 'particle size' of an irregularly shaped particle is somewhat ambiguous as it depends on both a definition of what that size represents and on the measurement technique. Therefore, referring back to the paragraph above, it is possible to quote diameter of a 'sphere' that corresponds to specific sizing technique, e.g.:

- Sieve diameter – particle passing through a square sieve aperture (the particle dimension is actually a sieve size).
- Stokes diameter – based on sedimentation velocity in fluid (if a sedimentation technique is used the particle dimension obtained can also be calculated and described as a free fall diameter or a drag diameter).
- Projected area diameter – related to the projected area of a particle (optical or electron) microscopy.

Table 15.1 Measurement technique conversion factors

To convert	Multiply by
Sieve diameter to projected area diameter	1.40
Sieve diameter to Stokes diameter	0.94
Projected area diameter to sieve diameter	0.71
Projected area diameter to Stokes diameter	0.67
Stokes diameter to sieve diameter	1.07
Stokes diameter to projected area diameter	1.50

15.3.1.2 Shape Factors

The ratio of a pair of listed diameters (often known as a shape factor) has been found to be fairly constant over quite wide size ranges for any one material which has been produced in the same way or derived from the same source. It is possible for instance to correlate analyses in which the coarser fraction of a material has been subjected to a sieve analysis and the sub-sieve fraction has been sized in some other way. Typical values of the ratios are given in Table 15.1 but these figures should be used with caution particularly when the particles are of extreme shape.

Another way to describe the shape of a particles is to compare the real particle to a sphere, here the term sphericity is defined as the ratio of the surface area of a sphere (with the same volume as the given particle) to the surface area of the particle.

$$\text{Sphericity} = \Psi = \frac{\pi^{1/3}(6V)^{2/3}}{A} \tag{15.9}$$

where

$V =$ the volume of the particle
$A =$ the surface area of the particle

As may be expected, the sphericity of a sphere is 1, and any particle which is not a sphere will have sphericity less than 1; for example, the sphericity of a cube = 0.806.

Table 15.2 gives descriptive terms for particulates in different particle size ranges.

15.3.1.3 Methods of Size Measurement

A brief summary of important and common particle size measurement techniques and instrumentation is given below and is restricted to 'dry' techniques; however, this is a highly specialised and rapidly evolving area, particularly in the sizing of nanoparticles where instrument manufacturers are often best placed to offer advice. The interested reader should make reference to specialist texts [11] or standards [14–52].

Table 15.2 Descriptive terms of particle size ranges

Descriptive term	Typical size range	Examples
Coarse solid	5–100 mm	Coal, aggregates, etc.
Granular solid	0.3–5 mm	Granulated sugar, rice
Coarse powder	100–300 μm	Table salt
Fine powder	10–100 μm	Icing sugar
Superfine powder	1–10 μm	Face powder
Nanoparticles	1–100 nm	'Quantum dots'

Sieves

Size analysis by sieves is perhaps the oldest technique of establishing the size distribution of a particulate material as well as being the most popular and least expensive. Each sieve is usually constructed from a weave of wires which leave a gap or aperture between the wires. Individual sieves can be described by their aperture size in mm or μm, or their mesh size (the number of openings per inch). The apertures are commonly square but can be round holes punched through a plate (electroformed sheets or perforated metal plates) and can be of sizes ranging from a few centimetres to 10 μm (dry: 125 mm to 20 μm; wet: 10 μm). The techniques works well for free flowing materials but problems of agglomeration can be experienced with fine powders. The use of a wet sieving procedure can help with agglomerating particles and may limit the damage of fragile particles. The procedure typically involves a 'stack' of sieve trays starting with the sieve with the largest holes or apertures at the top and then the next down has a smaller aperture in the ration of root two and the last tray or pan has no holes. The stack is then shaken or vibrated at a particular frequency and amplitude for a predetermined time. The distribution is then constructed from the mass of material retained on each tray. The results of the technique are dependent on the apertures, the shape of the particles, the details of the applied vibration and the length of time the stack is shaken.

Light

A popular sizing technique is laser diffraction where a collimated beam of mono chromatic light (from a laser) is projected on to a dilute suspension of particles. The suspension medium can be a liquid or a gas (Fig. 15.9).

The light is scattered (Fraunhofer or Mie scattering) with a pattern of maximum and minimum intensities at angles that are related to the size of the particle. The combined scattered light of the ensemble of particles falls on to in the focal plane of an array of detectors. The size distribution is back-calculated from the detector signals using Fraunhofer or Mie theory. With a suitable arrangement of lenses, particles in the range 2 mm–0.1 μm can be sized with this technique.

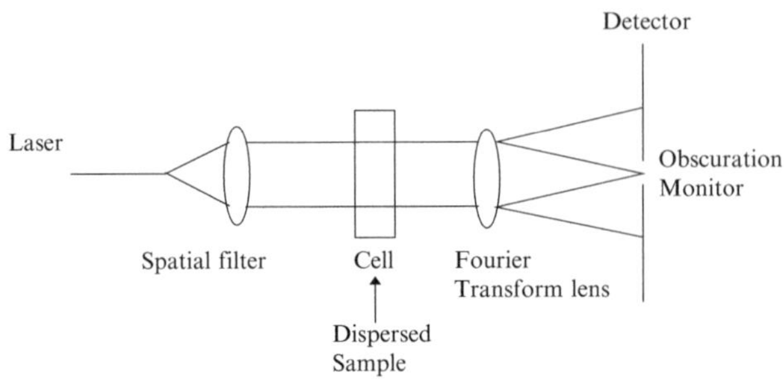

Fig. 15.9 Schematic of a laser diffraction instrument

ESZ

The electrical sensing zone (ESZ) technique is based on the electrical properties of a conductive fluid (an electrolyte) with suspended particles. The suspension is passed through an orifice and the change in conductivity as a particle passes the orifice is recorded as an electrical signal. The amplitude of this signal can be related to the size or volume of the particle passing through the orifice. The technique was developed by W. H. Coulter for counting blood cells and has now been extended to industrial powders with devices that can cover size range 1200 μm down to about 0.6 μm, where number and size distributions of particles can be determined. There are some limitations to the technique, for example, highly conducting particles, very porous particles, finding a suitable finding inorganic electrolytes in which to suspend particles and the required low particle concentration.

Optical Microscopy (Plus Image Analysis)

Optical microscopy is typically used with particles in the 0.8–150 μm size range. The technique has the advantage of being able to directly observe, examine and measure individual particles and can often be used where other techniques fail. Determination of the size of the particles is based on defined diameter measurements some of which are Feret, Martin and Projected Area diameters. This a very time consuming procedure if performed manually; however, image analysis techniques can be used to drastically cut the time taken to obtain particle size data (see example in Fig. 15.10).

There remain three problems which are difficult to overcome. The first is that typically relatively few particles are examined from the bulk which gives a difficulty of obtaining a very small representative sample for study. This can be overcome somewhat by automated sampling and image analysis allowing very many samples to be analysed. Second, the particles will naturally align in a preferred orientation on the microscope slide which may mean that the technique does not capture the true three-dimensional information of the particles (this can be overcome somewhat in dynamic image analysis, although, here too, some

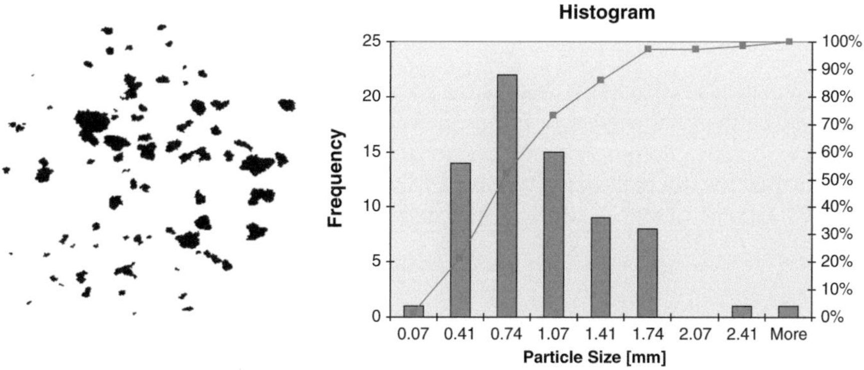

Fig. 15.10 Image and size analysis of carbon 'flakes'

alignment may occur depending on flow conditions). Thirdly there is the limitation imposed by the small depth of focus of the microscope optics.

Nano Sizers

A nanoparticle can be defined as a particle of any shape with dimensions in the 1×10^{-9} and 1×10^{-7} m range. There is a range of options from instrument manufacturers in the sizing at the nano-scale, for example, particle size for a range of less than a nanometre to several micrometres using techniques like dynamic light scattering (DLS), particle tracking, electron microscopy (TEM, SEM), atomic force microscopy (AFM), and Multi-Angle Static Light Scattering (MALS).

A range of techniques has been described above; however, each technique could potentially give a different 'equivalent size' for identical particles tested, as particle shape often significantly deviates from sphericity; also cloud properties depend upon different PSD characteristics. The choice should be made therefore on the technique which will provide the most useful information for the particle process or application under consideration. To borrow an example given in a lecture by the late Professor Brian Kaye on carbon black sizing:

1. If carbon black is to be used as a pigment then the opacity may be best assessed by a measurement based on the equivalent area of the particle.
2. If, however, the opacity is to be utilised as a 'smoke screen' as may be used in defence applications, then the light scattering property of the particle would be important and may be related to a particle size obtained by laser diffraction.
3. If the interest is the dispersal of this smoke screen or the health hazard of an aerosol, then the aerodynamic diameter is clearly applicable.

15.3.2 Bulk Properties

It is often the case that knowledge of the properties of individual particles is of less importance than knowledge of how a material behaves in bulk form. This is true when we look to design or specify materials handling or processing plant, where information on the bulk density or wall friction has a major influence on performance. A range of bulk properties and test methods are reviewed.

15.3.2.1 Bulk Density

Bulk density is probably one of the most common and widely used of the bulk characteristics [12]. It is used in hopper design to determine wall loading, it is used to size volumetric feeders, such as screws and rotary valves, it is used to estimate "flowability" and used in many other ways. It is rather unfortunate then that such a useful characteristic is not a constant for a given material. The bulk density of a material is simply the mass of material divided by the volume that it occupies. The density of the particles themselves can be taken as constant; however the complication comes because the amount of "space" between the particles depends on how the material has been handled before the measurement is taken. The volume that a unit mass of product can occupy can change by 50 % between the material being in a compressed state to being in a very loose state. Cement, for example, has a compacted bulk density of 1400 kg/m^3 and an aerated bulk density of 1000 kg/m^3. It is obviously important that the correct bulk density value is selected for any calculation.

The full expression for bulk density, ρ_b, is

$$\rho_b = \frac{Mass_{solids} + Mass_{spaces}}{Volume_{solids} + Volume_{spaces}} \tag{15.10}$$

For dry bulk solids, the void spaces would usually contain air or some other gas, the density of which can be taken as negligible compared to the density of the solid particles, so we can approximate

$$\rho_b = \frac{Mass_{solids}}{Volume_{total}} \tag{15.11}$$

We can relate this to another common characteristic, voidage or void fraction ε, which is the percentage of the total volume not occupied by particles:

$$\varepsilon = \frac{Volume_{spaces}}{Volume_{total}} \tag{15.12}$$

Again assuming air or gas in the void spaces and taking particle density as ρ_p, we can write:

$$\rho_b = \rho_p(1 - \varepsilon) \tag{15.13}$$

The range of values voidage can take can be illustrated by considering a static heap of mono-sized spheres. If the spheres are in a regular hexagonal packing the classic "cannon ball" stack, the voidage would be 26 %; whereas if they were in regular cubic packing the voidage would increase to 48 %. However this does not represent the loosest packing even for large smooth identical spheres. The cannon ball stack gives each ball six contact points, but simple static mechanics requires only two contact points below the centre of gravity of the ball for equilibrium. Therefore it is possible to have a stable structure with many fewer contact points and a resulting increase in voidage [3, 12]. If the particles are irregular in shape, have a size distribution and in some way cohere to one another the packing arrangement can be very loose and so the voidage can be very large.

Measurement of bulk density is in theory quite simple, and requires only a knowledge of material mass and volume and is generally based on one of two techniques:

The first is to weigh out a quantity of material using a simple balance and put this in to a calibrated cylinder in much the same way as you would a liquid. If the particulate material is poured into the cylinder, the volume taken up would be of the material in a loose or poured state; the associated bulk density would be commonly described as "poured bulk density". If this same cylinder is then tapped or dropped from a small height on to the bench several times, the volume would likely decrease and the new value of bulk density would be the "tapped bulk density". Similar techniques can be used to determine aerated bulk density from a fluidising column or compacted bulk density from a material placed under load.

The second technique is to fix the volume of the bulk material by filling a cup like vessel to overflowing and then level with a straight edge. This is then weighed on a balance and the bulk density calculated. This gets around some of the problems of trying to estimate the actual level of powder in a cylinder with a surface which is typically anything but flat and see through a glass that has become coated in powder. Typical density values for a few materials are presented in Table 15.3.

Table 15.3 Typical bulk density values for a few materials

Material	Tapped bulk density [kg/m^3]	Poured bulk density [kg/m^3]	Particle density [kg/m^3]
Iron powder	3410	3360	7200
Aluminium powder	1220	1095	2650
Cement	1400	1100	2700
Nylon pellets	680	680	1140

Some Definitions of Bulk Density Terms

Bulk density	the bulk density under defined conditions (non specific)
Poured density	'loose' bulk density; the minimum bulk density of a motionless powder; the bulk density of the particulate material when poured into a vessel or heap
Tapped density	bulk density of a powder bed formed in a container of stated and specified dimensions when subjected to vibration or tapping under known and stated conditions
Pressed or compacted bulk density	bulk density of a compact of material under a given load or pressure
Aerated density	bulk density of a powder bed formed in a container of stated and specified dimensions when subjected to minimum compression to ensure an open loosely packed volume
Fluidised bulk density	bulk density of a fluidised powder bed

One possible complication with bulk density measurements is the effect of the porosity of the particles themselves; in order to avoid ambiguity it is worthwhile stating whether the bulk density value is inclusive or exclusive of closed pores.

15.3.2.2 Flowability

Bulk density measurements have been used to give some qualitative prediction of the "flowability" or "handlability" of a bulk solid, that is, some estimate of the likely ease or difficulty in dealing with these materials. One such predictor is the often quoted Hausner ratio (HR) which is defined as:

$$HR = \frac{\rho_b tapped}{\rho_b poured} \tag{15.14}$$

Another close relative is Carr's Compressibility Index (*CI*):

$$CI[\%] = \left(1 - \frac{\rho_b poured}{\rho_b tapped}\right) \times 100 \tag{15.15}$$

This *CI* index ranks materials as:

5–15 %	free – flowing to excellent flow – granules
12–16 %	free – flowing to good flow – powders
18–21 %	fair to passable powdered granule flow
23–28 %	easy fluidizable powders – poor flow
28–35 %	cohesive powders – poor flow
33–38 %	cohesive powders – very poor flow
>40 %	cohesive powders – very very poor flow

Fig. 15.11 Poured angle of repose

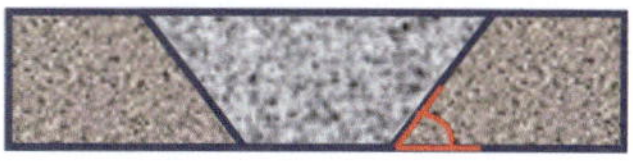

Fig. 15.12 Drained angle of repose

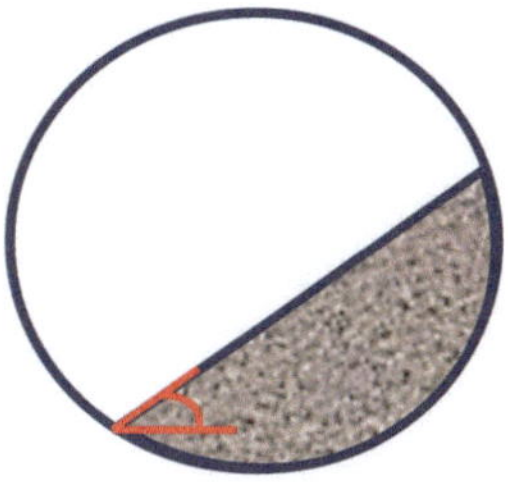

Fig. 15.13 Dynamic angle of repose

These relatively quick and easy measurements can be effective in giving some indication as to how powders will likely behave in solids handling plants but are by no means comprehensive and caution should be exercised if relying only on this information.

Angle of Repose

The angle of repose is another parameter that is used to determine the flowability of a bulk solid. The angle of repose is defined as the angle of the free surface of a heap of particulate material to the horizontal plane. We are faced with the same problem as we were with bulk density, the angle of repose is not a constant for a given material and is dependent on the method of heap formation. There are again two measurements which are commonly quoted: the poured angle of repose and the drained angle of repose. The poured angle of repose is the angle measured from a heap formed by pouring material on to a flat surface to the horizontal (Fig. 15.11). The drained angle of repose is the angle measured on the internal conical face of a material which has been formed when drained from a orifice of a flat bottomed container to the horizontal (Fig. 15.12). A third angle of repose which you may come across is the dynamic angle of repose which is the angle to the horizontal of the free surface formed in a relatively slowly rotating drum (Fig. 15.13).

There are several things we should be aware of when using angles of repose. The first is that the angle formed will depend on the details of the formation process. For example, the fall height for the poured angle or the orifice size for the drained angle will influence the angle. Therefore the angle measured is not independent of the measuring apparatus. The second is that a different angle will be found for the same material tested using the three different techniques (Table 15.4). The measurements can only be reliably made when using powders that are free flowing to slightly

Table 15.4 Angles of repose of three materials (based on [7])

Material	Shape	Poured angle	Drained angle	Dynamic angle
Tapioca	Spherical	30	37.5	32
Sand	Angular	37	39	36.5
Coal	Angular	37.5	41	34

cohesive and are fairly homogenous. Materials that are a material of components or have a wide size distribution will give angles that are difficult to determine and have low repeatability. There are also some uncertainties based on the fundamental physics of the problem, relating to stress history and avalanche behaviour, discussed by Duran [3].

Despite the difficulties noted above, the angle of repose in whatever form can be useful tool to rank materials. As a rough guide the relationship between the angle of repose and flowability often follows the structure below.

Angle of repose [degrees]	Flowability
25–30	Very free flowing
30–38	Free flowing
38–45	Fair flowing
45–55	Cohesive
>55	Very cohesive

This classification allows us to make some judgement on the likely flow behaviour of a material but has very limited use for equipment selection and design and in particular, it is a mistake to use the angle of repose in an estimate of the wall angle required for the converging section of a hopper. However, the angle of repose can be used in some cases to estimate the surcharge in a storage vessel or is useful in estimating the ground area requirements when forming a stockpile.

An interesting development of the rotating drum device is the Aero Flow device developed by Brian Kaye where the avalanche behaviour is studied using the covering and uncovering of elements of an array of photodiodes as the drum revolves. The signals from these detectors are processed to create a "fractal fingerprint" which is typical of the avalanching behaviour. The time between successive avalanche events is used to characterise the behaviour with poor flowing materials demonstrating more scatter in their avalanche times than freer flowing powders. The tests are relatively quick and the Aero Flow can be a useful tool in comparative measurements and quality control in some circumstances.

Descriptions, such as 'free flowing' or 'poor flowing' are subjective and only reflect a specific condition in particular circumstances. A food powder can appear to be 'free flowing' when it is loosely poured, but may settle to a very firm and stable condition when de-aerated or subject to compacting stresses. A dry, crystalline product e.g. salt will usually flow through a relatively small orifice, but have extreme reluctance to deform if damp or 'caked' due to the presence of tiny crystal

Fig. 15.14 A stable rat hole in a conical hopper

bridges binding particles together. The EU commissioned 'Strategic document for research in food powders' [4] examined the challenges facing manufacture with food powders and in particular the important aspects of characterisation, storage, feeding and processing within plant.

Among the many characteristics of a bulk powder perhaps the most important for flow are how the powder slides on a contact surface e.g. at the wall of a hopper, the face of a mixer blade, (wall friction), and the resistance of the bulk solid to deformation e.g. to squeeze through a hopper outlet, (its shear or failure strength). More information on characterization of properties of bulk solids, viz. the Mohr stress circle, compressive strength, flow functions and shear testers, in relation to the design of extruders and hoppers is presented in Chaps. 5 and 13.

Flow behaviour is not only dependent on the material characteristics – the equipment design also has a major influence. Steep hoppers with large outlet sizes often have flow benefits: they prevent arches and rat holes, they generate mass flow (no static or dead regions in the hopper during discharge – an important feature when handling food powders which may age and spoil) and avoid segregation.

By contrast, a 'rat hole' may form if the hopper slope is shallow and the powder is cohesive, so the effective storage capacity of the hopper is reduced. Additionally if the hopper is refilled whilst the rat hole remains then this has ramifications for product residence time which may affect its flow behaviour and its fitness for purpose if the powder has a shelf life and spoilage occurs (Fig. 15.14).

If the outlet is too small, an 'arch' may stop the flow altogether. To avoid these effects, the slope of wall, β_c and outlet size D_{crit} needed for reliable flow can be calculated using a variety of techniques described in detail elsewhere (see Chap. 13).

Results from this shear test on a number of materials undertaken by McGee [8–10] are presented in Fig. 15.15 in a form where the axes are factored by a numerical constant and gravity. Figure 15.15 includes the results of tests on a lactose and titanium dioxide presented in a space partitioned by lines of constant outlet size.

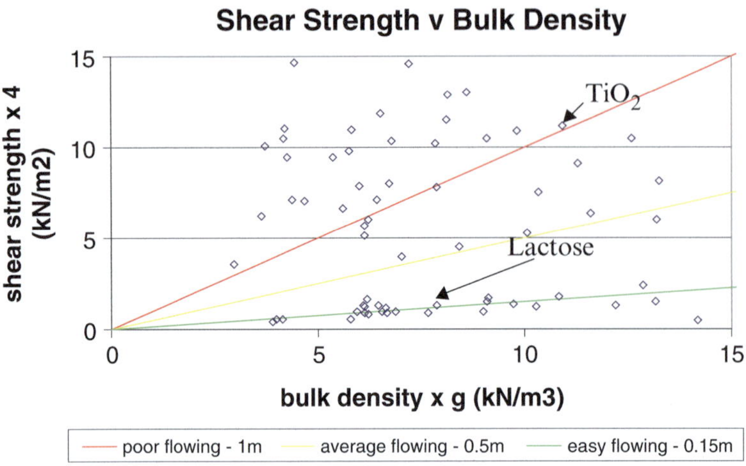

Fig. 15.15 Shear strength and bulk density plotted in a graph partitioned by lines of constant outlet size

This presentation helps characterise materials in a space where a sensitive plant parameter (hopper outlet size) defines flow boundaries. Note the partitioning lines identify constant outlet sizes as follows: – 1 m – poor flowing; 0.5 m – average and 0.15 m – easy flowing.

15.3.2.3 Spider Diagrams

To integrate some of the facets of characterisation of bulk solids a 'spider' diagram can used based on measured characteristics of wall friction (ϕ_w), shear strength (τ_s), bulk density (ρ_b), and a number of derived equipment characteristics such as; hopper wall angle for mass flow (β_c), outlet size (D_{crit}) and Hausner ratio (*H.R.*, ratio of tapped bulk density/loose bulk density).

To construct the spider diagram, a series of three concentric circles are divided by an axis for each of the characteristics. These axes intersect with the smallest diameter circle where that particular characteristic describes 'easy flow' with subsequent bigger diameter circles defining 'modest' or 'average' and 'poor flow'. Two idealised situations can then be presented, as in Fig. 15.16.

The diagrams can be more than qualitative if the data from the tests on the large number of materials (over 200) reported by McGee is used to define the 'easy', 'average' and 'poor' flow circles, Table 15.5.

Note that the bulk density axis is the reverse of the others because decreasing bulk density usually means poorer flow. A practical example is that most milling operations lower bulk density and worsen flowability of powders when they are stored.

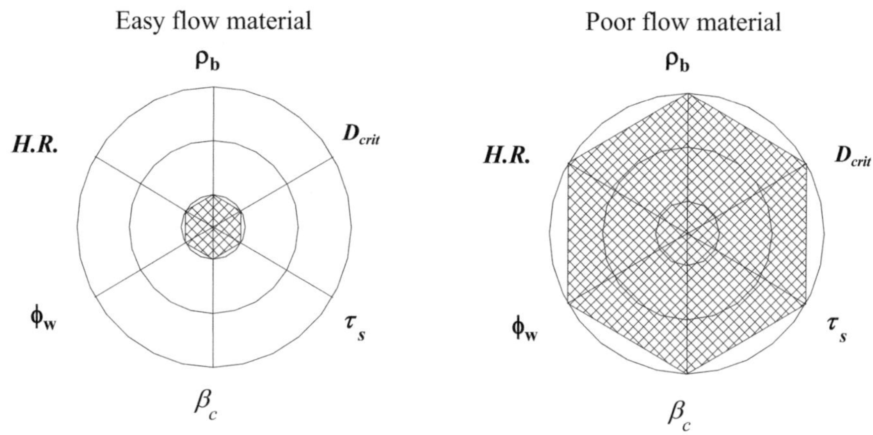

Fig. 15.16 Spider diagrams for ideal 'easy' flow and 'poor' flow material

Table 15.5 Parameters suggested by the tests reported by McGee [8]

Circle	Wall friction, ϕ_w (deg)	Bulk density, ρ_b (kg/m^3)	Shear strength, τ_s (N/m^2)	Hausner ratio, HR (–)	Outlet size, D_{crit} (cm)	Mass flow wall angle, β_c (deg)
Easy flow	20	1200	300	1.1	15	65
Average	25	800	1000	1.25	50	73
Poor flow	30	400	2000	1.5	100	80

Ideal 'Easy Flow' Material

A low friction material (<20°), which would mass flow in a conical hopper with a wall angle of 65° to the horizontal. It would have a high bulk density (around 1200 kg/m^3) but not be affected much by compaction or vibration so would have a low Hausner ratio (up to 1.1). Its low shear strength (maximum 300 N/m^2) coupled with the high bulk density would guarantee flow through a small outlet (<15 cm diameter). A practical example would be a free flowing grade of lactose with wall friction angle of 17° against stainless steel, shear strength 197 N/m^2, Hausner ratio of 1.1, rat hole diameter 9 cm and requiring a 64° wall angle for mass flow in a conical hopper. With a bulk density 867 kg/m^3 this particular example would have a small spike on the density axis of the spider diagram indicating a slight deviation from the ideal flow material.

Ideal 'Poor Flow' Material

A high friction material (>30°), which would barely mass flow in even the steepest conical hopper (>80°) (in fact probably require a Vee shaped hopper). It would

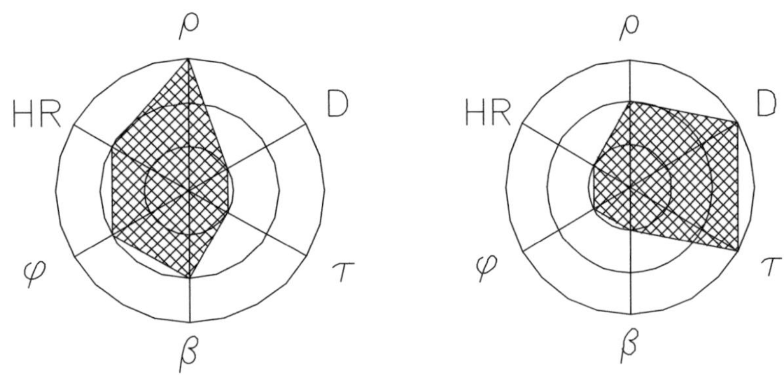

Fig. 15.17 Spider diagrams for Powder A and Powder B

have a low bulk density (about 400 kg/m^3), which would be significantly affected by compaction indicated by a high Hausner ratio (about 1.5). Its high shear strength (2000 N/m^2) coupled with the low bulk density would mean very large outlets (>100 cm) would be needed to ensure flow. A practical example would be a fine milled icing sugar with wall friction of 30.5° against stainless steel, bulk density of 540 kg/m^3, shear strength 2144 N/m^2, Hausner ratio of 1.49, rat hole diameter 149 cm and requiring a wall angle for mass flow in a conical hopper of 80° to the horizontal.

The technique when applied to two other examples highlights particular aspects of the 'profile' that merit special attention. Figure 15.17 shows the resultant diagram for two different powders, noted as Powder A and Powder B.

For Powder B most aspects for flow are good except the shear strength (τ) and outlet size (D). Practical application for handling this material required a good shaped Vee hopper with a long slot outlet to ensure reliable flow to process.

Powder A has low shear strength but aspects associated with wall friction and bulk density and more awkward. These produce a characteristic spider diagram quite different from the preceding example. To avoid flow difficulties with Powder A requires steep slopes on hoppers and chutes. The merits of comparing wall friction values against different surfaces e.g. stainless steel versus polyethylene or finishes e.g. 2 mill finish, mechanically polished or electropolished could be explored.

The spider diagram should be seen as a useful and versatile tool where the numeric values of the parameters can be 'tuned' for an indication of 'poor' or 'good' gained from heuristic knowledge on plant. Indeed an entirely different set of bulk parameters can be used as the axes if this provides better information on the materials behaviour in a particular process and plant.

15.4 Definitions, Abbreviations and Symbols

Equivalent projected area diameter	diameter of a circle having the same area as the particle's projection
Equivalent sieve diameter	equivalent volume diameter of a particle just passing through a square sieve aperture (near-mesh size).
Equivalent sphere	sphere that has the same property as the observed particle in relation to a given measurement principle
Equivalent volume diameter	diameter of a sphere having the same volume as the particle
Median size	particle size at the 50 % point of a cumulative PSD
Stokes diameter	diameter of a sphere that has the same density and based sedimentation rate as the particle under conditions of Stokes' law (viscous flow conditions).

AFM	atomic force microscopy
DLS	dynamic light scattering
ESZ	electrical sensing zone
MALS	multi-angle static light scattering
PSD	particle size distribution
PT	Particle tracking
SEM	scanning electron microscopy
s.r.s.	simple random selection
TEM	transmission electron microscopy

A	surface area of a particle
CI	Carr's Index
d	particle size
D	hopper outlet diameter
HR	Hausner Ratio
n	number of samples
s	standard deviation
$\hat{u}$	the measured mean
V	volume of a particle

β_c	hopper wall angle for mass flow
ε	voidage or void fraction
ρ_b	Bulk density
ρ_p	particle density
ϕ_w	wall friction angle
$\hat{\sigma}^2$	best estimate of the true variance
Ψ	sphericity
τ_s	shear stress

References

1. Allen, T., Khan, A.A.: Critical evaluation of powder sampling procedures. Trans. Inst. Chem. Eng. **238**, CE 108–112 (1970)
2. Bates, L.: In: McGlinchey, D. (ed.) Characterisation of Bulk Solids, ISBN 1405116242. Blackwell Publishing, Oxford 296pp (2005)
3. Duran, J.: Sands, Powders and Grains – An Introduction to the Physics of Granular Materials. Springer, New York (2000). ISBN 0387986561
4. Fitzpatrick, J., Ahrne, L. et al.: 'Strategic document for research on food powders'; EC 5th Framework Programme Accompanying measure contact QLK1-CT-**2001**-30172; http://www.foodpowders.net/documents/Strategic%20Document-Draft%204.3.pdf (2001)
5. Gy, P. Special Issue: 50 years of Pierre Gy's theory of sampling proceedings: first world conference on sampling and blending (WCSB1) tutorials on sampling: theory and practice. Chemometr. Intell. Lab. Syst. **74**(1) (2004)
6. Harnby, N.: Proceeding Third International Powder Technology and Bulk Solids Conference; Pub. Hayden (1976)
7. Heywood, H.: Proceedings of the 1st Particle Size Analysis Conference, Sept 1996, pp. 355–359 (1966)
8. McGee, E., McGlinchey, D.: 'Using wall friction data to design hopers' Proceedings of PARTEC 2004, Nuremberg, Germany, (2004)
9. McGee, E.: An investigation into characterisation of bulk solids and flow in hoppers. Ph.D. thesis, Glasgow Caledonian University (2005)
10. McGee, E., McGlinchey, D.: Using data from vertical shear cell tests to define flowability. Proceedings of the 5th International Conference for Conveying and Handling of Particulate Solids, Sorrento (2006)
11. Merkus, H.G.: Particle Size Measurements: Fundamentals, Practice, Quality (Particle Technology Series vol. 17). Springer, ISBN-10: 1402090153 (2009)
12. Merkus, H.G., Meesters, G.M.H. (eds.): Particulate Products: Tailoring Properties for Optimal Performance (Particle Technology Series vol. 19); Springer; ISBN-10: 3319007130, (2014)
13. Pitard, F.F.: Pierre Gy's Sampling Theory and Sampling Practice, Second Edition: Heterogeneity, Sampling Correctness, and Statistical Process Control, 2 edn. CRC Press, ISBN-10: 0849389178, (1993)

Standards[1]

14. ASTM E11—13 Standard Specification for Woven Wire Test Sieve Cloth and Test Sieves
15. CSN EN 933-1 (**2012**) Tests for Geometrical Properties of Aggregates—Part 1: Determination of Particle Size Distribution—Sieving Method
16. ISO 565, Test sieves—Woven metal wire doth and perforated plate—Nominal sizes of apertures
17. ISO 607, Surface active agents and detergents—Methods of sample division
18. ISO 902:**1976**(en) Aluminium oxide primarily used for the production of aluminium—Measurement of the angle of repose
19. ISO 2591, Test sieving
20. ISO 3165, Sampling of chemical products for industrial use—Safety in sampling

[1] Note: Search always the latest available version of a standard. They usually are revised at a regular basis.

21. ISO 3310-1:**2000** Test Sieves—Technical Requirements and Testing—Part 1: Test Sieves of Metal Wire Cloth
22. ISO 3944:**1992**(en) Fertilizers—Determination of bulk density (loose)
23. ISO 4324:**1977**(en) Surface active agents—Powders and granules—Measurement of the angle of repose
24. ISO 5311:**1992**(en) Fertilizers—Determination of bulk density (tapped)
25. ISO 6206, Chemical products for industrial use—Sampling—Vocabulary
26. ISO 6274:**1982** Concrete—Sieve analysis of aggregates.
27. ISO 6782:**1982**(en) Aggregates for concrete—Determination of bulk density
28. ISO 8213:**1986** Chemical products for industrial use—Sampling techniques—Solid chemical products in the form of particles varying from powders to coarse lumps
29. ISO 8398:**1989**(en) Solid fertilizers—Measurement of static angle of repose
30. ISO 9276-1:**1998**, Representation of results of particle size analysis—Part 1: Graphical representation
31. ISO 9276-2, Representation of results of particle size analysis—Part 2: Calculation of average particle sizes/diameters and moments from particle size distributions
32. ISO 9276-4:**2001**, Representation of results of particle size analysis—Part 4: Characterization of a classification process
33. ISO 9276-6:**2008**, Representation of results of particle size analysis—Part 6: Descriptive and quantitative representation of particle shape and morphology
34. ISO 11648-2:**2001**(en) Statistical aspects of sampling from bulk materials—Part 2: Sampling of particulate materials
35. ISO 13099-1:**2012**, Colloidal systems—Methods for zeta-potential determination—Part 1: Electroacoustic and electrokinetic phenomena
36. ISO 13099-2:**2012**, Colloidal systems—Methods for zeta-potential determination—Part 2: Optical methods
37. ISO 13317-1:**2001**, Determination of particle size distribution by gravitational liquid sedimentation methods—Part 1: General principles and guidelines
38. ISO 13319:**2007** Determination of Particle Size Distributions—Electrical Sensing Zone Method
39. ISO 13320:**2009** Particle Size Analysis—Laser Diffraction Methods
40. ISO 13322-1:**2004**, Particle size analysis—Image analysis methods—Part 1: Static image analysis methods
41. ISO 13322-2:**2006**(en) Particle size analysis—Image analysis methods—Part 2: Dynamic image analysis methods
42. ISO/TS 13762:**2001**, Particle size analysis—Small angle X-ray scattering method
43. ISO 14488:**2007**, Particulate materials—Sampling and sample splitting for the determination of particulate properties
44. ISO 14629:**2012**(en) Fine ceramics (advanced ceramics, advanced technical ceramics)—Determination of flowability of ceramic powders
45. ISO 14887:**2000**, Sample preparation—Dispersing procedures for powders in liquids
46. ISO/TS 17200:**2013** Nanotechnology -- Nanoparticles in powder form -- Characteristics and measurements
47. ISO 20998-1:**2006**, Measurement and characterization of particles by acoustic methods—Part 1: Concepts and procedures in ultrasonic attenuation spectroscopy
48. ISO 20998-2:**2013**(en) Measurement and characterization of particles by acoustic methods—Part 2: Guidelines for linear theory
49. ISO 21501-3:**2007** Determination of particle size distribution—Single particle light interaction methods—Part 3: Light extinction liquid-borne particle counter
50. ISO 22412:**2008**(en) Particle size analysis—Dynamic light scattering (DLS)
51. ISO 26824:**2013**(en) Particle characterization of particulate systems—Vocabulary
52. ISO/TS 27687, Nanotechnologies—Terminology and definitions for nano-objects—Nanoparticle, nanofibre and nanoplate

Index

© Springer International Publishing Switzerland 2016
H.G. Merkus, G.M.H. Meesters (eds.), *Production, Handling and Characterization of Particulate Materials*, Particle Technology Series 25,
DOI 10.1007/978-3-319-20949-4

MIX
Papier aus verantwortungsvollen Quellen
Paper from responsible sources
FSC® C105338

If you have any concerns about our products,
you can contact us on
ProductSafety@springernature.com

In case Publisher is established outside the EU,
the EU authorized representative is:
Springer Nature Customer Service Center GmbH
Europaplatz 3, 69115 Heidelberg, Germany

Printed by Libri Plureos GmbH
in Hamburg, Germany